Accessing Uncultivated Microorganisms

FROM THE ENVIRONMENT
TO ORGANISMS AND GENOMES
AND BACK

CONTENTS

Contributors *vii*
Preface *xi*

I. INTRODUCTION 1

1. Does Cultivation Still Matter?
Karsten Zengler
3

2. The Human Intestinal Microbiota and Its Impact on Health
Mirjana Rajilić-Stojanović, Willem M. de Vos, and Erwin G. Zoetendal
11

II. THE STATUS QUO: WHO IS OUT THERE? HOW TO DETERMINE MICROBIAL DIVERSITY 33

3. The Uncountables
William T. Sloan, Christopher Quince, and Thomas P. Curtis
35

4. The Missing Fungi: New Insights from Culture-Independent Molecular Studies of Soil
S. K. Schmidt, K. L. Wilson, A. F. Meyer, C. W. Schadt, T. M. Porter, and J. M. Moncalvo
55

5. The Diversity of Free-Living Protists Seen and Unseen, Cultured and Uncultured
David A. Caron and Rebecca J. Gast
67

6. Microbial Biogeography: Patterns in Microbial Diversity across Space and Time
Noah Fierer
95

7. The Least Common Denominator: Species or Operational Taxonomic Units?
Ramon Rosselló-Mora and Arantxa López-López
117

8. Measuring Diversity
Jed A. Fuhrman
131

9. Metagenomics as a Tool To Study Biodiversity
Karen E. Nelson
153

III. ARE MICROORGANISMS NONCULTURABLE OR NOT YET CULTURABLE? 171

10. New Cultivation Strategies for Terrestrial Microorganisms
Peter H. Janssen
173

11. Cultivation of Marine Symbiotic Microorganisms
Todd A. Ciche
193

12. Methods To Study Consortia and Mixed Cultures
Boran Kartal and Marc Strous
205

IV. DO WE HAVE TO CHANGE GEAR? NEW CULTIVATION APPROACHES AND NEW MOLECULAR APPROACHES COMBINED 221

13. Microbial Cell Individuality
Simon V. Avery
223

14. Nanomechanical Methods To Study Single Cells
Ramya Desikan, Laurene Tetard, Ali Passian, Ram Datar, and Thomas Thundat
245

15. Single-Cell Genomics
Martin Keller, Christopher W. Schadt, and Anthony V. Palumbo
267

16. How Many Genes Does a Cell Need?
Hamilton O. Smith, John I. Glass, Clyde A. Hutchison III, and J. Craig Venter
279

Index 301

CONTRIBUTORS

Simon V. Avery
School of Biology, Institute of Genetics, University of Nottingham,
University Park, Nottingham NG7 2RD, United Kingdom

David A. Caron
Department of Biological Sciences, University of Southern California,
3616 Trousdale Parkway, Los Angeles, CA 90089-0371

Todd A. Ciche
Department of Microbiology and Molecular Genetics, Michigan State University,
2215 Biomedical Physical Sciences Building, East Lansing, MI 48824-4320

Thomas P. Curtis
School of Civil Engineering and Geosciences, University of Newcastle upon Tyne,
Newcastle NE1 7RU, United Kingdom

Ram H. Datar
Biosciences Division, Oak Ridge National Laboratory,
1 Bethel Valley Road, Oak Ridge, TN 37831-6123

Ramya Desikan
Biosciences Division, Oak Ridge National Laboratory,
1 Bethel Valley Road, Oak Ridge, TN 37831-6123

Willem M. de Vos
Laboratory of Microbiology, Wageningen University,
Dreijenplein 10, 6703 HB Wageningen, The Netherlands

Noah Fierer
Department of Ecology & Evolutionary Biology,
Cooperative Institute for Research in Environmental Sciences,
University of Colorado, Boulder, CO 80305

Jed A. Fuhrman
Department of Biological Sciences and Wrigley Institute, University of Southern California, Los Angeles, CA 90089-0371

Rebecca J. Gast
Department of Biology, Woods Hole Oceanographic Institution, Woods Hole, MA 02543

John I. Glass
The J. Craig Venter Institute, 9704 Medical Center Drive, Rockville, MD 20850

Clyde A. Hutchison III
The J. Craig Venter Institute, 9704 Medical Center Drive, Rockville, MD 20850

Peter H. Janssen
AgResearch Limited, Palmerston North 4442, New Zealand

Boran Kartal
Department of Microbiology, Faculty of Science, Radbound University Nijmegen, Toernooiveld 1, 6525 ED Nijmegen, The Netherlands

Martin Keller
Biosciences Division, Oak Ridge National Laboratory, 1 Bethel Valley Road, Oak Ridge, TN 37831-603

Arantxa López-López
Marine Microbiology Group, Institut Mediterrani d'Estudis Avancats (IMEDEA, CSIC-UIB), E-07190 Esporles, Illes Balears, Spain

A. F. Meyer
Department of Ecology and Evolutionary Biology, University of Colorado, Boulder, CO 80309

J. M. Moncalvo
Department of Ecology and Evolutionary Biology, University of Toronto, and Department of Natural History, Royal Ontario Museum, Toronto, ON M5S 2C6, Canada

Karen E. Nelson
The J. Craig Venter Institute, 9704 Medical Center Drive, Rockville, MD 20850

Anthony V. Palumbo
Biosciences Division, Oak Ridge National Laboratory, 1 Bethel Valley Road, Oak Ridge, TN 37831-6035

Ali Passian
Biosciences Division, Oak Ridge National Laboratory, 1 Bethel Valley Road, Oak Ridge, TN 37831-6123

T. M. Porter
Department of Ecology and Environmental Biology, University of Toronto, Toronto, ON M5S 3B2, Canada

Christopher Quince
Department of Civil Engineering, University of Glasgow,
Glasgow G12 8LT, United Kingdom

Mirjana Rajilić-Stojanović
Laboratory of Microbiology, Wageningen University, Dreijenplein 10,
6703 HB Wageningen, The Netherlands

Ramon Rosselló-Mora
Marine Microbiology Group, Institut Mediterrani d'Estudis Avancats
(IMEDEA, CSICUIB), E-07190 Esporles, Illes Balears, Spain

Christopher W. Schadt
Biosciences Division, Oak Ridge National Laboratory,
1 Bethel Valley Road, Oak Ridge, TN 37831-6035

S. K. Schmidt
Department of Ecology and Evolutionary Biology, University of Colorado,
Boulder, CO 80309

William T. Sloan
Department of Civil Engineering, University of Glasgow,
Glasgow G12 8LT, United Kingdom

Hamilton O. Smith
The J. Craig Venter Institute, 9704 Medical Center Drive,
Rockville, MD 20850

Marc Strous
Department of Microbiology, Faculty of Science, Radboud University Nijmegen,
Toernooiveld 1, 6525 ED Nijmegen, The Netherlands

Laurene Tetard
Biosciences Division, Oak Ridge National Laboratory,
1 Bethel Valley Road, Oak Ridge, TN 37831-6123

Thomas Thundat
Biosciences Division, Oak Ridge National Laboratory,
1 Bethel Valley Road, Oak Ridge, TN 37831-6123

J. Craig Venter
The J. Craig Venter Institute, 9704 Medical Center Drive,
Rockville, MD 20850

K. L. Wilson
Department of Ecology and Evolutionary Biology, University of Colorado,
Boulder, CO 80309

Karsten Zengler
Department of Bioengineering, University of
California—San Diego, 9500 Gilman Drive,
La Jolla, CA 92093–0412

Erwin G. Zoetendal
Laboratory of Microbiology, Wageningen University,
Dreijenplein 10, 6703 HB Wageningen, The Netherlands

PREFACE

This book has its origin in a longtime fascination for the diversity of microorganisms and their sheer endless metabolic capabilities. This fascination was nurtured over the years by multiple colleagues and friends, which to name all would go far beyond the scope of this preface. However, the person who shaped my view about microorganisms the most and thus contributed to the text being written is Fritz Widdel. As a graduate student in his laboratory at the Max Planck Institute for Marine Microbiology in Bremen, I was not only introduced to principles of thermodynamics and biochemistry but learned to share his affection for isolating and growing microorganisms in the laboratory. As a horticulturist cares for his plants I learned from Fritz how to pamper different microorganisms, pay respect to their needs, and accept their various "temperaments and mood swings." Observing, understanding, and hopefully predicting these "behaviors" became a driving force for my scientific (ad)ventures. And although nowadays isolation and cultivation are considered by many microbiologists to be old-fashioned, they are still the pillars of microbiology.

Microbially mediated processes are at the center of all global biogeochemical cycles. Solid understanding of processes within a cell, at various levels, is paramount if we are going to apply this knowledge to an understanding of the environment, whether the habitat is the size of a sand grain or the whole ocean floor. Evaluating the microbial diversity and identifying individuals that are responsible for specific processes in the environment are fundamental and challenging tasks in microbial ecology and applied microbiology. It seems appropriate to emphasize the topic of accessing microorganisms and their role in the environment at a time when concomitant advances have been made in various research fields to fulfill these tasks. Researchers address these challenges from different angles, use different methods, and often have different perspectives. It seems that over the course of the last years, however, interdisciplinary research became one of the most successful approaches for ground-breaking discoveries in environmental microbiology. This book highlights various aspects of assessing and accessing microorganisms from the environment and of ways to elucidate basic biological concepts. It spans from critical evaluations on how to determine, define, and measure microbial diversity, including fungi and protists, discussing various cultivation approaches, as well as presenting molecular tool boxes, and reviewing genomic, proteomic, and metagenomic approaches, to single-cell methods, including the development of synthetic cells. I hope this book will inspire researchers and students from various fields to consider all aspects in the exciting challenge of the present microbiological era to understand and explain the diversity of microorganisms in their environment.

KARSTEN ZENGLER

INTRODUCTION

I

DOES CULTIVATION STILL MATTER?

Karsten Zengler

1

For more than a century, isolation and cultivation of microorganisms have been fundamental to elucidating the functioning of microorganisms and thus understanding their role in the environment. Bacterial cultures were first obtained when Robert Koch introduced growth media solidified by agar and the concept of pure culture more than 130 years ago (Koch, 1877). Shortly after, microbiologists achieved milestones in microbial physiology and microbial ecology by enriching and cultivating new microorganisms (Beijerinck, 1895; Cohn, 1875; Winogradsky, 1887). Despite these significant advances in microbial ecology, it became obvious only a few years after Koch's breakthrough that not all microorganisms could be cultivated in the laboratory and that the majority of cells visible by microscopy did not form visible colonies on plates (Conn, 1918). During the last three decades microbiologists realized, mainly through the introduction of molecular techniques, such as rDNA sequencing and more recently environmental sequencing approaches, that microbial diversity is sheerly enormous and that actually only a tiny fraction of organisms present in any given environment can be cultivated in the laboratory.

As applications of new molecular tools in microbial ecology were on the rise, interest in cultivation approaches to study microbial diversity declined. Isolation and cultivation of microorganisms became an almost lost skill. In addition, these approaches did seem to be particularly fascinating to students and young scientists. Worldwide, only a few laboratories continued isolating microorganisms from the environment following the tradition of the "Delft school" (van Niel, 1949). Interestingly, the "molecular revolution" in environmental microbiology, as it was coined by some authors, has recently initiated a revived interest in cultivation techniques. The reasons are quite obvious—with the application of various molecular tools to assess microbial diversity, evidence accumulated that the organisms in our culture collections might not be the ones of importance in the environment or that the few organisms that could be cultivated in the laboratory were not even the most abundant in nature. The conclusion was that the vast majority of microbes still remain reluctant to our isolation and cultivation attempts. Recently several

Karsten Zengler, Department of Bioengineering, University of California—San Diego, 9500 Gilman Drive, La Jolla, CA 92093–0412.

Accessing Uncultivated Microorganisms: from the Environment to Organisms and Genomes and Back
Edited by Karsten Zengler © 2008 ASM Press, Washington, DC

advances have been made to overcome cultivation biases and have spurred renewed interest in classical microbiology as well as in innovative isolation techniques (Keller and Zengler, 2004).

However, numerous biodiversity surveys revealed that even with these new approaches we are far from filling the void in understanding the role of microorganisms in the environment. So what does the future hold? Maybe new inventions in the field of cultivation will allow us to get a hand on almost every microorganism. Maybe substantial funding opportunities, similar to funding that was directed toward environmental sequencing efforts, would permit isolation of relevant microorganisms from the environment. Or maybe we have to live with the fact that despite numerous technological advances, we will still be outsmarted by the smallest of all living creatures. So how can we possibly understand the metabolism and function of certain microorganisms in nature and use this knowledge to understand the environment as a whole if we are not able to study the organism of interest in the laboratory? How can we prove that our hypotheses are correct if we, for example, cannot genetically manipulate an organism and confirm that a certain gene is encoding a certain protein that signs responsible for a certain function? Currently numerous advances have been made to study microorganisms from the environment without the need of having the organism in culture, all termed "culture independent." Many of those recent advances are targeting the genome, parts of the genome, or expressed genetic material of an organism. So, if we are already able to collect this information from a natural assemblage of microorganisms without having any microorganism in culture, why bother with cultivation and isolation attempts? Or, as the title of this chapter implicates, does cultivation still matter?

A tool box of methods has been developed over the last decades in order to study microbial diversity and to assign functions to specific organisms. The main goal of these approaches is not only to describe microbial diversity but ultimately to explain it. This would eventually enable us to control and manipulate microbial diversity for various applications and, in the end, to predict how an environment will react to changes. These methods in environmental microbiology can be divided into three major groups. The first is characterized by a "black box" approach, in which physical, chemical, and biological parameters of the environment are studied, e.g., by microelectrodes, labeling studies, rate measurements, gene surveys, etc. These approaches, which include a range of in situ methods, have the enormous advantage that effects on the environment can be directly measured. However, most of these techniques are not designed to link specific microbes to certain processes. Techniques intended to do so are the combinations of fluorescence in situ hybridization with microautoradiography (Ouverney and Fuhrman, 1999) and with secondary ion mass spectrometry (Orphan et al., 2001; Lechene et al., 2006). The second is the metagenomic approach (Handelsman, 2004), which can be regarded as the molecular "black box" approach. The aim of this method is to assess the genomic potential of an environment as a whole. This approach led to studies focusing on expressed gene products such as RNA profiling and metaproteome studies (Wagner et al., 2007; Maron et al., 2007). The goal of these studies is not only to determine the genetic potential of an environment but to find out which genes are expressed at a certain moment in time under certain conditions. The hope is that by understanding both the metagenome and metaproteome it will be feasible to make predictions about how certain microbes will affect the environment and how they will react to changing conditions. The third is a "bottom-up" approach in which single organisms are studied in great detail and the obtained information is subsequently extrapolated to conditions found in nature. This approach is aimed to assemble the microbial community by understanding one organism, or one process, at a time. This approach includes cultivation and isolation methods as well as recently available single-cell techniques. However, as mentioned earlier, many microorganisms of interest are recalcitrant to isolation. Also, single-cell microbiology of uncultivated microbes is still in its infancy. Hopefully, this is likely to change in the future.

To assess microbial diversity, most studies rely on various forms of rDNA sequence analysis. However, it is crucial to keep in mind that (i) physiology and metabolism of an organism cannot be extrapolated from its rDNA gene sequence and (ii) the small-subunit rRNA gene lacks resolution at the species level (Rosselló-Mora and Amann, 2001). It is also essential to recognize that even if diversity surveys at the species level will become possible in the future, different strains of the same species can have very different properties. This fact can become "painfully" clear if consumed food contains the disease-causing *Escherichia coli* strain O157:H7 instead of some of its harmless cousins. But whether this function of the organisms is of any "relevance" depends on the perspectives and questions addressed by the researcher. However, intraspecies diversity can be assessed by cumbersome DNA-DNA hybridization, multilocus sequence typing, or average nucleotide identity (Konstantinidis et al., 2006) but so far requires large DNA quantities or substantial genomic information. Then again, individual organisms in closest proximity with (likely) identical niches (e.g., two single *Beggiatoa* filaments) cannot only comprise two different genomes but also resemble two distinct 16S rRNA genes (Muβmann et al., 2007). Understanding the breadth and implications that can be made by biodiversity surveys based on rDNA sequences is therefore absolutely essential. It is also important to keep in mind that results in the laboratory cannot always be directly translated to the environment. This can have several reasons; for example, the organism in the laboratory and the one occupying that particular niche in nature are not identical. This, for example, can happen if the "wrong" strain has been isolated or if biogeographic differences exist between different strains (Peña et al., 2005), or if the isolate evolved over time in the laboratory (Cooper et al., 2003). It is noteworthy that isolation of strains, even from low diversity environments, will not necessarily result in identical strains; e.g., two strains of *Salinibacter ruber,* for which genomes had been sequenced, could not be re-isolated from the same environment (J. Antón, personal communication). It is not clear at this time whether these discrepancies are due to biases in culturability, unnoticed changes of the habitat, or extreme genome plasticity of the population. Cultivation of microorganisms in the laboratory over time periods that translate into hundreds and thousands of generations of the species kept will ultimately lead to strains (often unrecognized by the researcher) that are much better adapted to conditions provided in the laboratory than to the ones found in nature (Cooper et al., 2003). This adaptation is manifested by only slight mutations (indels) of the genome (Herring et al., 2006). One has to keep in mind that important parameters in the environment, such as the supply of nutrients or growth factors, as well interactions among organisms cannot, or only insufficiently, be simulated in the laboratory. It is also essential to take into account that most measurements are performed at a certain point in time but rarely over time scales that are of importance for the environment. For most environments, biodiversity patterns will not necessarily be stationary. Even minor changes to the environment, undetected by the investigator, can have broad effects on the community structure. Since microbial populations can show extreme heterogeneity at different levels (micro- and macrodiversity), it is crucial that we measure environmental parameters at the same scale we intend to resolve microbial diversity.

The importance of understanding interactions of microorganisms with their environment becomes immediately obvious to us when microorganisms directly interact with the human body. Adult humans harbor more bacterial cells in the gut system than there are cells in the body. The human gut, the "outside world" inside the body, represents an extremely diverse environment. Although we are far from a comprehensive understanding of such an extremely complex microbial community, substantial progress has been made in determining effects certain microbes of the gut microflora have on human health. Resolving microbial diversity as well as interactions and potential pathogenicity of microorganisms in the human gut is essential for understanding their role for human health. Mirjana Rajilić-Stojanović, Willem de Vos, and Erwin Zoetendal describe the recent discoveries, and their implications for human well-being, that have been made by studying this "familiar" environment in great detail.

But how can we actually measure microbial diversity if the numbers of individuals in any given sample is sheerly gargantuan? Data compiled so far by various molecular methods suggest that true diversity in environmental samples will be extremely difficult to quantify purely by experimental methods. To estimate the number of different taxa within a sample, it is not necessary to account for every single taxonomic unit within a community. Diversity estimators allow the extrapolation of observed data and the prediction of total community diversity. However, we have to keep in mind that the number of individuals within a sample will be extremely small compared to the total number of individuals within the community we aim to describe. In their chapter on the uncountables, Bill Sloan, Chris Quince, and Tom Curtis describe what mathematical diversity estimators and taxon-abundance distributions have been deployed to acquire a baseline map of microbial diversity.

Regarding total abundance, viruses are the most abundant biological entity in nature, and sequence-based surveys have revealed enormous phage diversity (Suttle, 2007; Casas and Rohwer, 2007). Recently it has been recognized that viruses can tweak the (bacterial) host metabolism by tapping into critical, rate-limiting steps of the host's metabolism; this could have broad implications for environmental "fitness" of the host population (Breitbart et al., 2007). However, most microbial diversity surveys have been focusing on bacteria and archaea. Since a comprehensive study of their diversity is already challenging (Ashby et al., 2007; Huber et al., 2007), many microbiologists tend to ignore eukaryotic microorganisms (fungi and protists) completely in their biodiversity surveys. However, a comprehensive understanding of any environment will not be possible without detailed analysis of diversity and interactions of its prokaryotic and eukaryotic counterparts. Fungi play an important ecological role, particularly in soil environments, and are key players in decomposition and cycling of nutrients. Although mycologists have been collecting and growing fungi for several hundred years, the 80,000 fungal species described so far represent only about 5% of the estimated total diversity. In their chapter on fungal diversity, Steve Schmidt and colleagues address the question "where are the missing fungi?"

Another group of organisms that have been known for several hundred years are the protists. First discovered by Hooke and van Leeuwenhoek more than three centuries ago through the invention of the microscope (Gest, 2007; http://ttp.nlm.nih.gov), protists resemble vastly expanding groups of organisms; this was made evident by applying 18S rDNA sequencing. Protists play essential ecological roles in any aquatic environment, spanning from freshwater ecosystems to the world oceans, where photosynthetic members of these highly diverse taxa provide the base of the food web. Heterotrophic protists consume other microorganisms and are key players mediating the decomposition and remineralization processes of organic matter, and thus play a vital role in benthic-pelagic coupling. David Caron and Rebecca Gast review the current state of these extremely diverse taxa of eukaryotic microbes and their ecological roles in nature.

As new molecular techniques and methods become available, it is not only possible to assess the microbial diversity of a given environment at a certain point in time but to study spatial patterns of microbial diversity within a sample or over geographic regions and to monitor changes of these patterns over time. Understanding these patterns for different microbial habitats and determining what factors are involved in shaping microbial diversity in these habitats will allow establishment of theories and principles of microbial biogeography that may or may not match the ones established for macroorganisms, i.e., plants and animals. In his chapter, Noah Fierer gives a broad overview of the biogeography of uncultivated microbes.

One crucial point in all biodiversity studies is how we define the units that we intend to study. Various terms such as phylotype, ribotype, ecotype, geotype, genotype, and operational taxonomic unit, are used throughout the literature to describe microbial diversity, unfortunately not always correctly. Ramon Rosselló-Mora and Arantxa López-López review studies of microbial community structures and dynamics with the intention of providing recommendations on how terms should be used correctly to avoid confusion.

There are many ways to assess the microbial diversity of an environment. Depending on the scope of a study, which includes the coverage, sensitivity, resolution, and also the cost efficiency, microbial diversity can be measured by a large variety of available methods. The approaches can range from microscopy and cultivation, to analysis of cell components, to community metagenomic approaches. Each method has its advantages and drawbacks and should be chosen wisely based on the question that we seek to answer. Jed Fuhrman illustrates in his chapter the manifoldness of these approaches and their relevancy in biodiversity studies.

Molecular approaches have been designed to capture the collective uncultivated majority of microorganisms from an environmental sample because attempts to isolate the entire microbial community of an environmental sample have not been successful for complex habitats. Community sequencing (metagenomics) and community proteomics (metaproteomics) are two approaches that take the collective information (genomic and proteomic) of a microbial community. The vast amount of the obtained information is subsequently analyzed in silico, relying on single-genome data as scaffolds. To overcome the problem of integrating "missing data," including sequences that do not have close relatives in the database, new in silico tools have been designed. In her chapter, Karen Nelson reviews the developments in single- and community-genome and proteome analysis.

Microbial communities can be extremely diverse. This is especially the case for soil microbial communities. Not surprisingly, researchers have discovered many clades of so far uncultivated microorganisms while analyzing the genetic information obtained from analyzing soil samples by various methods. Peter Janssen describes in his chapter the methods and techniques successfully used to isolate some of these until then uncultivated microbes. He reviews topics that should be considered regarding the isolation of organisms, including the choice of medium, carbon and energy source, pH, and temperature. In particular, the chapter focuses on aspects that are often not considered during cultivation, such as salt components of the medium, the choice of gelling agents and glassware, the size of sample and inoculum, the time of incubation, and how colonies are being detected. Altogether the chapter provides detailed and resourceful advice for successful cultivation approaches.

The isolation of a single organism is often not feasible for several reasons. The importance of synthrophic interaction or dependence on factors produced by other organisms has been known for a long time but is increasingly recognized as a key factor of "unculturability." Boran Kartal and Marc Strous describe in their chapter that enrichment cultures, grown under stable and close to natural conditions, e.g., in a chemostat, can still be studied in great detail. They present a toolbox of techniques that make it possible to not only identify key members of enrichment cultures but to determine the niche of the enriched microorganism, calculate kinetic parameters, determine intermediates, resolve the proteins responsible for specific activities, and link the protein to a gene sequence. As an example of the study of a recently discovered pathway (anaerobic ammonium oxidation), they used this combined approach to study anaerobic ammonium-oxidizing bacteria within a mixed culture. These bacteria became by far the best understood members of the phylum *Planctomycetes*.

Interaction (i.e., symbiosis) between different organisms can bear new characteristics and therefore permit utilization of new niches. Among higher organisms, such as plants and animals, symbioses with microorganisms are extensive and have recently been attributed to being essential for host physiology (Cash et al., 2006). Symbioses resulting in a great range of fitness exist between prokaryotes and viruses (phages), between different prokaryotes, and between prokaryotes and eukaryotes. Todd Ciche reviews in his chapter the various forms of symbiosis and describes procedures to cultivate marine symbiotic microorganisms to fulfill Koch's postulates. Since isolation of microorganisms is not always possible, he adapts Koch's postulates for symbiosis.

Another strategy to study microorganisms in the environment without obtaining them as a pure or mixed culture is to focus on single cells. However, it is crucial to keep in mind that a clonal pure culture consists of individual cells that can exhibit phenotypic heterogeneity. Simon Avery discusses in his chapter the latest developments in microbial cell individuality. Individual cells within a genetically uniform population can have different levels of gene transcription and translation, leading to diverse subpopulations that are genetically identical. In addition to genotypic heterogeneity (through mutation and genome rearrangements), phenotypic heterogeneity provides a population with a dynamic way to enhance fitness without having to manifest changes on a genome level. However, in conventional studies heterogeneity at the single-cell level is typically masked by data that averages across millions of cells in a population.

Techniques involving mechanical and laser manipulation, flow cytometry, or development of microfluidic chips have been designed to study individual cells within a pure culture or to obtain single cells from an environmental sample. In comparison to a classical serial dilution approach, which yields the most dominant organism in a sample, these techniques allow capturing single cells and even targeting low-abundance organisms from a sample. Recently, several advances have been made that now allow the study of structural and chemical properties with high sensitivity and resolution at a single-cell level. Ramya Desikan, Laurene Tetard, Ali Passian, Ram Datar, and Thomas Thundat review in their chapter the latest advances in scanning probe microscopy and related techniques that allow one to obtain microscopic images at a resolution of a tenth of a nanometer. In their detailed review, they discuss techniques that include scanning tunneling microscopy, atomic force microscopy, near-field scanning optical microscopy, photon scanning tunneling microscopy, photonic force microscopy, scanning near field ultrasonic holography, scanning electrochemical microscopy, and others. Some of these techniques can sense chemical, physical, and biological parameters with high sensitivity and selectivity.

New advances also have been made in single-cell genomics, which now allow amplifying and subsequently sequencing the genome of single (or very few) cells. Martin Keller, Christopher Schadt, and Anthony Palumbo describe the latest developments in this field. In particular, they discuss the three standard methods that are used to amplify genomic DNA—primer extension preamplification, degenerate oligonucleotide primed PCR, and multiple displacement amplification—and demonstrate that the latter gives the highest yields. Yet, a significant sequence bias still exists during amplification, which subsequently causes variations in genome coverage. However, if this technique is combined with targeted capturing of single cells, it will complement environmental metagenomic data by linking specific genes to the corresponding microorganisms. While the metagenomic approach is casting a wide net and is trying to sequence an environment as a whole, this "polony" sequencing approach is aiming to stitch the metagenome together one genome at a time.

Sequencing of a complete genome allows, in combination with information on physiology and metabolism, creation of metabolic networks and prediction of functions in silico. However, after more than a decade of complete genome sequencing (Fleischman et al., 1995), we are still not able to understand the most fundamental levels of life. How many and which genes are necessary for an organism under a certain growth condition; in other words, "How many genes does a cell need?" To answer this question, different methods have been applied to reduce the genome size or to create a minimal genome of a cell. One approach, which became feasible after costs for oligonucleotide synthesis had dropped dramatically, consisted of chemically synthesizing a genome and subsequently transferring it into a cell envelope. This has recently been performed successfully by transplanting genomic DNA from *Mycoplasma mycoides* into *Mycoplasma capricolum* cells (Lartigue et al., 2007). Ham Smith, John Glass, Clyde Hutchison, and Craig Venter describe the latest developments in the field of synthetic cells in their chapter.

Let us return to the question of this chapter—does cultivation still matter? Diverse applications of molecular tools produced vast amounts of information from microorganisms even without having them in culture. However, pure cultures or enrichments of microorganisms will always have their significance in microbiology. For a number of questions it is still essential to grow organisms in the laboratory that can then be studied under defined conditions. This not only allows obtaining detailed physiological information but makes controlled genetic manipulation of those organisms possible. Molecular approaches and isolation and cultivation of microorganisms should not be exclusive. Instead, cultivation should be used to benefit metagenomic and metaproteomic studies through information obtained from "domesticated" microorganisms and can provide new scaffolds for accurate genome assembly. Vice versa, information from metagenomic studies can help isolate previously uncultivated microorganisms; this was demonstrated, for example, by the cultivation of *Leptospirillum ferrodiazotrophum*, the key nitrogen-fixing organism in an acidophilic microbial community (Tyson et al., 2005). It is likely that not all microbes will be accessible as defined cultures in the laboratory by using current technology. Therefore, the development and combination of innovative techniques to study uncultivated microorganisms, ideally in their natural environment, will be essential to advance our understanding of microbial physiology and ecology and will shed light on these most diverse creatures on our planet.

REFERENCES

Ashby, M. N., J. Rine, E. F. Mongodin, K. E. Nelson, and D. Dimster-Denk. 2007. Serial analysis of ribosomal DNA and the unexpected dominance of rare members of microbial communities. *Appl. Environ. Microbiol.* **73:**4532–4542.

Beijerinck, W. M. 1895. Über *Spirillum desulfuricans* als Ursache von Sulfatreduktion. *Zentralblatt Bakteriol.* **1:**1–9, 49–59, and 104–114.

Breitbart, M., L. R. Thompson, C. A. Suttle, and M. B. Sullivan. 2007. Exploring the vast diversity of marine viruses. *Oceanography* **20:**135–139.

Casas, V., and F. Rohwer. 20007. Phage metagenomics. *Methods Enzymol.* **421:**259–268.

Cash, H. L., C. V. Whitham, C. L. Behrendt, and L. V. Hooper. 2006. Symbiotic bacteria direct expression of an intestinal bactericidal lectin. *Science* **313:**1126–1130.

Cohn, F. 1875. Untersuchungen über Bakterien. *Beitr. Biol. Pflanz.* **1:**127–224.

Conn, H. J. 1918. The microscopic study of bacteria and fungi in soil. *N. Y. Agr. Exp. Sta., Tech. Bull.* **64:**3–20.

Cooper, T. F., D. E. Rozen, and R. E. Lenski. 2003. Parallel changes in gene expression after 20,000 generations of evolution in *Escherichia coli*. *Proc. Natl. Acad. Sci. USA* **100:**1072–1077.

Fleischmann, R. D., M. D. Adams, O. White, R. A. Clayton, E. F. Kirkness, A. R. Kerlavage, C. J. Bult, J. F. Tomb, B. A. Dougherty, J. M. Merrick, R. D. Fleischmann, M. D. Adams, O. White, R. A. Clayton, E. F. Kirkness, A. R. Kerlavage, C. J. Bult, J. F. Tomb, B. A. Dougherty, J. M. Merrick, K. McKenney, G. Sutton, W. FritzHugh, C. Fields, and J. C. Venter. 1995. Whole-genome random sequencing and assembly of *Haemophilus influenzae* Rd. *Science* **269:**496–512.

Gest, H. 2007. Fresh views of the 17th-century discoveries by Hooke and van Leeuwenhoek. *Microbe* **2:**483–488.

Handelsman, J. 2004. Metagenomics: application of genomics to uncultured microorganisms. *Microbiol. Mol. Biol. Rev.* **68:**669–685.

Herring, C. D., A. Raghunathan, C. Honisch, T. Patel, M. K. Applebee, A. J. Joyce, T. J. Albert, F.

R. Blattner, D. van den Boom, C. R. Cantor, and B. Ø. Palsson. 2006. Comparative genome sequencing of *Escherichia coli* allows observation of bacterial evolution on a laboratory timescale. *Nat. Genet.* **38:**1406–1412.

Huber, J. A., D. B. M. Welch, H. G. Morrison, S. M. Huse, P. R. Neal, D. A. Butterfield, and M. L. Sogin. 2007. Microbial population structures in the deep marine biosphere. *Science* **318:**97–100.

Keller, M., and K. Zengler. 2004. Tapping into microbial diversity. *Nat. Rev. Microbiol.* **2:**141–150.

Koch, R. 1877. Untersuchungen über Bakterien VI. Verfahren zur Untersuchung, zum Conservieren und Photographieren. *Beitr. Biol. Pflanz.* **2:**399–434.

Konstantinidis, K. T., A. Ramette, and J. M. Tiedje. 2006. Toward a more robust assessment of intraspecies diversity, using fewer genetic markers. *Appl. Environ. Microbiol.* **72:**7286–7293.

Lartigue, C., J. I. Glass, N. Alperovich, R. Pieper, P. P. Parmar, C. A. Hutchison III, H. O. Smith, and J. C. Venter. 2007. Genome transplantation in bacteria: changing one species to another. *Science* **317:**632–638.

Lechene, C., F. Hillion, G. McMahon, D. Benson, A. M. Kleinfeld, J. P. Kampf, D. Distel, Y. Luyten, J. Bonventre, D. Hentschel, K. M. Park, S. Ito, M. Schwartz, G. Benichou, and G. Slodzian. 2006. High-resolution quantitative imaging of mammalian and bacterial cells using stable isotope mass spectrometry. *J. Biol.* **5:**20.

Maron, P. A., L. Ranjard, C. Mougel, and P. Lemanceau. 2007. Metaproteomics: a new approach for studying functional microbial ecology. *Microb. Ecol.* **53:**486–493.

Muβmann, M., F. Z. Hu, M. Richter, D. de Beer, A. Preisler, B. B, Jørgensen, M. Huntemann, F. O. Glöckner, R. Amann, W. J. H. Koopman, R. S. Lasken, B. Janto, J. Hogg, P. Stoodley, R. Boissy, and G. D. Ehrlich. 2007. Insights into the genome of large sulfur bacteria revealed by analysis of single filaments. *PLoS Biol.* **5:**e230.

Orphan, V. J., C. H. House, K.-U. Hinrichs, K. D. McKeegan, and E. F. DeLong. 2001. Methane-consuming *Archaea* revealed by directly coupled isotopic and phylogenetic analysis. *Science* **293:**484–487.

Ouverney, C. C., and J. A. Fuhrman. 1999. Combined microautoradiography-16S rRNA probe technique for determination of radioisotope uptake by specific microbial cell types *in situ*. *Appl. Environ. Microbiol.* **65:**1746–1752.

Peña, A., M. Valens, F. Santos, S. Buczolits, J. Antón, P. Kämpfer, H. J. Busse, R. Amann, and R. Rosselló-Mora. 2005. Intraspecific comparative analysis of the species *Salinibacter ruber*. *Extremophiles* **9:**151–161.

Rosselló-Mora, R., and R. Amann. 2001. The species concept for prokaryotes. *FEMS Microbiol. Rev.* **25:**39–67.

Suttle, C. A. 2007. Marine viruses—major players in the global ecosystem. *Nat. Rev. Microbiol.* **5:**801–812.

Tyson, G. W., I. Lo, B. J. Baker, E. E. Allen, P. Hugenholtz, and J. F. Banfield. 2005. Genome-directed isolation of the key nitrogen fixer *Leptospirillum ferrodiazotrophum* sp. nov. from an acidophilic microbial community. *Appl. Environ. Microbiol.* **71:**6319–6324.

van Niel, C. B. 1949. The "Delft School" and the rise of general microbiology. *Microbiol. Mol. Biol. Rev.* **13:**161–174.

Wagner, M., H. Smidt, A. Loy, and J. Zhou. 2007. Unravelling microbial communities with DNA-microarrays: challenges and future directions. *Microb. Ecol.* **53:**498–506.

Winogradsky, S. 1887. Über Schwefelbacterien. *Botanische Zeitung* **45:**489–507, 513–523, 529–539, 545–559, 569–576, 585–594, and 606–610.

THE HUMAN INTESTINAL MICROBIOTA AND ITS IMPACT ON HEALTH

Mirjana Rajilić-Stojanović, Willem M. de Vos, and Erwin G. Zoetendal

2

Microbial ecosystems are found all over the world and are involved in the cycling of the elements that are crucial for our lives. Without microbes, existence of other life forms on Earth is impossible. On the other side, microbes are often linked to diseases. The relationship between our health and the influence of microbes has interested researchers for centuries and continues to be a highly active research area. The first researcher who discovered the involvement of microbes in our health was Antonie van Leeuwenhoek. In 1681 he examined, during his illness, his watery excrement under the microscope and saw that it contained much more small "animals" (probably *Giardia lamblia*) than normal stool (Dobell, 1920). Since then, many microbes have been isolated from all types of ecosystems, resulting in nearly 600 registered culture collections in 66 countries and even more private collections (http://wdcm.nig.ac.jp/hpcc.html). Despite all cultivation attempts for over a century to isolate microbes from a wide variety of ecosystems, it is generally accepted that we can only cultivate a minority of all existing microbes (Rappé and Giovannoni, 2004). This was well demonstrated by nucleic acid-based studies that have indicated that the majority of microbes in any ecosystem is different from those obtained by cultivation. The main reason for this cultivation anomaly is caused by the fact that we do not know the conditions microbes require for growth and that we introduce stresses by implementing cultivation procedures, which, for example, include oxygen exposure and breaking symbiotic interactions. After the discovery that ribosomal RNA (rRNA) is present in every cell and that its nucleotide sequence can be used for phylogenetic classification (Woese, 1987; Woese et al., 1990), there was an enormous increase in studies using approaches based on the sequence diversity of small-subunit (SSU) rRNA and its encoding gene that detected and identified many uncultured microbes. Currently, more than 400,000 SSU rRNA sequences are available in the DNA databases, and this is far more than what can be found for any other gene (http://www.arb-silva.de).

Although microbes are everywhere and, in theory, they all may have an impact on our health, microbial diversity is so enormous that it

Mirjana Rajilić-Stojanović, Willem M. de Vos, and Erwin G. Zoetendal, Laboratory of Microbiology, Wageningen University, Dreijenplein 10, 6703 HB Wageningen, The Netherlands.

Accessing Uncultivated Microorganisms: from the Environment to Organisms and Genomes and Back
Edited by Karsten Zengler © 2008 ASM Press, Washington, DC

is hardly possible to describe all of them. Therefore, we focus this chapter on the microbes residing in our gastrointestinal (GI) tract that are collectively termed the microbiota, since these have the most intimate interaction with our body and, therefore, have a prominent impact on our health.

THE HUMAN GI TRACT AND ITS MICROBIOTA

The human GI tract is the system of organs that is involved in the conversion and absorption of food, thus supplying our body with energy and nutrient sources. To do this efficiently, the GI tract has a specific anatomy, which includes the stomach (divided into pylorus and antrum), the small intestine (divided into duodenum, jejunum, and ileum), the large intestine (divided into cecum; ascending, transverse, descending, and sigmoid colon; and rectum), and the anus (Fig. 1). This approximately 7-meter-long system of organs with a surface of approximately 300 square meters in adults (similar to that of a tennis court) (Bengmark, 1998) is continuously exposed to the outside environment and, therefore, challenged by its dangers, including microbes that are intruded or ingested. Therefore, the GI tract also has, besides an energy and nutrient-harvesting role, an important defensive role. Involved in this defensive role are the low pH in the stomach, the coverage of the complete GI tract with a mucous layer, and the harboring of an army of cells involved in our immune system. Another player in the GI tract that is involved in our well-being is our commensal microbiota, which is distributed along the entire GI tract with density and diversity increasing from the stomach to the colon. The GI microbiota is involved in the conversion of indigestible food components (Cummings and Macfarlane, 1997; MacFarlane et al., 1986), production of essential vitamins and cofactors (Albert et al., 1980; Ramotar et al., 1984), and limiting colonization by pathogens via colonization resistance (van der Waaij, 1989).

It is evident that the microbiota in our GI tract plays a pivotal role in our health and that we cannot live without a microbiota. However, our understanding of the GI microbiota and its role in health and disease is still very limited. Many factors have contributed to our lack of knowledge about such an important ecosystem with which we are constantly interacting. One of the main problems is that most parts of the human GI tract are not accessible without invasive procedures. In fact, the small intestine is the

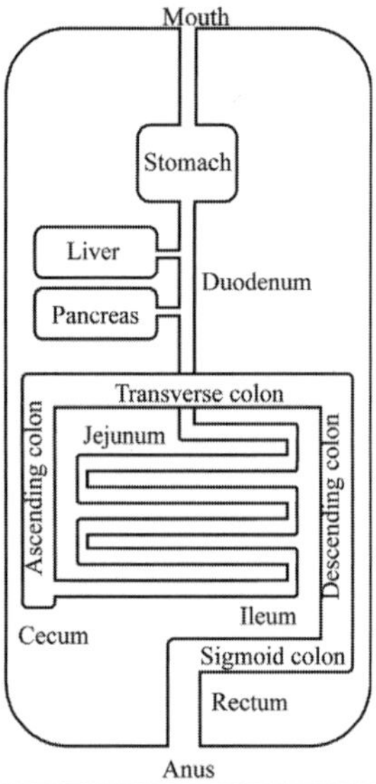

GI segment	Length, cm	Passage, time, h	Density of microbiota, cells/ml(g)
Stomach	12	2-6	10^0-10^4
Duodenum	25		10^4-10^5
Jejunum	160	3-5	10^5-10^7
Ileum	215		10^7-10^8
Caecum	6		
Ascending colon	15		
Transverse colon	50	10-20	10^{10}-10^{11}
Descending colon	25		
Sigmoid colon	40		
Rectum	18	1	10^{10}-10^{11}

FIGURE 1 Approximate dimensions of the human GI tract and corresponding microbiota density (Freitas, 1999; Guyton, 1985; Justesen et al., 1984; Moore and Holdeman, 1974a).

first part where host and microbes meet and where most conversion of food components takes place, but it is the least studied region because of its inaccessibility. As a result, most of our current knowledge about the microbiota is based on the analysis of the feces. Furthermore, the GI microbiota is an extremely complex ecosystem, whose density exceeds that of any other reported ecosystem (Bäckhed et al., 2005) and whose composition varies highly between individuals. Therefore, after more than 100 years of study, the GI microbiota remains insufficiently described, even at the composition level.

Describing the diversity of a microbial ecosystem is only the first, yet indispensable, step in advancing its understanding. Reliable insight into the diversity of the GI microbiota is essential for providing a reference framework to study its dynamics in time and space, analyze its functions, and characterize host-microbe interactions. However, as is discussed below, there are many challenges to overcome before we completely understand what is present in the GI tract, what each member is doing, and how microbes are related to our health status.

DIVERSITY OF THE GI MICROBIOTA—STATE OF THE ART

During the past century many microbes from the human GI tract have been described through isolation procedures and subsequent characterization as well as culture-independent approaches, mainly based on SSU rRNA gene analysis. In 2005 Bäckhed and coworkers stated in a seminal review paper that of 55 known bacterial phyla only 8 can be found in the human GI tract (Bäckhed et al., 2005). However, 2 years later, both figures should be reconsidered. Depending on the SSU rRNA gene database used, the number of existing bacterial phyla ranges from 31 to 88 (DeSantis et al., 2006), and analysis of all presently available data indicates that members of at least 10 bacterial phyla, 1 archaeal phylum, and 1 eukaryotic phylum are inhabitants of the human GI tract (Fig. 2). That these reconsiderations are needed

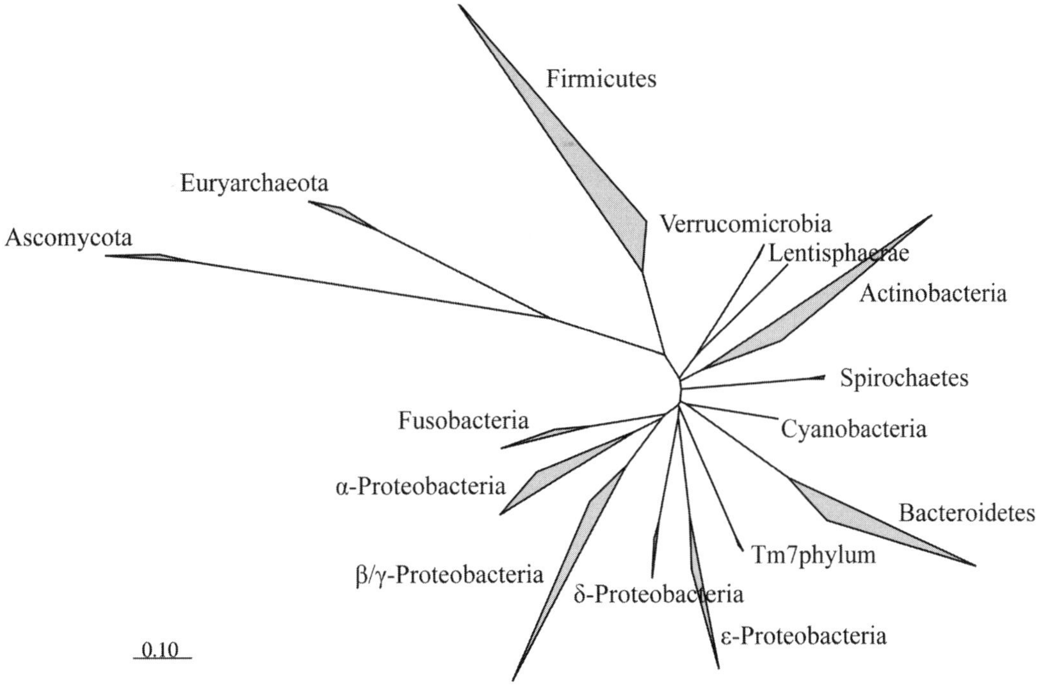

FIGURE 2 SSU rRNA-based phylogenetic tree representing the diversity of the human GI microbiota. Each phylum is presented as a distinct phylogenetic group except for the *Proteobacteria*, which are presented by four phylogenetic groups. The reference bar indicates 10% sequence divergence.

argues for the reliability of the present SSU rRNA gene databases and indicates that the description of the diversity of this ecosystem has not yet been brought to an end, but rather is a continuous process.

The recent revival of the interest in the human GI microbiota has triggered numerous studies and has resulted in an exceptionally dynamic field of research. Many studies have aimed at better description of this remarkably relevant ecosystem, using both traditional cultivation approaches (Bakir et al., 2006; Duncan et al., 2006; Mohan et al., 2006; Song et al., 2006) and also molecular-based approaches (Bik et al., 2006; Eckburg et al., 2005). Traditional culturing has, up to now, resulted in a partial description of the diversity of the GI microbiota. Nevertheless, the recent successful attempts to obtain novel isolates are encouraging as they brought into light a number of quantitatively, and potentially functionally, relevant GI microbes (for recent review, see Duncan et al., 2007). Together with these novel isolates, the hitherto cultivable diversity of GI microbiota consists of 442 bacterial, 3 archaeal, and 17 eukaryotic species (Rajilić-Stojanović, 2007). However, these kinds of figures become continuously outdated because of the constant reports of novel intestinal inhabitants (e.g., a recent description of another bacterial species, *Enterococcus caccae*) (Carvalho et al., 2006). Despite the fact that the cultivation-based approaches underestimate the true diversity of the human GI microbiota, systematic integration of results presented in almost 300 scientific publications has revealed that the number of known, fully characterized GI inhabitants is within the range of the most frequently cited diversity prediction for this ecosystem of 400 to 500 species. Application of culture-independent approaches has revealed that the microbial ecosystem in the human GI tract is more complex, and the majority of recent reports estimate the species richness of this ecosystem as 1,000 (Nicholson et al., 2005; Noverr and Huffnagle, 2004; Phillips, 2006). However, analysis of GI microbes that have either been identified by culture-dependent or culture-independent approaches demonstrated that the number of presently reported GI phylotypes exceeds this recent estimation of 1,000 organisms, whereas the predicted species richness suggests that the number of species that can inhabit the human GI tract will be measured in thousands (Rajilić-Stojanović et al., 2007). From over 1,200 presently detected inhabitants of the human GI tract, only about 12% were recovered by application of both molecular and cultivation-based approaches, while the majority of GI microbiota (~75%) was detected solely in SSU rRNA-based surveys. Despite the large discrepancy between the GI microbiota portrayals assessed by cultivation-dependent and -independent approaches, both approaches clearly show that most of the intestinal inhabitants belong to three bacterial phyla—*Firmicutes*, *Bacteroidetes*, and *Actinobacteria*.

The diverse and abundant *Firmicutes*, *Bacteroidetes*, and *Actinobacteria* are accompanied in the GI tract by several distantly related microbial groups, including the members of the phylum candidate TM7 (Rajilić-Stojanović, 2007). The TM7 phylum represents a newly recognized, widely distributed group of yet uncultured bacteria (Hugenholtz et al., 2001). Members of this group had previously been detected in a number of ecosystems including soil (Podar et al., 2007) and bioreactors (Thomsen et al., 2002), and also in the human oral cavity (Kazor et al., 2003; Paster et al., 2001), the mouse intestine (Salzman et al., 2002), the rumen (Tajima et al., 2000), and the beetle intestine (Egert et al., 2003). Therefore, its detection in the human intestinal tract does not seem to represent an environmental contamination, although its abundance and function in this ecosystem remain unknown. Taking into account this recent report of the TM7 phylum, it appears that 4 out of 10 presently known intestinal bacterial phyla have been detected within the past decade. The only hitherto known cyanobacterial phylotype of intestinal origin was reported in 2005 (Eckburg et al., 2005). This phylotype was only detected as an SSU rRNA sequence that could not be classified further than the order *Chroococcales*.

Furthermore, an isolate representing the previously uncultured intestinal members belonging to the *Verrucomicrobia* was reported in 2004 (Derrien et al., 2004), while the only hitherto known human intestinal member of the *Lentisphaera* was isolated in 2003 (Cho et al., 2004; Zoetendal et al., 2003). The reason for the delayed detection of these organisms by cultivation-based techniques is the special nutritional demands of those microbes. The fact that they escaped detection by SSU rRNA-based techniques might be explained by their relatively low abundance and the limitations of "universal" PCR primers. The design of SSU rRNA-based primers is largely dependent on sequence knowledge, which is exponentially increasing; at the same time, it has been shown that many previously designed primers/probes do not have the claimed specificity or universality (Horz et al., 2005). Another explanation for the ongoing detection of novel phylotypes is the fact that low abundant phylotypes are often not captured in SSU rRNA gene libraries. Other methods, such as the recently developed serial analysis of rRNA genes, could be fruitful in the detection of these rare phylotypes (Ashby et al., 2007).

From the present pace of detecting novel phyla, the presence of other, not yet acknowledged, distantly related groups can be anticipated. However, newly detected SSU rRNA gene sequences rarely represent distinct branches in the phylogenetic tree as the majority of the recovered sequences cluster within three phyla—*Firmicutes, Bacteroidetes,* and *Actinobacteria* (Bik et al., 2006; Eckburg et al., 2005; Ley et al., 2006). Although the completion of the description of the GI microbiota is a long-term objective, examples of novel isolates, such as *Bacteroides dorei* (Bakir et al., 2006), whose SSU rRNA gene sequence has been reported in nine cultivation-independent studies, or *Clostridium glycyrrhizinilyticum* (Sakuma et al., 2006), which corresponds to two previously reported SSU rRNA gene sequences, are suggesting that the complete description of the intestinal microbiota is a challenging but, in the end, feasible goal. The fact that the proportion of novel phylotypes reported in some studies appears discouraging (e.g., Eckburg et al., 2005) should be regarded with caution as the number of detected (novel) phylotypes largely depends on the phylotype cutoff value. As previously reported, within the same clone library, the number of distinct phylotypes ranges from 148 to 643 when the phylotype defining cutoff value is changed from 97 to 99% (Venter et al., 2004). The choice of cutoff value seems to be a rather ambiguous issue that depends on the author's decision. It is known that the sequence of different copies of an SSU rRNA in the same genome is rarely below 99% although it can substantially vary (Acinas et al., 2004). Moreover, this variation increases between different strains of the same species but rarely reaches values of 97% (Klappenbach et al., 2001; Sacchi et al., 2005). Hence, it is evident that SSU rRNA gene sequences alone will not provide absolute answers about the diversity of the human GI microbiota. This insight could be reached after isolation and characterization of all members of this ecosystem, which, at this stage, is a highly ambitious goal.

IMPLEMENTATION OF NOVEL CULTIVATION STRATEGIES IN GI TRACT RESEARCH

As indicated above, there is a huge gap between the existing microbial diversity in the human GI tract and the diversity represented in the cultivated fraction. Since SSU rRNA and its corresponding genes do not provide any physiological data or information about the metabolic potential of the uncultured phylotypes, getting them into culture will be a great challenge for microbiologists. This is essential since cultured representatives can be physiologically characterized and used as model microbes in several in vitro or animal model studies. The importance of these types of studies has been demonstrated by interactions between host and microbes in a mono-associated mouse model that was colonized by *Bacteroides thetaiotaomicron* (Hooper and Gordon, 2001; Hooper et al., 1999). These studies demonstrated that *B. thetaiotaomicron* was able to modulate the expression of host genes with a variety of physiological functions, including nutrient absorption and immune responses, and that *B. thetaiotaomicron* was able to adapt its lifestyle according to the nutrient availability in the intestine. Despite the undisputable value of information that was obtained in such reductionist approaches, it should be kept in mind that these studies provide only a first glance at the importance of the intestinal microbiota since the complex ecosystem is oversimplified by using a single species (which does not represent more than 1% of the total community).

It has to be mentioned that systematic isolation using undefined rich media might not be the wisest approach since it has, up to now, resulted in the isolation of similar phylotypes. For example, the numerous members of the *Proteobacteria* are frequently obtained by traditional isolation procedures whereas they are often rare and present in low numbers in the GI tract (Chiesa et al., 1993; Müller, 1986). Although it cannot be ruled out that these microbes are relevant for the numerous functions of microbiota as a whole, more valuable information about this complex ecosystem would be achieved with targeted isolation and characterization of microbial groups of particular interest. An example of the cultivation that targeted butyrate producers (Barcenilla et al., 2000; Duncan et al., 2004) enabled the isolation of *Roseburia* spp. (Duncan et al., 2002a) and the reclassification of *Fusobacterium prausnitzii* into the novel genus of *Faecalibacterium* (Duncan et al., 2002b). These findings appeared to be highly relevant, as those organisms are abundant in the feces (Hold et al., 2003; Suau et al., 2001).

Another example of a targeted cultivation approach is the use of defined media using specific carbon and energy sources. These types of media likely represent the conditions at the end of the colon much better than the routinely used rich media, since conditions at the end of the GI tract are harsh where only the complex and difficult to degrade carbohydrates are left. A recent example is the isolation of *Akkermansia muciniphila* through use of basal media containing pork gastric mucin as the sole carbon and energy source (Derrien et al., 2004). *A. muciniphila* belongs to the phylum *Verrucomicrobia* and was commonly detected in feces from healthy individuals by SSU rRNA-based approaches. Quantification of *A. muciniphila* by SSU rRNA targeted fluorescent in situ hybridization and quantitative PCR demonstrated that it is a common member of the human GI tract with a mean abundance of 1.3% of the total fecal community (Derrien, 2007). Interestingly, a similar approach using cellobiose as the sole carbon and energy source has resulted in the isolation of *Victivallis vadensis* (Zoetendal et al., 2003), which belongs to the recently described phylum *Lentisphaerae* (Cho et al., 2004). These studies indicated that use of basal media containing specific carbon sources may lead to the isolation of novel isolates from the human GI tract.

These targeted approaches will definitely help microbiologists in the isolation of novel microbes from the human GI tract. However, it must be realized that not all microbes reach the end of the large intestine alive. It was recently shown that bacterial cells in feces or in pure culture, especially when stress conditions are applied, can be enumerated and divided into active, injured, and dead cell populations by flow cytometry (FCM) with live/dead staining probes (Apajalahti et al., 2003; Ben-Amor et al., 2002; Ben-Amor et al., 2005). When *Bifidobacterium* spp. are exposed to increasing bile concentrations in pure culture, the fraction of injured and dead cells subsequently increased. Interestingly, the injured cells are characterized as cells that are detected with both live and dead staining probes and that dedicated cultivation procedures are needed to have them grow in culture (Ben-Amor et al., 2002). By using the viability FCM approach to screen fecal samples it has been demonstrated that up to one-third of all fecal bacteria were scored as dead (Apajalahti et al., 2003; Ben-Amor et al., 2002; Ben-Amor et al., 2005). Remarkably, the composition of bacteria in the live and dead fractions is overrepresented by different phylotypes as demonstrated by fluorescence-activated cell sorting of these groups and subsequent cloning and sequencing of the SSU rRNA genes. Bacteria related to known butyrate-producing bacteria predominated in the live fraction, while bacteria affiliated with *Bacteroides*, *Ruminococccus*, and *Eubacterium* were more abundant in the dead fractions (Ben-Amor et al., 2005). These are important findings, since they link phylogenetic information of the GI bacteria to activity. These findings also suggest that some of the GI phylotypes can never be isolated from the feces as they are not alive anymore when they leave the body. The fact that fecal microbiota does not comprehensively represent what can be found in other sites of the GI tract is supported by finding only representatives of GI cyanobacteria and α-proteobacteria in studies where diversity of mucosal samples from the upper GI tract was analyzed (Eckburg et al., 2005; Wang et al., 2005). Therefore, it is likely that such bacteria could be isolated from samples obtained from other regions in the human GI tract.

In addition to the use of targeted cultivation approaches to get the uncultured microbes into culture, high-throughput cultivation is an alternative. The idea behind high-throughput cultivation is to cage single microbes by encapsulation in agarose, allow them to form microcolonies in the environmental samples, and subsequently sort these microcolonies by FCM (Zengler et al., 2002). The benefit of this approach is that microbial interactions, such as nutrient exchanges, remain during cultivation, since the separation of the microcolonies will be performed after cultivation. Another high-throughput cultivation technique that has recently been described is the Anopore system, which allows the simultaneous cultivation of thousands of microbial cells on a chip (Ingham et al., 2006). The Anopore system allows simultaneous growth of an enormous number of microbes; this is relevant for characterization of the highly dense and diverse microbiota within the GI tract. It has previously been noted that one of the major limitations for describing the GI microbiota in full by cultivation is its complexity. When a single sample is processed, an enormous number of different isolates can be obtained, all of which are physically difficult to fully describe (Moore and Holdeman, 1974b). Hence, recent high-throughput cultivation technologies could overcome this hurdle, and major improvements in this area could be made in the near future.

All the approaches mentioned above provide microbiologists a wide range of opportunities when bringing currently uncultured microbes into culture. However, even with those approaches at hand, isolation of so far uncultured microbes seems almost an endless process given the interindividual variation and complexity of the GI microbiota. To shed light on which role so far uncultured groups of microbes could play in our health, the continuous application of culture-independent approaches is therefore crucial.

THE GI MICROBIOTA OF HEALTHY INDIVIDUALS

Since microbiologists have acknowledged that cultivation will provide only a partial picture of the true diversity, the application of culture-independent approaches to study microbial diversity has increased exponentially during the past two decades. The most applied approaches are cloning and subsequent sequencing of SSU rRNA genes, fluorescent in situ hybridization (FISH), and fingerprinting approaches such as denaturing gradient gel electrophoresis (DGGE). When focusing on the ecosystem that resides in the human GI tract, these approaches have already provided a phylogenetic framework of the diversity and novel insights into its ecology. In addition, it has been observed that the predominant bacterial composition in feces of healthy adult individuals is relatively stable over time (Franks et al., 1998; Seksik et al., 2003; Tannock et al., 2000; Vanhoutte et al., 2004; Zoetendal et al., 2001; Zoetendal et al., 1998). In contrast to healthy adults, however, severely disturbed and/or unstable fecal microbiotas can be correlated with humans with GI tract disorders, such as Crohn's disease (CD) (Seksik et al., 2003), ulcerative colitis (UC) (Andoh et al., 2007), and intestinal bowel syndrome (IBS) (Mättö et al., 2005; Maukonen et al., 2006). Although cultivation data already demonstrated that community shifts occur with aging and that these changes happen especially in newborn babies and elderly people (Mitsuoka, 1992), culture-independent data have shown that during life the microbial diversity becomes more complex and that the cultivability of the microbiota decreases over time (J. Doré, personal communication).

A promising approach for novel insights into the complex GI microbiota is the phylogenetic microarray that enables a qualitative and quantitative high-throughput analysis of targeted ecosystems. The great potential of this method was illustrated by DeSantis et al. (2007), who showed that an SSU rRNA gene-based phylogenetic microarray had superior diagnostic power for the analysis of microbial community structure over a clone library approach. Phylogenetic microarrays are in general based on SSU rRNA-targeted diagnostic oligonucleotide probes that allow the detection of microorganisms at different levels of phylogenetic resolution (Guschin et al., 1997; Loy and Bodrossy, 2006; Smoot et al., 2005; Wagner et al., 2007; Zhou, 2003). At this moment, ecosystem-specific microarrays are available for a wide range of complex microbial ecosystems, including the human intestine and the oral cavity (Palmer et al., 2006; Rajilić-Stojanović, 2007; Smoot et al., 2005).

The direct coupling of microbiota profiling and phylogenetic analysis facilitates efficient data extraction. Phylogenetic microarrays can be valuable for assessing all kinds of relevant information, for example, assessing the difference between metabolically active and inactive bacteria by analyzing DNA- and RNA-based profiles of the microbiota. Other fingerprinting techniques, such as DGGE and related techniques, have been previously applied for such analysis (Tannock et al., 2004; Zoetendal et al., 1998). Identification of different groups, observed as DGGE patterns, requires the construction of clone libraries, making the analysis laborious and time consuming. Assessment of those profiles by using a phylogenetic microarray for a fecal sample of an individual showed a more than twofold difference between the hybridization signals of more than 20 phylogenetic groups (Fig. 3). The microbial landscape obtained by either DNA or RNA was found to be tremendously different, as *Actinobacteria* were composing 1% of the total bacteria (based on the DNA profile) and 18% of the active bacteria (based on the RNA profile), whereas *Bacteroidetes* were found to make up 30% of the total (based on the DNA profile) and only 17% of the active bacteria (based on the RNA profile) (Fig. 3 B, C). In a previous DGGE-based study, *Collinsella* spp. and *Bifidobacterium* spp. were identified as the most active (Tannock et al., 2004), which is in agreement with our findings, as those groups showed a remarkably increased signal in hybridization with RNA compared to the hybridization with DNA (40- and 27-fold difference, respectively) (Rajilić-Stojanović, 2007). That the difference of the other 18 phylogenetic groups was not detected with DGGE illustrates the lower sensitivity of this technique in comparison to the phylogenetic microarrays.

Although the major factor for the observed difference in the DNA versus RNA microbiota profile is differential activity of different microbial groups, it should be kept in mind that other factors are contributing to the observed discrepancies. One of the relevant factors is the copy number of rRNA operons that differs between microbes. The number of SSU rRNA genes in the genome has been determined or estimated for a dozen of GI microbes (Fogel et al., 1999; Klappenbach et al., 2001) and averages around three copies of this gene. However, *Bacteroides* and *Streptococcus* spp. have an average of six SSU rRNA gene copies, which will lead to overestimation of these groups by DNA-based quantification. This factor might be even more relevant since the number of genome copies per cell is related to the growth rate (Button and Robertson, 2001), and from various studies (e.g., Ben-Amor et al., 2005) it is known that different bacterial groups are not equally active along the GI tract.

Recently, the first studies that employ phylogenetic microarrays for the analysis of GI microbiota were presented, and the results of

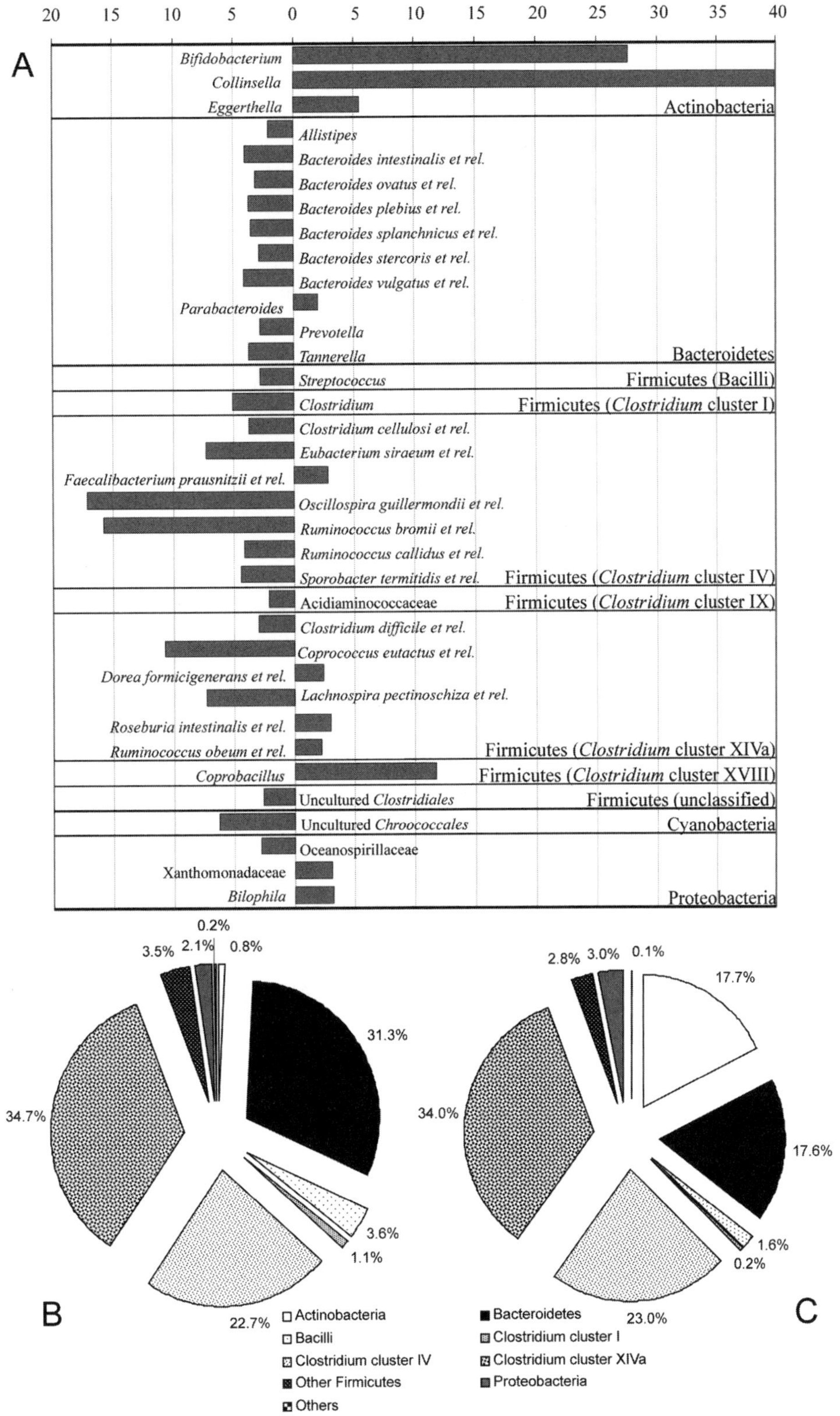

FIGURE 3 (A) Ratio of the average hybridization signals obtained by the analysis of fecal DNA and RNA. Less active phylogenetic groups (higher hybridization signals with DNA) are presented as left-oriented bars, while active phylogenetic groups (higher hybridization signal with RNA) are presented with right-oriented bars. Relative composition of the total microbiota (B) and active microbiota (C).

these studies have illustrated the power of the applied, holistic approach. Analysis of the development of microbiota of 14 newborn babies for over a year showed it to be individual-specific since the first day of birth (Palmer et al., 2007), confirming earlier studies (Favier et al., 2002). Furthermore, this study also demonstrated that the microbiota of infants is unexpectedly chaotic in early months, but it follows, in all babies, a similar pattern of development toward an adult-like microbiota, which is approached already at age 1. Despite our current inability to identify driving forces of these mechanisms, it is astonishing that an opportunistic, initial colonization of each newborn baby develops into a relatively stable and similar ecosystem at later age. In another study, it has been shown for the first time that different subpopulations do not follow the same pattern of change during temporal variations (Rajilić-Stojanović, 2007). While members of the *Firmicutes* were found to actively respond to environmental stimuli, *Bacteroidetes* and *Actinobacteria* have exhibited an astonishing level of stability within a life span of 4 years. This finding allows speculation that the human GI microbiota differs from other ecosystems, such as bioreactors, in which unchanged metabolic activity can be preserved, despite notable changes in the community composition (Fernandez et al., 2000), and that because of the constant and intimate interactions between the human host and microbes a subset of the microbiota seems to resist environmental triggering. The molecular mechanisms of the interactions between host and intestinal microbes are beginning to be understood (Haller, 2006; Rakoff-Nahoum et al., 2004), and newly presented results that showed specific preservation of *Bifidobacterium* and *Bacteroides* spp. are suggesting the presence of specific receptors that may recognize these organisms (Rajilić-Stojanović, 2007).

The first applications of phylogenetic microarrays have just suggested the importance this technology will have for revealing the truths about the GI microbiota and its relation with the host. However, this technology, as any other, has its limitations, which should be acknowledged. These approaches depend on the isolation of DNA and subsequent PCR amplification of SSU rRNA genes, which could be biased by irregular cell lysis and primer specificity, respectively. These are general drawbacks and are difficult to avoid. Furthermore, it may be considered that the capacity of the phylogenetic microarray approaches is limited by the fact that they detect only dominant groups. A powerful technique that can provide insights into the less abundant groups is quantitative PCR (qPCR), which can detect and enumerate subdominant populations with a sensitivity as little as 10^3 cells per gram of feces (~0.000001% of the total community) (Matsuda et al., 2007). Numerous group- and species-specific qPCR assays allowed quantification of a significant proportion of the human GI inhabitants (Matsuki et al., 2002; Matsuki et al., 2004; Rinttilä et al., 2004). On the other hand, specific PCR approaches focusing on less dominant or rare populations can also be combined with phylogenetic microarray analysis. This will provide insight into the diversity and relative abundance within these subdominant populations, which could be very useful in studying the population dynamics of probiotics or pathogens in the GI tract, as they often are not present in high numbers in the gut.

THE GI MICROBIOTA AND DISEASE

The GI microbiota plays a crucial role in our health by providing essential components for our body. On the other hand, microbes in the GI tract can also cause several diseases. Pathogens such as *Clostridium difficile*, *Campylobacter* spp., and members of the *Enterobacteriaceae* are frequently associated with acute diarrhea (Aranda-Michel and Giannella, 1999). These and other well-known GI pathogens, such as *Helicobacter pylori* and *Listeria monocytogenes*, have been well characterized in the laboratory. In addition to those pathogens, bacteria belonging to our GI microbiota are also suspected to play a relevant role in the development of several intestinal disorders. In general, these disorders have a complex and not yet defined etiology in which, besides the microbiota, genetic and environmental factors are implicated. For example, in a mouse model, colon cancer has been associated with the higher production of carcinogenic compounds by the microbiota, such as hydrogen sulfide and secondary bile acids (McGarr et al., 2005). In addition, alterations of hydrogen-consuming bacteria have been observed in patients with IBS (King et al., 1998; Mättö et al., 2005). However, associating these types of functional alterations to specific microbiota patterns in these diseases is very complicated. Part of this problem includes our inability to cultivate all members of the microbiota. Since on average 80% of the GI microbes have not been cultured, it is impossible to formulate hypotheses about the role of these uncultured microbes in health and disease as they are only known by their (partial) SSU rRNA sequence. Another major factor that complicates studying the association between microbes and a certain GI disorder is the individual-specific composition of the GI microbiota (Ley et al., 2006; Zoetendal et al., 1998). In each study, different individuals are recruited for the analysis of the microbiota composition, and this certainly has an impact on the results. Moreover, there is little consistency between the studies with respect to the target microbes or target molecules, which makes the comparison between different studies even more complicated. Nevertheless, correlations have been suggested between the composition and activity of the microbiota and a variety of clinical statuses. For example, intestinal bowel diseases, such as CD, UC, and IBS, have often been associated with a rather unstable microbiota composition, which contrasts with the stable composition in healthy individuals (Andoh et al., 2007; Mättö et al., 2005; Maukonen et al., 2006; Seksik et al., 2003). However, the presence of an unstable microbiota makes it even more difficult to pinpoint microbial groups that are correlated with these diseases, since it requires multiple samples and dedicated statistics. High-throughput analysis, such as the approaches described above, could offer new possibilities to tackle this problem.

UC and IBS are complex GI disorders that are affected by environmental and genetic factors (Bengtson et al., 2006; Gwee, 2005; Halfvarson et al., 2003; Sood et al., 2003). IBS, similar to UC in remission, is characterized by low-grade inflammation and functional disturbances in the intestine, and both pathologies are strongly associated with the GI microbiota (Haller, 2006). The results of the high-throughput profiling of fecal samples showed that the microbiota of both IBS and UC patients has a distinct composition when compared to that of corresponding healthy controls (Fig. 4) (Rajilić-Stojanović, 2007). A general observation was that members of the *Firmicutes* were the most dramatically affected in both aberrations, which is in line with our finding that this phylogenetic group is more strongly affected than the other microbial groups by environmental changes (Rajilić-Stojanović, 2007). Correlated with the severity of the disease, the microbiota composition of UC patients was found to be significantly different from that of healthy adults ($P = 0.002$), while the difference in the microbiota composition between IBS patients and corresponding controls showed a strong trend in relation to the health status ($P = 0.157$). The *Firmicutes*, which were found to be the most dramatically affected in both IBS and UC, are also the most diverse and abundant taxonomic group within the GI microbiota. Furthermore, the GI *Firmicutes* primarily consist of uncultured organisms, the functionality of which is unknown. Therefore, the accurate rationalization of the observed trends is possible for only few microbial groups, such as *Roseburia* spp, which were among the *Firmicutes* that were decreased in UC patients. This reduced abundance of *Roseburia* spp., which are well-established butyrate producers (Duncan et al., 2002a), could be correlated with the recently shown reduction of butyrate levels in feces of UC patients (Marchesi et al., 2007). In contrast, this genus was found to be increased in diarrhea-predominant IBS patients, which adds to the controversy about butyrate's effect on human health (Macfarlane et al., 1994; Scheppach and Weiler, 2004). Although

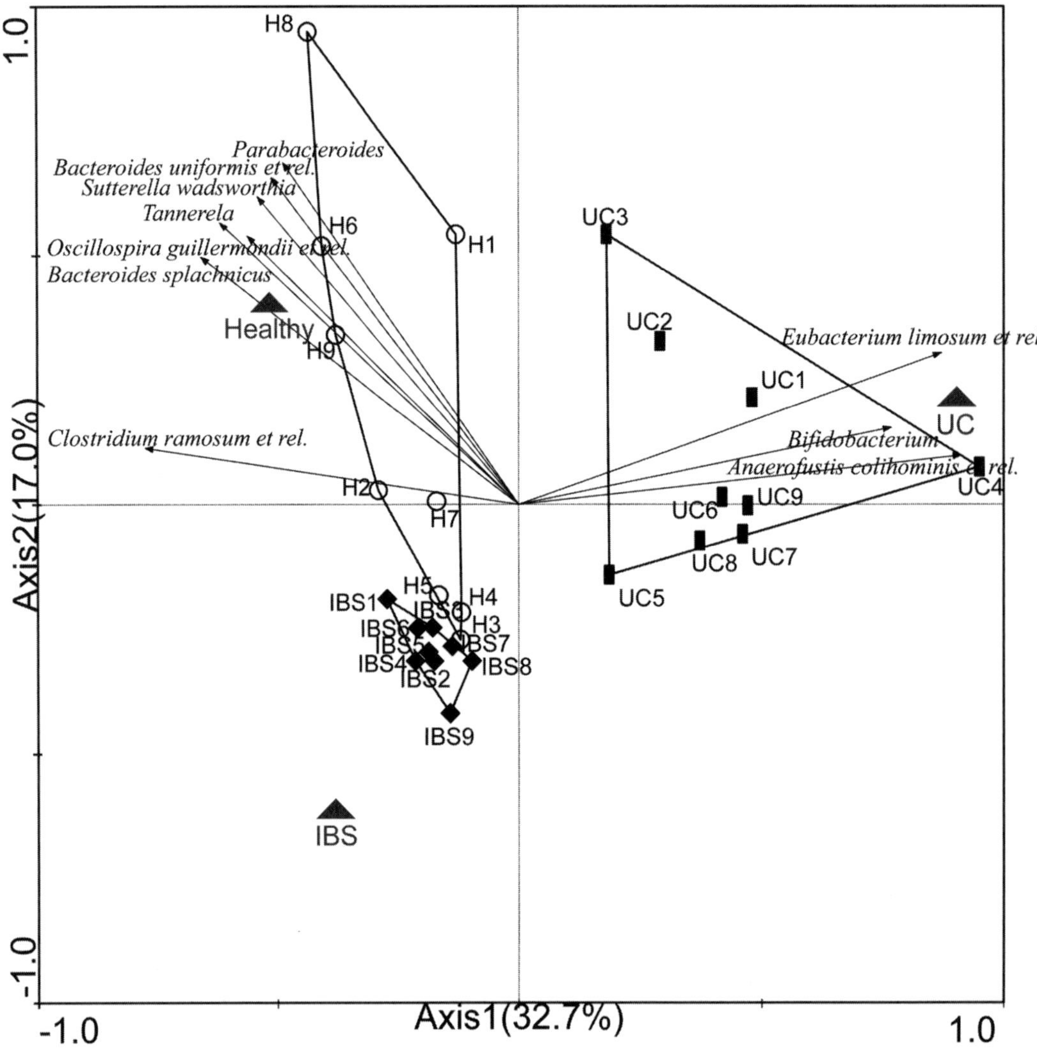

FIGURE 4 PCA triplot depicting the microbiota of healthy adults (○) (D1–D9), patients with IBS (◆) (IBS1–IBS9), and patients with UC (■) (UC1–UC9). Percentage values at the axes indicate contribution of the principal components to the explanation of total variance in the dataset; phylogenetic groups named after cultivated representative that contributed at least 60% to the explanatory axis used in the plot are presented as vectors, while centroides of the plot describe health status of subjects.

butyrate for a long time was considered an important microbial metabolite that promotes GI cell growth (Roediger, 1980), some novel findings indicated that butyrate can also have a necrotic effect on colonic cell lines (Matsumoto et al., 2006).

The reduced diversity of the microbiota in UC patients has previously been associated with an inflamed mucosa (Ott et al., 2004; Sepehri et al., 2007). This reduction seems to be accompanied by the reduction of the total number of microbes in relation to some of the disease. It was shown by molecular-based enumeration that the abundance of fecal bacteria in UC patients was reduced from 1.16×10^{11} cells per gram (characteristic for healthy individuals) to 3.51×10^{10} cells per gram (Ben-Amor, 2004). However, it cannot be ruled out that the reduction of the microbial diversity is not as strongly affected as reported since it was demonstrated by targeting only presently acknowledged bacterial species. It has been shown that the widely applied universal bacterial probes, such as EUB 338, do not obtain a significant hybridization signal when hybridized against some of the intestinal bacteria, such as the abundant and widely distributed intestinal bacterium *A. muciniphila* (Derrien, 2007). However, as the presence of *A. muciniphila* was not acknowledged until 2004 (Derrien et al., 2004), the sequence of this microbe could not be taken into account during EUB 338 probe design in 1990 (Amann et al., 1990). Similarly, it is possible that other not yet identified intestinal inhabitants are omitted when the total bacteria are analyzed by universal bacterial probes, and consequently lead to the conclusion that microbial diversity is reduced. This seems to be even more likely to be case with the microbiota of patients with digestive diseases, as it was shown that a quantitatively relevant fraction of the microbiota of UC and CD patients was composed of bacteria that could not be identified as members of any presently known phylogenetic group of intestinal bacteria (Sokol et al., 2006). Therefore, the impact of these unknown bacteria remains speculative.

In line with UC patients, the microbiota in CD patients is also characterized by a significant reduction in *Firmicutes*, together with an increase in *Proteobacteria* and *Bacteroidetes* (Gophna et al., 2006; Sokol et al., 2006). This suggests that the number of *Firmicutes* may be considered as a biomarker for intestinal bowel diseases. Obtaining more insights into the function of these *Firmicutes* is therefore of utmost importance.

Another novel finding obtained with the phylogenetic microarray analysis in relation to intestinal disorders concerns the increased heterogeneity among the microbial composition of both IBS and UC patients (Rajilić-Stojanović, 2007). In the case of IBS, this finding was in line with the systematization of IBS patients into three distinct groups based on their clinical symptoms, and UC also has been suggested as a heterogeneous disease (Sartor, 2006).

The GI microbiota has also been associated with diseases that are not directly linked to the intestine or its mucosal surface. A decrease in *Bifidobacterium* and *Enterococcus* species and increased levels of clostridia have been associated with allergies (Kirjavainen et al., 2001; Noverr and Huffnagle, 2005). Similarly, a higher incidence of *Clostridium histolyticum*-like bacteria (members of the *Clostridium* cluster II) has been observed in children with autism (Parracho et al., 2005). Moreover, a recent paper by van Tongeren and colleagues (2005) indicated that the highly frail elderly have a significantly reduced number of *Lactobacillus* and/or *Enterococcus* species, members of the *Bacteroidetes* phylum, and *F. prausnitzii* in their feces in comparison to the healthier elderly, while the number of *Enterobacteriaceae* increased. Last but not least, a decreased *Bacteroidetes* population was observed in obese individuals (Ley et al., 2006). Remarkably, this population increased during weight loss when the subjects were following different low-calorie diets.

Altogether, these observations point to an important role of our intestinal microbiota in health and disease. Major questions that need to be addressed include whether there is a causal relation between the presence of specific microbiota and the clinical symptoms and, if so, what mechanisms underlie the observed effects. To answer these questions, the functions of the GI microbiota and its interaction with the host must be elucidated. One way to approach this includes the application of meta-analytic approaches in GI research.

META-ANALYSIS OF THE GI MICROBES

Despite the value of SSU rRNA-based approaches in finding links between microbes and a certain health status, these approaches do not provide information about the functions since these cannot be extracted from SSU rRNA data. Moreover, a correlation between the presence of a microbe and a disease status does not explain whether the microbe is the cause or the result of the disorder. Therefore, studies focusing on the genetic potential or activity of uncultured microbes are needed to shed light on potential roles of microbes in health and disease.

One way to gain insight into potential functions of microbes is by performing metagenomic approaches. Metagenomics, which means the study of collected genomes from an ecosystem, can be used to study the phylogenetic, physical, and functional properties of the microbial communities (Handelsman, 2004). The first metagenomic approach that focused on the human GI tract described the diversity of the viral community in human feces (Breitbart et al., 2003), which consisted of approximately 1,200 viral genotypes. Remarkably, the majority of viral sequences in this collection showed highest sequence similarity to phages known to infect gram-positive bacteria, which is in agreement with SSU rRNA gene-based studies. The first prokaryotic metagenomic studies from GI samples of humans have recently been published, and we expect to see an explosive increase in metagenomic data within the coming years (Gill et al., 2006; Jones and Marchesi, 2007; Manichanh et al., 2006). The first comparative metagenomic approach with fecal samples from humans indicated that *Firmicutes* are significantly reduced in complexity in CD patients compared to healthy subjects (Manichanh et al., 2006), which is in line with previous observations as discussed above. The metagenomic approach described by Gill and colleagues (Gill et al., 2006) demonstrated that many genes in the GI microbiota represent essential functions, such as vitamin production, that often have been ascribed to the microbiota.

A major limitation of metagenomics is that the number of different genes within a community is enormous and that it is almost impossible at the moment to get reasonable coverage for complex ecosystems such as the human GI tract. In addition, the presence of a gene in an ecosystem does not mean that it is functionally important. Therefore, metagenomics should be considered as a catalog for activity-based approaches, such as metatranscriptomics or metaproteomics. This approach has been successfully applied for an acid mine drainage ecosystem, a low diversity environment, and resulted in a description of metabolic networks in this ecosystem (Ram et al., 2005).

Besides coupling meta-analytic approaches, potential functions of microbes can also be studied by function-driven analyses (Handelsman, 2004). Recently an intracellular approach to screen metagenomic libraries for clones that induced or inhibited quorum sensing was developed (Williamson et al., 2005). In addition, a substrate-induced gene-expression screen has been developed to identify genes in a groundwater metagenome that were induced by aromatic hydrocarbons (Uchiyama et al., 2005). However, one should keep in mind that the control and expression of these genes in their original hosts could be very different. Recently, a high-throughput phenotyping approach was used to screen GI bacterial metagenomic clones for capacities to modulate the growth of epithelial cells in vitro (Gloux et al., 2007). The authors demonstrated that this approach may allow the identification of potential novel mechanisms of host-microbe interactions in the GI tract. The application of this type of screening could be useful to screen for beneficial or harmful functions in the GI microbiota. Subsequently, the expression for the corresponding genes can be monitored in situ and serve as potential biomarkers for intervention studies.

Metagenomics combined with other meta-analytic approaches will provide a wealth of information, and this requires continuous developments and improvements in high-throughput sequencing technologies and bioinformatics. However, there is also a current need for more genome sequences from a wide variety of GI isolates, as those form the basis for functional categorization of genes. For example, the members of *Clostridium* cluster IV and *Clostridium* cluster XIVa are the most abundant *Firmicutes* in the human GI tract, from which no genome sequence has been determined yet. Therefore, initiatives to determine the genome sequence of GI tract members that represent the major microbial phyla, such as the Human Gut Microbiome Initiative (http://www.genome.gov/Pages/Research/Sequencing/SeqProposals/HGMISeq.pdf), are very valuable since they will enable researchers to construct metabolic networks from data obtained by meta-analytic approaches. In addition, novel genomics approaches such as combining microbial cell sorting and subsequent complete genome amplification and sequencing offer great possibilities to obtain genome sequences of uncultured microbes (Podar et al., 2007; Raghunathan et al., 2005). However, as the diversity in the gut is enormous, individual- and location-specific genome sequencing of selected and representative single isolates has to be complemented by sequence analysis of metagenomes from a variety of locations in the GI tract from different individuals.

CONCLUDING REMARKS AND FUTURE PERSPECTIVES

This chapter summarized the current status of our knowledge on GI microbiota diversity and focused on the recent observations with respect to the association of uncultured microbes and different health statuses. Fingerprinting techniques, such as DGGE, and other SSU rRNA-based approaches have played a major role in the molecular revolution of our view on complex ecosystems such as the GI tract. Their application has revealed many relevant observations of the human GI microbiota, including the conclusion that the microbiota is individual-specific, affected by the host genotype, and stable over time (Franks et al., 1998; Seksik et al., 2003; Tannock et al., 2000; Vanhoutte et al., 2004; Zoetendal et al., 2001; Zoetendal et al., 1998). However, the high-resolution profiling that can be performed with phylogenetic microarrays has several advantages over the well-established fingerprinting by DGGE (Neufeld et al., 2006). Therefore, it is to be expected that phylogenetic microarrays will prevail over other fingerprinting techniques in future diagnostic surveys of the intestinal and other ecosystems.

The implementation of high-throughput technologies, such as phylogenetic arrays, is very promising in getting more detailed insights into the population dynamics of the GI tract ecosystems as they allow the simultaneous analysis of thousands of microbes in a single analysis. This may enable us to link uncultured microbes to a certain health status, and this will help us form hypotheses about the role of these microbes in a certain disease. However, even the targeted 1,000 phylotypes, which are currently represented on the most comprehensive microarray for the human GI microbiota—Human Intestinal Tract Chip (HITChip)—represent only a fraction of the total microbial diversity, as thousands of novel distinct SSU rRNA sequences are expected to be found in samples of human GI origin (Rajilić-Stojanović, 2007). Despite the fact that the generated phylogenetic profiles provide diagnostic information even about unknown phylotypes, the further improvement of the HITChip, and other similar phylogenetic microarrays (Palmer et al., 2006), is dependent on the discovery of novel GI inhabitants. Therefore, the accuracy and comprehension of phylogenetic microarray analysis will largely benefit from the reports of novel SSU rRNA sequences from human GI clone libraries (Bik et al., 2006; Eckburg et al., 2005; Leitch et al., 2007; Ley et al., 2006)and also from the cultivation of the novel GI inhabitants (Carvalho et al., 2006).

Applications of FISH have enabled the first insights into the morphology and physiology of uncultured organisms, such as members ofTM7 phylum candidate (Hugenholtz et al., 2001; Thomsen et al., 2002). In addition to the quantification of microbial groups in complex ecosystems, which has been exceptionally fruitful (Lay et al., 2005; Mueller et al., 2006; Sokol et al., 2006), FISH has also other potential. Its combination with fluorescence-activated cell sorting is a promising approach for determining relevant data on the complex ecosystems, such as analysis of alive, injured, and dead subsets of GI bacteria (Ben-Amor et al., 2005; Vaughan et al., 2005). The development of functional gene targeted CPRINC-FISH (Kenzaka et al., 2005) or RING-FISH (Zwirglmaier et al., 2004) in combination with cell sorting has a great potential for detection of microbial subpopulations that are responsible for relevant conversions. Similar to targeted cultivation (Duncan et al., 2004), this approach would enable identification of relevant microbial species. Furthermore, functional genes have been shown to have similar discriminative power as housekeeping genes, such as SSU rRNA, which can be used for identification and detection of novel phylotypes of a particular guild (Wagner et al., 2005). Therefore, it can be speculated that relevant functional genes will form the basis for future generations of microarrays that couple analysis of diversity and activity.

Finally, functional analyses are the ultimate goal of the research on the human GI microbiota. Reductionist approaches, which analyze simplified microbial communities in animal or in vitro models, have revolutionized the perception of the microbial impact on the metabolism of the host. The pioneering work of J. Gordon and coworkers, which has shown that microbial colonization improves nutritional and defensive functions of the host (Hooper and Gordon, 2001; Hooper et al., 1999), was further expanded by including other microbial species (Derrien, 2007; Sonnenburg et al., 2006), and also by analyzing the impact of the host on the microbe (de Vries, 2006; Sonnenburg et al., 2006). Furthermore, reductionist approaches of analyzing different cultured cells provide insight into new genes and functions of individual organisms; this is a milestone for the global approaches, such as metagenomics. Although metagenomic analysis has shown their power for revealing the coding potential of the microbiome, the genome of the entire microbial ecosystem (Gill et al., 2006; Manichanh et al., 2006), further functional studies, such as metaproteomics, have been reported but show limitations in our predictive capacity (Klaassens et al., 2006). Hence, there is a great need to integrate all available reductionist and global, cultivation- and molecular-based approaches to finally describe and understand our micro companions throughout our journey on planet Earth.

REFERENCES

Acinas, S. G., L. A. Marcelino, V. Klepac-Ceraj, and M. F. Polz. 2004. Divergence and redundancy of 16S rRNA sequences in genomes with multiple rrn operons. *J. Bacteriol.* **186:**2629–2635.

Albert, M. J., V. I. Mathan, and S. J. Baker. 1980. Vitamin B_{12} synthesis by human small intestinal bacteria. *Nature* **238:**781–782.

Amann, R. I., B. J. Binder, R. J. Olson, S. W. Chisholm, R. Devereux, and D. A. Stahl. 1990. Combination of 16S rRNA-targeted oligonucleotide probes with flow cytometry for analyzing mixed microbial populations. *Appl. Environ. Microbiol.* **56:**1919–1925.

Andoh, A., S. Sakata, Y. Koizumi, K. Mitsuyama, Y. Fujiyama, and Y. Benno. 2007. Terminal restriction fragment length polymorphism analysis of the diversity of fecal microbiota in patients with ulcerative colitis. *Inflamm. Bowel. Dis.* **13:**955–962.

Apajalahti, J. H., A. Kettunen, P. H. Nurminen, H. Jatila, and W. E. Holben. 2003. Selective plating underestimates abundance and shows differential recovery of bifidobacterial species from human feces. *Appl. Environ. Microbiol.* **69:**5731–5735.

Aranda-Michel, J., and R. A. Giannella. 1999. Acute diarrhea: a practical review. *Am. J. Med.* **106:**670.

Ashby, M. N., J. Rine, E. F. Mongodin, K. E. Nelson, and D. Dimster-Denk. 2007. Serial analysis of rRNA genes and the unexpected dominance of rare members of microbial communities. *Appl. Environ. Microbiol.* **73:**4532–4542.

Bäckhed, F., R. E. Ley, J. L. Sonnenburg, D. A. Peterson, and J. I. Gordon. 2005. Host-bacterial mutualism in the human intestine. *Science* **307:**1915–1920.

Bakir, M. A., M. Sakamoto, M. Kitahara, M. Matsumoto, and Y. Benno. 2006. *Bacteroides dorei* sp. nov., isolated from human faeces. *Int. J. Syst. Evol. Microbiol.* **56:**1639–1643.

Barcenilla, A., S. E. Pryde, J. C. Martin, S. H. Duncan, C. S. Stewart, C. Henderson, and H. J. Flint. 2000. Phylogenetic relationships of butyrate-producing bacteria from the human gut. *Appl. Environ. Microbiol.* **66:**1654–1661.

Ben-Amor, K. 2004. Ph.D. thesis. Microbial ecophysiology of the human intestinal tract: a flow cytometric approach, p.166. Laboratory of Microbiology, Wageningen University, Wageningen, The Netherlands.

Ben-Amor, K., P. Breeuwer, P. Verbaarschot, F. M. Rombouts, A. D. L. Akkermans, W. M. De Vos, and T. Abee. 2002. Multiparametric flow cytometry and cell sorting for the assessment of viable, injured, and dead bifidobacterium cells during bile salt stress. *Appl. Environ. Microbiol.* **68:**5209–5216.

Ben-Amor, K., H. G. Heilig, H. Smidt, E. E. Vaughan, T. Abee, and W. M. de Vos. 2005.

Genetic diversity of viable, injured, and dead fecal bacteria assessed by fluorescence-activated cell sorting and 16S rRNA gene analysis. *Appl. Environ. Microbiol.* **71:**4679–4689.

Bengmark, S. 1998. Ecological control of the gastrointestinal tract. The role of probiotic flora. *Gut* **42:**2–7.

Bengtson, M. B., T. Ronning, M. H. Vatn, and J. R. Harris. 2006. Irritable bowel syndrome in twins: genes and environment. *Gut* **55:**1754–1759.

Bik, E. M., P. B. Eckburg, S. R. Gill, K. E. Nelson, E. A. Purdom, F. Francois, G. Perez-Perez, M. J. Blaser, and D. A. Relman. 2006. Molecular analysis of the bacterial microbiota in the human stomach. *Proc. Natl. Acad. Sci.* **103:**732–737.

Breitbart, M., I. Hewson, B. Felts, J. M. Mahaffy, J. Nulton, P. Salamon, and F. Rohwer. 2003. Metagenomic analyses of an uncultured viral community from human feces. *J. Bacteriol.* **185:**6220–6223.

Button, D. K., and B. R. Robertson. 2001. Determination of DNA content of aquatic bacteria by flow cytometry. *Appl. Environ. Microbiol.* **67:**1636–1645.

Carvalho, M. D. G. S., P. L. Shewmaker, A. G. Steigerwalt, R. E. Morey, A. J. Sampson, K. Joyce, T. J. Barrett, L. M. Teixeira, and R. R. Facklam. 2006. *Enterococcus caccae* sp. nov., isolated from human stools. *Int. J. Syst. Evol. Microbiol.* **56:**1505–1508.

Chiesa, C., L. Pacifico, F. Nanni, A. M. Renzi, and G. Ravagnan. 1993. *Yersinia pseudotuberculosis* in Italy. Attempted recovery from 37,666 samples. *Microbiol. Immunol.* **37:**391–394.

Cho, J. C., K. L. Vergin, R. M. Morris, and S. J. Giovannoni. 2004. *Lentisphaera araneosa* gen. nov., sp. nov, a transparent exopolymer producing marine bacterium, and the description of a novel bacterial phylum, *Lentisphaerae*. *Environ. Microbiol.* **6:**611–621.

Cummings, J. H., and G. T. Macfarlane. 1997. Colonic microflora: nutrition and health. *Nutrition* **13:**476.

Derrien, M. 2007. Ph.D. thesis. Mucin utilisation and host interactions of the novel intestinal microbe *Akkermansia muciniphila*, p. 164. Laboratory of Microbiology, Wageningen University, Wageningen, The Netherlands.

Derrien, M., E. E. Vaughan, C. M. Plugge, and W. M. de Vos. 2004. *Akkermansia muciniphila* gen. nov., sp. nov., a human intestinal mucin-degrading bacterium. *Int. J. Syst. Evol. Microbiol* **54:**1469–1476.

DeSantis, T. Z., E. L. Brodie, J. P. Moberg, I. X. Zubieta, Y. M. Piceno, and G. L. Andersen. 2007. High-density universal 16S rRNA microarray analysis reveals broader diversity than typical clone library when sampling the environment. *Microb. Ecol.* **53:**371–383.

DeSantis, T. Z., P. Hugenholtz, N. Larsen, M. Rojas, E. L. Brodie, K. Keller, T. Huber, D. Dalevi, P. Hu, and G. L. Andersen. 2006. Greengenes, a chimera-checked 16S rRNA gene database and workbench compatible with ARB. *Appl. Environ. Microbiol.* **72:**5069–5072.

de Vries, M. C. 2006. Ph.D. thesis. Analyzing global gene expression of *Lactobacillus plantarum* in the human gastrointestinal tract, p. 160. Laboratory of Microbiology, Wageningen University, Wageningen, The Netherlands.

Dobell, C. A. 1920. The discovery of the intestinal protozoa of man. *Proc. R. Soc. Med.* **13:**1–15.

Duncan, S. H., R. I. Aminov, K. P. Scott, P. Louis, T. B. Stanton, and H. J. Flint. 2006. Proposal of *Roseburia faecis* sp. nov., *Roseburia hominis* sp. nov. and *Roseburia inulinivorans* sp. nov., based on isolates from human faeces. *Int. J. Syst. Evol. Microbiol.* **56:**2437–2441.

Duncan, S. H., G. L. Hold, A. Barcenilla, C. S. Stewart, and H. J. Flint. 2002a. *Roseburia intestinalis* sp. nov., a novel saccharolytic, butyrate-producing bacterium from human faeces. *Int. J. Syst. Evol. Microbiol.* **52:**1615–1620.

Duncan, S. H., G. L. Hold, H. Harmsen, C. S. Stewart, and H. J. Flint. 2002b. Growth requirements and fermentation products of *Fusobacterium prausnitzii*, and a proposal to reclassify it as *Faecalibacterium prausnitzii* gen. nov., comb. nov. *Int. J. Syst. Evol. Microbiol.* **52:**2141–2146.

Duncan, S. H., P. Louis, and H. J. Flint. 2004. Lactate-utilizing bacteria, isolated from human feces, that produce butyrate as a major fermentation product. *Appl. Environ. Microbiol.* **70:**5810–5817.

Duncan, S. H., P. Louis, and H. J. Flint. 2007. Cultivable bacterial diversity from the human colon. *Lett. Appl. Microbiol.* **44:**343–350.

Eckburg, P. B., E. M. Bik, C. N. Bernstein, E. Purdom, L. Dethlefsen, M. Sargent, S. R. Gill, K. E. Nelson, and D. A. Relman. 2005. Diversity of the human intestinal microbial flora. *Science* **308:**1635–1638.

Egert, M., B. Wagner, T. Lemke, A. Brune, and M. W. Friedrich. 2003. Microbial community structure in midgut and hindgut of the humus-feeding larva of *Pachnoda ephippiata* (*Coleoptera: Scarabaeidae*). *Appl. Environ. Microbiol.* **69:**6659–6668.

Favier, C. F., E. E. Vaughan, W. M. de Vos, and A. D. L. Akkermans. 2002. Molecular monitoring of succession of bacterial communities in human neonates. *Appl. Environ. Microbiol.* **68:**219–226.

Fernandez, A. S., S. A. Hashsham, S. L. Dollhopf, L. Raskin, O. Glagoleva, F. B. Dazzo, R. F. Hickey, C. S. Criddle, and J. M. Tiedje. 2000. Flexible community structure correlates with stable community function in methanogenic bioreactor communities perturbed by glucose. *Appl. Environ. Microbiol.* **66:**4058–4067.

Fogel, G. B., C. R. Collins, J. Li, and C. F. Brunk. 1999. Prokaryotic genome size and SSU rDNA copy number: estimation of microbial relative abundance from a mixed population. *Microb. Ecol.* **38:**93–113.

Franks, A. H., H. J. M. Harmsen, G. C. Raangs, G. J. Jansen, F. Schut, and G. W. Welling. 1998. Variations of bacterial populations in human feces measured by fluorescent in situ hybridization with group-specific 16S rRNA-targeted oligonucleotide probes. *Appl. Environ. Microbiol.* **64:**3336–3345.

Freitas, R. A. 1999. *Navigational Alimentography.* Landes Bioscience, Georgetown, TX.

Gill, S. R., M. Pop, R. T. Deboy, P. B. Eckburg, P. J. Turnbaugh, B. S. Samuel, J. I. Gordon, D. A. Relman, C. M. Fraser-Liggett, and K. E. Nelson. 2006. Metagenomic analysis of the human distal gut microbiome. *Science* **312:**1355–1359.

Gloux, K., M. Leclerc, H. Iliozer, R. L'Haridon, C. Manichanh, G. Corthier, R. Nalin, H. M. Blottière, and J. Doré. 2007. Development of high-throughput phenotyping of metagenomic clones from the human gut microbiome for modulation of eukaryotic cell growth. *Appl. Environ. Microbiol.* **73:**3734–3737.

Gophna, U., K. Sommerfeld, S. Gophna, W. F. Doolittle, and S. J. O. Veldhuyzen van Zanten. 2006. Differences between tissue-associated intestinal microfloras of patients with Crohn's disease and ulcerative colitis. *J. Clin. Microbiol.* **44:**4136–4141.

Guschin, D. Y., B. K. Mobarry, D. Proudnikov, D. A. Stahl, B. E. Rittmann, and A. D. Mirzabekov. 1997. Oligonucleotide microchips as genosensors for determinative and environmental studies in microbiology. *Appl. Environ. Microbiol.* **63:**2397–2402.

Guyton, A. C. 1985. *Anatomy and Physiology.* Saunders College Pub., Philadelphia, PA.

Gwee, K. A. 2005. Irritable bowel syndrome in developing countries—a disorder of civilization or colonization? *Neurogastroent. Motil.* **27:**317–324.

Halfvarson, J., L. Bodin, C. Tysk, E. Lindberg, and G. Jarnerot. 2003. Inflammatory bowel disease in a Swedish twin cohort: a long-term follow-up of concordance and clinical characteristics. *Gastroenterology* **124:**1767.

Haller, D. 2006. Intestinal epithelial cell signalling and host-derived negative regulators under chronic inflammation: to be or not to be activated determines the balance towards commensal bacteria. *Neurogastroenterol. Motil.* **18:**184–199.

Handelsman, J. 2004. Metagenomics: application of genomics to uncultured microorganisms. *Microbiol. Mol. Biol. Rev.* **68:**669–685.

Hold, G. L., A. Schwiertz, R. I. Aminov, M. Blaut, and H. J. Flint. 2003. Oligonucleotide probes that detect quantitatively significant groups of butyrate-producing bacteria in human feces. *Appl. Environ. Microbiol.* **69:**4320–4324.

Hooper, L. V., and J. I. Gordon. 2001. Commensal host-bacterial relationships in the gut. *Science* **292:**1115–1118.

Hooper, L. V., J. Xu, P. G. Falk, T. Midtvedt, and J. I. Gordon. 1999. A molecular sensor that allows a gut commensal to control its nutrient foundation in a competitive ecosystem. *Proc. Natl. Acad. Sci. USA* **96:**9833–9838.

Horz, H. P., M. E. Vianna, B. P. F. A. Gomes, and G. Conrads. 2005. Evaluation of universal probes and primer sets for assessing total bacterial load in clinical samples: general implications and practical use in endodontic antimicrobial therapy. *J. Clin. Microbiol.* **43:**5332–5337.

Hugenholtz, P., G. W. Tyson, R. I. Webb, A. M. Wagner, and L. L. Blackall. 2001. Investigation of candidate division TM7, a recently recognized major lineage of the domain bacteria with no known pure-culture representatives. *Appl. Environ. Microbiol.* **67:**411–419.

Ingham, C. J., M. van den Ende, P. C. Wever, and P. M. Schneeberger. 2006. Rapid antibiotic sensitivity testing and trimethoprim-mediated filamentation of clinical isolates of the *Enterobacteriaceae* assayed on a novel porous culture support. *J. Med. Microbiol.* **55:**1511–1519.

Jones, B. V., and J. Marchesi. 2007. Transposon-aided capture (TRACA) of plasmids resident in the human gut mobile metagenome. *Nat. Methods* **4:**55–61.

Justesen, T., O. H. Nielsen, and P. A. Krasilnikoff. 1984. Normal cultivable microflora in upper jejunal fluid in children without gastrointestinal disorders. *J. Pediatr. Gastroenterol. Nutr.* **3:**683–686.

Kazor, C. E., P. M. Mitchell, A. M. Lee, L. N. Stokes, W. J. Loesche, F. E. Dewhirst, and B. J. Paster. 2003. Diversity of bacterial populations on the tongue dorsa of patients with halitosis and healthy patients. *J. Clin. Microbiol.* **41:**558–563.

Kenzaka, T., S. Tamaki, N. Yamaguchi, K. Tani, and M. Nasu. 2005. Recognition of individual genes in diverse microorganisms by cycling primed in situ amplification. *Appl. Environ. Microbiol.* **71:**7236–7244.

King, T. S., M. Elia, and J. O. Hunter. 1998. Abnormal colonic fermentation in irritable bowel syndrome. *Lancet* **352:**1187.

Kirjavainen, P. V., E. Apostolou, T. Arvola, S. J. Salminen, G. R. Gibson, and E. Isolauri. 2001. Characterizing the composition of intestinal microflora as a prospective treatment target in infant allergic disease. *FEMS Immunol. Med. Microbiol.* **32:**1.

Klaassens, E. S., W. M. de Vos, and E. E. Vaughan. 2006. A metaproteomics approach to study the functionality of the microbiota in the human infant gastrointestinal tract. *Appl. Environ. Microbiol.*:AEM.01921–01906.

Klappenbach, J. A., P. R. Saxman, J. R. Cole, and T. M. Schmidt. 2001. rrndb: the ribosomal RNA operon copy number database. *Nucleic Acids Res.* **29:**181–184.

Lay, C., L. Rigottier-Gois, K. Holmstrom, M. Rajilic, E. E. Vaughan, W. M. de Vos, M. D. Collins, R. Thiel, P. Namsolleck, M. Blaut, and J. Doré. 2005. Colonic microbiota signatures across five northern European countries. *Appl. Environ. Microbiol.* **71:**4153–4155.

Leitch, E. C., A. W. Walker, S. H. Duncan, G. Holtrop, and H. J. Flint. 2007. Selective colonization of insoluble substrates by human faecal bacteria. *Environ. Microbiol.* **9:**667–679.

Ley, R. E., P. J. Turnbaugh, S. Klein, and J. I. Gordon. 2006. Microbial ecology: human gut microbes associated with obesity. *Nature* **444:**1022–1023.

Loy, A., and L. Bodrossy. 2006. Highly parallel microbial diagnostics using oligonucleotide microarrays. *Clin. Chim. Acta* **363:**106.

Macfarlane, G. T., J. H. Cummings, and C. Alison. 1986. Protein degradation by human intestinal bacteria. *J. Gen. Microbiol.* **132:**1647–1656.

Macfarlane, G. T., G. R. Gibson, and S. Macfarlane. 1994. Short chain fatty acid and lactate production by human intestinal bacteria grown in batch and continuous culture. *In* H. J. Binder, J. H. Cummings, and K. H. Soergel (ed.), *Short Chain Fatty Acids.* Kluwer Publishing, London, United Kingdom.

Manichanh, C., L. Rigottier-Gois, E. Bonnaud, K. Gloux, E. Pelletier, L. Frangeul, R. Nalin, C. Jarrin, P. Chardon, P. Marteau, J. Roca, and J. Doré. 2006. Reduced diversity of faecal microbiota in Crohn's disease revealed by a metagenomic approach. *Gut* **55:**205–211.

Marchesi, J. R., E. Holmes, F. Khan, S. Kochhar, P. Scanlan, F. Shanahan, I. D. Wilson, and Y. Wang. 2007. Rapid and noninvasive metabonomic characterization of inflammatory bowel disease. *J. Proteome Res.* **6:**546–551.

Matsuda, K., H. Tsuji, T. Asahara, Y. Kado, and K. Nomoto. 2007. Sensitive quantitative detection of commensal bacteria by rRNA-targeted reverse transcription-PCR. *Appl. Environ. Microbiol.* **73:**32–39.

Matsuki, T., K. Watanabe, J. Fujimoto, Y. Miyamoto, T. Takada, K. Matsumoto, H. Oyaizu, and T. Ryuichiro. 2002. Development of 16S rRNA-gene-targeted group-specific primers for the detection and identification of predominant bacteria in human feces. *Appl. Environ. Microbiol.* **68:**5445–5451.

Matsuki, T., K. Watanabe, J. Fujimoto, T. Takada, and R. Tanaka. 2004. Use of 16S rRNA gene-targeted group-specific primers for real-time PCR analysis of predominant bacteria in human feces. *Appl. Environ. Microbiol.* **70:**7220–7228.

Matsumoto, T., T. Hayasak, Y. Nishimura, M. Nakamura, T. Takeda, Y. Tabuchi, M. Obinata, T. Hanawa, and H. Yamada. 2006. Butyrate induces necrotic cell death in murine colonic epithelial cell MCE301. *Biol. Pharm. Bull.* **29:**2041–2045.

Mättö, J., L. Maunuksela, K. Kajander, A. Palva, R. Korpela, A. Kassinen, and M. Saarela. 2005. Composition and temporal stability of gastrointestinal microbiota in irritable bowel syndrome—a longitudinal study in IBS and control subjects. *FEMS Immunol. Med. Microbiol.* **43:**213–222.

Maukonen, J., R. Satokari, J. Mättö, H. Söderlund, T. Mattila-Sandholm, and M. Saarela. 2006. Prevalence and temporal stability of selected clostridial groups in irritable bowel syndrome in relation to predominant faecal bacteria. *J. Med. Microbiol.* **55:**625–633.

McGarr, S. E., J. M. Ridlon, and P. B. Hylemon. 2005. Diet, anaerobic bacterial metabolism, and colon cancer:a review of the literature. *J. Clin. Gastroenterol.* **39:**98–109.

Mitsuoka, T. 1992. Intestinal flora and aging. *Nutr. Rev.* **50:**438–446.

Mohan, R., P. Namsolleck, P. A. Lawson, M. Osterhoff, M. D. Collins, C.-A. Alpert, and M. Blaut. 2006. *Clostridium asparagiforme* sp. nov., isolated from a human faecal sample. *Syst. Appl. Microbiol.* **29:**292.

Moore, W. E. C., and L. V. Holdeman. 1974a. Human fecal flora:the normal flora of 20 Japanese-Hawaiians. *Appl. Microbiol.* **27:**961–979.

Moore, W. E. C., and L. V. Holdeman. 1974b. Special problems associated with the isolation and identification of intestinal bacteria in fecal flora studies. *Am. J. Clin. Nutr.* **27:**1450–1455.

Mueller, S., K. Saunier, C. Hanisch, E. Norin, L. Alm, T. Midtvedt, A. Cresci, S. Silvi, C. Orpianesi, M. C. Verdenelli, T. Clavel, C. Koebnick, H.-J. F. Zunft, J. Doré, and M. Blaut. 2006. Differences in fecal microbiota in different European study populations in relation to age, gender, and country:a cross-sectional study. *Appl. Environ. Microbiol.* **72:**1027–1033.

Müller, H. E. 1986. Occurrence and pathogenic role of *Morganella-Proteus-Providencia* group bacteria in human feces. *J. Clin. Microbiol.* **23:**404–405.

Neufeld, J. D., W. W. Mohn, and V. de Lorenzo. 2006. Composition of microbial communities in hexachlorocyclohexane (HCH) contaminated soils from Spain revealed with a habitat-specific microarray. *Environ. Microbiol.* **8:**126–140.

Nicholson, J. K., E. Holmes, and I. D. Wilson. 2005. Gut microorganisms, mammalian metabolism and personalized health care. *Nat. Rev. Microbiol.* **3**:431–438.

Noverr, M. C., and G. B. Huffnagle. 2004. Does the microbiota regulate immune responses outside the gut? *Trends Microbiol.* **12**:562–568.

Noverr, M. C., and G. B. Huffnagle. 2005. The "microflora hypothesis" of allergic diseases. *Clin. Exp. Allergy* **35**:1511–1520.

Ott, S. J., M. Musfeldt, D. F. Wenderoth, J. Hampe, O. Brant, U. R. Folsch, K. N. Timmis, and S. Schreiber. 2004. Reduction in diversity of the colonic mucosa associated bacterial microflora in patients with active inflammatory bowel disease. *Gut* **53**:685–693.

Palmer, C., E. M. Bik, D. B. Digiulio, D. A. Relman, and P. O. Brown. 2007. Development of the human infant intestinal microbiota. *PLoS Biol.* **5**:e177.

Palmer, C., E. M. Bik, M. B. Eisen, P. B. Eckburg, T. R. Sana, P. K. Wolber, D. A. Relman, and P. O. Brown. 2006. Rapid quantitative profiling of complex microbial populations. *Nucleic Acids Res.* **34**:e5.

Parracho, H. M. R. T., M. O. Bingham, G. R. Gibson, and A. L. McCartney. 2005. Differences between the gut microflora of children with autistic spectrum disorders and that of healthy children. *J. Med. Microbiol.* **54**:98–991.

Paster, B. J., S. K. Boches, J. L. Galvin, R. E. Ericson, C. N. Lau, V. A. Levanos, A. Sahasrabudhe, and F. E. Dewhirst. 2001. Bacterial diversity in human subgingival plaque. *J. Bacteriol.* **183**:3770–3783.

Phillips, M. L. 2006. Interdomain interactions: dissecting animal-bacterial symbioses. *BioScience* **56**:376–381.

Podar, M., C. B. Abulencia, M. Walcher, D. Hutchison, K. Zengler, J. A. Garcia, T. Holland, D. Cotton, L. Hauser, and M. Keller. 2007. Targeted access to the genomes of low-abundance organisms in complex microbial communities. *Appl. Environ. Microbiol.* **73**:3205–3214.

Raghunathan, A., H. R. Ferguson, C. J. Bornarth, W. Song, M. Driscoll, and R. S. Lasken. 2005. Genomic DNA amplification from a single bacterium. *Appl. Environ. Microbiol.* **71**:3342–3347.

Rajilić-Stojanović, M. 2007. Ph.D. thesis. Diversity of the human gastrointestinal microbiota: novel perspectives from high throughput analyses, p. 216. Laboratory of Microbiology, Wageningen University, Wageningen, The Netherlands.

Rajilić-Stojanović, M., H. Smidt, and W. M. de Vos. 2007. Diversity of the human gastrointestinal tract microbiota revisited. *Environ. Microbiol.* **9**: 2125–2136.

Rakoff-Nahoum, S., J. Paglino, F. Eslami-Varzaneh, S. Edberg, and R. Medzhitov. 2004. Recognition of commensal microflora by toll-like receptors is required for intestinal homeostasis. *Cell* **118**:229.

Ram, R. J., N. C. Verberkmoes, M. P. Thelen, G. W. Tyson, B. J. Baker, R. C. Blake, M. Shah, R. L. Hettich, and J. F. Banfield. 2005. Community proteomics of a natural microbial biofilm. *Science* **308**:1915–1920.

Ramotar, K., J. M. Conly, H. Chubb, and T. J. Louie. 1984. Production of menaquinones by intestinal anaerobes. *J. Infect. Dis.* **150**:213–218.

Rappé, M. S., and S. J. Giovannoni. 2004. The uncultured microbial majority. *Annu. Rev. Microbiol.* **57**:369–394.

Rinttilä, T., A. Kassinen, E. Malinen, L. Krogius, and A. Palva. 2004. Development of an extensive set of 16S rDNA-targeted primers for quantification of pathogenic and indigenous bacteria in faecal samples by real-time PCR. *J. Appl. Microbiol.* **97**:1166–1177.

Roediger, W. E. W. 1980. Role of anaerobic bacteria in the metabolic welfare of the colonic mucosa in man. *Gut* **21**:793–798.

Sacchi, C. T., D. Alber, P. Dull, E. A. Mothershed, A. M. Whitney, G. A. Barnett, T. Popovic, and L. W. Mayer. 2005. High level of sequence diversity in the 16S rRNA genes of *Haemophilus influenzae* isolates is useful for molecular subtyping. *J. Clin. Microbiol.* **43**:3734–3742.

Sakuma, K., M. Kitahara, R. Kibe, M. Sakamoto, and Y. Benno. 2006. *Clostridium glycyrrhizinilyticum* sp. nov., a glycyrrhizin-hydrolysing bacterium isolated from human faeces. *Microbiol. Immunol.* **50**:481–485.

Salzman, N. H., H. de Jong, Y. Paterson, H. J. M. Harmsen, G. W. Welling, and N. A. Bos. 2002. Analysis of 16S libraries of mouse gastrointestinal microflora reveals a large new group of mouse intestinal bacteria. *Microbiology* **148**:3651–3660.

Sartor, R. B. 2006. Mechanisms of disease: pathogenesis of Crohn's disease and ulcerative colitis. *Nat. Clin. Pract. Gastroenterol. Hepatol.* **3**:390–407.

Scheppach, W., and F. Weiler. 2004. The butyrate story: old wine in new bottles? *Curr. Opin. Clin. Nutr. Metab. Care.* **7**:563–567.

Seksik, P., L. Rigottier-Gois, G. Gramet, M. Sutren, P. Pochart, P. Marteau, R. Jian, and J. Doré. 2003. Alterations of the dominant faecal bacterial groups in patients with Crohn's disease of the colon. *Gut* **52**:237–242.

Sepehri, S., R. Kotlowski, C. N. Bernstein, and D. O. Krause. 2007. Microbial diversity of inflamed and noninflamed gut biopsy tissues in inflammatory bowel disease. *Inflamm. Bowel Dis.* **13**: 675–683.

Smoot, L. M., J. C. Smoot, H. Smidt, P. A. Noble, M. Könneke, Z. A. McMurry, and D. A. Stahl. 2005. DNA microarrays as salivary diagnostic tools for characterizing the oral cavity's microbial community. *Adv. Dent. Res.* **18:**6–11.

Sokol, H., P. Seksik, L. Rigottier-Gois, C. Lay, P. Lepage, I. Podglajen, P. Marteau, and J. Doré. 2006. Specificities of the fecal microbiota in inflammatory bowel disease. *Inflamm. Bowel Dis.* **12:**106–111.

Song, Y., E. Kononen, M. Rautio, C. Liu, A. Bryk, E. Eerola, and S. M. Finegold. 2006. *Alistipes onderdonkii* sp. nov. and *Alistipes shahii* sp. nov., of human origin. *Int. J. Syst. Evol. Microbiol.* **56:**1985–1990.

Sonnenburg, J. L., C. T. Chen, and J. I. Gordon. 2006. Genomic and metabolic studies of the impact of probiotics on a model gut symbiont and host. *PLoS Biol.* **12:**e413.

Sood, A., V. Midha, N. Sood, A. S. Bhatia, and G. Avasthi. 2003. Incidence and prevalence of ulcerative colitis in Punjab, North India. *Gut* **52:**1587–1590.

Suau, A., V. Rochet, A. Sghir, G. Gramet, S. Brewaeys, M. Sutren, L. Rigottier-Gois, and J. Doré. 2001. *Fusobacterium prausnitzii* and related species represent a dominant group within the human fecal flora. *Syst. Appl. Microbiol.* **24:**139–145.

Tajima, K., S. Arai, K. Ogata, T. Nagamine, H. Matsui, M. Namakura, R. I. Aminov, and Y. Benno. 2000. Rumen bacterial community transition during adaptation to high-grain diet. *Anaerobe* **6:**273–284.

Tannock, G. W., K. Munro, R. Bibiloni, M.A. Simon, P. Hargreaves, P. Gopal, H. Harmsen, and G. Welling. 2004. Impact of consumption of oligosaccharide-containing biscuits on the fecal microbiota of humans. *Appl. Environ. Microbiol.* **70:**2129–2136.

Tannock, G. W., K. Munro, H. J. M. Harmsen, G. W. Welling, J. Smart, and P. K. Gopal. 2000. Analysis of the fecal microflora of human subjects consuming a probiotic product containing *Lactobacillus rhamnosus* DR20. *Appl. Environ. Microbiol.* **66:**2578–2588.

Thomsen, T. R., B. V. Kjellerup, J. L. Nielsen, P. Hugenholtz, and P. H. Nielsen. 2002. In situ studies of the phylogeny and physiology of filamentous bacteria with attached growth. *Environ. Microbiol.* **4:**383–391.

Uchiyama, T., T. Abe, T. Ikemura, and K. Watanabe. 2005. Substrate-induced gene-expression screening of environmental metagenome libraries for isolation of catabolic genes. *Nat. Biotechnol.* **23:**88–93.

van der Waaij, D. 1989. The ecology of the human intestine and its consequences for overgrowth by pathogens such as *Clostridium difficile*. *Ann. Rev. Microbiol.* **43:**69–87.

van Tongeren, S. P., J. P. Slaets, H. J. Harmsen, and G. W. Welling. 2005. Fecal microbiota composition and frailty. *Appl Environ Microbiol* **71:**6438–6442.

Vanhoutte, T., G. Huys, E. Brandt, and J. Swings. 2004. Temporal stability analysis of the microbiota in human feces by denaturing gradient gel electrophoresis using universal and group-specific 16S rRNA gene primers. *FEMS Microbiol. Ecol.* **48:**437–446.

Vaughan, E. E., H. G. H. J. Heilig, K. Ben-Amor, and W. M. de Vos. 2005. Diversity, vitality and activities of intestinal lactic acid bacteria and bifidobacteria assessed by molecular approaches. *FEMS Microbiol. Rev.* **29:**477.

Venter, J. C., K. Remington, J. F. Heidelberg, A. L. Halpern, D. Rusch, J. A. Eisen, D. Wu, I. Paulsen, K. E. Nelson, W. Nelson, D. E. Fouts, S. Levy, A. H. Knap, M. W. Lomas, K. Nealson, O. White, J. Peterson, J. Hoffman, R. Parsons, H. Baden-Tillson, C. Pfannkoch, Y.-H. Rogers, and H. O. Smith. 2004. Environmental genome shotgun sequencing of the Sargasso Sea. *Science* **304:**66–74.

Wagner, M., A. Loy, M. Klein, N. Lee, N. B. Ramsing, D. A. Stahl, M. W. Friedrich, and R. L. Jared. 2005. Functional marker genes for identification of sulfate-reducing prokaryotes. *Methods. Enzymol.* **397:**469–489.

Wagner, M., H. Smidt, A. Loy, and J. Zhou. 2007. Unravelling microbial communities with DNA-microarrays: challenges and future directions. *Microb. Ecol.* **53:**498–506.

Wang, M., S. Ahrne, B. Jeppsson, and G. Molin. 2005. Comparison of bacterial diversity along the human intestinal tract by direct cloning and sequencing of 16S rRNA genes. *FEMS Microbiol. Ecol.* **54:**219.

Williamson, L. L., B. R. Borlee, P. D. Schloss, C. Guan, H. K. Allen, and J. Handelsman. 2005. Intracellular screen to identify metagenomic clones that induce or inhibit a quorum-sensing biosensor. *Appl. Environ. Microbiol.* **71:**6335–6344.

Woese, C. R. 1987. Bacterial evolution. *Microbiol. Rev.* **51:**221–271.

Woese, C. R., O. Kandler, and M. L. Wheelis. 1990. Towards a natural system of organisms: proposal for the domains Archaea, Bacteria, and Eucarya. *Proc. Natl. Acad. Sci. USA* **87:**4576–4579.

Zengler, K., G. Toledo, M. Rappé, J. Elkins, E. J. Mathur, J. M. Short, and M. Keller. 2002. Cultivating the uncultured. *Proc. Natl. Acad. Sci. USA* **99:**15681–15685.

Zhou, J. 2003. Microarrays for bacterial detection and microbial community analysis. *Curr. Opin. Microbiol.* **6:**288.

Zoetendal, E. G., A. D. Akkermans, W. M. Akkermans-van Vliet, A. J. G. M. de Visser, and W. M. de Vos. 2001. The host genotype affects the bacterial community in the human gastrointestinal tract. *Microb. Ecol. Health Dis.* **13:**129–134.

Zoetendal, E. G., A. D. Akkermans, and W. M. de Vos. 1998. Temperature gradient gel electrophoresis analysis of 16S rRNA from human fecal samples reveals stable and host-specific communities of active bacteria. *Appl. Environ. Microbiol.* **64:**3854–3859.

Zoetendal, E. G., C. M. Plugge, A. D. L. Akkermans, and W. M. de Vos. 2003. *Victivallis vadensis* gen. nov., sp. nov., a sugar-fermenting anaerobe from human faeces. *Int. J. Syst. Evol. Microbiol.* **53:**211–215.

Zwirglmaier, K., W. Ludwig, and K. H. Schleifer. 2004. Recognition of individual genes in a single bacterial cell by fluorescence in situ hybridization—RING-FISH. *Mol. Microbiol.* **51:**89–96.

THE STATUS QUO: WHO IS OUT THERE? HOW TO DETERMINE MICROBIAL DIVERSITY

II

THE UNCOUNTABLES

William T. Sloan, Christopher Quince, and Thomas P. Curtis

3

The microbial world is immense. There are more than 10^{30} individual microorganisms that we know about. It is also ancient, having evolved over 3.5 billion years (Nisbet and Sleep, 2001). With the vagaries of evolution acting over such a long time on such a massive number of individuals, it is hardly surprising that we are discovering many different and new microbial taxa in almost every environment we investigate. How many new species await discovery in specific environments, and globally, is an important question. Microbial communities are held to be reservoirs for new drugs and metabolic processes (Rondon et al., 2000) and are central to health, sustainable cities, agriculture, and most of the world's geochemical cycles. Will shifts in diversity brought about, for example, by climate change or pollution affect the size of the reservoirs or functioning of the communities and jeopardize the services they provide for us? Such questions are almost impossible to address without some baseline map of microbial diversity.

William T. Sloan and Christopher Quince, Department of Civil Engineering, University of Glasgow, G12 8LT, United Kingdom. *Thomas P. Curtis*, School of Civil Engineering and Geosciences, University of Newcastle upon Tyne, Newcastle, NE1 7RU, United Kingdom.

Modern molecular methods that allow for in situ characterization of microbial communities appear to be telling us that some environments are very much more diverse than others and are beginning to reveal patterns in the global distribution of taxa. For example, the pioneering research of Torsvik et al. (1990) used DNA reassociation kinetics to infer that there may be as many as 10,000 different taxa in a gram of soil. This was the first culture-independent analysis of soil bacteria. The method measures the rate at which single-stranded denatured DNA in a sample reassociates. This is then calibrated using the reassociation rate of a reference genome to give an estimate of diversity. In most current studies, rather than use whole genomes, bacterial biodiversity is defined by the number of different 16S rRNA genes in a sample. Methods for gathering this information have evolved rapidly from PCR amplification, cloning, and then sequencing to massively parallel sequencing where the cloning step can be missed out and an enormous number of sequences can be collated (e.g., Sogin et al., 2006). When biodiversity is based on differences in the 16S rDNA, then a unit of diversity becomes arbitrary and the term "species" is probably inappropriate. Typically the operational taxonomic unit in

Accessing Uncultivated Microorganisms: from the Environment to Organisms and Genomes and Back
Edited by Karsten Zengler © 2008 ASM Press, Washington, DC

diversity studies is defined as 3% dissimilarity between 16S rRNA gene sequences (e.g., Schloss and Handelsman, 2006); this is the convention we have adopted here. Recently the empirical evidence for very high soil diversity has been augmented by very rich catalogs of 16S rDNA gene sequences from various soils. Roesch et al. (2007), using 454 pyrosequencing, identified 5,543 different ribotypes from a Canadian boreal forest soil. Puzzlingly, there is evidence to suggest that the global distribution of soil microbial diversity will not necessarily reflect the biogeographic patterns that we have come to accept for larger organisms. For example, the received wisdom from classical ecology is that biodiversity is highest at the equator and decreases as one moves toward the poles, where the diversity is negligible. Neufeld and Mohn (2005) found an unexpectedly high reservoir of microbial diversity in Arctic soils that appears to contradict this norm and that may be vulnerable to climate change. To date, the most concerted effort to provide a map of global microbial biodiversity has been in the world's oceans. It appears that the experimentally determined marine microbial diversity for the upper oceans is far lower than that in soils. Hagstrom et al. (2002) published an analysis of the rate of accumulation of new sequence based on the rate of accession of new sequences to GenBank. The picture was clear; the rate of discovery appears to be reducing. The implication drawn from this is that diversity in the upper oceans is small; the inventory appeared to be nearing completion at about 1,117 unique sequences. Data from the ongoing Global Ocean Survey appear to reinforce this finding (Rusch et al., 2007). Although the diversity of the upper oceans may ultimately prove to be low, we show here that the sampling effort required to access all the genetic diversity they hold may be enormous. The deep oceans by contrast appear to be much richer; Huber et al. (2007) identified more than 30,000 unique 16S rDNA sequences from two low-temperature diffuse deep sea flow vents. Thus microbial ecologists seem to be slowly drawing conclusions on patterns in microbial biogeography and arriving at a consensus on the relative, if not the absolute, diversity of different environments.

The conclusions and consensus are compromised, however, by the fact that the true diversity in any environmental sample is extremely difficult to quantify through purely experimental methods. It is not currently possible to characterize all genes in a sample, and thus diversity must be inferred. Microbial ecologists are not unique in having to infer the characteristics of communities from incomplete samples; it is a problem that cuts across many disciplines in biological and social sciences. However, the extent of the problem in microbial ecology differs. Irrespective of the particular molecular method used, the disparity between sample size and community size far exceeds that experienced in most other scientific disciplines. The considerable technical sophistication and skill required to collect, analyze, and correctly enumerate microbial populations in environmental samples can sometimes obscure just how small, in relative terms, samples are. Take, for example, a large clone library of say 500 clones derived from a 1-g soil sample; the soil sample itself may contain as many as 10^9 individual organisms, and applied microbial ecologists will generally be interested in the services provided by the communities at a scale somewhat larger than a single 1-g sample. By analogy, when there are currently 6×10^9 humans in the world, a single sample of a few hundred individuals is unlikely to be sufficient to characterize the global distribution of any human traits unless it is extremely homogeneous. This disparity between sample size and community size is enormous and far greater than those normally encountered in biological or social sciences. Consequently patterns are perceived through a sparse, often distorted (Sloan et al., 2007) map of the microbial world.

Given that our modus operandi in microbial ecology is to work with small samples, how should we proceed to estimate the underlying community diversity? Several methods have been proposed in the literature, all of which use data on the relative abundance of the taxonomic units in environmental samples (for a review, see Bohannan and Hughes, 2003). They differ in the degree to which the raw abundance data is supplemented with additional assumptions and data. Nonparametric methods use the observed sample data alone to extrapolate to community diversity. With parametric methods the additional assumptions come in the form of a mathematical model, which has to be justified in terms of how well it fits the observed data, how it conforms to the conceptual model of the biology, and how generic it is. If a hierarchy of diversity estimators existed, then the top spot would probably go to biologically informed parametric diversity estimators because, when properly calibrated and validated, they have the potential to predict diversity and community structure in ecosystems that are subject to change using minimal observed data. Unfortunately, there is currently no consensus on the most appropriate models, although researchers are beginning to whittle away at a suite of potential models for various environments (Hong et al., 2006). Thus, when sample sizes are small and there is no knowledge of the underlying model that can be brought to bear, then "discretion is the better part of valor," and it is more informative to give a conservative estimate of diversity on the basis of a nonparametric estimator.

THE SCALE OF THE PROBLEM

Before descending into the details of diversity estimators, it is worth graphically reinforcing the scale of the undersampling problem facing microbiologists. It is the distribution of taxon abundances in a community that determines diversity. A good knowledge of this distribution will yield an accurate estimate of the diversity. However, the disparity of scale between samples and the communities they aim to represent means that the sample and community distributions can be very different indeed. To demonstrate this, Sloan et al. (2007) derived the sample distribution for 200 individuals (equivalent to clones in a 16S rDNA gene library) selected at random from large populations (10^{12} individuals) in which taxon abundances are distributed in four ways (Fig. 1), two of which have been previously proposed as plausible theoretical distributions (Finlay and Clarke, 1999; Curtis et al., 2002) (Fig. 1a, b) and two of which have no biological basis and could be considered ridiculous (Fig. 1c, d). All the sample distributions have a very similar shape, which is redolent of the distribution of taxon abundances in real 16S rRNA gene clone libraries. Thus, for example, the fact that clone abundance distributions look like the tail end of a lognormal distribution does not mean that the taxon in the larger community is distributed lognormally (although it might be). Descriptors such as diversity indices, taxon bundance distributions, and similarity indices have their roots in the ecology of macroorganisms, which are easier to observe and rely on a fairly complete census of the organisms at a particular site. What is clear from Fig. 1 is that for microbial communities these descriptors may differ significantly between sample and community. When very high-throughput sequencing becomes routinely available to microbial ecologists, a complete census of a sample may become possible (Sogin et al., 2006; Huber et al., 2007) although we show below that the sequencing effort even for a moderately diverse community would be enormous. Improved molecular methods will only, however, take us so far; the number of individuals in samples, even when we can identify all of them, will still be very small in comparison to those in the microbial communities as a whole. For example, a 1-g sample is small if one's aspirations are to characterize an entire field of soil. It will always be necessary to infer larger-scale descriptors of community structure from very small samples, which requires consideration of sampling effects.

NONPARAMETRIC ESTIMATION OF DIVERSITY

A distinction is drawn here between nonparametric diversity indices and diversity estimators. Diversity indices, such as the Shannon index or Simpson's index, offer a convenient

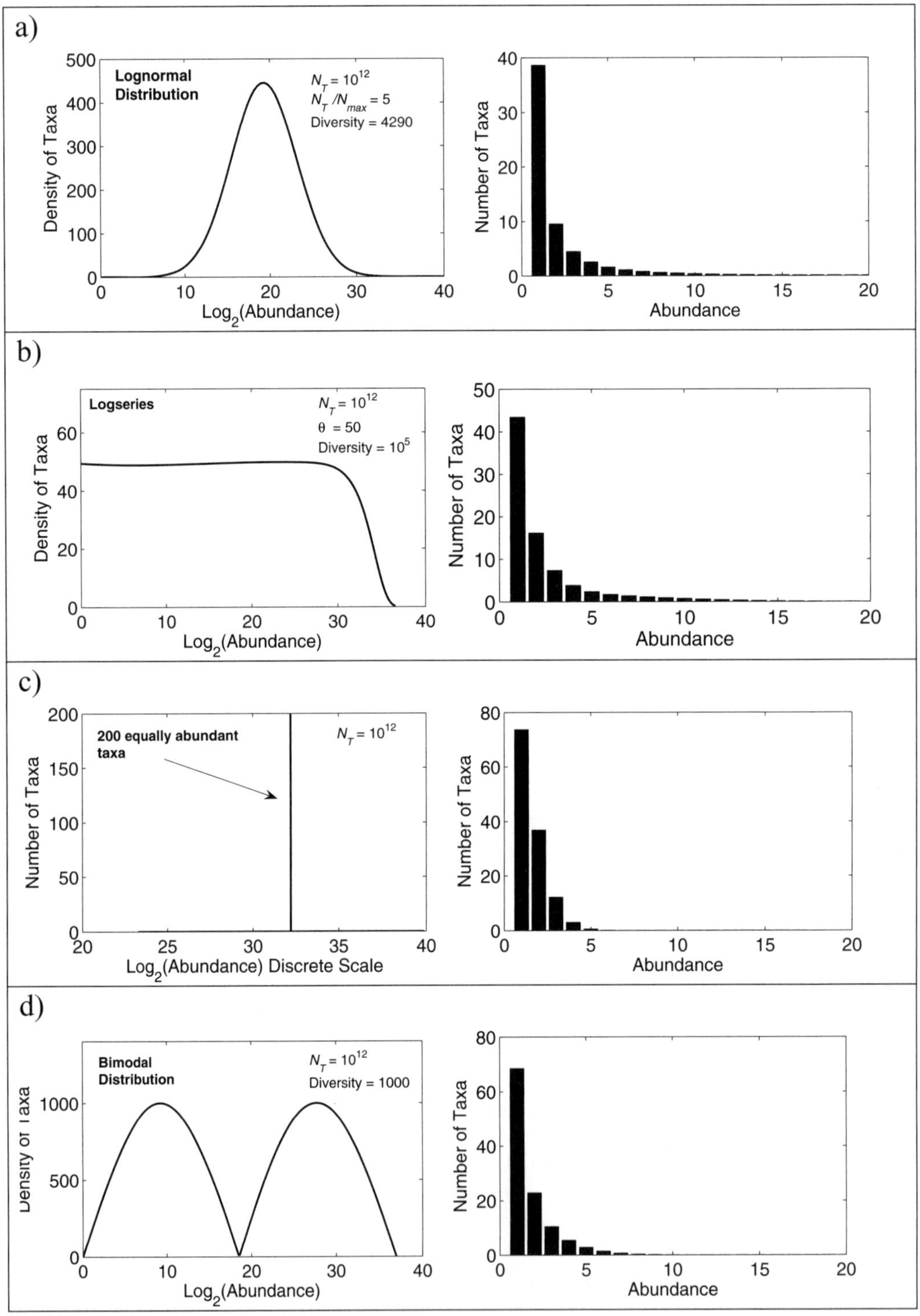

FIGURE 1 Distribution of taxon abundances in communities of 10^{12} individuals and in small samples of 200 individuals for (a) a lognormally distributed community; N_T/N_{max} is the ratio of the total number of individuals to the number of individuals belonging to the most abundant taxon, which can be used to index richness (Curtis et al., 2002); (b) a logseries distributed community; θ is one of the parameters of the lognormal that can be used as an index to species richness (Hubbell, 2001); (c) a community where 200 taxa are equally abundant, and (d) a bimodal distribution (redrawn from Sloan et al., 2007).

one-stop statistic that summarizes the shape and richness of the sample distribution. They have been picked up from mainstream ecology and are used extensively in microbial ecology for comparing community composition. The fact that they actually summarize sample distributions and do not extrapolate to community diversity is often overlooked. For example, in Fig. 1 the Simpson's index of the sample distributions ranges from between 0.97 and 0.98, which reflects the typically high proportion of rare taxa, whereas for the underlying community distribution it ranges from approximately 0.999 for the 200 equally abundant taxa and the logseries to 0.95 for the lognormal. Simpson's index for a sample is defined to be the probability that two individuals selected at random are from a different operational taxonomic unit and is given by:

$$D = 1 - \sum_{i=1}^{S} \frac{N_i(N_i - 1)}{L(L-1)} \quad (1)$$

where N_i is the abundance of the ith taxa, L is the total number of individuals in the sample, and S is the number of different taxa in the sample. So if there were only one taxon with abundance L, then $D = 0$, whereas if there were L different taxa with abundance 1, then $D = 1$. So $0 \leq D \leq 1$ is an index to sample diversity and for the distributions in Fig. 1 it tells us about the number of taxa in the underlying community. This and other indices are extensively used and compared by a number of authors (e.g., Kassen et al., 2000; Carney et al., 2004). Here, however, having noted that they summarize the countable rather than infer the uncountable and having registered the generic problem that they do not necessarily reflect the shape of the underlying community distribution, we leave diversity indices.

Nonparametric diversity estimators are bulletproof. In using them, microbiologists render themselves immune from criticism on any assumptions they might have introduced. Those engaged in the microbial biodiversity research will know that currently it is the subject of polemic, and assumptions will be attacked no matter how biologically inspired they might be. So the use of nonparametric diversity estimators can win debates. But how close do they get to the true diversity?

There are many nonparametric diversity estimators from which to choose in mainstream ecology, the majority of which are reviewed in Colwell and Codington (1994). Nonparametric estimators are usually devised to yield a lower bound on the diversity (Chao, 1987; Chao and Lee, 1993), and they all give slightly different results. The discrepancies between the diversity estimates they provide are a function of the underlying distribution, which is often obscure in environmental samples. Thus, for the most part, debating the relative merits of various nonparametric estimators for microbial communities is futile unless there is sufficient information to infer the distribution. When sample sizes become sufficiently large to inform the debate, it becomes possible to apply parameteric estimates. So choose a good nonparametric, stick with it, and use the diversity estimates for intercomparisons.

Take, for example, Chao's estimator, which is ingenious in its use of the observed data. Chao (1987) reasoned that the relative abundance of the rare taxa in a sample gives the most information on the taxa that might be absent. Letting f_i be the number of species that are observed i times, then f_0 is the number of species that are not observed, and the true diversity is given by:

$$S_T = S + f_0 \quad (2)$$

where S is the number of species observed in the sample. Using a piece of mathematical analysis, she proved that

$$E(f_0) \geq \left(\frac{L-1}{L}\right) \frac{(E(f_1))^2}{2E(f_2)} \quad (3)$$

where $E(\cdot)$ means "the expected value of" and L is the sample size. Cleverly, this inequality holds independent of the underlying distribution. Therefore, for large L, a lower bound estimator for the total diversity is

$$\hat{S}_T = S + \frac{f_1^2}{2f_2} \quad (4)$$

Furthermore, she derived an expression for the variance in the estimate,

$$Var(\hat{S}_T) = f_2\left(\frac{1}{4}\left(\frac{f_1}{f_2}\right)^4 + \left(\frac{f_1}{f_2}\right)^3 + \frac{1}{2}\left(\frac{f_1}{f_2}\right)^2\right) \quad (5)$$

So given an observed taxon-abundance distribution, Chao's estimator is easy to apply. It was popularized in ecology in a series of papers by Colwell (e.g., Colwell and Coddington, 1994). Hughes et al. (2001) applied Chao's estimator to 16S rDNA sequence data from clones in samples from the human gut, aquatic mesocosms, and Scottish soils. They demonstrated how the diversity estimates and their variance changed with sample size. Sample size does not appear in the formulas above; thus the change in estimates reflects the fact that the distribution in rare taxa changes with increasing sample size, which reinforces the observations made earlier in Fig. 1. Low variance in the estimate of diversity does not necessarily reflect accuracy. An indication that large samples are required to obtain a good estimate with Chao's estimator can be gained from inspecting equation 6. The maximum of S_T would occur when there is one doubleton and the rest are singletons, in which case,

$$\hat{S}_T = S + \frac{(S-1)^2}{2} = \frac{S^2+1}{2} \quad (6)$$

Therefore, if the number in the sample is L, then the maximum possible prediction of diversity is $\frac{(L-1)^2+1}{2}$, regardless of how many taxa are out there. So one can expect Chao's estimator to underestimate until the sample size is, at the very minimum, the square root of twice the true diversity.

Schloss and Handelsman (2006) explored the use of nonparametric estimators by simulation. They found that for a sample with a species richness of 5,000 it could take anything from 18,000 to 40,000 clones to determine the true species richness if the bacteria have a lognormal distribution but a mere 150 clones, if the distribution is even (Fig. 2). They further demonstrated the effects of sample size in

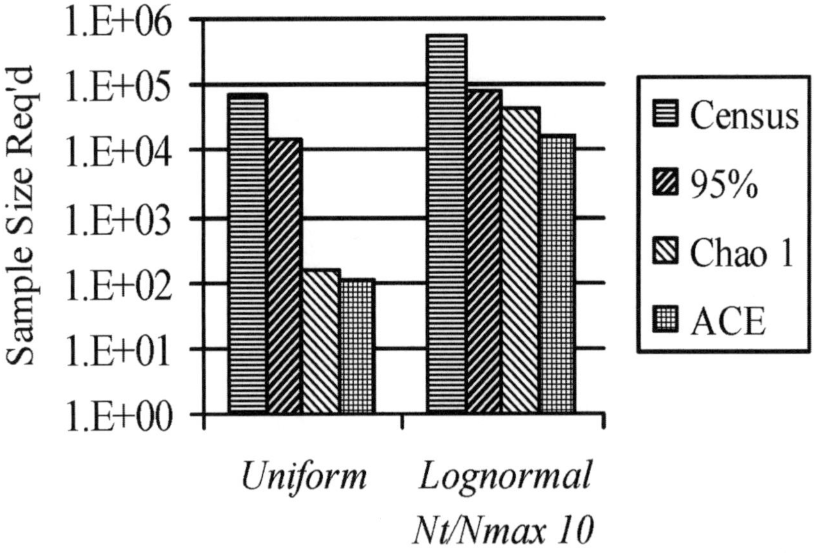

FIGURE 2 The sample sizes required to correctly characterize a sample with a diversity of 5,000, undertaking a complete census, a 95% census, or using nonparametric methods. Note that if the diversity is uniform, nonparametric estimators are very efficient. However, if the diversity is lognormally distributed, a very large sample is required to obtain the correct answer. The simulations are described in more detail in Schloss and Handelsmann (2006), and this figure appeared in Curtis et al. (2007).

an analysis of local and global biodiversity data, which showed unequivocally that nonparametric estimates are a function of sample size and that we do not yet have sufficient data at a local or global scale to estimate richness using these estimators. The global analysis was undertaken by analyzing the sequences in the RDP-II database. While this stratagem is subject to many caveats, in addition to the question of sample size, it does give a lower limit to our estimates of global diversity of about 35,000 (based on species discrimination at 97% sequence identity) and 325,000 (based on species discrimination at 99% sequence identity). We know that the global diversity of prokaryotes is greater than this, but we do not know how much greater (Curtis et al., 2006). This uncertainty at small and large scale is not the fault of the estimators; they are clever and very useful mathematical tools, not magic wands.

Occasionally an unusual taxon abundance data set appears where all the sequences detected in a 16S rRNA gene clone library are different (Borneman and Triplett, 1997). This tells us little about how well sampled the community is (Hughes et al., 2001), but it is possible to estimate how likely the observation is as a function of the underlying diversity. Lunn et al. (2004) argued fairly intuitively that the likelihood that diversity is high increases as more unique species are observed. The probability of observing only unique clones is not just a function of the number of species in the community, it is also affected by the abundance distribution of those species; if the underlying distribution is uneven, then the common taxa are likely to appear more regularly and one is unlikely to observe only unique clones unless the number of species is very high indeed. In practice, we cannot be sure what the underlying distribution is. However, Lunn et al. (2004) showed that if we assume all taxa are equally abundant, then we can guarantee a conservative estimate of the probability of observing all singletons:

$$\Pr(\text{all singletons}) = \frac{N_T^{L-1}(S_T - 1)!(N_T - L)!}{S_T^{L-1}(S_T - L)!(N_T - 1)!}$$

where L is the sample size, N_T is the number of individuals in the community, and S_T is the number of species in the community. This approach was used to analyze data from the clone library of 100 singletons obtained from Amazonian soil (Borneman and Triplett, 1997). They showed that for 100 singletons it would be very unlikely ($P = 0.006$) if the soil diversity was less than 10^3, and quite unlikely ($P = 0.6$) if the diversity was less than 10^4, and probable ($P = 0.95$) if the diversity was about 10^5. Such samples are, of course, not common, and there is some doubt about the truly flat nature of the data set that inspired the work (Schloss and Handelsman, 2006). Lunn's estimator like Chao's gives a lower bound on the diversity, and also like Chao's estimator, the unequivocal nature of the reasoning is attractive. The estimates of Lunn et al. (2004) also serve to hint at the possibility that the diversity in soils at least could be very much larger than nonparametric estimators would suggest if the distribution of taxon abundances was very uneven.

Assume a Distribution

In the 1980s and 1990s microbial ecologists were beginning to marvel at the diversity of organisms that was being revealed through new molecular methods. It was widely recognized that the extent of microbial diversity in any environmental sample had never been experimentally determined (it still has not), and some commentators believed bacterial diversity to be beyond practical calculation. Data sets such as the completely flat clone abundance libraries of Borneman and Triplett (1997) suggested that in some environments, such as soils, pursuing a purely experimental route to determining diversity would be arduous if not futile. However, Curtis et al. (2002) pointed out that it is not necessary to count every species in a community to estimate the number of different taxa therein. It is sufficient to estimate the area under the taxon-abundance distribution for that environment (Fig. 3), which may seem obvious now. The nature of the taxon-abundance distribution then becomes critical. Curtis et al. had little empirical evidence in support of any particular distribution and, as was shown in

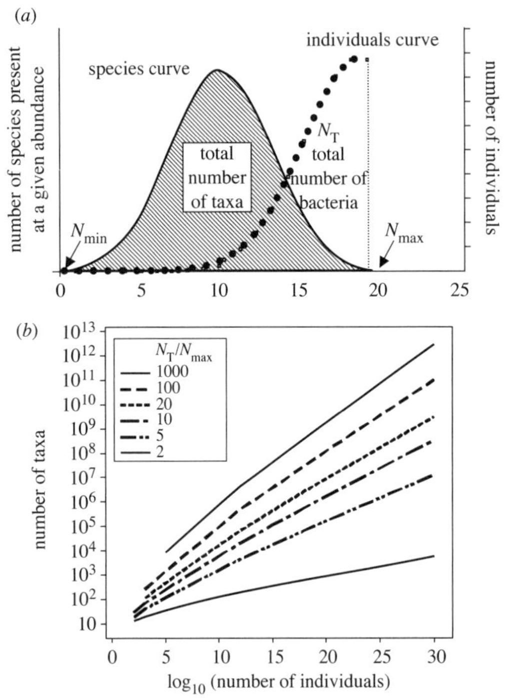

FIGURE 3 A "quick and dirty" way to estimate diversity by assuming a distribution. (a) The total number of taxa in a community with a lognormal species abundance curve is simply the area under that curve (called the species curve). The individuals curve is the number of species at each abundance (the species curve) multiplied by their abundance (the *x*-axis). There is therefore a mathematical relationship between the area under a species area curve, the number of individuals *NT* (the area under the individuals curve), and the maximum and the minimum abundance (N_{max} and N_{min}). (b) The relationship, over 30 orders of magnitude in population size, for various ratios of NT/N_{max} by assuming that N_{min} is equal to 1 (Curtis et al., 2002). This figure appeared in Curtis et al. (2007).

Fig. 1, even what might have seemed like a large clone library may not contain enough information to distinguish between distributions. Therefore, they speculated that the lognormal distribution might be a suitable candidate distribution because it is thought to characterize communities that exhibit highly dynamic and random growth (MacArthur, 1960; May, 1975). They were able to show that the diversity of prokaryotic communities may be related to the ratio of two variables: the total number of individuals in the community N_T and the abundance of the most abundant members of that community N_{max}. Both of these variables could be measured with some confidence in microbial communities, and the average value of N_T/N_{max} in multiple samples will reflect the ratio for the community. They added a further assumption that either the least abundant taxon has an abundance of 1 or Preston's canonical hypothesis (Preston, 1961) is valid. Consequently, they estimated the bacterial diversity on a small scale (oceans 160 per milliliter; soil 6,400 to 38,000 per gram; sewage works 70 per milliliter). Their estimates were significantly higher than the lower bounds given by nonparametric methods. For example, Chao's estimator gives a lower bound of 223 to 300 taxa in an anaerobic digester based on clone abundance data (Godon et al., 1997), whereas Curtis et al. (2002) estimated there might be more than 9,000. The assumptions regarding the shape of the underlying community taxon-abundance distribution is undoubtedly open to criticism (Bohannan and Hughes, 2003); however, what this article served to do was to focus attention on determining what the underlying taxon-abundance distribution is by demonstrating that it is fundamental in determining the extent of prokaryotic diversity.

What Is the Shape of the Distribution?

Clearly, merely assuming the shape of the distribution is ultimately unsatisfactory; proof is required. Curtis et al. (2002) selected the lognormal distribution based on an unproven, but plausible argument, on the ecological mechanisms that give rise to it, but there is a plethora of alternative distributions that have been shown to describe the taxon-abundance distributions of macroorganisms (e.g., May, 1975; Hubbell, 2001). They could potentially have made a case for at least some of them. Recognizing this, several researchers have investigated a range of different parametric taxon-abundance distributions. Rather than verify the underlying ecological mechanisms that give rise to a particular distribution, the prevailing approach is to fit a variety of distributions to the observed data and argue their merits based on some measure of goodness of fit (Hong et al., 2006). However, we have already seen that it is difficult to distinguish different community-level abundance distributions on the basis of small clone libraries (Fig. 1). Thus some method of enumerating more of the DNA in samples is required. Unfortunately, rapid DNA fingerprinting methods like denaturing gradient gel electrophoresis or terminal restriction fragment polymorphism, which analyze more of the genes in environmental samples, will not suffice. This is because bands on a denaturing gradient gel electrophoresis or peaks in a terminal restriction fragment polymorphism only appear when the relative abundance of the DNA exceeds a given threshold. The exact value of this threshold is rarely established, but it has been estimated to be approximately 1% of the total DNA (Cocolin et al., 2000). So fingerprinting methods have no chance of detecting rare taxa. Indeed, it has been shown that they often only highlight the very common taxa, and consequently, it is only the tail of the taxon-abundance distribution that can be seen. This partial view of the community composition can completely obscure patterns in biodiversity (Woodcock et al., 2006).

Inferring the Shape by DNA Reassociation

A more promising avenue for obtaining large samples of the genes from the environment was thought to be the method of reassociation kinetics originally pioneered by Torsvik et al. (1990). In this method purified DNA is sheared into fragments and then denatured, and the time taken for the single-stranded DNA to reassociate is measured. The rate of change in concentration of single-stranded DNA had previously been shown to follow second-order kinetics. So if $C(t)$ is the concentration, then

$$\frac{dC}{dt} = -kC^2 \tag{7}$$

where k is the rate of reassociation, which means that

$$\frac{C}{C_0} = \frac{1}{(1 + kC_0t)} \tag{8}$$

where C_0 is the concentration at time $t = 0$. The graph of this relationship is usually referred to as the Cot curve. By calibrating the value of k such that the observed Cot curve and that predicted by equation 8 coincide, it is possible to estimate the rate of reassociation of DNA from a mixed community of bacteria. To convert this to an estimate of diversity, the reassociation rate, k_r, of DNA from a monoculture of bacteria with known genome length, usually *Escherichia coli*, is determined with the same procedure. It is then, quite reasonably, assumed that there are no repeated fragments in any one genome and therefore the reassociation rate of any given fragment, k_f, is simply,

$$k_f = \frac{k_r L_r}{l} \tag{9}$$

where L_r is the length of the reference genome in nucleotide pairs and l is the length of a fragment. Under the assumptions that all taxa are equally abundant, no fragments are repeated, and all fragments are of the same length, then the calibrated reassociation rate of the mixed DNA can also be related to the total number of nucleotides (metagenome length) by

$$k_f = \frac{kL}{l} \tag{10}$$

Thus, by assuming the rate at which fragments reassociate is constant, then by combining equations 9 and 10 we get

$$\frac{k_r}{k} = \frac{L}{L_r} \tag{11}$$

Under the assumption that all taxa are equally abundant and have the same genome length, L/L_r is approximately the diversity S_T and hence

$$S_T = \frac{k_r}{k} \tag{12}$$

The research of Torsvik et al. with this technique has given rise to the now widely accepted paradigm of "10,000 bacterial species per gram soil." Gans et al. (2005) pointed out that the method almost certainly grossly underestimates the genetic diversity because of the assumption that all bacterial species in a sample are equally abundant. This is an inherently conservative assumption, because if, for example, the reassociation rate for the DNA in the environmental sample was half that of the reference rate, then the minimum number of genome equivalents that could give rise to this is two, but it more probably arises from a mixture of genomes. In an attempt to rectify this, Gans et al. (2005) recast equation 8 such that the Cot curve is a function of the underlying distribution of taxon abundances. They then extracted the taxon-abundance curves by calibrating against observed Cot curves from pristine and contaminated soil in Germany. By integrating the taxon-abundance curves, they were able to extrapolate to estimate the diversity. The estimates were staggeringly high; as many as 10^6 different prokaryotes per gram of soil. However, subsequent analyses by Volkov et al. (2006) and Bunge et al. (2006) suggested that it was not possible to distinguish between different distributions using Cot curves and that the variance in diversity estimates was too high to draw any conclusions about the underlying diversity. Nonetheless, the analysis of Gans et al. (2005) served to refocus attention on determining the shape of the underlying taxon-abundance distribution (Curtis and Sloan, 2005). Furthermore, the uncertainty in the abundance distribution extracted from Cot curves should not detract from the utility of reassociation kinetics in its original form for estimating the diversity from hyperdiverse environments. Soils, for example, may be so diverse that even the most common taxa in the community sit at a relative abundance of less than 0.001. Such low abundances preclude any meaningful analysis of diversity using fingerprinting methods because the abundances of all taxa would sit below the detection threshold. Similarly, a huge clone library would be required to confidently estimate the abundance of even the very common taxa. With reassociation kinetics there is no threshold, and if fragments are rare, they just take longer (potentially a very long time) to bump into complementary strands. By allowing half the single-stranded DNA to anneal before calibrating a reassociation rate, a good lower bound on diversity is obtained using half the DNA in the sample.

Fitting a Shape Using Larger Samples

The challenge of obtaining a more representative sample of the relative abundances of microbial taxa in environmental samples is, in a large part, being addressed by high-throughput sequencing technologies (e.g., Sogin et al., 2006). The advent of these technologies has the potential to sweep aside many of the methods for estimating diversity that have been outlined above. As the sample sizes grow, then the need to supplement data with assumptions and employ novel, often idiosyncratic, ways of extrapolating to microbial community composition diminishes. Instead, straightforward robust methods of fitting models to the observed taxon-abundance distributions can be employed. By way of an example, take the ongoing Global Ocean Survey (GOS) being conducted by the Sorcerer II expedition. To date, surface water samples from 41 mostly marine locations have been collected along a 2000-km transect starting on the north Atlantic seaboard and crossing into the Pacific via the Panama canal. The microbial DNA in these samples was shotgun sequenced (Rusch et al. 2007). This is where the DNA is randomly broken up into numerous small segments, which are sequenced by using the chain termination method to obtain reads. Multiple overlapping reads for the target DNA are obtained by performing several rounds of this fragmentation and sequencing. A wealth of information can be derived from matching overlapping ends of different reads to assemble them into a contiguous sequence, not least information on the genetic diversity of the samples. For the GOS data set, Rusch et al. (2007) managed to assemble 4,125 partial or full 16S rRNA genes from 7,068 reads. When clustered at 97% similarity, these produced 811 ribotypes. This represents a random sample that is much larger than that previously accumulated in, for example, clone libraries, that is at a higher resolution than fingerprinting methods, and where there is a finite probability of sampling rare taxa. Sogin et al. (2006) also report that the PCR biases that bedevil any molecular analysis of diversity are much reduced when the PCR amplicons are short. Perhaps there is sufficient information in these 16S rDNA data sets to fit a species-abundance distribution with some confidence and extrapolate to the expected total diversity.

The aim is to fit an underlying model of the taxon-abundance distribution for the community by using observations of abundance in the sample. The inverse problem of generating realizations of a sample distribution given a perfect knowledge of the abundance of taxa in the community is much simpler; we know that the sample distribution is a multivariate hypergeometric distribution (Feller, 1950), or if the community is sufficiently large that we can assume sampling with replacement, then the distribution is multinomial. On the basis of the sample distribution being multinomial, Woodcock et al. (2007) were able to systematically adjust the parameters of taxon-abundance distributions for waterborne bacterial communities residing in a set of tree holes so that simulated and observed sample taxon-abundance distributions matched. However, a more systematic approach that not only leads to an estimate of the parameters of the community taxon-abundance distribution but also gives estimates of the uncertainty in parameter values and uncertainty in the estimates of diversity comes about through a formal consideration of the likelihood function. This quantifies the probability of observing the data given the parameters of the underlying distribution. A single, best set of parameters can be determined by maximizing this probability, i.e., maximum likelihood estimation. However, the likelihood function contains more information than this; by using a trivial application of Baye's theorem, where no prior knowledge of the parameters is assumed, the likelihood function can be shown to be proportional to the distribution of the parameters given the data. This may seem like a subtle distinction, but this information can be exploited by sampling from the likelihood function to generate a realization of the probability distribution for the parameter values given the data. The latter, usually referred to as the "posterior distribution," completely characterizes the model fit. For instance, if we are interested in a particular parameter, we would examine its marginal posterior distribution, its distribution integrating over the other parameter values. Similarly, quantities of interest can be averaged over the posterior distribution; these are known as "posterior means." Of particular interest is the posterior mean of the deviance, which can be used to compare goodness of fit when two models have the same number of parameters; smaller values indicate a better fit (Spiegelhalter et al., 2002). The deviance is defined as minus two times the log-likelihood it can be viewed as a generalization of the chi-squared statistic. Details on the application of Baye's theorem, or the sampling algorithms like MCMC (Gelman, 2003), are beyond the scope of an article on microbial diversity. Besides, these are generic methods, which increasingly form parts of commercial computer packages. What is critical to the successful applications of the method is being able to evaluate the likelihood function for a particular taxon-abundance distribution and parameter set. This is generically useful in fitting taxon-abundance distributions to data derived from any molecular method. It was first described by Chao and Bunge (2002) for the particular case of gamma distributed abundances and subsequently applied to bacterial clone data using a range of different distributions by Hong et al. (2006). The approach has never previously been applied to a data set as rich as the GOS rDNA data.

Again, suppose we know precisely how many individuals there were in the *i*th species in say community λ_i. Then provided the sample size is reasonably large, to a good approximation, the number of times that species will appear in a sample obeys a Poisson distribution (Feller, 1950), with mean $\mu\lambda_i$, where μ is the sample frequency. The sampling frequency is the probability that any given individual is sampled, which is simply the number of individuals in the sample divided by the number of individuals in the community ($\mu = L/N_T$). Hence, the probability that the ith species is observed n times would be $\frac{e^{-\mu\lambda_2}(\mu\lambda_i)^n}{n!}$. Unfortunately, we do not know λ_i, the abundance of the ith species in the community. However, suppose that we have good reason to believe the abundances in the community are randomly drawn from some generic underlying parametric taxon-abundance distribution, $T(\lambda|\theta)$, where λ is abundance and θ is a low-dimensional vector of parameters that ultimately we want to estimate. Then, since each species enters the sample independently, the probability that a given species is observed n times, P_n, is independent of the other observations, and is given by a "compound Poisson" distribution (Bulmer, 1974; Etienne and Olff, 2005),

$$P_n = \int_0^\infty \frac{e^{-\mu\lambda}(\mu\lambda)^n}{n!} T(\lambda \mid \boldsymbol{\theta}) d\lambda \qquad (13)$$

The probability of observing a particular set of abundances is the product of the P_ns for each species. However, in reality we do not know the abundances of labeled species; rather we have a species-abundance data set $f = (f_1, f_2, \ldots, f_L)$, consisting of the number of species observed with $k = 1, 2, .., L$ individuals, where some of the f_ks will be zero. In addition, the number of unobserved species is given by $S_T - S$. Therefore, since the labels are arbitrary, we must multiply the product by the number of sets of labeled abundances that give the same f, which can be obtained from combinatorics. The combined likelihood function is then

$$P(\mathbf{f} \mid S_T, \boldsymbol{\theta}, \mu) = P_0^{S_T - S} \prod_{k=1}^{L} P_k^{f_k} \frac{S_T!}{(S_T - S)! \prod_{k=1}^{L} f_k!}. \qquad (14)$$

To determine the likelihood, we need the probabilities P_n, which are defined in equation 13 and depend on the taxon-abundance distribution, its parameters, θ, and the sampling frequency, μ. Remember μ is the number of individuals in the sample L divided by N_T, the number of individuals in the community. This is a problem because we rarely know what constitutes a community. In the case of the GOS data, is it a few liters of sea water or the entire sea? A simple transformation, however, lets us absorb the dependence of P_n on the community size into a set of rescaled parameters, θ'. Letting $x = \mu\lambda$, most taxon-abundance distributions are invariant, so that $T(\lambda|\theta)d\lambda = T(x|\theta')dx$, then

$$P_n = \int_0^\infty \frac{e^{-x} x^n}{n!} T(x \mid \boldsymbol{\theta}') dx \qquad (15)$$

This equation, together with equation 14, defines the likelihood of the data for an arbitrary assumed taxon-abundance distribution as a function of a set of parameters, θ' and the diversity S_T, in a way that does not require prior knowledge of the community size.

The likelihood function has been deliberately defined for some arbitrary abundance distribution to highlight the fact that currently, for microbial communities, the choice of the distribution to model taxon-abundance data is subjective. We have not had sufficient data for long enough to draw any conclusions about whether certain distributions are consistently best in certain environments. This will only be achieved by fitting a wide range of distributions to many large samples from many different environments and comparing how well they fit, or alternatively, and perhaps more satisfactorily, deriving the appropriate distribution on the basis of biological mechanisms that can be independently validated. However, sticking for the moment to the process of fitting a taxon-abundance distribution, then the integral in equation 15 can be evaluated explicitly for simple cases like the exponential or gamma distribution. Below we demonstrate fitting the lognormal distribution to the GOS 16S rRNA gene data and contrast it with the fit of the inverse Gaussian distribution. Both distributions were taken as candidate models for microbial community taxon abundance by Hong et al. (2006). The lognormal has the form:

$$T(\lambda \mid \boldsymbol{\theta}) = \frac{1}{\sqrt{2\pi V}} e^{\frac{(\ln \lambda - M)^2}{2V}} d\lambda \qquad (16)$$

with two parameters, $\theta = (M, V)$, which under rescaling become $\theta' = (M + \ln\mu, V)$. The inverse Gaussian distribution has form:

$$T(\lambda \mid \theta) = \sqrt{\frac{\alpha}{2\pi\lambda^3}} e^{\frac{\alpha(\lambda-\beta)^2}{2\beta^2\lambda}} d\lambda \qquad (17)$$

Like the lognormal, the inverse Gaussian has two parameters, $\theta = (\alpha, \beta)$, which when rescaled become $\theta' = (\mu\alpha, \mu\beta)$. For both distributions, the integral has to be performed numerically.

Thus, having defined the likelihood function, we can generate the posterior distribution of model parameters, including the community diversity S_T, and completely characterize the model fit. The reason for ploughing through some of the mathematics of the approach is to convince that its rigor yields meaningful estimates on the uncertainty associated with any estimate of diversity. If sample sizes are small, then that uncertainty is so large as to make the use of parametric methods like this almost meaningless, other than to say we have little idea of what the diversity in the environment is. In this case, nonparametric methods such as Chao's estimator, which yields a lower bound on the diversity, are more useful. The true utility of the rigorous approach to fitting a parametric model to taxon-abundance distribution becomes apparent when the number of individuals in the sample gets large. Then the variance in the posterior distributions drops, and one can distinguish which distributions fit the data best and, as Hong et al. (2007) show for a microbial data set, obtain bounds on the diversity given a set of reasonable parametric distributions.

Application to a Marine Microbial Plankton Data Set

The observed distribution of taxon abundances in 16S rRNA genes of marine microbial plankton data from the Global Ocean Survey (Rusch et al., 2007) was introduced as a precursor to the description on fitting data to larger samples. Comprising a random sample of 4,125 16S rRNA genes, it certainly represents a large sample in comparison to, for example, a typical clone library. What can we glean about the underlying diversity by applying our rigorous parametric approach? Sampling, using the MCMC method (Gelman, 2003), from the likelihood functions for the lognormal and the inverse Gaussian taxon-abundance distributions outlined above, we can infer the diversity estimates for these abundance data. The results are shown in Table 1 together with a lower bound for the diversity obtained from Chao's estimator. The two distributions used in Table 1 generated diversity estimates that were consistent, in that their confidence intervals overlapped, and these estimates were considerably larger than that obtained from the nonparametric Chao's estimator. The mean deviance was significantly lower for the lognormal taxon-abundance distribution, suggesting that more weight should be attached to its predictions. We tried using other distributions, including the gamma and mixed exponential, but the fit was very poor, and this was seen as grounds for discarding them. Therefore, on the basis of a lognormal distribution one might conclude that the aggregate marine microbial surface plankton community in the region of the Americas contains between 2,000 and 3,500 species, with about 2,500 species the most probable prediction. This low diversity in the upper ocean contrasts with the number of different taxa found in the deep ocean around flow vents (Huber et al., 2007).

A graphical representation of how well the distributions fit the sample taxon-abundance distribution is given in Fig. 4. It shows the

TABLE 1 Diversity estimates from the data of Rusch et al. (2007)[a]

Method	S_T	E[D]
Chao	**1,038**	NA[b]
Lognormal	1,998, **2,634**, 3,664	368.94
Inverse Gaussian	1,695, **2,049**, 2,401	391.72

[a]Results are shown for the Chao estimator and parametric methods assuming a lognormal and inverse Gaussian taxon-abundance distribution. The observed number of ribotypes for this data set was 811. The predictions are denoted S_T. For the parametric methods, median diversity estimates are given in bold face with upper and lower 95% confidence intervals, and posterior mean deviance, E[D], is also given as a measure of goodness of fit.

[b]NA, not available.

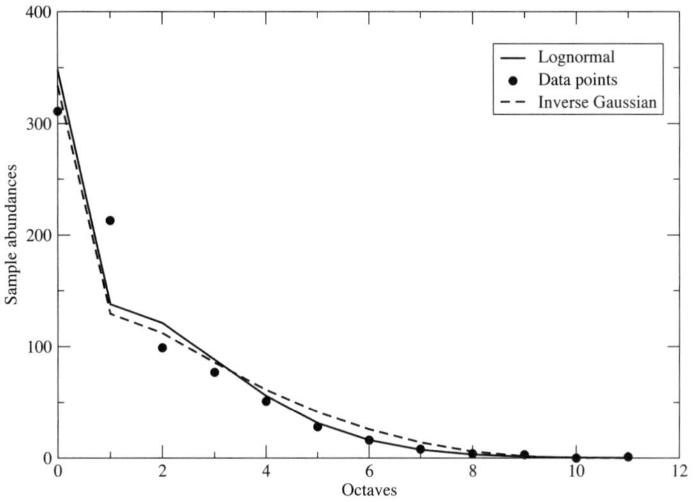

FIGURE 4 Expected sample abundances obtained using the lognormal (solid line) and inverse Gaussian distributions (dashed line). The predictions are posterior means as explained in the text. Actual data points are solid circles.

expected number of species with a given abundance, $S_T P_n$, against the actual numbers in the sample. Following the convention set in the ecology of macroorganisms, the data points have been aggregated into "octaves," all data points with abundances greater than 2^{n-1} and smaller than or equal to 2^n are placed in the nth octave and their frequencies summed. From this graph we see that the lognormal distribution fits both the start and the tail of the distribution well, but not the middle, notably underestimating the number of doubleton species. The inverse Gaussian, as we would expect from the mean deviance values, fits less well than the lognormal but produces fairly similar predictions.

How Hard Do You Have To Sample To Capture the Diversity of a Gene?

The diversity estimates suggest that Rusch et al. (2007) identified about 30% of the species that comprise the marine microbial plankton. To do this required 7,068 16S rDNA shotgun sequencing reads (more than the 4,125 genes identified since multiple reads at approximately 900 bp could contribute to a single 16S rRNA gene of ~1600 bp). An important question is how many reads targeted at the 16S rRNA gene would be required to identify all microbial species present? The standard approach to this question is to first generate a rarefaction curve. For a sample of L individuals, the rarefaction curve gives the expected number of different taxa in random subsamples (without replacement) of sizes 1 to $L - 1$, which can be determined from a straightforward sampling formula (Tipper, 1979). This rarefaction curve is then extrapolated to sample sizes larger than that observed. Typically this is done by fitting a saturating curve such as the Michaelis-Menten equation to the rarefaction curve (e.g., Roesch et al., 2007). The dotted line in Fig. 5 shows the Michaelis-Menten equation fitted to the rarefaction curve of the GOS data set (Rusch et al., 2007). The inset of this figure gives the actual rarefaction curve as a dashed line. This curve saturates quickly to just over a 1,000 species, which might suggest that current sequencing levels can easily reach all the species present in the biota.

The problem with this approach is the extrapolation; there is no basis for believing that any particular curve extrapolated out to a region for which there is no data will be accurate. A much better method can be developed using the parametric taxon-abundance distribution fitting outlined in the previous section. In inferring the diversity we also infer the parameters of the taxon-abundance distribution. We can therefore generate a synthetic community with each species abundance sampled from the taxon-abundance distribution. The sampling formula (Tipper, 1979) can then

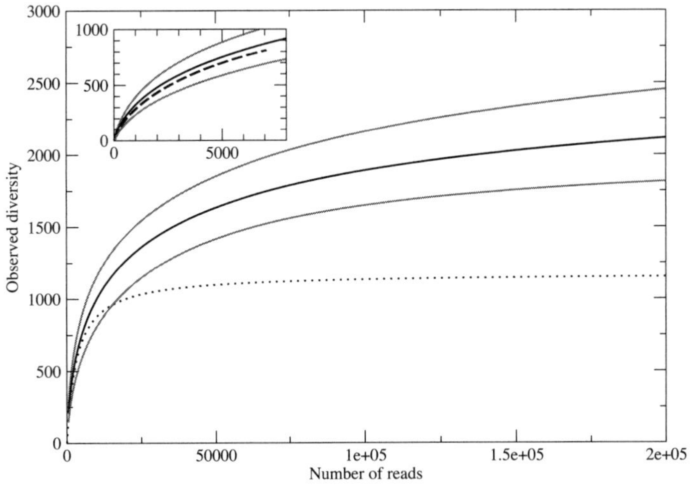

FIGURE 5 Estimates of observed diversity as a function of number of 16S rDNA reads for the data set of Rusch et al. (2007). The dotted line gives a curve generated by fitting a Michaelis-Menten equation to the rarefaction curve. The solid line is the median observed diversity from generating artificial communities with parameters obtained by sampling from the likelihood, Equation (3), assuming a log-normal taxon-abundance distribution. The gray lines give 95% confidence intervals. The inset graph gives the same data magnified near the origin together with the actual rarefaction curve (dashed line).

be applied to calculate the expected diversity observed at any level of sampling for that community. Repeating this for many synthetic communities gives a distribution of expected observed diversities. Figure 3 summarizes this for the data set of Rusch et al. (2007), where the solid line gives the median expected diversity and the gray lines the confidence intervals for 30,000 communities assembled by using parameters sampled from the likelihood of Equation (3). We used the lognormal taxon-abundance distribution to assemble these communities, because it fitted the data better than the inverse Gaussian. This method requires an estimate of the sampling frequency μ to obtain the untransformed parameters of the taxon-abundance distribution. We estimated that Rusch et al. (2007) sampled from a pool of 10^{12} individuals and since each individual contains roughly two reads of 16S rDNA, this gave $\ln(\mu) = \ln(7{,}068/2 \times 10^{12}) = -19.46$. This estimate for μ and the community assembly method itself is validated by the inset of Fig. 5, from which we see that the actual rarefaction curve (dashed line) lies within the confidence intervals of our observed diversity estimates. The difference between the extrapolated rarefaction curve and that generated by community assembly in Fig. 3 is striking. The curve from community assembly has still not saturated after 200,000 reads with roughly 20% of the predicted species diversity not yet observed. The reason for this can be found by looking at the parameter values predicted for the lognormal distribution. The parameter V lies with 95% confidence in the range 4.12 to 7.18 with a median of 5.42. Taking the median value for V, we expect that the natural logarithm of the population size of the most abundant species will be about twice the square root of 5.42 larger than that of the rare species. This corresponds to an abundance that is 2 orders of magnitude larger. Consequently, accessing these rare species requires exhaustive sampling. Extrapolating beyond the end of the graph in Fig. 5, then the median reads required to sample 90% of the species is 551,000 if the diversity is 2,634 taxa as predicted by the lognormal distribution. This may be considered feasible if one targets the 16S rRNA gene. But the large number of reads required highlights that attempting to sample all the diversity by any method based on random sampling, however high the throughput achieved, will be challenged, not so much by the absolute diversity of microbial taxa in samples, but by the uneven distribution of taxon abundances. It will be interesting to see prediction on the sampling effort required to capture 90% of the different taxa in soils (Roesch et al., 2007) or the deep sea, where the distribution of taxon abundances appears to be very even indeed (Sogin et al., 2006; Huber et al., 2007). Then, even when one targets a small section of the bacterial genome, the sampling effort required for an exhaustive census may be enormous.

How Hard Do You Have To Sample To Capture the Total Genetic Diversity?

The rationale for studying bacterial diversity given at the beginning of this chapter was the prospect of these uncharted taxa being a reservoir of new drugs and metabolic processes. In this case, inferring diversity from 16S rDNA sequences is interesting only insofar as it tells us how much remains to be uncovered. Even, in the oceans, if we can accumulate a random sample of 551,000 genes from targeting the 16S rRNA gene and thus capture the majority of the diversity in this gene, this will tell us nothing about novel metabolic pathways; the genes that make up the whole genome might though. It is for this reason that the GOS shotgun sequencing is being conducted on the metagenome without targeting specific genes. This then begs the question of how many reads would be required to capture all the genetic diversity. It is possible to estimate this if we make assumptions about the length of a typical marine bacterial genome, then the length of a typical fragment (read length), and use the taxon-abundance distribution fitted using the abundance of different 16S rDNA sequences.

Formally we want to know the expected number of different genomes that we can construct for a given number of reads under some assumption on how the taxon abundances are distributed. From the preceding analysis it seems reasonable to assume that the taxon-abundance distribution is that obtained by fitting to 16S rDNA sequences. So let $\{\lambda_i\}_{i=1}^{S_T}$ be a single realization of the abundance of taxa in the community. First consider a single particular fragment of the ith species. If M is the number of fragments per genome, then the probability that the fragment appears in any single read, $\tilde{P}_i$, is

$$\tilde{P}_i = \frac{\textit{Number of these fragments in the sample}}{\textit{Total number of fragments}} = \frac{\lambda_i}{N_T M} \qquad (18)$$

Now if the total number of fragments in the community, $N_T M$, is much larger than the number of fragments being sequenced, K, then the sequencing procedure approximates sampling with replacement. Letting the random variable X_i be the number of times our fragment of interest from the ith species appears in K reads, then, strictly speaking, X_i is binomially distributed, but for a large K this can be approximated by a Poisson distribution with expected number of observations $\mu_i = K\tilde{P}_i$, so

$$P(X_i = m) = \frac{\mu_i^m e^{-\mu_i}}{m!} \qquad (19)$$

Hence the probability that you see the fragment more than C times is

$$P(X_i \geq C) = 1 - \big(P(X_i = 0) + P(X_i = 1) + \ldots + P(X_i = C-1)\big) = 1 - \sum_{m=1}^{C-1} \frac{\mu_i^m e^{-\mu_i}}{m!} \qquad (20)$$

To construct the whole genome at least C times, then you have to sample each of its component fragments C times. Letting the random variable Y_i be the number of times that the genome of ith species can be constructed from the K reads, then

$$P(Y_i \geq C) = \left(1 - \sum_{m=1}^{C-1} \frac{\mu_i^m e^{-\mu_i}}{m!}\right)^M \qquad (21)$$

In our idealized situation where all the fragments are of equal length, you would only have to construct each genome once to accumulate all of the genetic diversity. Furthermore, we do not care if it appears more than once; provided the number of reads remains much smaller than the total pool of fragments, then the expected number of genomes, E_R, in K reads is simply

$$E_R = \sum_{i=1}^{S_T} P(Y_i \geq 0) = \sum_{i=1}^{S_T} \left(1 - e^{-\mu_i}\right)^M \qquad (22)$$

This can be calculated for any realization of the taxon-abundance distribution and thus equation 22 gives a relationship between K and E_R. This relationship will be unaffected by having multiple copies of rRNA genes per genome because, provided there are the same number of copies in all the genomes, the relative abundance distribution of $\{p_i\}_{i=1}^{S_T}$ will be the same as the relative abundance of genes.

For the GOS data set (Rusch et al., 2007), if we assume that the marine bacterial genomes are relatively short and comprise approximately 1.5×10^6 base pairs and that the average number of base pairs per fragment is approximately 900, then the average number of fragments in a typical genome is $M = 1{,}667$. Here we estimate the expected number of genomes that one can construct for realizations of taxon-abundance distributions drawn from the posterior distributions of our Bayesian fitting of the lognormal distribution. This is done for a variety of read numbers ranging from 10^6 to 10^{10}. The relationship between the median expected number of genomes that can be constructed along with the 95% confidence limits is plotted against number of reads in Fig. 6. It is immediately apparent that the sequencing effort to catalog the genetic diversity, even in the sea, is enormous. With the 7 million reads in the existing GOS data set we would be lucky to construct one genome. Indeed, Rusch et al. (2007) were only able to assemble one full genome. To have a chance of assembling the 2,225 genomes that would represent 90% of the median genetic diversity would require 9.193×10^9 reads. Thus, with at least 1,000 times more sequencing effort, Rusch et al. (2007) conducted approximately 6.4×10^6 reads and did not manage to assemble one complete genome, which suggests that our assumptions on fragment lengths and average genome lengths lead to an extremely conservative estimate of the sequencing effort.

Hagstrom et al. (2001) suggested that we might be getting close to a full census with 1,117 unique sequences for the oceans. But from our analysis, the sequencing effort required to gather genetic information about the rare taxa in the community could make an exhaustive catalog extremely difficult to achieve.

BEYOND DIVERSITY ESTIMATES

One could be forgiven for thinking that the technical challenges of estimating prokaryote diversity have obscured microbial ecologists' vision of why they are studying it. In most papers, and indeed in this chapter, the first few paragraphs pay lip service to the wealth of information that might be revealed if only we knew the genetic makeup of microbial communities. The microbial world is undoubtedly a vast resource, but the field of molecular microbial ecology is so new and the scientific hurdles in the way of exploiting all this genetic diversity are currently rather daunting and ill-defined. Microbial ecologists have prudently shied away from being too explicit on how this resource might be tapped. Part of the fascination in

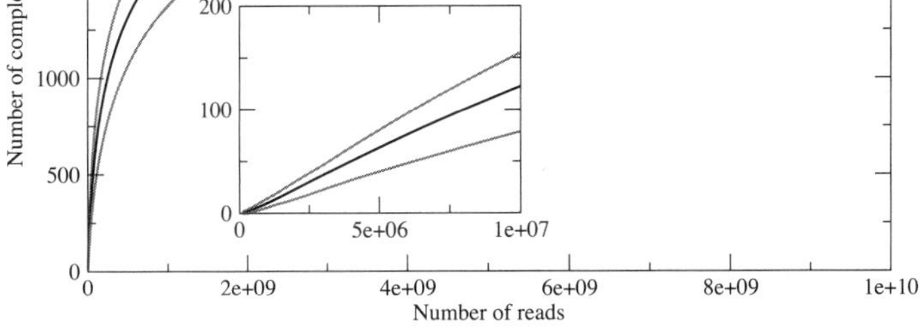

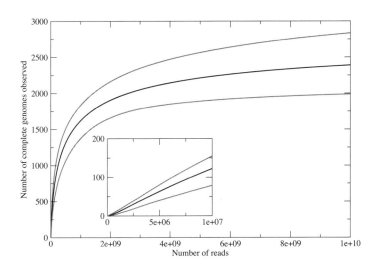

FIGURE 6 Estimates of number of genomes that can be assembled as a function of fragments read for the data set of Rusch et al. (2007). The solid line is the median expected number of genomes for the artificial communities used in Fig. 3. The gray lines give 95% confidence intervals. The inset graph gives the same data magnified near the origin. Rusch et al. (2007) conducted approximately 6.4×10^6 reads and assembled only a single genome.

studying microbial diversity is just how little we do know about it. With more than 10^{30} individuals that we know about (Whitman et al., 1998), there are 10^9 times more bacteria on Earth than there are stars in the universe (Curtis and Sloan, 2005). One needs no rationale to ponder on the extent of the microbial universe. We need merely to recognize it for what it is: an immense and unexplored frontier in science—a frontier of literally astronomical dimensions and of astonishing complexity (the simplest microbe is arguably more complex than any star). At this early stage in exploring the microbial world by using molecular methods the information and complexity can appear overwhelming. It is for this reason that the abstraction of mathematical descriptions of patterns in the data is so important. The mathematical diversity estimators and taxon-abundance distributions described in this chapter provide us with crude maps of the microbial world that will help guide future exploration and direct resources. The advent of high-throughput sequencing technologies has meant that sample sizes have increased dramatically in the recent past, and there is no reason to suppose that the technological revolution will slow. Thus, the pool of genetic information and the rate at which we acquire it are set to grow. This makes quantifying the biodiversity and biogeography (Rusch et al. 2007) of bacteria a realistic prospect. However, in the near future there is much work still to be done. It appears that for some environments we are coming close to a complete census of at least some bacterial genes such as 16S rRNA. However, the rarefaction curves presented above suggest that as we move toward a complete census in any environment, the ocean surface being a prime example, there is diminution in the new sequences returned for sequencing effort. This is simply a consequence of the fact that the distribution of taxon abundances is uneven. The small number of very abundant taxa will obscure the larger number of moderately rare or very rare organisms, and therefore the chance of picking up sequences from the rarer taxa by chance is low. The chance of sequencing all the component fragments of a rare genome is extremely slim and hence the sequencing effort in cataloging all the genetic diversity is several orders of magnitude greater than for single genes. For hyperdiverse environments, like soils, the task of even pinning down how many different taxa there are will likely require larger samples than have been routinely available. Cataloging the diversity of even a single gene in these environments will require a massive sequencing effort and attempting to catalog the genetic diversity is a frightening but foreseeable prospect.

Unfortunately, the patterns in bacterial biodiversity may not be stationary. Climatic and anthropogenic changes to some environments, at least, will conspire to ensure that bacterial community structure will change. Our censuses capture only instances in time in a transient process of community assembly and disturbance. Should we commit to continually survey bacterial diversity in all environments to account for this? This would be a tall order and might ultimately be futile. New communities of microorganism form on a daily basis in, for example, the gut of a child newly born, a waste reactor newly built, a new leaf, or a new root. The successful functioning of these communities is vital to us, and hence understanding their diversity may be critical. However, it appears that there is not a single characteristic community composition in, for example, the guts of infants (Curtis and Sloan, 2005) or in ostensibly identical bioreactors (Roeselers et al., 2006). These local communities do not form by magic; initially, at least, bacteria must be drawn from some source community that comprises all the bacteria that could colonize (Sloan et al., 2006). Thus the experimental determination of broad-scale patterns in bacterial diversity provides a map for us to pursue a deeper understanding of community composition and the interaction between local and global diversity. To achieve this understanding we need to go beyond the considerable challenge of enumerating and cataloging diversity and attempt to calibrate and verify mathematical models that encapsulate the ecological processes that shape microbial biodiversity.

ACKNOWLEDGMENTS

We thank Rampal Etienne for the algorithm that was used to calculate the compound Poisson lognormal distribution. We also thank Aaron Halpern and Doug Rusch for kindly providing the GOS sample abundance distribution and for discussion that aided the interpretation of those data.

REFERENCES

Bohannan, B. J. M., and J. Hughes. 2003. New approaches to analyzing microbial biodiversity data. *Curr. Opin. Microbiol.* **6:**282–287.

Borneman, J., and E. W. Triplett. 1997. Molecular microbial diversity in soils from eastern Amazonia: evidence for unusual microorganisms and microbial population shifts associated with deforestation. *Appl. Environ. Microbiol.* **63:**2647–2653.

Bulmer, M. G. 1974. Fitting poisson lognormal distribution to species-abundance data. *Biometrics* **30:**101–110.

Bunge, J., S. S. Epstein, and D. G. Peterson. 2006. Comment on "Computational improvements reveal great bacterial diversity and high metal toxicity in soil." *Science* **313:**918.

Carney, K. M., P. A. Matson, and B. J. M. Bohannan. 2004. Diversity and composition of tropical soil nitrifiers across a plant diversity gradient and among land-use types. *Ecol. Lett.* **7:**684–694.

Chao, A. 1987. Estimating the population-size for capture recapture data with unequal catchability. *Biometrics* **43:**783–791.

Chao, A., and J. Bunge. 2002. Estimating the number of species in a Stochastic abundance model. *Biometrics* **58:**531–539.

Chao, A., and S. M. Lee. 1993. Estimating population-size for continuous-time capture-recapture models via sample coverage. *Biometrical Journal* **35:**29–45.

Cocolin, L., L. F. Bisson, and D. A. Mills. 2000. Direct profiling of the yeast dynamics in wine fermentations. *FEMS Microbiol. Lett.* **189:**81–87.

Colwell, R. K., and J. A. Coddington. 1994 Estimating terrestrial biodiversity through extrapolation. *Philos. Trans. R. Soc. London, Ser. B* **345:**101–118.

Curtis, T., W. T. Sloan, and J. Scannell. 2002. Modelling prokaryotic diversity and its limits. *Proc. Natl. Acad. Sci. USA* **99:**10494-–10499.

Curtis, T. P., I. M. Head, M. Lunn, S. Woodcock, P. D. Schloss, and W. T. Sloan. 2006 What is the extent of prokaryotic diversity? *Philos. Trans. R. Soc. London, Ser. B* **361:**2023–2037.

Curtis, T. P., and W. T. Sloan. 2005. Exploring microbial diversity—a vast below. *Science* **309:**1331–1333.

Etienne, R. S., and H. Olff. 2005. Confronting different models of community structure to species-abundance data: a Bayesian model comparison. *Ecol. Lett.* **8:**493–504.

Feller, W. 1950. *An Introduction to Probability Theory and Its Applications.* Wiley, New York, NY.

Finlay, B. J., and K. J. Clarke. 1999. Ubiquitous dispersal of microbial species. *Nature* **400:**828–828.

Gans, J., M. Wolinsky, and J. Dunbar. 2005. Computational improvements reveal great bacterial diversity and high metal toxicity in soil. *Science* **309:**1387–1390.

Gelman, A. 2003. A Bayesian formulation of exploratory data analysis and goodness-of-fit testing. *Int. Stat. Rev.* **71:**369–382.

Godon, J. J., E. Zumstein , P. Dabert, F. Habouzit, and R. Moletta. 1997. Molecular microbial diversity of an anaerobic digestor as determined by small-subunit rDNA sequence analysis. *Appl. Environ. Microbiol.* **63:**2802–2813.

Hagstrom, A., T. Pommier, F. Rohwer, K. Simu, W. Stolte, D. Svensson, and U. L. Zweifel. 2002. Use of 16S ribosomal DNA for delineation of marine bacterioplankton species. *Appl. Environ. Microbiol.* **68:**3628–3633.

Hong, S. H., J. Bunge, S. O. Jeon, and S. S. Epstein. 2006. Predicting microbial species richness. *Proc. Natl. Acad. Sci. USA* **103:**117-–122.

Hubbell, S. P. 2001. *The Unified Neutral Theory of Biodiversity and Biogeography.* Princeton University Press, Princeton, NJ.

Huber, J A., D. B. M. Welch, H. G. Morrison, S. M. Huse, P. R Neal, D. A. Butterfield, and M. L. Sogin. 2007. Microbial population structures in the deep marine biosphere. *Science* **318:**97–100.

Hughes, J. B., J. J. Hellmann, T. H. Ricketts, and B. J. M. Bohannan. 2001. Counting the uncountable: statistical approaches to estimating microbial diversity. *Appl. Environ. Microbiol.* **67:**4399–4406.

Kassen, R., A. Buckling, G. Bell, and P. B. Rainey. 2000. Diversity peaks at intermediate productivity in a laboratory microcosm. *Nature* **406:**508–512.

Lunn, M., W. T. Sloan, and T. P. Curtis. 2004. Estimating bacterial diversity using Flat clone libraries and sampling concepts. *Environ. Microbiol.* **6:**1081–1086.

MacArthur, R. 1960. On the relative abundance of species. *The American Naturalist* **874:**25–36.

May, R. M. 1975. Patterns of species abundance and diversity, p. 81–120. *In* M. L. Cody and J. M. Diamond (ed.), *Ecology and Evolution of Communities.* Harvard University Press, Cambridge, MA.

Neufeld, J. D., and W. W. Mohn. 2005. Unexpectedly high bacterial diversity in arctic tundra relative to boreal forest soils, revealed by serial analysis of ribosomal sequence tags. *Appl. Environ. Microbiol.* **71:**5710–5718.

Nisbet, E. G., and N. H. Sleep. 2001. The habitat and nature of early life. *Nature* **409:**1083–1091.

Preston, F. W. 1961. The canonical distribution of commonness and rarity. *Ecology* **43:**185–215.

Roesch, L. F., R. R. Fulthorpe, A. Riva, G. Casella, A. K. M. Hadwin, A. D. Kent, S. H. Daroub, F. A. O. Camargo, W. G. Farmerie, and E. W. Triplett. 2007. Pyrosequencing enumerates and contrasts soil microbial diversity. *ISME Journal* **1:**283–290.

Roeselers, G., B. Zippel, M. Staal, M. van Loosdrecht, and G. Muyzer. 2006. On the reproducibility of microcosm experiments—different community composition in parallel phototrophic biofilm microcosms. *FEMS Microbiol. Ecol.* **58:**169–178.

Rondon, M. R., P. R. August, A. D. Bettermann, S. F. Brady, T. H. Grossman, M. R. Liles, K. A. Loiacono, B. A. Lynch, I. A. MacNeil, C. Minor, C. L. Tiong, M. Gilman, M. S. Osburne, J. Clardy, J. Handelsman, and R. M. Goodman. 2000. Cloning the soil metagenome: a strategy for accessing the genetic and functional diversity of uncultured microorganisms. *Appl. Environ. Microbiol.* **66:**2541–2547.

Rusch, D. B., A. L. Halpern, G. Sutton, K. B. Heidelberg, S. Williamson, S. Yooseph, D. Wu, J. A. Eisen, J. M. Hoffman, K. Remington, K. Beeson, B. Tran, H. Smith, H. Baden-Tillson, C. Stewart, J. Thorpe, J. Freeman, C., Andrews-Pfannkoch, J. E. Venter, K. Li, S. Kravitz, J. F. Heidelberg, T. Utterback, Y. H. Rogers, L. I. Falcon, V. Souza, G. Bonilla-Rosso, L. E. Eguiarte, D. M. Karl, S. Sathyendranath, T. Platt, E. Bermingham, V. Gallardo, G. Tamayo-Castillo, M. R. Ferrari, R. L. Strausberg, K. Nealson, R. Friedman, M. Frazier, and J. C. Venter. 2007. The Sorcerer II Global Ocean Sampling expedition: Northwest Atlantic through Eastern Tropical Pacific. *PloS Biol.* **5:**398–431.

Schloss, P. D., and J. Handelsman. 2006. Toward a census of bacteria in soil. *PloS Comput. Biol.* **2:**786–793.

Sloan, W. T., S. Woodcock, M. Lunn, I. M. Head, and T. P. Curtis. 2007. Modeling taxa-abundance distributions in microbial communities using environmental sequence data. *Microb. Ecol.* **53:** 443–455.

Sloan, W. T., S. Woodcock, M. Lunn, I. M. Head, S. Nee, and T. P. Curtis. 2006. The roles of immigration and chance in shaping prokaryote community structure. *Environ. Microbiol.* **8:**732–740.

Sogin, M. L., H. G. Morrison, J. A. Huber, D. M. Welch, S. M. Huse, P. R. Neal, J. M. Arrieta, and G. J. Herndl. 2006. Microbial diversity in the deep sea and the underexplored "rare biosphere." *Proc. Natl. Acad. Sci. USA* **103:**12115–12120.

Spiegelhalter, D. J., N. G. Best, B. R. Carlin, and A. van der Linde. 2002. Bayesian measures of model complexity and fit. *J. R. Stat. Soc. Ser. B Stat. Methodol.* **64:**583–616.

Tipper, J. C. 1979. Rarefaction and rarefiction—use and abuse of a method in paleoecology. *Paleobiology* **5:**423–434.

Torsvik, V., J. Goksoyr, and F. L. Daae. 1990. High diversity in DNA of soil bacteria. *Appl. Environ. Microbiol.* **56:**782–787.

Volkov, I., J. R. Banavar, and A. Maritan. 2006. Comment on "Computational improvements reveal great bacterial diversity and high metal toxicity in soil." *Science* **313:**918.

Whitman, W. B., D. C. Coleman, and W. J. Wiebe. 1998. Prokaryotes: the unseen majority. *Proc. Natl. Acad. Sci. USA* **95:**6578–6583.

Woodcock, S., T. Bell, C. J. Van der Gast, M. Lunn, T. P. Curtis, I. M. Head, and W. T. Sloan. 2007. Neutral assembly of bacterial communities. *FEMS Microbiol. Ecol.* **62:**171–180.

Woodcock, S., T. P. Curtis, I. M. Head, M. Lunn, and W. T. Sloan. 2006. Taxa-area relationships for microbes: the unsampled and the unseen. *Ecol. Lett.* **9:**805–812.

THE MISSING FUNGI: NEW INSIGHTS FROM CULTURE-INDEPENDENT MOLECULAR STUDIES OF SOIL

S. K. Schmidt, K. L. Wilson, A. F. Meyer, C. W. Schadt, T. M. Porter, and J. M. Moncalvo

4

Fungi are an often-neglected group of microorganisms, taking a back seat to bacteria and archaea with many microbiologists. Long considered to be a subdivision of plants, kingdom *Eumycota* is the closest neighbor of *Animalia* in the *Eukarya* (Baldauf and Palmer, 1993). The fungi are one of the most numerous and diverse groups of organisms on Earth with over 80,000 described species (Kirk et al., 2001) and an estimated 1.5 million yet to be identified (Hawksworth, 2001). Although this number may rise or fall depending on how a fungal species is defined (Moncalvo, 2005), the number of undiscovered fungi is truly enormous. However, these estimates are based largely on rates of discovery of new species that usually fall within known genera and families. Unlike bacteria and archaea, relatively few higher-level evolutionary lineages have been discovered except via splitting and regrouping of already described fungal groups. If higher-level diversity remains undiscovered, and recent studies reviewed here suggest that it does, estimates of "undiscovered" diversity could be largely underestimated.

Where are the missing fungi? Numerous understudied habitats for fungal diversity are thought to exist in association with insects, forest canopies, plant roots, and plant leaves in the tropics and other biomes (Arnold et al., 2000; Gardes and Bruns, 1996; Blackwell and Jones, 1997; Frohlich and Hyde, 1999; Longcore, 2005). However, the largest "undiscovered country" for fungal diversity is the world of soil where microbial niches abound at both microspatial and temporal scales (Schmidt et al., 2007). Broad molecular surveys of the soil biome, from tropics to polar regions and from temperate grasslands to alpine tundra, have revealed a rich new view of the bacterial and archaeal diversity, regularly adding new clades that are deeply branched often to the phylum level (Rappé and Giovannoni, 2003; Janssen, 2006; Schloss and Handelsman, 2006; Robertson et al., 2005). Molecular surveys of soil fungi have received much less attention despite the fact that fungi dominate decomposition and nutrient cycles in many soils. For example, tun-

S.K. Schmidt, K. L. Wilson, and A. F. Meyer, Department of Ecology and Evolutionary Biology, University of Colorado, Boulder, CO 80309. *T. M. Porter*, Department of Ecology and Evolutionary Biology, University of Toronto, Toronto, ON M5S 3B2, Canada. *C. W. Schadt*, Biosciences Division, Oak Ridge National Laboratory, Oak Ridge, TN 37831-6038. *J. M. Moncalvo*, Department of Ecology and Evolutionary Biology, University of Toronto, and Department of Natural History, Royal Ontario Museum, Toronto, ON M5S 2C6, Canada.

Accessing Uncultivated Microorganisms: from the Environment to Organisms and Genomes and Back
Edited by Karsten Zengler © 2008 ASM Press, Washington, DC

dra and coniferous forest ecosystems have far more fungal biomass than bacterial and archaeal biomass, especially during times of the year (e.g., fall, winter, and spring) that have not been intensively sampled in traditional studies of microbial diversity (Lipson et al., 2002; Schadt et al., 2003; Schmidt et al., 2007).

Mycologists have been collecting in the field and growing fungi in culture for several hundred years. Collecting fungal fruiting bodies and lichens in various habitats has yielded many thousands of species but is biased toward those fungi that produce discernible fruiting bodies. Growing fungi in culture with various sampling techniques has yielded additional thousands of species but is biased toward fungi that can be tamed in the laboratory and fungi that are relatively fast growing. Though these techniques combined have yielded most of the 80,000 described species of fungi, this still represents only about 5% of the estimated total. The question remains, "where are the missing fungi?"

THE BENEFITS OF SOIL GENE LIBRARIES

The ability to sequence fungal DNA from soil samples has opened up a new realm of possibilities for identifying new fungal groups. Community genomic DNA can be extracted from soil and selected genes amplified using PCR. Cloning and sequencing of these amplified genes can create a "gene library" of fungal sequences. Sequences in a new gene library can be tentatively identified by comparing them to known sequences. However, phylogenetic methods are needed to truly understand the relationship of new sequences to those in databases. This approach can be used to identify fungi that were resident in the soil, independent of the biases of traditional culturing and collection methods. Of course, any library created in this manner is only a partial representation of all fungi present in the soil being studied. A clone library approximates a random sample of fungi in the soil, and is likely to be biased toward some groups of fungi over others based on variables in each step of the procedure used. Even with this cautionary note, the creation of clone libraries from soil is a tool that should give us the ability to look more deeply into the fungal world, something akin to looking at the night sky with the naked eye and then being given a powerful telescope.

Of course, a new gene library that is created from soil samples is only as good as the master reference library that can be used for comparison. Fortunately for fungal phylogenetics a new body of research has provided an excellent reference tool. NSF has funded a large project to study several genes across many families of fungi for a more robust phylogenetic picture of the fungi. This Assembling the Fungal Tree of Life (AFTOL) initiative has definitively clarified the phylogenetic tree for all known fungi (Hibbett et al., 2002; Lutzoni et al., 2004; James et al., 2006). Before AFTOL, there was much confusion regarding the exact relationship of the various fungal taxonomic and phylogenetic groups. AFTOL confirmed that *Ascomycota* and *Basidiomycota* are monophyletic, while *Chytridiomycota* and *Zygomycota* are polyphyletic and more basal (James et al., 2006). AFTOL also confirmed that arbuscular mycorrhizal fungi (the *Glomeromycota*) (Schüßler et al., 2001) should be considered to be a separate phylum (Lutzoni et al., 2004). The AFTOL work makes any discussion of fungal relationships easier and more definitive.

There is still a major problem in the comparison of phylogenetic trees created by soil libraries. The most common genes used to identify bacteria, archaea, and fungi are the genes that code for ribosomal RNA. These genes are fairly well conserved, yet can provide the ability to identify down to the species level. Three rDNA regions are commonly used to identify fungi: the 18S small subunit (SSU); the internal transcriber spacer sequences ITS-1 and ITS-2; and the 28S large subunit (LSU). However, two problems arise when new libraries are created with sequences from only one of these regions: First, using only one region makes it impossible to directly compare libraries from different gene regions, and second, short sequences limit our ability to make definitive conclusions about similarities and differences among clones and correct phylogenetic placement. AFTOL used six genes, which provides a good base for reference. However, when a new soil library reveals a new and unique fungal clade, it is difficult to find support for the new group when other studies use a different gene to create their library.

THE PITFALLS OF CLONE LIBRARIES

Even though molecular techniques have proved powerful in elucidating microbial diversity and revealing novel groups (Landeweert et al., 2003; Schadt et al., 2003; He et al., 2005; O'Brien et al., 2005; Vandenkoornhuyse et al., 2002), they have also proved to have significant limitations. For instance, DNA extraction, primer and cloning biases, and unequal rDNA operon copy number between taxa can result in an incomplete or distorted picture of the community studied (Howlett et al., 1997; von Wintzingerode et al., 1997; Frostegard et al., 1999; Lueders and Friedrich, 2003; Kurata et al., 2004). Some of these biases can be minimized by using the lowest annealing temperature that would produce visible amplification product and by pooling PCR reaction products before cloning (Polz and Cavanaugh, 1998; Ishii and Fukui, 2001; Acinas et al., 2005).

Several issues need be addressed in any claim of discovery of novel taxonomic lineages, including undetected chimeric sequences, long-branch attraction artifacts, incomplete molecular sampling of described fungi, and poorly annotated sequences in the public databases (Bridge et al., 2003; Vilgalys, 2003; Berney et al., 2004; Cavalier-Smith, 2004). Several tools can be used to detect chimeras, for instance, Chimera Check (http://rdp.cme.msu.edu) (Cole et al., 2003), Bellerophon analysis (http://foo.maths.uq.edu.au/~huber/bellerophon.pl) (Huber et al., 2004), Pintail and Mallard (Ashelford et al., 2006), and secondary structure modeling and partial tree analysis using ARB (http://www.arb-home.de). However, the detection of chimeras remains a difficult and imprecise activity, which often mandates subjective "judgment calls" (Ashelford et al., 2005, 2006). The use of multiple tools and a conservative standard for excluding suspicious clones should minimize the claim of novelty based on spurious sequences.

Long-branch attraction phenomena can be minimized by using methods that take into account different rates of substitutions among lineages and adding more taxa that are related to those with long branches (Bergsten, 2005). In addition, the ongoing deposition of rDNA sequences in public databases increases the probability of finding slowly evolving sequences closely related to any possible novel sequence (Berney et al., 2004). Such sequences help break up long branches and are useful for distinguishing chimeras from legitimately new clones. The number of sequences of known and described fungi being deposited in public databases has accelerated recently, particularly through the work of the AFTOL consortium (James et al., 2006; Lutzoni et al., 2004; http://aftol.org). The work of AFTOL and others has minimized the risk of misidentifying a sequence as novel simply because it represents an unsequenced but already described taxon. Finally, the problem of poorly identified or annotated sequences in the public databases (Ashelford et al., 2006; Bridge et al., 2003; Hugenholtz and Huber, 2003; Vilgalys, 2003) being used as guide sequences for phylogenetic trees showing novel groups can be minimized by only using sequences from recent authoritative publications of fungal phylogenies (James et al., 2006; Lutzoni et al., 2004).

NOVEL SOIL FUNGI

Below, we review the new fungal lineages that have been discovered in various soil libraries over the past 10 years. Most soil libraries to date have discovered new fungi only at the genus or species level. There are, however, several exciting exceptions. The first involves several new clades of *Chytridiomycota* discovered in alpine soils. The second is a deeply branching, novel clade in the phylum *Ascomycota* that was first identified several years ago, also in alpine soil. This clade has since been found in a number of other soil libraries and appears to represent a new subphylum of fungi within the *Ascomycota*. We will spend the bulk of our discussion on these two new discoveries after reviewing other important but unrelated clades that have also been discovered in soil libraries.

Ascomycota

Schadt et al. (2003) used libraries of LSU rDNA from a tundra soil in three seasons to identify three new, deeply branching *Ascomycota* clades that had no previously identified members. The largest clade (Group I) was found predominantly in the summer library, whereas another large new clade (Group II) was predominant in both the under-snow and snow melt libraries. Figure 1 depicts a tree showing the relationship of these new clades to previously known groups within the *Ascomycota*. Discovery of these large and previously unrecognized major groups of fungi sparked a flurry of controversy, mostly centering on whether they were actually chimeras. However, a number of independent studies have since shown that Soil Clone Group I (SCGI) is widespread in nature, occurring in numerous soils from Central and North America, Europe, and Australia (Porter et al., 2008).

Phylogenetically, SCGI branches deeply between the *Ascomycota* subphyla of *Saccharomycotina* and *Taphrinomycotina*, tentatively identifying it as a new subphylum (Porter et al., 2008). *Taphrinomycotina* typically have a simple and rudimentary morphology where a yeast stage is common and the sexual stage, when known, is characterized by unitunicate asci and lacks a well-formed ascomata except in *Neolecta* (Landvik et al., 2003). Lifestyles range from obligate plant pathogens, animal pathogens, to saprobic fission yeasts (Nishida and Sugiyama, 2004; Liu and Hall, 2004). The *Saccharomycotina* are composed of about 800 recognized species with basal forms that are filamentous, others that are true budding yeasts, and more derived forms that show filamentous-yeast dimorphism. Most of these forms are free-living, but at least one genus is known to contain animal and human pathogens (*Candida*), and one genus contains plant pathogens (*Eremothecium*) (Anderson et al., 2003a; Liu and Hall, 2004). The sexual stage is characterized by the lack of ascomata and the formation of unitunicate asci. Because SCGI falls between these two subphyla, it could be hypothesized to have characteristics that are similar but could also be completely different.

Significantly, SCGI does not seem to have been overlooked because it is a rare member of soil communities. On the contrary, Porter et al. (2008) found SCGI to be widespread and common within soils from locations in Colorado, Costa Rica, and Canada with relative abundances within overall fungal libraries of between 9 and 38% of the total number of clones sequenced. Likewise, Schadt et al. (2003) found that 16% of all clones from Colorado alpine soils (across three seasons) fell into this unique clade. These findings indicate that SCGI may play an important role in the soil system, but as of yet we can only speculate as to its morphology, physiology, and ecological niche.

The work of Porter et al. (2008) also has allowed us to find SCGI in previously published clone libraries that employed DNA regions other than the LSU used by Schadt et al. (2003). Porter et al. (2008) were able to sequence a large fragment (>2.5 kbp) that included the SSU, ITS, and most of the LSU regions, thus allowing linkage of the original LSU sequences of Schadt et al. (2003) to studies that employed ITS and SSU sequences. Comparison of SCGI clones from each location suggests little geographic overlap among subclades within this large group, indicating that we have barely glimpsed the true diversity of this subphylum in studies to date. Table 1 lists the studies that have found SCGI members.

Figure 1 also indicates another interesting new clade designated as Group II. Though not as deeply branching as Group I, Group II appears to be a diverse clade that is nested within the *Ascomycota* subphylum *Pezizomycotina*. The analyses of Schadt et al. (2003) placed Group II in the midst of *Pezizomycotina* with good support. Figure 1 identifies a Group III

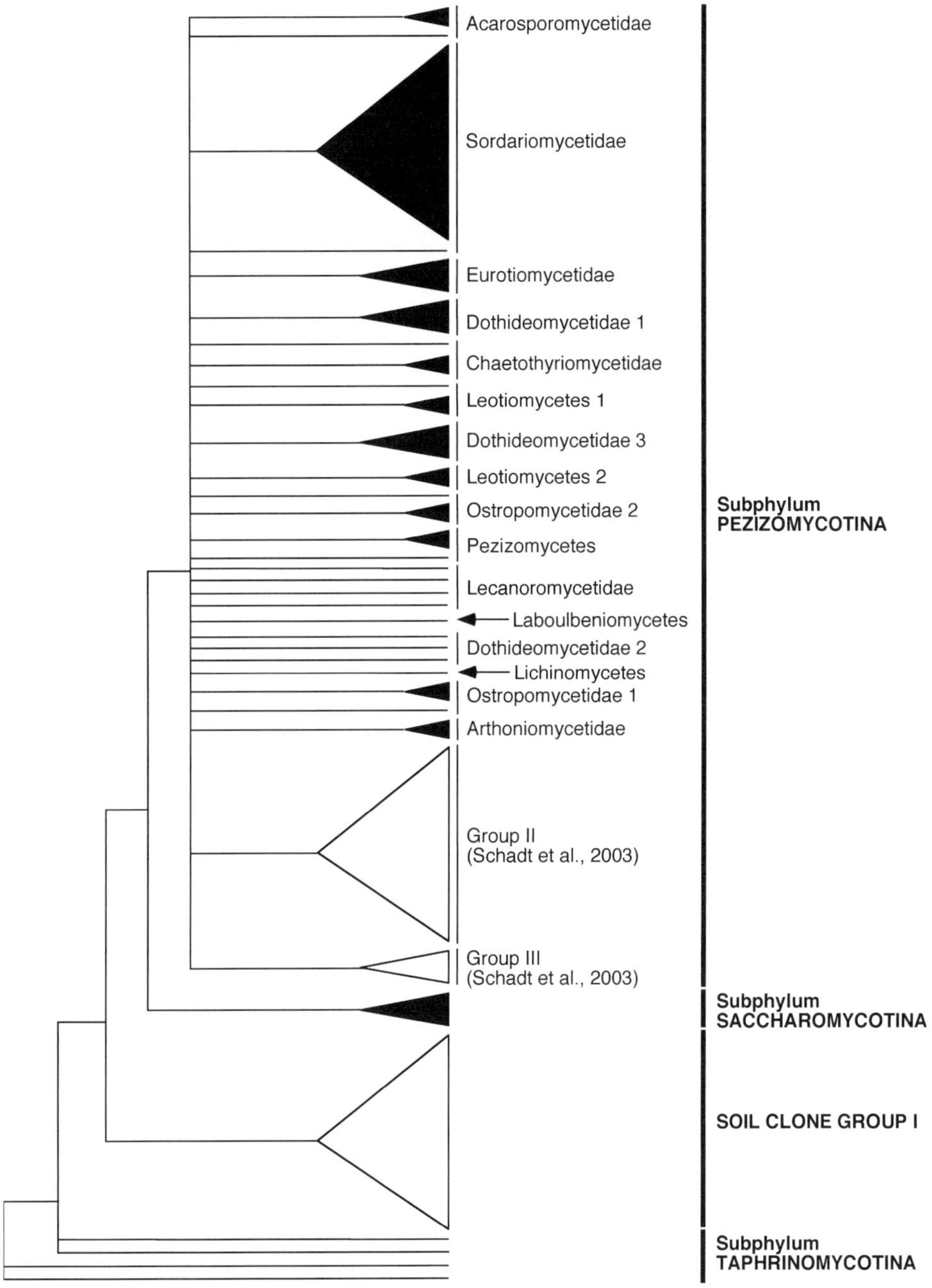

FIGURE 1 Composite tree from a Bayesian analysis of LSU rRNA genes showing the relative position of Soil Clone Group I (Schadt et al., 2003; Porter et al., 2008) between the *Taphrinomycotina* and *Saccharomycotina* subphyla, and Groups II and III within the *Pezizomycotina* subphylum as reported by Schadt et al. (2003).

clade that is distinct from Group II and is also likely part of *Pezizomycotina*. Neither of these clades has been positively identified in other studies, though this may be an issue of different gene regions being used for sequencing as discussed above.

TABLE 1 Biogeographic distribution of SCGI in rDNA clone libraries of soils[a]

Authors	Year	Soil	Location
Schadt et al.	2003	Tundra dry meadow soil	Colorado, U.S.A.
Porter et al.	2008	Western hemlock forest soil	British Columbia, Canada
Porter et al.	2008	Eastern hemlock forest soil	Ontario, Canada
Porter et al.	2008	Montane pine-fir forest soil	Colorado, U.S.A.
Porter et al.	2008	Treeline spruce-fir soil	Colorado, U.S.A.
Porter et al.	2008	Forest oxisoil	Costa Rica
Izzo et al.	2005	Mycorrhizal root tips, mixed conifer forest	California, U.S.A.
Pringle et al.	2000	Fungal glomalean spore from grassland	North Carolina, U.S.A.
Rosling et al.	2003	Mycorrhizal root tips, boreal forest	Sweden
Chen and Cairney	2002	Sclerophyll forest soil	New South Wales, Australia
Menkis et al.	2005	Mycorrhizal root tips, forest soil	Lithuania
Anderson et al.	2003	Temperate grassland soil	Scotland
O'Brien et al.	2005	Coniferous and hardwood forest soil	North Carolina, U.S.A.
Meyer and Schmidt	2007[b]	Under snow talus and forest alpine soil	Colorado, U.S.A.
Jumpponen et al.	2005	Tallgrass prairie	Kansas, U.S.A.
Anderson et al.	2003	Grassland soil	Scotland

[a]Identification of SCGI in previously published libraries was revealed by the work of Porter et al. (unpublished) who sequenced a large fragment of SCGI DNA from environmental soil samples and was able to link clone libraries that were based on smaller and nonoverlapping DNA sequences (SSU, ITS, or LSU).

[b]Unpublished data.

Chytridiomycota and *Zygomycota*

The *Chytridiomycota* (chytrids) are the only fungi with flagellate zoospores. Phylogenetic reconstruction indicates that they are the most basal fungi although they are not monophyletic but rather intermixed with the *Zygomycota* (James et al., 2006). Although chytrids are usually thought of as aquatic organisms, they are also found in many terrestrial environments, including soils from grasslands, forests, and tundra. While relatively few chytrids have been found in fungal soil libraries, some deeply divergent new chytrid groups have been recovered from high-altitude soils that are seasonally saturated with water during snow melt and also from recent intensive culturing efforts using seasonally wet and cold high-altitude soils from around the world.

Meyer and Schmidt (unpublished data) reported the presence of at least one major novel group of chytrids in clone libraries from barren high-altitude (4,000 meters above sea level) soils. Although their Bayesian tree contains the closest nBLAST matches of all of their clones, one clade does not show any known close relative. This clade (TAL 6) falls with strong support outside any known chytrid group at the ordinal level (Fig.2).

The high-altitude soils studied by Meyer and Schmidt (unpublished data) also contain other unusual, possibly anaerobic chytrids. One of the clones (T3 BA7) (Fig. 2) from this site groups with uncultured clone LEM010. The LEM010 clone originated in the anoxic sediment of a freshwater lake (Dawson and Pace, 2002), suggesting this novel chytrid may be a strict anaerobe of sediments and saturated soils. Another chytrid clone (T3 BF8), from the same study, groups strongly with an unusual chytrid, *Hyaloraphidium curvatum*. *H. curvatum* was first described as a colorless alga (Ustinova et al., 2000), but molecular analysis revealed it as a member of the *Monoblepharidales* (Bullerwell et al., 2003). It produces no zoospores and has no rhizoids (Ustinova et al., 2000).

Recent intensive culturing efforts using high-alpine soils have yielded other novel chytrids. Simmons (2007) amassed nine new pure cultures of soil chytrids that group with the unusual aquatic chytrid *Chytriomyces angularis*. This species has resisted previous phylogenetic placement in the known chytrid

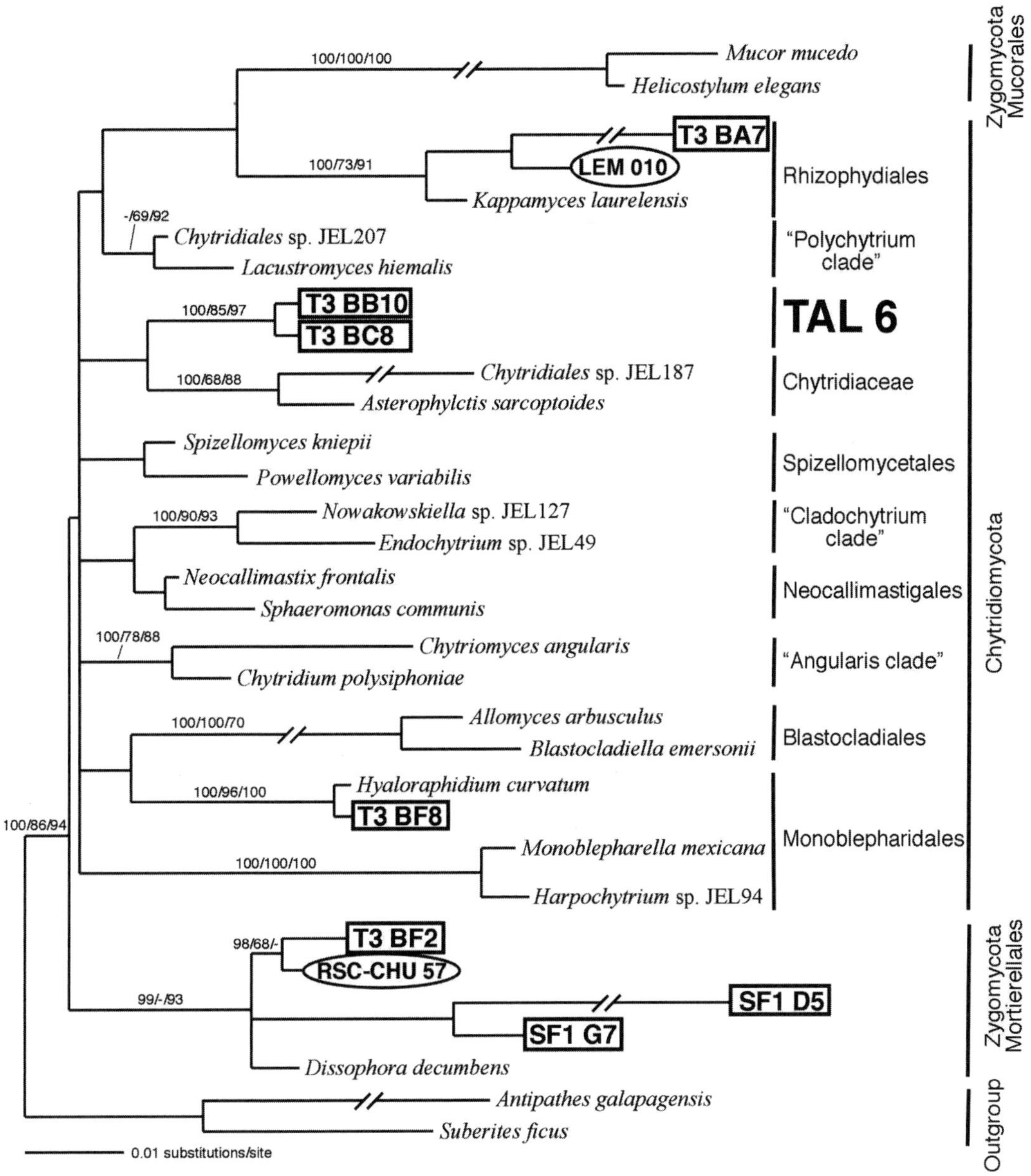

FIGURE 2 Phylogenetic tree of environmental sequences of the *Chytridiomycota* and *Zygomycota* from high-altitude soil clone libraries. This tree consists of library clones (in rectangles), nBLAST matches of the clones, and selected guide sequences, and was constructed using Bayesian methods. Clone sequences of uncultured organisms that are nBLAST matches of the high-altitude soil clones are ovals. Support values are Bayesian posterior probabilities, and bootstraps using parsimony and neighbor joining methods (Bayes/MP/NJ). Selected posterior probabilities are those greater than 95%; selected bootstraps are those recovered greater than 50%. The TAL 6 clade represents novel chytrid sequences at the ordinal level. Two other novel lineages are also shown, T3 BA7 and T3 BF2. Zygomycete clones SF1 D5 and SF1 G7, found under snow in forest soils in Colorado, are members of a recently described snow mold consortium (Schmidt et al., 2007, 2008).

orders (James et al., 2000). Results of the Bayesian and maximum parsimony analyses of nucSSU and partial nucLSU rDNA sequences grouped the nine isolates and *C. angularis* in a monophyletic clade within the *Chytridiomycota*. Transmission electron microscopy studies have further verified that these isolates had ultrastructure similar to that of *C. angularis*. Thus, molecular and ultrastructural data support the establishment of the new order *Lobulomycetales*, which contains the family *Lobulomycetaceae*, *C. angularis,* and the new soil chytrids described by Simmons (2007).

Figure 2 also identifies what could be a new *Zygomycota* clade. SF1 D5 and SF1 G7, both identified in under-snow forest soils in Colorado, are members of the subalpine snow mold consortium recently described by Schmidt et al. (2007, 2008). A clone that is probably related to this group was also detected in the winter, but not the summer, tundra clone libraries of Schadt et al. (2003). Psychrophilic fungi related to this group have also been described from Arctic and Antarctic soils (Bergero, 1999; Wynn-Williams, 1996). Several of these fungi have been isolated in pure culture and showed remarkably rapid growth rates at subzero temperatures in the laboratory. On the basis of their growth kinetics and the high levels of nutrients available under late-winter snow packs, Schmidt et al. (2008) hypothesize that these fungi are *r*-selected opportunists that fill an important but transiently available niche in many seasonally snow-covered ecosystems. Remarkably, their growth kinetics also match quite well with field-observed exponential increases in CO_2 fluxes through late-winter snow packs in alpine and subalpine environments (Brooks et al., 1997; Monson et al., 2006).

A potentially novel group of basal fungi that clusters independently from the major groups of chytrids and zycomycetes analyzed by James et al. (2006) has been identified using LSU rDNA from temperate forest soil in Canada (Porter et al., unpublished). This group may be related to the novel chytrid group reported by Schmidt et al. (2008) using SSU rDNA isolated from Colorado soil; however, this relationship remains to be tested.

THE TIP OF THE ICEBERG?

We believe that the soil studies described above have barely scratched the surface in terms of describing the diversity of fungi. The use of environmental genomic techniques shows promise in detecting new fungal clades and expanding our general knowledge of fungal diversity. There should be no doubt that additional studies will discover new, deeply branching clades of fungi and many more species in existing clades, especially if these studies are conducted in diverse geographic and ecological niches and at various times of the year. The use of long sequencing reads, containing SSU, ITS, and LSU regions of the fungal genome, should be more generally used to maximize the information obtained and enhance our ability to make comparisons between studies.

Every new gene library from soil done so far contains many unclassifiable sequences that arguably represent new species or clades at the genus level or above, even though most of these studies were not carried out with the intent of discovering new fungi. In addition to the studies discussed above, potential novel clades are hinted at in the recent work of several groups (Anderson and Campbell, 2003a; Anderson et al., 2003b; He et al., 2005; Jumpponen and Johnson, 2005; Landeweert et al., 2003; Lynch and Thorn, 2006; Malosso et al., 2006; Menkis et al., 2005; Selbmann et al., 2005; Taylor et al., 2007; Vandenkoornhuyse et al., 2002). However, because most of these studies did not further delve into phylogenetically verifying the depth of divergence of their sequences, further synthetic studies are needed to truly know how many novel fungal groups are represented in the current fungal databases. It is intriguing, however, that some of these published studies may also be detecting the novel *Ascomycota* clade (SCGI) discussed above.

In the studies presented above, the discovery of SCGI subphylum in *Ascomycota* is a significant addition to the diversity of fungi in general. Groups II and III that seem to belong in the

Ascomycota subphyla *Pezizomycotina* need to be confirmed and may also add significantly to our knowledge of fungal diversity. The new chytrids described in this chapter hint at far more diversity in chytrids that need additional exploration. It is also noteworthy that environmental DNA studies conducted to date have not recovered unknown deeply divergent groups of *Basidiomycota*, which composes the bulk of fungi forming structures detectable with the naked eye. This stresses the fact that fungi are mostly microorganisms and that detection and assessment of their global diversity necessitate the use of microbiological tools similar to those used in studies of bacteria and archaea.

Finally, it is important to flesh out the fungal family tree as quickly as possible. Given the reality of global climate change, there is a high probability that many of the missing fungi will go extinct before we even know that they ever existed. This is especially apparent given that many of the recently discovered groups of ascomycetes and chytrids discussed in this chapter come from seasonally cold soils that are especially vulnerable to current warming trends. These fungi may play pivotal roles in ecological cycles of nature and may produce unique antibiotics and enzymes of potential importance to humanity. Ongoing efforts to characterize, and especially to culture, these fungi are therefore of paramount importance.

The missing fungi are an open frontier, beckoning to us, as researchers, to explore and discover.

ACKNOWLEDGMENTS

This work was supported by the NSF Microbial Observatories program (MCB-0455606), the NSF Biotic Surveys and Inventories Program (DEB-0426116), and the National Geographic Society (#7535-03).

REFERENCES

Acinas, S. G., R. Sarma-Rupavtarm, V. Klepac-Ceraj, and M. F. Polz. 2005. PCR-induced sequence artifacts and bias: insights from comparison of two 16S rRNA clone libraries constructed from the same sample. *Appl. Environ. Microbiol.* **71:**8966–8969.

Anderson, I. C., C. D. Campbell, and J. I. Prosser. 2003a. Diversity of fungi in organic soils under a moorland—Scots pine (*Pinus sylvestris* L.) gradient. *Environ. Microbiol.* **5:**1121–1132.

Anderson, I. C., C. D. Campbell, and J. I. Prosser. 2003b. Potential bias of fungal 18S rDNA and internal transcribed spacer polymerase chain reaction primers for estimating fungal biodiversity in soil. *Environ. Microbiol.* **5:**36–47.

Arnold, A. E., Z. Maynard, G. S. Gilbert, P. D. Coley, and T. A. Kursar. 2000. Are tropical fungal endophytes hyperdiverse? *Ecol. Lett.* **3:**267–274.

Ashelford, K. E., N. A. Chuzhanova, J. C. Fry, A. J. Jones, and A. J. Weightman. 2005. At east 1 in 20 16S rRNA sequence records currently held in public repositories is estimated to contain substantial anomalies. *Appl. Environ. Microbiol.* **71:**7724–7736.

Ashelford, K. E., N. A. Chuzhanova, J. C. Fry, A. J. Jones, and A. J. Weightman. 2006. New screening software shows that most recent large 16S rRNA gene clone libraries contain chimeras. *Appl. Environ. Microbiol.* **72:**5734–5741.

Baldauf, S. L., and J. D. Palmer. 1993. Animal and fungi are each other's closest relatives: congruent evidence from multiple proteins. *Proc. Natl. Acad. Sci. USA* **90:**11558-–11562.

Bergero, R., M. Girlanda, G. C. Varese, D. Intili, and A. M. Luppi. 1999. Psychrooligotrophic fungi from Arctic soils of Franz Joseph Land. *Polar Biol.* **21:**361–368.

Bergsten, J. 2005. A review of long-branch attraction. *Cladistics* **21:**163–193.

Berney, C., J. Fahrni, and J. Pawlowski. 2004. How many novel eukaryotic 'kingdoms'? Pitfalls and limitations of environmental DNA surveys. *BMC Biology* **2:**13.

Blackwell, M., and K. Jones. 1997. Taxonomic diversity and interactions of insect-associated ascomycetes. *Biodiversity Conserv.* **6:**689–699.

Bridge, P. D., P. J. Roberts, B. M. Spooner, and G. Panchal. 2003. On the unreliability of published DNA sequences. *New Phytologist* **160:**43–48.

Brooks, P. D., S. K. Schmidt, and M. W. Williams. 1997. Winter production of CO_2 and N_2O from Alpine tundra: environmental controls and relationship to inter-system C and N fluxes. *Oecologia* **110:**403–413.

Bullerwell, C. E., L. Forget, and B. F. Lang. 2003. Evolution of monoblepharidalean fungi based on complete mitochondrial genome sequences. *Nucleic Acids Res.* **31:**1614–1623.

Cavalier-Smith, T. 2004. Only six kingdoms of life. *Proc. R. Soc. London, Ser. B.* **271:**1251–1262.

Chen, D. M., and J. W. G. Cairney. 2002. Investigation of the influence of prescribed burning on ITS profiles of ectomycorrhizal and other soil fungi at three Australian sclerophyll forest sites. *Mycol. Res.* **106:** 532–540.

Cole, J. R., B. Chai, T. L. Marsh, R. J. Farris, Q. Wang, S. A. Kulam, S. Chandra, D. M. McGarrell, T. M. Schmidt, G. M. Garrity, and J. M. Tiedje. 2003. The Ribosomal Database Project (RDP-II): previewing a new autoaligner that allows regular updates and the new prokaryotic taxonomy. *Nucleic Acids Res.* **31:**442–443.

Dawson, S. C., and N. R. Pace. 2002. Novel kingdom-level eukaryotic diversity in anoxic environments. *Proc. Natl. Acad. Sci. USA* **99:**8324–8329.

Frohlich, J., and K. D. Hyde. 1999. Biodiversity of palm fungi in the tropics: are global fungal diversity estimates realistic? *Biodiversity Conserv.* **8:**977–1004.

Frostegard, A., S. Courtois, V. Ramisse, S. Clerc, D. Bernillon, F. Le Gall, P. Jeannin, X. Nesme, and P. Simonet. 1999. Quantification of bias related to the extraction of DNA directly from soils. *Appl. Environ. Microbiol.* **65:**5409–5420.

Gardes, M., and T. D. Bruns. 1996. Community structure of ectomycorrhizal fungi in a *Pinus muricata* forest: above- and below-ground views. *Can. J. Bot.* **74:**1572–1583.

Hawksworth, D. L. 2001. The magnitude of fungal diversity: the 1.5 million species. *Mycol. Res.* **105:**1422–1432.

He, J. Z., Z. H. Xu, and J. Hughes. 2005. Analyses of soil fungal communities in adjacent natural forest and hoop pine plantation ecosystems of subtropical Australia using molecular approaches based on 18S rRNA genes. *FEMS Microbiol. Lett.* **247:**91–100.

Hibbett, D., A. Lutzoni, D. McLaughlin, J. Spatafora, and R. Vilgalys. 2002. Assembling the Fungal Tree of Life (AFTOL) Project Summary. http://ocid.nacse.org/research/deephyphae/htmls/AFTOL_sum_DH.html.

Howlett, B. J., B. D. Rolls, and A. J. Cozijnsen. 1997. Organisation of ribosomal DNA in the ascomycete *Leptosphaeria maculans*. *Microbiol. Res.* **152:**261–267.

Huber, T., G. Faulkner, and P. Hugenholtz. 2004. Bellerophon, a program to detect chimeric sequences in multiple sequence alignments. *Bioinformatics* doi:10.1093/bioinformatics/bth226.

Hugenholtz, P., and T. Huber. 2003. Chimeric 16S rDNA sequences of diverse origin are accumulating in the public databases. *Int. J. Syst. Evol. Microbiol.* **53:**289–293.

Ishii, K., and M. Fukui. 2001. Optimization of annealing temperature to reduce bias caused by a primer mismatch in multitemplate PCR. *Appl. Environ. Microbiol.* **67:**3753–3755.

Izzo, A., J. Agbowo, and T. D. Bruns. 2005. Detection of plot-level changes in ectomycorrhizal communities across years in an old-growth mixed-conifer forest. *New Phytologist* **166:**619–630.

James T. Y., D. Porter, C. A. Leander, R. Vilgalys, and J. E. Longcore. 2000. Molecular phylogenetics of the *Chytridiomycota* supports the utility of ultrastructural data in chytrid systematics. *Can. J. Bot.* **78:**336–350.

James, T. Y., F. Kauff, C. L. Schoch, P. B. Matheny, V. Hofstetter, C. J. Cox, G. Celio, C. Gueidan, E. Fraker, J. Miadlikowska, H. T. Lumbsch, A. Rauhut, V. Reeb, A. E. Arnold, A. Amtoft, J. E. Stajich, K. Hosaka, G. H. Sung, D. Johnson, B. O'Rourke, M. Crockett, M. Binder, J. M. Curtis, J. C. Slot, Z. Wang, A. W. Wilson, A. Schussler, J. E. Longcore, K. O'Donnell, S. Mozley-Standridge, D. Porter, P. M. Letcher, M. J. Powell, J. W. Taylor, M. M. White, G. W. Griffith, D. R. Davies, R. A. Humber, J. B. Morton, J. Sugiyama, A. Y. Rossman, J. D. Rogers, D. H. Pfister, D. Hewitt, K. Hansen, S. Hambleton, R. A. Shoemaker, J. Kohlmeyer, B. Volkmann-Kohlmeyer, R. A. Spotts, M. Serdani, P. W. Crous, K. W. Hughes, K. Matsuura, E. Langer, G. Langer, W. A. Untereiner, R. Lucking, B. Budel, D. M. Geiser, A. Aptroot, P. Diederich, I. Schmitt, M. Schultz, R. Yahr, D. S. Hibbett, F. Lutzoni, D. J. McLaughlin, J. W. Spatafora, and R. Vilgalys. 2006. Reconstructing the early evolution of fungi using a six-gene phylogeny. *Nature* **443:**818–822.

Janssen, P. H. 2006. Identifying the dominant soil bacterial taxa in libraries of 16S rRNA and 16S rRNA genes. *Appl. Environ. Microbiol.* **72:**1719–1728.

Jumpponen, A., and L. C. Johnson. 2005. Can rDNA analyses of diverse fungal communities in soil and roots detect effects of environmental manipulations— a case study from tallgrass prairie. *Mycologia* **97:**1177–1194.

Kirk, P. M., P. F. Cannon, and J. C. David (ed.). 2001. *Ainsworth and Bisby's Dictionary of the Fungi*, 9th ed. CAB International, Wallingford, United Kingdom.

Kurata, S., T. Kanagawa, Y. Magariyama, K. Takatsu, K. Yamada, T. Yokomaku, and Y. Kamagata. 2004. Reevaluation and reduction of a PCR bias caused by reannealing of templates. *Appl. Environ. Microbiol.* **70:**7545–7549.

Landeweert, R., P. Leeflang, T. W. Kuyper, E. Hoffland, A. Rosling, K. Wernars, and E. Smit. 2003. Molecular identification of ectomycorrhizal mycelium in soil horizons. *Appl. Environ. Microbiol.* **69:**327–333.

Landvik, S., T. K. Schumacher, O. E. Eriksson, and S. T. Moss.. 2003. Morphology and ultrastructure of *Neolecta* species. *Mycol. Res.* **107:**1021–1031.

Lipson, D. A., C. W. Schadt, and S. K. Schmidt. 2002. Changes in soil microbial community structure and function in an alpine dry meadow following spring snowmelt. *Microb. Ecol.* **43:**307–314.

Liu, Y. J., and B. D. Hall. 2004. Body plan evolution of ascomycetes, as inferred from an RNA polymerase II, phylogeny. *Proc. Natl. Acad. Sci. USA* **101:**4507–4512.

Longcore, J. E. 2005. Zoosporic fungi from Australian and New Zealand tree-canopy detritus. *Aust. J. Bot.* **53:**259–272.

Lueders, T., and M. W. Friedrich. 2003. Evaluation of PCR amplification bias by terminal restriction fragment length polymorphism analysis of small-subunit rRNA and mcrA genes by using defined template mixtures of methanogenic pure cultures and soil DNA extracts. *Appl. Environ. Microbiol.* **69:**320–326.

Lutzoni, F., F. Kauff, C. J. Cox, D. McLaughlin, G. Celio, B. Dentinger, M. Padamsee, D. Hibbett, T. Y. James, E. Baloch, M. Grube, V. Reeb, V. Hofstetter, C. Schoch, A. E. Arnold, J. Miadlikowska, J. Spatafora, D. Johnson, S. Hambleton, M. Crockett, R. Shoemaker, G. H. Sung, R. Lucking, T. Lumbsch, K. O'Donnell, M. Binder, P. Diederich, D. Ertz, C. Gueidan, K. Hansen, R. C. Harris, K. Hosaka, Y. W. Lim, B. Matheny, H. Nishida, D. Pfister, J. Rogers, A. Rossman, I. Schmitt, H. Sipman, J. Stone, J. Sugiyama, R. Yahr, and R. Vilgalys. 2004. Assembling the fungal tree of life. Progress, classification and evolution of subcellular traits. *Am. J. Bot.* **91:**1446–1480.

Lynch, M. D. J., and R. G. Thorn. 2006. Diversity of basidiomycetes in Michigan agricultural soils. *Appl. Environ. Microbiol.* **72:**7050–7056.

Malosso, E., I. S. Waite, L. English, D. W. Hopkins, and A. G. O'Donnell. 2006. Fungal diversity in maritime Antarctic soils determined using a combination of culture isolation, molecular fingerprinting and cloning techniques. *Polar Biol.* **29:**552–561.

Menkis, A., R. Vasiliauskas, A. F. S. Taylor, J. Stenlid, and R. Finlay. 2005. Fungal communities in mycorrhizal roots of conifer seedlings in forest nurseries under different cultivation systems, assessed by morphotyping, direct sequencing and mycelial isolation. *Mycorrhiza* **16:**33–41.

Moncalvo, J. M. 2005. Molecular systematics—major fungal phylogenetic groups and fungal species concepts. p. 1–33 *In* J. P. Xu (ed.), *Evolutionary Genetics of Fungi*. Horizon Scientific Press, Norfolk, United Kingdom.

Monson, R. K., D. L. Lipson, S. P. Burns, A. A. Turnipseed, A. C. Delany, M. W. Williams, and S. K. Schmidt. 2006. Winter forest soil respiration controlled by climate and microbial community composition. *Nature* **439:**711–714.

Nishida, H., and J. Sugiyama. 1994. *Archiascomycetes*: detection of a major new lineage within the *Ascomycota*. *Mycoscience* **35:**361–366.

O'Brien, H. E., J. L. Parrent, J. A. Jackson, J.-M. Moncalvo, and R. Vilgalys. 2005. Fungal community analysis by large-scale sequencing of environmental samples. *Appl. Environ. Microbiol.* **71:**5544–5550.

Polz, M. F., and C. M. Cavanaugh. 1998. Bias in template-to-product ratios in multitemplate PCR. *Appl. Environ. Microbiol.* **64:**3724–3730.

Porter, T. M., C. W. Schadt, L. Rizvi, A. P. Martin, S. K. Schmidt, L. Scott-Denton, R. Vilgalys, and J.-M. Moncalvo. Widespread occurrence and phylogenetic placement of a soil clone group adds a prominent new branch to the fungal tree of life. *Mol. Phylogenet. Evol.*, in press.

Pringle, A., J.-M. Moncalvo, and R. Vilgalys. 2000. High levels of variation in ribosomal DNA sequences within and among spores of a natural population of the arbuscular mycorrhizal fungus *Acaulospora colossica*. *Mycologia* **92:**259–268.

Rappé, M. S., and S. J. Giovannoni. 2003. The uncultured microbial majority. *Annu. Rev. Microbiol.* **57:**369–394.

Robertson, C. E., J. K. Harris, J. R. Spear, and N. R. Pace. 2005. Phylogenetic diversity and ecology of environmental archaea. *Curr. Opin. Microbiol.* **8:**638–642.

Rosling, A., R. Landeweert, B. D. Lindahl, K.-H. Larsson, T. W. Kuyper, A. F. S. Taylor, and R. D. Finlay. 2003. Vertical distribution of ectomycorrhizal fungal taxa in a podzol soil profile. *New Phytologist* **159:**775–783.

Schadt, C. W., A. P. Martin, D. A. Lipson, and S. K. Schmidt. 2003. Seasonal dynamics of previously unknown fungal lineages in tundra soils. *Science* **301:**1359–1361.

Schloss, P. D., and J. Handelsman. 2006. Toward a census of bacteria in soil. *PloS Comp. Biol.* **2:**786–793.

Schmidt, S. K., E. K. Costello, D. R. Nemergut, C. C. Cleveland, S. C. Reed, M. N. Weintraub, A. F. Meyer, and A. M. Martin. 2007. Biogeochemical consequences of rapid microbial turnover and seasonal succession in soil. *Ecology* **88:**1379–1385.

Schmidt, S. K., K. L. Wilson, M. M. Gebauer, A. F. Meyer, and A. J. King. Phylogeny and ecophysiology of opportunistic "snow molds" from a subalpine forest ecosystem. *Microb. Ecol.*, in press.

Schübler, A., D. Schwarzott, and C. Walker. 2001. A new fungal phylum, the *Glomeromycota*: phylogeny and evolution. *Mycol. Res.* **105:**1413–1421.

Selbmann, L., G. S. de Hoog, A. Mazzaglia, E. I. Friedmann, and S. Onofri. 2005. Fungi at the edge of life: cryptoendolithic black fungi from Antarctic desert. *Stud. Mycol.* **51:**1–32.

Simmons, D. R. 2007. Systematics of the *Lobulomycetales*, a new order within the *Chytridiomycota*. University of Maine, Orono, ME.

Taylor, L., I. C. Harriott, J. Long, and K. O'Neill. 2007. TOPO TA is A-OK: a test of phylogenetic bias in fungal environmental clone library construction. *Environ. Microbiol.* doi:10.1111/j:1462-2920.

Ustinova, I., L. Krienitz, and V. A. R. Huss. 2000. *Hyaloraphidium curvatum* is not a green alga, but a lower fungus; *Amoebidium parasiticum* is not a fungus, but a member of the DRIPs. *Protist* **151:**253–262.

Vandenkoornhuyse, P., S. L. Baldauf, C. Leyval, J. Straczek, and J. P. W. Young. 2002. Extensive fungal diversity in plant roots. *Science* **295:**2051.

Vilgalys, R. 2003. Taxonomic misidentification in public DNA databases. *New Phytologist* **160:**4–5.

von Wintzingerode, F., U. B. Gobel, and E. Stackebrandt. 1997. Determination of microbial diversity in environmental samples/ pitfalls of PCR-based rRNA analysis. *FEMS Microbiol. Rev.* **21:**213–229.

Wynn-Williams, D. D. 1996. Antarctic microbial diversity: the basis of polar ecosystem processes. *Biodiversity Conserv.* **5:**1271–1293.

THE DIVERSITY OF FREE-LIVING PROTISTS SEEN AND UNSEEN, CULTURED AND UNCULTURED

David A. Caron and Rebecca J. Gast

5

Single-celled eukaryotic organisms (protists) comprise an enormous collection of evolutionarily diverse species that collectively play fundamental and dominant ecological roles in most ecosystems on our planet. Protistan species range in size from the minute alga *Ostreococcus* sp. that is less than 1 μm in diameter, to heterotrophic amoeboid oceanic forms such as the radiolarian *Collozoum longiforme* that forms gelatinous colonies in excess of 1 m in length (Swanberg and Harbison, 1980; Courties et al., 1994). Within this 6 orders of magnitude in size, photosynthetic protists constitute much of the base of the food web in freshwater ecosystems and in the world ocean, whereas heterotrophic protists (aka protozoa) play essential roles as consumers of bacteria, photosynthetic protists, and other microorganisms, as links in pelagic and benthic food webs, and as mediators of decomposition and nutrient remineralization (Sherr and Sherr, 2002; Calbet and Landry, 2004; Sherr et al., 2007).

The study of these predominantly microscopic forms began with the invention and application of the earliest microscopes to samples of water more than three centuries ago by Robert Hooke and Antonie van Leeuwenhoek. In a series of letters to the Royal Society of London, van Leeuwenhoek described many minute "animalcules" in a wide variety of environments. These early studies revealed an incredible morphological diversity among the "infusoria," a now-defunct terminology that referred to the microscopic algae and protozoa inhabiting pond water. Studies of marine ecosystems followed and, collectively, resulted in the production of some of the most notable tomes from aquatic environments on microbial eukaryote taxa by the middle of the 18th century and early 19th century (Stokes, 1878; Leidy, 1879; Haeckel, 1887; Calkins, 1901; Lohmann, 1902; Gran, 1912; Fauré-Fremiet, 1924; Schewiakoff, 1926; Kofoid and Skogsberg, 1928).

These early works described many of the largest and most morphologically distinctive forms of protists and established the morphology-based taxonomy that is still considered the "gold standard" for describing these taxa. Significant compilations of the morphological descriptions of protists were realized despite the sometimes dubious interpretations and interpolations of morphological

David A. Caron, Department of Biological Sciences, University of Southern California, 3616 Trousdale Parkway, Los Angeles, CA 90089-0371. *Rebecca J. Gast*, Department of Biology, Woods Hole Oceanographic Institution, Woods Hole, MA 02543.

Accessing Uncultivated Microorganisms: from the Environment to Organisms and Genomes and Back
Edited by Karsten Zengler © 2008 ASM Press, Washington, DC

characters that the microscopes of the day could not resolve. The development of improved preservation and staining techniques since that time, and particularly the application of electron microscopy during the 1970s and 1980s, provided a wealth of morphological information to improve the taxonomic criteria used for the description and identification of these species (Lee et al., 2000).

This chapter concerns the "seen and unseen," "cultured and uncultured" protists. There are basically three categories of species included within these headings. Protists exist that have been observed, described morphologically, and established in laboratory cultures. Understandably, most of what we know regarding the ecology and physiology of these species has been ascertained from the study of the "seen and cultured" species that are amenable to laboratory culture. Surprisingly, however, there are also numerous protists that have been observed and described morphologically but have never been cultured. At least some of these "seen but uncultured" specimens are well-known taxa that have resisted attempts over many generations to bring them into culture. And finally, numerous studies within the last decade have reported DNA sequences obtained directly from environmental samples. Many of these sequences do not correspond to any morphologically described, cultured species of protists. Very little is known regarding the morphologies or ecologies of these microbial eukaryotes. These "unseen and uncultured" taxa constitute an area of very active research at this time.

A brief overview of these three categories of protistan species will be provided. It is a difficult undertaking for at least two reasons. First, there is tremendous breadth in form and function among the microbial eukaryotes that are circumscribed by the historically useful but now seemingly arbitrarily defined term "protist." It is difficult to generalize about such a huge number and diversity of organisms. Presently, estimates of the total number of extant species of protists are hotly debated in the literature. Much of this argument revolves around differences in the species concept applied to protists and the amount of intraspecies variability in taxonomic criteria that one is willing to accept. Given these uncertainties, some investigators have speculated that the total number of protistan species may be as few as 20,000 (Finlay and Fenchel, 1999), while others believe that the ciliates alone (one of several major lineages within the alveolates) may contain 30,000 species (Foissner, 1999). This does not include the many extinct protistan "species" known only from their fossilized remains (e.g., a considerable literature exists for foraminiferan tests obtained from deep-sea sediment cores).

Second, higher-level phylogenetic schemes for protistan taxa are presently in a state of upheaval. Older phylogenetic characterizations of single-celled eukaryotes grouped these species within a single kingdom, the *Protista*, but this superficial grouping has given way to a succession of schemes that include several different phylum- and kingdom-level groupings for these taxa (Cavalier-Smith, 1998; Adl et al., 2005; Shalchian-Tabrizi et al., 2006), and reorganization of the proposed evolutionary relationships among these groups (Schlegel, 1991; Schlegel, 1994; Baldauf, 2003; Simpson and Roger, 2004). There is still considerable debate regarding the number and evolutionary relationships among many lineages of protists, and new hypotheses of the phylogeny of protistan taxa are continuing to emerge at a brisk pace. Indeed, numerous monophyletic assemblages of single-celled eukaryotes exist that have still not been placed into a proper phylogenetic framework (Patterson, 1999). This rapidly changing landscape of the hypothesized phylogenetic relationships among protists complicates syntheses of information on these protistan groups.

The extensive reorganization of protistan phylogeny that has been undertaken in recent years stems in large part from the insufficiency of morphological criteria for characterizing the evolutionary relationships among some of these species and the tremendous amount of new taxonomic information to fill this gap that DNA sequences are providing. DNA sequences have become particularly useful for establishing the phylogenetic relationships among the minute protists that often possess few distinctive morphological features. This information has greatly aided studies of protistan evolution. Ironically, DNA sequences have also complicated the picture. Sequencing of small-subunit ribosomal DNA (18S rDNA, srDNA, or SSU DNA) from environmental samples has yielded a large number of sequences from apparently "unknown, undescribed" protistan taxa. The breadth and identities of these previously undetected (and still largely uncultured) taxa are only beginning to emerge. In addition, sequence information is still lacking for many well-described taxa, and problems exist regarding the quality of some of the sequences, particularly those obtained from environmental samples (Berney et al., 2004). As a consequence, modern phylogenetic schemes for protistan taxa within the Tree of Life (Fig. 1) will continue to change as more information on unknown or poorly characterized taxa becomes available (Baldauf, 2003; Simpson and Roger, 2004).

SEEN AND CULTURED: THE STATE OF OUR PRESENT KNOWLEDGE

There are certainly many hundreds of protistan species that have been observed and morphologically described from natural environments, isolated, and subsequently cultured in the laboratory. However, the curation of protistan cultures has not been as organized historically or as well supported financially as the curation of bacterial isolates. Therefore, the number of protistan species that are available through public and private culture collections has waxed and waned, and the mix of species available has changed through time. Most protists in culture are algal taxa and are maintained in public culture collections, while a small percentage of heterotrophic forms are present in public and private collections. The total number is a small fraction of the many thousands of protistan taxa that have been observed and described but do not presently reside in culture collections, and it is a minute number relative to the bacterial strains in extant culture repositories.

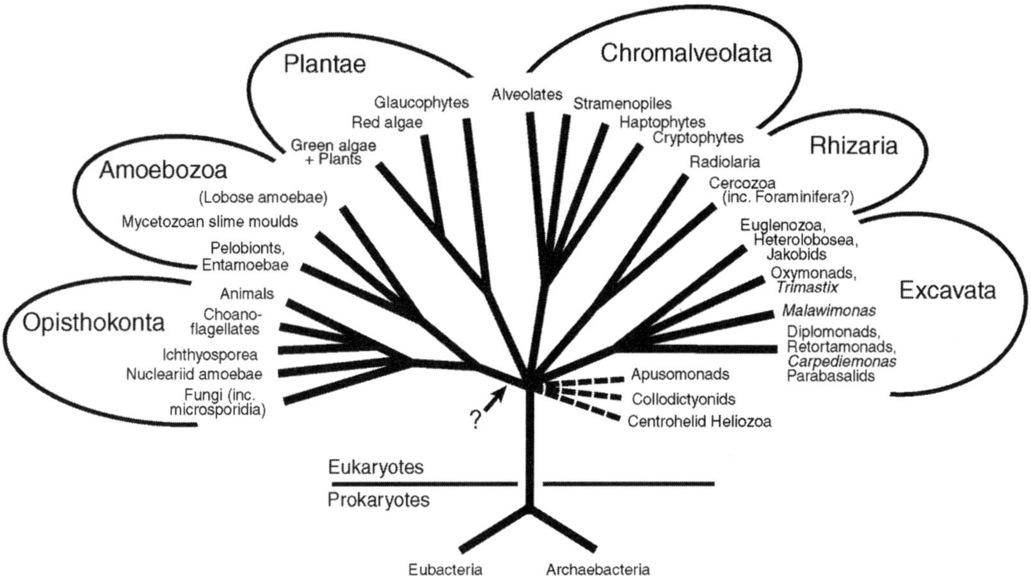

FIGURE 1 One recent proposition for the evolutionary relationships among living eukaryotic organisms. Note that protistan taxa dominate the many eukaryotic lineages of organisms with respect to diversity and evolutionary breadth. From S. L. Baldauf, D. Bhattacharya, J. Cockrill, P. Hugenholtz, J. Pawlowski, and A. G. B. Simpson. The tree of life: an overview, p. 43–75. *In* J. Cracraft and M. J. Donoghue (ed.), *Assembling the Tree of Life,* chapter 4. Oxford University Press, Oxford, United Kingdom, 2004.

There are numerous reasons for the paucity of cultured protists relative to the number of bacterial cultures that are available to the scientific community at any given time. These reasons generally relate to the considerable effort required for establishing and retaining protistan cultures. Present approaches for cryogenic preservation of protists are not adequately developed for many species, and those species must be sustained as live cultures. Maintenance of live protistan cultures typically requires a substantially greater amount of handling and care than most bacterial cultures. This is particularly true for heterotrophic protists that require live prey. The latter species presuppose that cultures of appropriate prey species are available (often bacteria or photosynthetic protists). Also, relatively few protistan species can be grown on solid media commonly employed for bacteria. The need to provide liquid cultures increases both the space required for protistan cultures and lessens the convenience of media preparation and storage. In addition, while most protists will reproduce indefinitely by binary fission, many possess sexual life stages that are not easily accommodated in laboratory culture. Long-term culture of some species necessitates knowledge of compatible mating types and/or life histories. Thus, many clonal strains of protists "fatigue" and stop growing after some initial period of success. Some protistan species may also require the presence of other species for the production of specific nutritional supplements (e.g., some amino acids, vitamins), and therefore the typical method of establishing clonal isolates excludes these organisms.

For the reasons noted above, there are many more "seen and cultur*able*" protistan taxa than "seen and cultur*ed*." It is not really appropriate to include the former specimens among the "unculturable" taxa because no exhaustive effort has been undertaken to bring as many specimens as possible into culture. The number of cultured forms could be substantially higher than is presently available (particularly for heterotrophic forms) if a concerted effort was carried out to generate and maintain these cultures, or if more effective long-term storage protocols were developed and applied.

These caveats of culture maintenance aside, we have learned a tremendous amount of information from the protistan taxa that have been brought into culture regarding their diversity, life histories, gross morphology, ultrastructure, nutrition, biochemistry, metabolism, physiology, ecology, evolution, and biogeochemical activities (Patterson and Larsen, 1991; Hausmann and Hülsmann, 1996; Falkowski and Raven, 1997; Lee et al., 2000). These advances are too numerous to recount in any detail in this chapter.

Cultured protists serve as the cornerstone for investigative studies of the biology of single-celled eukaryotes. Numerous species of photosynthetic and heterotrophic protists have been employed as "lab rats," model systems that have been used for many years to study everything from the physiological effects of pharmaceuticals to the acquisition and genetic control of organelles (e.g., chloroplast acquisition). For example, the facultatively phototrophic/heterotrophic protist *Euglena* (Fig. 2A) has been used for decades as a model system for studying chloroplast development (Hill et al., 1966). Many protistan taxa exhibit a high degree of physiological flexibility, making them interesting and versatile subjects for study.

A great deal of effort has been expended describing and culturing species of protists that are the cause of human illness (e.g., *Giardia, Plasmodium, Acanthamoeba, Trypanosoma, Cryptosporidium, Toxoplasma, Entamoeba*). Cultures and detailed procedures exist for maintaining many of these species in the laboratory. In addition, the genomes of most of these species have been or are being sequenced in an effort to elucidate their physiologies. This work will further efforts to provide appropriate conditions for long-term culture of these species and, hopefully, uncover new strategies for preventing and combating human infection.

Few protistan taxa have achieved a similar level of attention as the pathogenic/parasitic forms. Extensive research programs (and genome sequencing) have been advanced for some very well known free-living species that are readily amenable to laboratory culture and manipulation (e.g., *Tetrahymena*), but only a handful of "ecologically relevant" protistan taxa have been the subject of extensive genetic studies to date, compared to a few hundred bacterial/archaeal genomes completed or in progress. In part, the present bias against protists is due to the large genome sizes of most eukaryotes. Although single-celled, some protistan genomes are exceptionally large with some dinoflagellate genomes estimated at 215,000 Mbp compared to viral genomes of about 10^5 bp, bacterial genomes in the range of 10^6 bp, and the human genome at 3,000 Mbp (Hackett et al., 2004). Nevertheless, characterization of a few free-living protists has begun (Scala et al., 2002; Armbrust et al., 2004; Robbens et al., 2005), and this list of taxa will certainly lengthen as sequencing costs diminish and as transformation protocols for protists improve (Zaslavskaia et al., 2000).

Ecological studies of cultured protists have included representatives from most of the major lineages of protists and have detailed the nutritional aspects of these species, their elemental stoichiometries, feeding behaviors and rates, growth rates, and growth efficiencies. Our understanding of the biogeochemical significance and activities of these species has been ascertained largely through the manipulation and experimental examination of cultured protists. This information is vital because it places constraints on the abilities of these organisms. Using this information, ecologists have generated biogeochemical and food web models that significantly improved our understanding of the activities of photosynthetic and heterotrophic protists in natural communities and thus lend better insight into how aquatic communities function (Daniels et al., 2006; Richardson et al., 2006).

Specimens from many (but not all) major lineages of protists have been brought into culture. For example, all of the phyla listed in Fig. 1 possess representative taxa that have been cultured. However, some of these lineages are much more completely represented than others. Among the stramenopiles, for example, the diatoms are perhaps one of the best known morphologically described group of species, and also a group from which many species have been successfully established in the laboratory. Successful cultures of many planktonic, benthic, and even symbiotic diatoms (Lee et al., 1980) have been established and are available through public culture collections (e.g., the CCMP; The Provasoli-Guillard National Center for Culture of Marine Phytoplankton, http://ccmp.bigelow.org/). The morphologies of the silica frustules of these organisms have served as the primary criteria for the identification of these species since their discovery (Fig. 2L, M), and these features also have been extensively used in micropaleontological studies. The distinctive features of the frustules, and the relatively robust nature of many of these species, make the diatoms one of the best characterized groups of ecologically relevant protists.

In addition to the diatoms, there are a wide variety of stramenopile algae (e.g., chrysophyte algae) that have been observed in marine and freshwater ecosystems and for which there are well-established cultures. These include commonly encountered freshwater algae such as species from the genera *Dinobryon* and *Ochromonas*, as well as many other genera. In addition to these photosynthetic forms, there are a variety of heterotrophic stramenopiles that have also been cultured (e.g., *Paraphysomonas*, *Spumella*). Specimens from these genera can be readily cultured from most marine and freshwater ecosystems. As a consequence, a great deal of information for these species regarding feeding behavior, growth rates, growth efficiencies, and nutrient remineralization has been derived from laboratory studies (Fenchel, 1982a and b; Goldman and Caron, 1985; Boenigk et al., 2005; Boenigk et al., 2006).

Photosynthetic dinoflagellates within the alveolate lineage are also well represented in protistan culture collections. This group consists of a large number of species that are morphologically conspicuous and ecologically important (Fig. 2E). Dinoflagellates play important ecological roles in aquatic ecosystems.

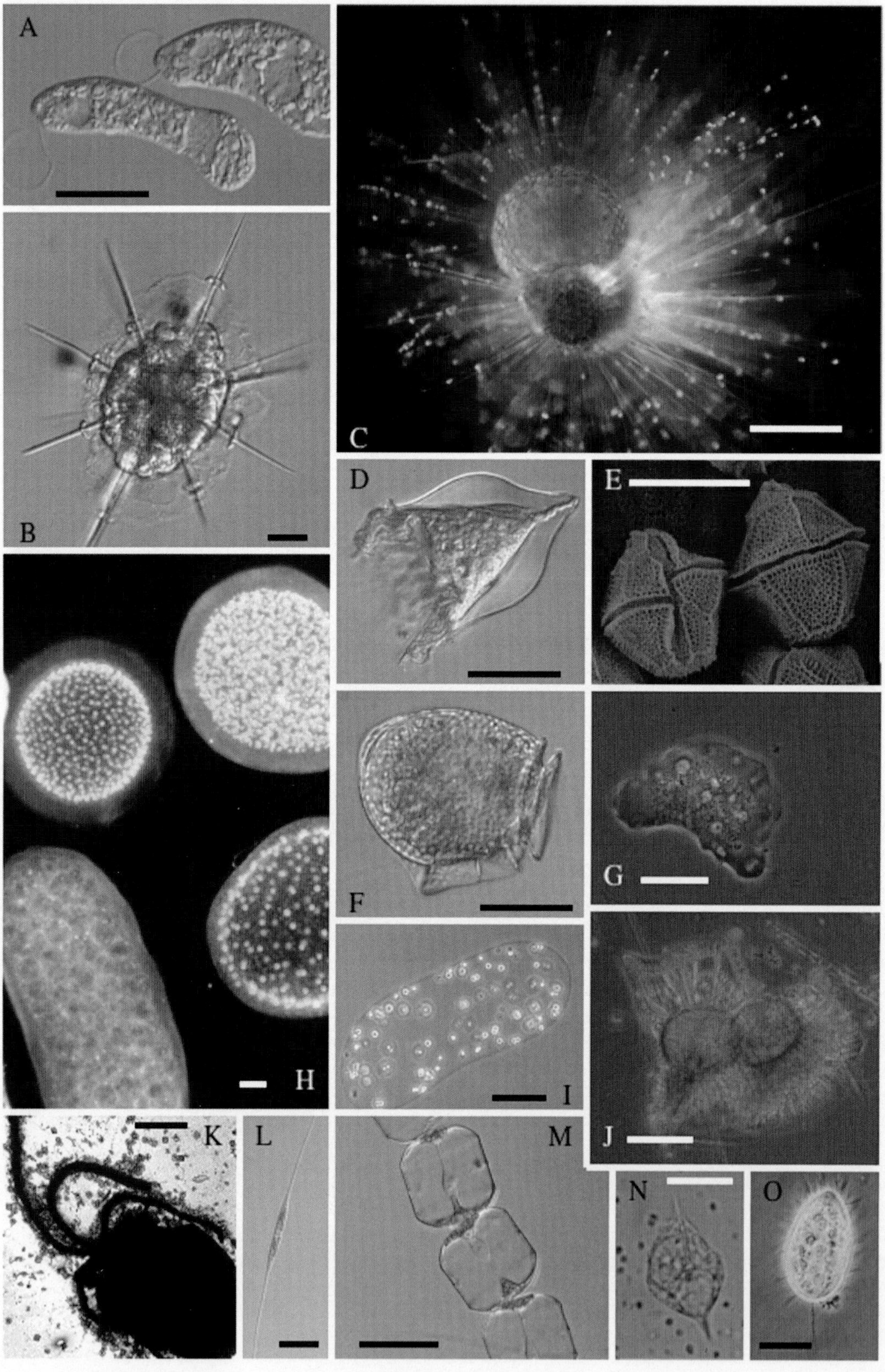
A
B
C
D
E
F
G
H
I
J
K
L
M
N
O

Based in large part on the study of cultured specimens, we know a great deal about the physiology of photosynthetic dinoflagellates and their contribution to total primary productivity of pelagic ecosystems (Falkowski and Raven, 1997). In addition, numerous dinoflagellates produce powerful toxins that constitute threats to human health and ecosystem function (Anderson and Ramsdell, 2005). Heterotrophic (apochlorotic and mixotrophic) dinoflagellates also exist. Phagotrophic dinoflagellates are important consumers of diatoms and other microbes in planktonic ecosystems (Lessard, 1991), but fewer of these species have been cultured, likely due to the lack of suitable prey items available in culture (Fig. 2F and J). Some of these species are highly specialized with respect to the microbes that are acceptable as their prey (see next section).

The ciliates are also important bacterivorous and herbivorous alveolates that occur in the plankton and benthos, and many cultures of these species have been established (Fig. 2D, O). Collectively, the feeding activities of phagotrophic dinoflagellates and ciliates account for a large fraction of the mortality of primary producers and bacteria in aquatic ecosystems (Sanders et al., 1992; Sherr and Sherr, 2002; Calbet and Landry, 2004). The ciliates have remained a monophyletic group within the alveolates throughout the period of phylogenetic reorganization that has engulfed protistology in recent years.

The remaining amoeboid and flagellated protists that have been established in laboratory cultures occur across what is now recognized as a wide range of taxonomic groups, often with very distant phylogenetic affinities (e.g., Fig. 1). Amoeboid form was once used to group protists exhibiting this type of morphology. Lobose amoebae (and other small naked and testate amoebae), foraminifera, radiolaria, and acantharia were grouped together based on their shared character of the pseudopodium (Fig. 2B, C, G, and H). This character has proven to be less phylogenetically significant than originally hypothesized. More recent classifications have grouped some amoebae into a single kingdom, the *Amoebozoa* (Baldauf et al., 2000), but this kingdom (and more recent categorizations) does not include many other protists exhibiting amoeboid form and locomotion (Fig. 1).

FIGURE 2 Examples of morphologically described and commonly cultured protists ("seen and cultured") and morphologically described but uncultured protists ("seen but uncultured"). (A) Differential interference contrast micrograph of *Euglena gracilis*, a facultatively phototrophic/heterotrophic protist (from D. J. Patterson). *Euglena* species have been cultured for many years. (B) A phase-contrast micrograph of a symbiont bearing antarctic acantharian. Neither the host nor the symbionts have been cultured. (C) Dark-field photomicrograph of the planktonic foraminiferan, *Globigerinoides sacculifer,* and its intracellular dinoflagellate symbionts. Planktonic foraminifera have never been cultured, the symbionts of *G. sacculifer* (*Gymnodinium beii*) have been. (D) An antarctic tintinnid ciliate photographed using differential interference contrast microscopy. Tintinnids have been isolated and cultured from numerous marine ecosystems. (E) Scanning electron micrograph of *Lingulodinium polyedrum*, a cultured, red-tide dinoflagellate. (F) Light micrograph of an antarctic species of *Dinophysis*. Species of this genus have only recently been brought into laboratory culture (see text). (G) Phase-contrast micrograph of a lobose amoeba maintained in enrichment cultures. (H) Dark-field photomicrograph of several species of symbiont-bearing colonial radiolaria. Central capsules of the radiolaria are visible as small dots within the pseudopodial networks. Dinoflagellate symbionts give the capsules a yellowish green color. The dinoflagellate symbiont, *Scrippsiella nutricula*, has been cultured, but the hosts have never been. (I) Phase-contrast micrograph of *Phaeocystis antarctica*, a cultured prymnesiophyte alga that forms mucilaginous colonies. Individual cells are the small dots embedded in the colony matrix. (J) *Ornithocercus* sp., a heterotrophic dinoflagellate with episymbiotic cyanobacteria, visualized using phase-contrast microscopy. The host has not been cultured, but the cyanobacterium has. (K) Negatively stained, transmission electron micrograph of *Pyramimonas* sp., a cultured antarctic prasinophyte. (L) Phase micrograph of *Cylindrotheca* sp., an antarctic diatom. (M) Phase-light micrograph of an unknown, uncultured antarctic diatom. (N) Light micrograph of the heterotrophic ebriid, *Hermesinum* sp. Ebriids have not been cultured. (O) The commonly cultured ciliate, *Uronema marina*, with ingested prey (the pelagophyte alga, *Aureococcus anophagefferens*), photographed using light microscopy. The pelagophyte has also been cultured and can be used to maintain cultures of this ciliate. Marker bars are 20 μm (A, B, F, L), 300 μm (C, H), 40 μm (D, E, I, M), 10 μm (G, J, N, O), and 2 μm (K).

Similarly, the term "flagellate" refers to a general morphological feature, the flagellum, that was used in older classifications of the protists (Lee et al., 1985). It is now recognized that photosynthetic and heterotrophic flagellated protists occur throughout a very wide range of protistan kingdoms.

Nevertheless, many of the phylogenetically far-flung amoeboid or flagellated taxa share a significant degree of similarity in their overall forms, motilities, diets, and distributions. For example, cultured species of choanoflagellates (opisthokonts), stramenopiles, cercozoa, euglenozoa, and excavates can all be grown in the laboratory using bacteria as prey. Presumably these species play similar ecological roles in nature. Moreover, the geographical distributions of these species overlap greatly in nature, and many of these species (if not all, as argued by some investigators) appear to be cosmopolitan in their distributions (Finlay, 2002).

SEEN—BUT UNCULTURED

Perhaps surprisingly, there are many protistan taxa that have been morphologically described based on microscopical observations but have never been brought into laboratory culture. There is often a general belief by those outside the field that microbial eukaryotic morphotypes that are easily visualized by microscopy are also easily cultured. This is a common fallacy for prokaryotes as well as microbial eukaryotes. Despite many decades of attempts, these "seen but uncultured" species of protists have eluded successful cultivation for one reason or another. Some of these species provide the most striking examples where a concerted but unsuccessful effort has been exerted to bring protistan species into culture. This result is in contrast to the many more species that have not been brought into culture simply because no concerted effort has been expended, as noted above. Relatively large size, delicate composition, poorly characterized nutrition, or complex life cycles have made a number of protistan species poorly suited for growth and reproduction in the laboratory, and will continue to be significant issues for establishing long-term cultures of these species.

The larger rhizarian protists (formerly known as sarcodine protists and presently classified as cercozoa) such as the radiolaria, acantharia, and planktonic foraminifera appear to fulfill one or more of these criteria. On the positive side, some benthic foraminifera have been brought into culture (Lee and Anderson, 1991; Lee et al., 1991). On the other hand, no planktonic foraminiferan, acantharian, or radiolarian has been cultured through successive generations in the laboratory (Fig. 2B, C, and H). They are exemplary of the most fastidious protists, a fact that also explains the paleontological usefulness of fossil shell assemblages of some of these taxa. Notably, specimens from many of these larger planktonic sarcodine species can be reared from juvenile life stages up to the release of vast numbers of swarmer cells, which are presumed to be gametes (Bé et al., 1983), but culture through successive generations has not been realized.

These relatively large, single-celled (but sometimes colonial) organisms have long generation times (several days to months), complex life histories, and nutrition that changes during ontogeny. Significant advances have been made in their general biology by rearing individual specimens from juvenile stages to adult size and swarmer formation (Anderson, 1980; Bé et al., 1983; Hemleben et al., 1988), but some aspects of their nutrition or required environmental conditions have prevented completion of their life cycles in the laboratory. This situation is rather remarkable given the substantial amount of effort that has been expended to culture these species, and it is unfortunate given the important role they play in paleoclimatological research. Collectively, the larger sarcodine protists form an abundant and ecologically important group of organisms in all tropical and subtropical oceanic environments (Caron et al., 1995; Michaels et al., 1995), and they have been well known and described for more than a century (Haeckel, 1887). Despite their protistological fame and considerable ecological importance, the factors that are lacking for successful culture of these species remain enigmatic.

It has long been speculated that the inability to re-create appropriate environmental conditions (light, temperature, pressure) during specific developmental stages of individuals may explain the lack of success in culturing the larger planktonic sarcodines. Acantharia, radiolaria, and planktonic foraminifera are predominantly oceanic species, and it has been speculated that they may undergo significant vertical migrations during ontogeny (Hemleben et al., 1988). The changing environmental parameters that would take place during ontogenetic migration have not been elucidated because the depth at which the earliest juvenile stages reside has not been determined. In addition, many basic aspects of reproduction remain in question for these species (e.g., if the swarmers released by adults are diploid zoospores or true gametes that require fertilization). The lack of compatible mating types might explain reproductive failure if the swarmers are haploid gametes. Finally, the prey of these heterotrophic species also must change markedly during ontogenetic development because these individuals grow from cells several micrometers in diameter to mature individuals with pseudopodial networks that can exceed 1-cm diameters for solitary species, or form gelatinous colonies more than 1 m in length. Many of these species also possess intracellular symbiotic algae that are acquired sometime during early ontogeny (Caron and Swanberg, 1990) (Fig. 2B, C, and H). It has been demonstrated that normal vegetative growth of at least some of these large protists is dependent on establishment and maintenance of these symbioses (Bé et al., 1982). Successful culture of these species might depend on providing the appropriate algae to the protistan host at the appropriate life stage in addition to other nutritional and environmental factors.

The acantharia, radiolaria, and planktonic foraminifera may represent the extreme situation of morphologically well-documented protistan taxa that have not been successfully cultured because there may exist multiple reasons for our failure in culturing these specimens (nutrition, environmental conditions, compatible mating types). For other protistan taxa, nutritional requirements or dependencies on specific trophic relationships among microbial taxa appear to constitute the main obstacles in preventing the establishment of long-term cultures of morphologically well-defined taxa. For example, species of the heterotrophic dinoflagellate genus *Dinophysis* are cosmopolitan and highly conspicuous components of coastal and oceanic plankton communities globally (Fig. 2F). These species have special significance in that they are capable of the production of okadaic acid, an occasional cause of diarrhetic shellfish poisoning. The ornate *Dinophysis* species appear to feed on a variety of organisms, but they acquire and harbor cryptophyte chloroplasts (a process referred to as kleptoplastidy) that they obtain from prey and maintain in a functional state (Stoecker et al., 1989; Stoecker et al., 1991; Stoecker, 1999). This obligate trophic relationship between *Dinophysis* species and cryptophytes has long been hypothesized as a reason for our inability to culture these important dinoflagellates. *Dinophysis* species have been maintained for some time in water samples containing mixed natural microbial assemblages, but long-term cultures of these species have been problematic.

Only recently has success in culturing species of *Dinophysis* been reported in the literature (Park et al., 2006) using the ciliate *Myrionecta rubra* as food for *Dinophysis acuminata*. *M. rubra* is a planktonic ciliate that consumes cryptophytes and can attain tremendous abundances in nature resulting in "red tides" in some coastal marine ecosystems (Crawford, 1989). The ciliate is itself a functional mixotroph (like *Dinophysis*) and uses kleptoplastidy to acquire the chloroplasts of ingested cryptophytes. It has been demonstrated recently that *M. rubra* can consume the cryptophyte *Geminigera cryophila* and retain the alga's chloroplasts in a functional state (Johnson et al., 2007). Apparently, *D. acuminata* is able to feed on the ciliate and retain the cryptophyte chloroplasts that the ciliate has obtained by kleptoplastidy from the alga, while growth of the dinoflagellate was not supported by the cryptophyte alone. This complex trophic relationship exemplifies the difficulties faced in providing the proper conditions (in this case, the appropriate prey) for the culture of some morphologically well-known species of protists, and may explain why *Dinophysis* species are commonly observed but have not been cultured until recently. A better understanding of the complex trophic interactions among microbes such as this one should allow more species to be brought into culture.

In general, heterotrophic dinoflagellates have not been as amenable to culture as photosynthetic dinoflagellates or other heterotrophic flagellate taxa. This disparity is not completely clear but may relate in part to feeding specialization among the dinoflagellates or dependency on specific associations with other organisms (Gordon et al., 1994; Jeong, 1999). *Dinophysis* is an excellent example of the degree of specialization that may be present among the heterotrophic dinoflagellates. Other heterotrophic dinoflagellates that are morphologically conspicuous but conspicuously absent from culture collections include taxa such as the ornate oceanic *Ornithocercus,* which possesses symbiotic cyanobacteria that attach to the outside of the dinoflagellate (Fig. 2J). The biochemical interactions in which these species participate with their symbionts, or perhaps the delicate nature of these oceanic dinoflagellates, may hold the key to successful culture of these dinoflagellates (Schnepf and Elbrächter, 1992).

Parasitic dinoflagellates (and perhaps related alveolate groups; see next section) present particularly difficult and rather specific problems for establishing cultures. These species require suitable hosts for infection, and thus the culture of these parasites is dependent on the availability of cultures of the hosts (unless storage protocols for the symbionts can be worked out). The range of suitable hosts is still poorly known for most of these parasites, and in other cases the host species themselves have not been cultured. For example, the presumed hosts for some parasitic dinoflagellates and other uncultured alveolates are radiolaria, which, as noted above, have not been brought into long-term culture (Gast, 2006; Dolven et al., 2007). Establishment and maintenance of cultures of these latter alveolate taxa would require a continuous source of hosts collected from the field.

Considerable success and important ecological information on the biology and ecology of parasitic dinoflagellates have been obtained for the one taxonomic group that has been relatively amenable to culture (Coats and Park, 2002; Park et al., 2004). Species within the genus *Amoebophrya* are known to parasitize and kill several photosynthetic dinoflagellate species as well as affect a variety of life processes of their hosts (Park et al., 2002a; Park et al., 2002b). The availability of suitable hosts that are readily amenable to laboratory culture has enabled research on *Amoebophrya.*

Examples of morphologically well-described heterotrophic flagellate taxa that are not dinoflagellates and have not been cultured also exist. For example, the Ebridia form a morphologically conspicuous and occasionally abundant group of heterotrophic flagellates that have not been brought into long-term laboratory culture. The species are identified by their distinctive siliceous skeletons (Fig. 2N). Species of *Hermesinum* and *Ebria* form sporadic blooms in coastal ecosystems where they have been demonstrated to consume small phytoplankton (Caron et al., 1989). Ebriids have been shown to have phylogenetic affinity to the cercozoa (Hoppenrath and Leander, 2006).

Photosynthetic flagellate taxa have many cultured representatives, as noted in the previous section. Mixotrophic behavior, usually a reference to combined photosynthetic and phagotrophic behavior, occurs among a large number of these species (Sanders and Porter, 1988). Mixotrophic algae are believed to prey on other microbes to obtain specific macro- or micronutrients (e.g., nitrogen, phosphorus, vitamins, trace metals). For example, many phytoplankton are known to have specific vitamin requirements (Croft et al., 2006) or other growth factors that are provided in most culture media. Presumably, co-occurring bacteria in nature are the source of these growth factors either directly via phagotrophy by the algae or indirectly by release and subsequent uptake by the algae from the surrounding environment. Elimination of these bacteria from cultures of phytoplankton may exacerbate the problems of establishing cultures of many phytoplankton species or may cause difficulties in obtaining axenic (unispecies) cultures. Nevertheless, these nutritional dependencies do not appear to be insurmountable problems for establishing cultures of a large number of photosynthetic protists. Indeed, only recently have the phagotrophic capabilities of some important phototrophic dinoflagellates become known (Jeong et al., 2005). For most species it is not known how heterotrophy by mixotrophic algae might contribute to their nutrition and how much this might explain the failure to culture some of the morphologically conspicuous but presently uncultured species of photosynthetic protists.

In addition to these potential problems with the culture of phagotrophic algae, there are heterotrophic species of protists that employ kleptoplastidy to the point where they apparently are in the process of permanently acquiring the sequestered chloroplasts from their photosynthetic prey, and thus their nutrition represents transitional states from a heterotrophic mode of nutrition to a phototrophic one. Two examples have been recently reported (Gast et al., 2007; Okamoto and Inouye, 2007). In the first case, a heterotrophic dinoflagellate that preys on *Phaeocystis antarctica* and retains its chloroplasts has been cultured for months in the laboratory. In the second case, a kathablepharid flagellate has been noted to retain chloroplasts from the prasinophyte genus *Nephroselmis*. The kathablepharid species has not yet been successfully cultured (Slapeta et al., 2006b). It is probable that complex trophic dependencies of this sort are responsible for some of the difficulties with culturing mixotrophic and kleptoplastidic protists.

UNSEEN—AND UNCULTURED

Environmental microbiologists have become increasingly aware during the last two decades that total microbial diversity in the ocean is far greater than has been revealed by traditional methodologies of microscopy and culture. The 1990s witnessed an explosion in the number of bacterial lineages that were reported based on novel gene sequences obtained from environmental samples, and that has led to the unavoidable conclusion that we have cultured only a fraction of the bacteria present in nature. This knowledge has now been developed into theoretical conceptualizations of abundance and species richness in natural microbial communities (Fig. 3A), and we now know that these communities are characterized by tremendous species richness, with large numbers of rare phylotypes (Schloss and Handelsman, 2004; Sogin et al., 2006). It is also acknowledged that since a large number of these sequence phylotypes do not match any cultured bacterial strains, many bacterial taxa have resisted traditional culture methods.

A similar community structure is now being recognized for microbial eukaryotes (Fig. 3B). In fact, a much greater degree of analogy to bacterial community structure has been observed for protistan communities than was originally expected. These assemblages also contain a tremendous diversity of taxa (based on the occurrence of unique "operational taxonomic units"), with most taxa present at very low abundances. Moreover, sequencing of DNA libraries from natural ecosystems has revealed the presence of significant numbers of previously undetected, undescribed microbial eukaryote taxa (see below). This latter finding implies that there has existed a significant bias in the methods that have been employed to observe and culture protists from natural ecosystems, and there is mounting evidence for that speculation. For example, *Paraphysomonas imperforata* is a heterotrophic chrysomonad with a cosmopolitan distribution (Finlay and Clarke, 1999). This species often dominates the bacterivorous flagellate assemblage that develops in natural seawater samples enriched with organic substrates. It has been repeatedly cultured from a variety of environments, and as a consequence, it has been employed as a model system for examining the physiological capabilities of small

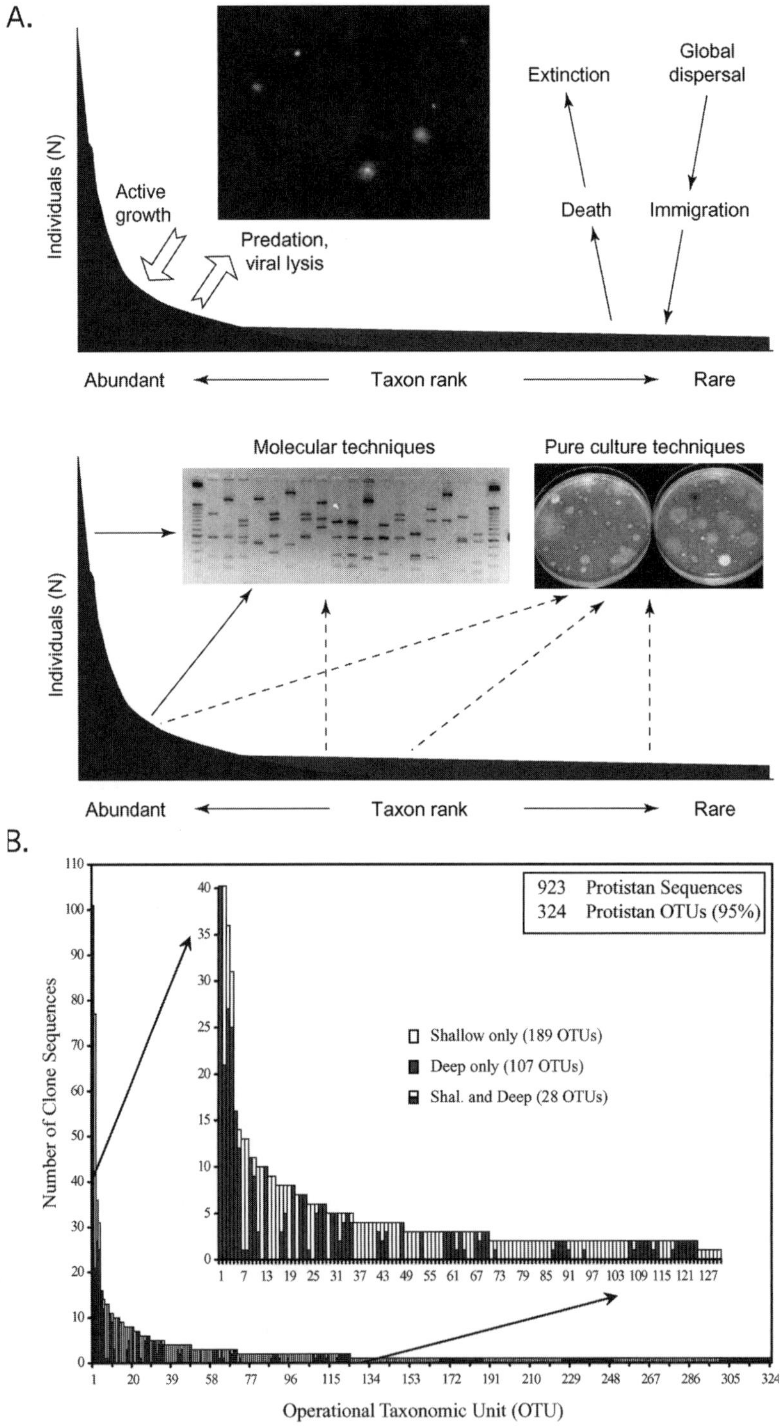

FIGURE 3 (A) Rank-abundance curve of microbial taxa in an idealized microbial (bacterial) community and (B) an actual rank-abundance curve for a natural protistan assemblage. The inset in panel B shows an enlargement of the rank-abundance curve for the most commonly encountered phylotypes. Panel A is from Pedrós-Alió (2006); panel B is from Countway et al., 2007.

bacterivorous flagellates in nature (Caron et al., 1985; Goldman et al., 1985; Andersen et al., 1986; Choi and Peters, 1992; Peters et al., 1996).

More recently, however, studies employing fluorescence in situ hybridization (FISH) have observed that this species, while present in many ecosystems, generally constitutes only a minor fraction of the total number of heterotrophic flagellates (Lim et al., 1999). In short, *P. imperforata* appears to be a "weed" species that flourishes under the enriched conditions of laboratory culture. That is not to say that "rare" taxa cannot play important roles in nature, but these findings imply that the ecological significance of *P. imperforata* as a dominant component of the bacterivore assemblage is unlikely in most ecosystems. That misconception is an artifact of its dominance in enrichment cultures. These results bear striking similarity to the model of bacterial community structure proposed by Pedrós-Alió (2006), and they support the speculation that methods employed to culture protistan species yield taxa that often do not represent the ecologically dominant taxa in nature (Fig. 3A).

At first glance, the occurrence of DNA sequences in environmental libraries that are not represented among sequences obtained from cultured protistan taxa is not surprising. As noted in the previous sections, only a small fraction of the total protistan diversity that is known to exist in nature is represented in culture collections, regardless of the species richness that one is willing to accept (Finlay and Fenchel, 1999; Foissner, 1999). Culture attempts have not been exhaustive, and most protists in nature have not been brought into culture simply because no concerted effort has been undertaken to do so. Also, some morphologically conspicuous species have resisted all attempts to culture them. Therefore, clone libraries might be expected to contain taxa for which we do not have cultured, sequenced representatives. Nevertheless, the sheer number of unseen and therefore undescribed taxa indicated by novel environmental DNA sequences, and the significant phylogenetic distances between many of these taxa and the nearest morphologically described protistan taxon, is remarkable.

It has been noted that at least some of the "unseen, uncultured" protistan taxa may not represent truly unique taxa, but instead may be a result of methodological artifacts. The potential problems include the presence of undetected chimeric sequences in databases and the misplacement of rapidly evolving sequences in molecular phylogenies. Berney et al. (2004) noted that a significant number of environmental sequences may in fact be chimeric sequences. On the basis of their analyses, the authors challenged the existence of undiscovered kingdom-level diversity within microbial eukaryotes and argued for verifying environmental sequences with sequences obtained from whole specimens. Similarly, Slapeta et al. (2006a) cautioned that the use of environmental sequences can be problematic when no reference organisms have been sequenced. In addition to methodological problems, environmental sequences (e.g., 18S rDNA sequences) lacking any accompanying morphological information are now accumulating in public databases at a rapid pace. While these sequences are useful for examining the geographical distributions of these phylotypes, they do not aid in identifying the morphotypes from which they originated.

It is also worth noting that "culture independent" does not mean "unbiased." Copy number for rRNA genes appears to vary considerably among protistan species. High copy number might favor the detection of certain taxa in natural samples (e.g., alveolates with large genome sizes). Moreover, the results obtained from sequencing genes amplified directly from environmental samples can vary considerably depending on the primer combinations employed. Stoeck et al. (2006) examined three libraries constructed from the same sample using different primer sets and found relatively little overlap between the phylotypes obtained from the three libraries. The result of this latter study tends to counterbalance the concerns of Berney et al. (2004) in that *some* of these numerous novel phylotypes must be derived from true novel lineages of microbial eukaryotes.

Despite issues of bias and potential error, there appears to be a substantial number of sequences that represent novel lineages of microbial eukaryotes for which no morphotypes have yet been described. Two major protistan groups have been the focus of much of the recent work to elucidate unseen, uncultured protistan lineages: the alveolates and the stramenopiles. Analyses of environmental samples from a wide variety of ecosystems have indicated the presence of phylogenetically distinct lineages within these large groupings. In addition, clone libraries constructed from assemblages of minute eukaryotes from marine and freshwater ecosystems have revealed the presence of a great deal of previously undetected diversity within *known* taxonomic groups.

The alveolates contain three well-known groups of protists: the ciliates, the dinoflagellates, and the apicomplexans. All three of these lineages have been well represented in environmental clone libraries, and as noted above, are also well represented among the "seen and cultured" protistan taxa (particularly the ciliates and dinoflagellates). Many of the "unseen, uncultured" DNA sequences from environmental samples have grouped with the known alveolates, but in two lineages that are distinct from the latter three groups (Fig. 4). These lineages have been designated "marine alveolates Group I" and "marine alveolates Group II" in the literature, and with a few notable exceptions, appear to contain no cultured or described species at this time. Group I alveolates have been found throughout the world and currently contain no named species. Group II alveolates also appear to be well supported as a unique lineage, but this group contains two genera that have been described: *Amoebophrya*, which is a parasite of several photosynthetic dinoflagellates (Coats et al., 1996), and *Hematodinium*, a parasite of blue crabs (Gruebl et al., 2002).

The Group I and II alveolates were first noted by López-García et al. (2001a, 2001b) from deep-sea ecosystems, and have since been detected from a wide variety of marine environments but not freshwater (Stoeck et al., 2003; Romari and Vaulot, 2004; Lovejoy et al., 2006; Stoeck et al., 2006; Countway et al., 2007). Groisillier et al. (2006) recently summarized sequence information relating to these two alveolate lineages and concluded that the Group II sequences may correspond to the *Syndiniales* of the dinoflagellates because the only two described genera showing phylogenetic affinity to this lineage (*Amoebophrya, Hematodinium*) are both from that dinoflagellate order (Fig. 4). If true, these findings would question the "novelty" of this group of alveolates but would not detract from the fact that we still know little about the overall breadth of species within this group or their ecologies. The morphotypes associated with the Group I taxa remain unknown at this time.

There has also been considerable speculation on the ecology of the "unseen, uncultured" protistan taxa in general, and within the novel alveolate lineages specifically. Stoeck et al. (2007) constructed both ribosomal DNA and RNA clone libraries of natural water samples in an effort to examine the fraction of the eukaryote community that was metabolically active (using the presence of sequences in clone libraries constructed from RNA as an indication of active cells). The authors noted that sequences of Group I alveolates were detected in rDNA libraries from the anoxic waters of the Mariager Fjord, but rRNA sequences of these taxa were not detected. The lack of Group I alveolate representation in the RNA-based library led the researchers to hypothesize that the ecology of these organisms might be tied to aerobic environments. These studies are predicated on empirical observations (by electron microscopy and by FISH) that have shown that growing protists contained greater numbers of ribosomes than starved cells (Fenchel, 1982b; Lim et al., 1993).

Moreira and López-García (2003) suggested that many of the novel sequences observed in environmental samples might be protistan species that are parasites of higher organisms. This conclusion was based on a phylogenetic analysis of small-subunit ribosomal DNA

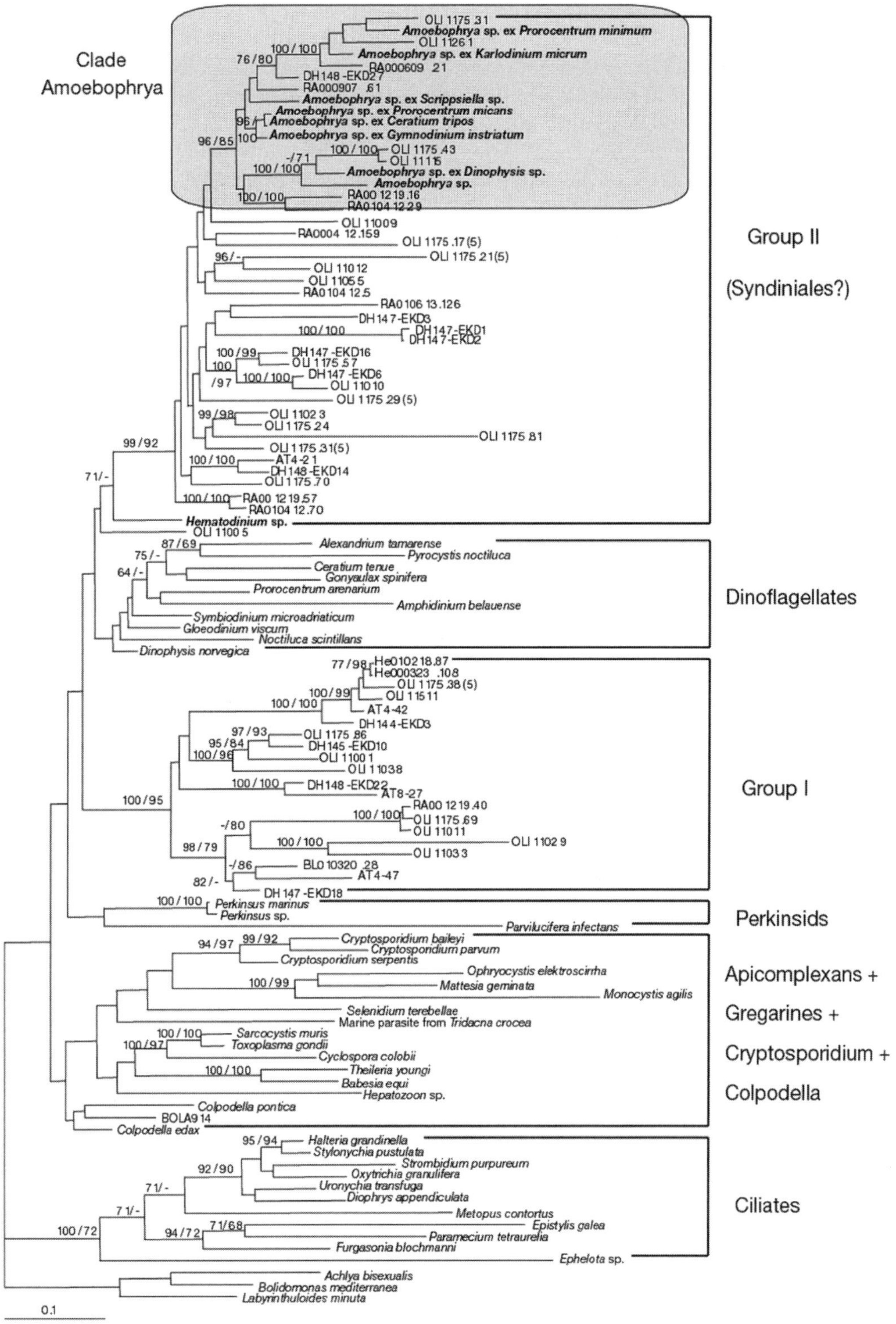

FIGURE 4 A proposed phylogenetic tree indicating the placement of major clades of "undescribed, uncultured" protistan taxa within the alveolates. Taken from Groisellier et al., 2006.

sequences from samples collected in the vicinity of hydrothermal vents in the Pacific and Atlantic oceans (Fig. 5). The authors noted the prevalence of phylotypes for which the closest known identified taxa were parasitic protists, including the gregarines (apicomplexans), other apicomplexans, *Syndiniales* (closely allied to the Group II alveolate sequences), and the kinetoplastids. The authors speculated that the localized high abundances of benthic fauna at vents might allow successful transmission of parasites among these inhabitants. This speculation is appealing because a parasitic life cycle might also explain why these species have not yet been recognized from direct microscopical analyses of water samples. If these species are present as nonvegetative life stages in the water column (e.g., cysts), they might well have remained undetected in past microscopical studies. Parasites would also not likely be recovered through traditional enrichment culture methods.

As yet, there has been little direct proof that the "unseen, uncultured" protistan taxa are indeed parasites. However, circumstantial evidence points in this direction for at least some of these organisms. Dolven et al. (2007) noted that some sequences of intracellular associates of several radiolarian taxa showed closest phylogenetic affinity to the sequences within the alveolate Groups I and II. These sequences could not be readily attributed to photosynthetic symbionts that also occurred in some specimens, and the authors noted that a parasitic mode of nutrition for these organisms could explain their results. A similar conclusion was reached by Gast (2006) in a separate study examining intracellular associates of the radiolorian *Thalassicola nucleata*.

The stramenopiles are an ecologically diverse group of protists that include many photosynthetic and heterotrophic forms. Within this group, Massana and coworkers (2002, 2004) have identified several novel independent lineages of MAST cells (for *ma*rine *st*ramenopiles) (Fig. 6). Their phylogenetic analyses indicate that these novel lineages were most probably heterotrophic. Subsequent studies demonstrating the growth of MAST cells in darkened incubation vessels, and the absence of chloroplasts in FISH-probed MAST cells have led the authors to conclude that these lineages represent small, heterotrophic, bacterivorous flagellates (Massana et al., 2006a; Massana et al., 2006b).

In addition to wholly novel lineages, clone libraries have indicated the presence of a much greater species diversity of minute (picoplankton), morphologically nondescript protists than has previously been realized. An example of the extensive genetic diversity detected in these very small protists can be found in the picoprasinophyte genus *Micromonas*. Sequence analysis of both environmental clone libraries and cultured representatives has led to the identification of at least five lineages within this globally distributed genus (Guillou et al., 2004; Slapeta et al., 2006b; Lovejoy et al., 2007). High diversity has also been observed within assemblages of minute chlorophytes in freshwater ecosystems (Fawley et al., 2004; Fawley et al., 2005). *Telonemia* is another example of an underestimated group of small flagellates. It has recently been proposed that these flagellates should be elevated to phylum level based on traditional descriptions and sequences of four genes of two species in the heterotrophic genus *Telonema* (Shalchian-Tabrizi et al., 2006). The genus was described nearly a century ago, and although only two species have been described to date, both have been brought into culture. Sequences related to these two species have been obtained from a number of geographically disparate environmental samples (Shalchian-Tabrizi et al., 2007), and this information has been employed to suggest that these sequences represent a distinct clade apparently related to the chromist lineages Cryptophyta and Haptophyta.

The abundance and diversity of these species have led some to speculate that minute size in a wide range of phylogenetically distinct algal taxa represents an example of convergent evolution (Potter et al., 1997). More importantly, the recognition of this immense diversity among the tiniest of single-celled organisms has led to renewed efforts to bring these species into culture. New classes of marine picoplanktonic algae have recently been described,

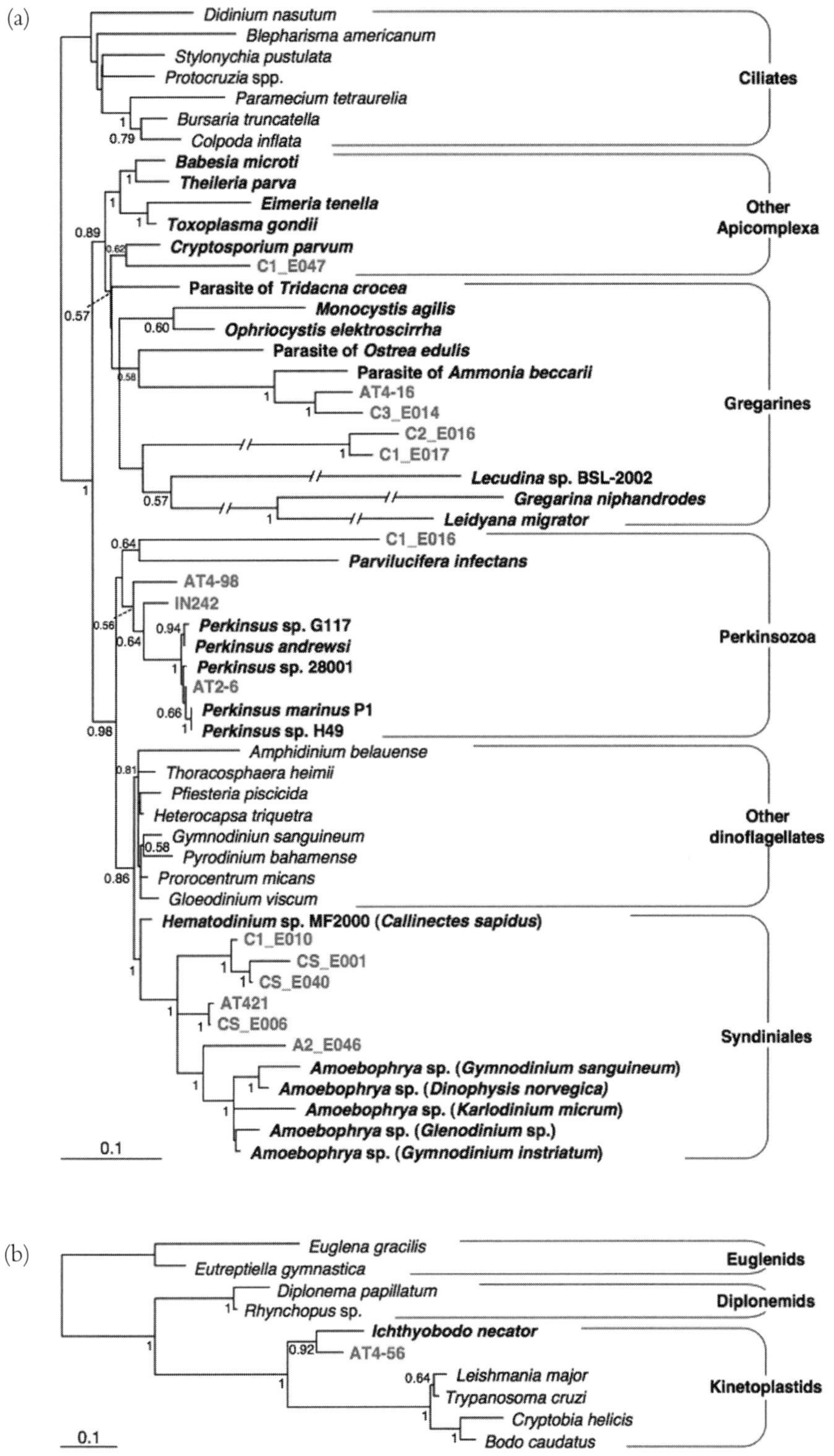

FIGURE 5 A proposed phylogenetic tree noting the association of several novel protistan clades within parasitic protistan lineages. Taken from Moreira et al., 2003.

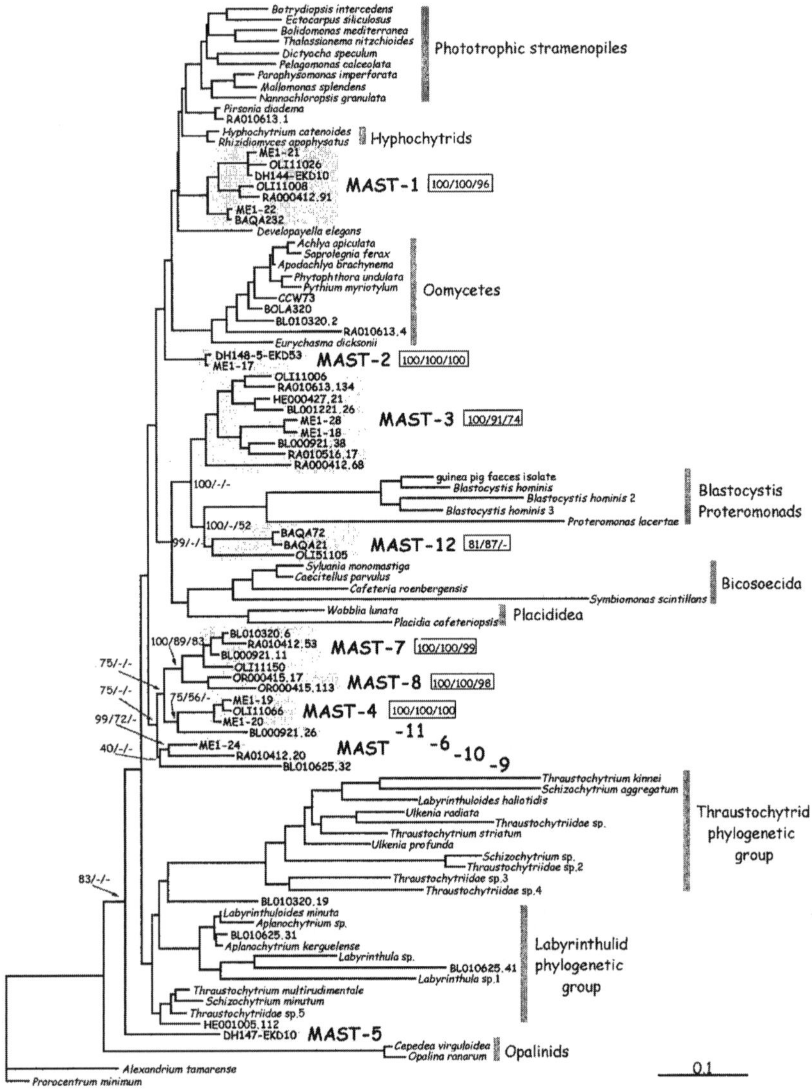

FIGURE 6 A proposed phylogenetic tree showing the placement of numerous novel marine stramenopile (MAST) phylotypes. Taken from Massana et al., 2004.

including the *Pelagophyceae* (Andersen et al., 1993), *Bolidophyceae* (Guillou et al., 1999a), and *Pinguiophyceae* (Kawachi et al., 2002), and a new algal group, the picobiliphytes, has been proposed (Not et al., 2007).

MOVING FORWARD: CULTURING THE UNCULTURED

A number of approaches have been, and will continue to be, useful in identifying the many unseen, unidentified taxa present in natural ecosystems. These methods will also aid in the eventual establishment and identification of cultures of presently "unculturable" species. These techniques include a mixture of classical and modern approaches.

Trial-and-error culture approaches continue to be a mainstay of protistological research. Isolation of single cells either by dilution-extinction or by physically picking cells from samples has been effective for minute, numerically dominant taxa and morphologically distinctive taxa, respectively. The transfer of isolated cells into a variety of culture media provides a reasonable chance of establishing cultures of many protists. These simple, traditional approaches are largely responsible for our present collection of cultured species. Success in this work obviously requires some general knowledge of the nutrition of the target cells (e.g., phototrophy versus heterotrophy), so these attempts will improve dramatically as the "needs" of the protistan taxa are characterized. An excellent example of progress using this approach is the breakthroughs that have occurred in the culture of many microalgal taxa as vitamin and trace metal requirements of these species have become known (Croft et al., 2006).

Physical separation of cells from other microbial taxa, and the establishment of cultures using these isolated cells, is a time-honored tradition in phytoplankton and protozoan research. It has regained popularity in recent years because of an increased recognition of the importance of minute "pico" algae and protozoa (Moon-van der Staay et al., 2001; Guillou et al., 2004; Massana et al., 2004; Romari and Vaulot, 2004). This approach has also recently led to the isolation and characterization of the smallest photosynthetic and heterotrophic eukaryotic cells in marine and freshwater ecosystems (Guillou et al., 1999a; Guillou et al., 1999b; Fawley et al., 2005).

The basic premise of single-cell isolation has remained unchanged, but the technology for accomplishing it has not. Flow cytometric sorting of minute protists, particularly phototrophs present at subdominant abundances, and therefore not easily accessed by dilution-extinction approaches, is becoming a popular means of obtaining cells for culture and other uses. Phototrophic protists are autofluorescent by virtue of their photosynthetic pigments. Cells that can be uniquely identified by their autofluorescence can be sorted directly and used to establish cultures. This information is a prime determinant of the basic culture conditions required for the sorted cells. Phagotrophic protists have also been identified through the combined use of fluorescent vacuole markers and flow cytometery (Carvalho and Graneli, 2006). Information on the potential nutritional strategy of the target taxa can aid greatly in subsequent culture attempts.

Flow cytometry also has been combined with FISH to label cells specifically and attach a fluorescent tag to target cells for sorting. FISH is particularly useful with probes designed based on information for "unseen" taxa whose sequences have been obtained from environmental samples (Massana et al., 2002), or single cells that have been isolated and sequenced (Ruiz Sebastian and O'Ryan, 2001; Takano and Horiguchi, 2005). Flow cytometry and FISH facilitate morphological characterization of the cells and can be used in combination to obtain information on the spatial distribution of uncultured taxa. Such information can provide insight into the ecology of these species (environmental conditions, trophic activities, etc.) that in turn provides guidance for designing culture media and choosing environmental conditions for establishing cultures.

An example of the cycle of discovery is exemplified by the recent description of a proposed new algal group, the picobiliphytes (Not et al., 2007). These minute phytoplankton were first detected as an independent lineage within phylogenetic analyses of environmental clone libraries. Subsequent studies using FISH allowed sorting and visualization of these cells by flow cytometry. Further analyses have inferred the presence of phycobiliprotein-containing plastids in these cells. Isolation and culturing have yet to be accomplished, but attempts are under way.

Isolation of single cells, FISH, flow cytometry, or other means can be very useful for linking sequence information for "unseen" taxa with specific morphotypes, which in turn can provide information on appropriate culture conditions. For example, Kolodziej and Stoeck (2007) employed FISH and scanning electron microscopy to identify MAST cells in a natural sample as 5-μm, heterotrophic flagellates. The use of FISH may also help identify ecological relationships that permit the inference of nutritional requirements of previously uncultured taxa. For example, the novel alveolate clades may be composed of protistan parasites and commensals. Linking parasitic taxa to their hosts could shed light on the nutritional requirements of the parasites. Understanding the benefit(s) provided by hosts would enable the design of strategies for culturing their dependent microfauna.

Linking sequence information and morphology is essential because (i) morphology is still the "gold standard" for protistan taxonomy and (ii) protistology is moving beyond the exclusive use of morphological criteria for the description of protists. DNA sequence information and ecological activity (e.g., physiological ability, trophic activity) are being combined with morphological information in an effort to make interpretation of the activity of protists in nature more effective (Modeo et al., 2003; Karpov et al., 2006). Also, the inadequacy of traditional taxonomic schemes to catalog and describe bacterial and archaeal diversity, and the need to reconcile those schemes with the emerging wealth of genetic information on these species, has been clearly recognized (Buckley and Roberts, 2007). Protistan taxonomy stands at a similar crossroad in that the traditional taxonomy based on morphological criteria is inadequate to describe fully the breadth of form and function manifested by this diverse group of species.

Genomics and metagenomics are being used as approaches to learn more about the biochemistry and metabolism of uncultured species of bacteria and archaea in order to better design methods for culturing these species. These approaches are not yet generally feasible for natural protistan assemblages because of the sizes of the genomes for most eukaryotic species. However, the application of whole genome/multiple displacement amplification to individual cells of interest is an approach that is becoming more feasible with new product assessment strategies (Raghunathan et al., 2005; Podar et al., 2007). This approach has been applied to dinoflagellates that possess large quantities of DNA per cell (Bolch, 2001; Edvardsen et al., 2003; Ki et al., 2004) and to radiolaria that have exceptionally large cell size (Yuasa et al., 2006). The results may provide vital information for designing culture schemes for uncultured species.

Future endeavors to culture novel protists from nature will undoubtedly focus on extreme or unique chemical ecosystems. These environments provide particularly fruitful ecosystems for pursuing the discovery and culture of previously undescribed, uncultured protistan species because they have been sparingly examined to date and have provided some of the most unique 18S rRNA genes yet observed. For example, environmental clone libraries from unique chemical habitats such as anoxic basins have indicated the presence of substantial, and highly novel, uncultured protistan diversity (Dawson and Pace, 2002; Stoeck and Epstein, 2003; Stoeck et al., 2003; Luo et al., 2005; Takishita et al., 2005; Jeon et al., 2006; Stoeck et al., 2006). Similarly, deep-sea ecosystems (López-García et al., 2001a; López-García et al., 2001b; Edgcomb et al., 2002; Countway et al., 2007; López-Garcia et al., 2007) and high-latitude ecosystems (Gast et al., 2004; Lovejoy et al., 2006) appear to harbor unique protistan fauna. These and other extreme ecosystems (Amaral Zettler et al., 2003) may constitute "low-lying fruit" for the culture of unique protistan types. This and the examination of much more mundane environments for taxa that are unexpected in those ecosystems (Holzmann et al., 2003) will undoubtedly yield exciting new discoveries and unique protists with novel ecologies.

CONCLUDING REMARKS

Central themes for ecological research on protistan assemblages at this time are the extent of novel protistan phylotypes, estimates of protistan diversity, and the geographical distribution of protistan species. As detailed above, much of the work accomplished thus far with natural assemblages has focused on the discovery, description, and cataloging of previously undescribed protistan types. This work will continue for the foreseeable future until we have developed a much clearer understanding of the extent of microbial eukaryotic diversity. The use of culture-independent approaches (i.e., genetic markers) has become particularly effective for expanding our knowledge because of the ability to detect presently unknown/uncultured taxa, the extreme taxonomic breadth of eukaryotic microbes, and the independence afforded by molecular biological approaches from the extensive information required by traditional, morphology-based, taxonomic schemes.

The overarching goal of this research, however, is not merely a "stamp collecting" endeavor. Ultimately, this work is intended to develop a thorough understanding of the functional significance of the considerable diversity that characterizes natural protistan assemblages. Fundamental, long-standing questions still remain as to how protistan assemblages are organized, maintained, and function. Do these communities represent random collections of species, or is species composition strongly affected by local environmental (or biological) parameters? Are communities of higher diversity more resistant to changes in ecosystem function than communities of lower diversity? Is diversity directly related to functional resilience following disturbance? Does ecological redundancy exist within these communities (i.e., are there multiple protistan taxa that can fulfill the same ecological role)? Answers to these questions will require more than a mere list of the species that are present in an ecosystem. They will require detailed physiological information on a great many protistan taxa, as well as experimental studies to test hypotheses involving community structure and response, trophic relationships, and biogeochemical function (McGrady-Steed et al., 1997; Naeem and Shibin, 1997; Griffiths et al., 2004; Girvan et al., 2005).

Cultures of ecologically important species will provide the fodder for this investigative work. Studies of cultured protists help establish the behavioral and physiological constraints imposed by environmental conditions on the survival and growth of these species and thus limits on their ecological and biogeochemical roles in nature (Rose and Caron, 2007). Studies of species in culture are also essential for establishing the growth requirements and efficiencies of microbes, their trophic interactions, and life histories. In turn, this information will improve the predictability of models of microbial activity.

Finally, environmental genomics has recently emerged as a powerful approach for understanding and investigating microbial community function (DeLong, 2004). There is great potential in these approaches. However, with these approaches has come the misunderstanding by some that we no longer need to culture organisms. The contention has been that simply by having genomic information, one can look directly into natural ecosystems and interrogate cells about what they are doing. In fact, this is far too simplistic. One must understand how microbes can respond physiologically under a given set of environmental conditions in order to be able to properly interpret genetic information obtained directly from natural microbial communities. That understanding can only be gained from carefully controlled experiments in which those responses are characterized under various environmental conditions and situations. In short, cultures are required to understand how genetic potential translates into metabolism and physiology. It is important to add that genomic information may provide useful insights for culturing some presently uncultured protists, but only through the study of cultured species can we hope to obtain a basic, mechanistic understanding of how organisms work at the organismal level, and thereby derive predictive capability regarding how they can and will respond to changing environmental conditions.

REFERENCES

Adl, S. M., A. G. B. Simpson, M. A. Farmer, R. A. Andersen, O. R. Anderson, J. R. Barta, S. S. Bowser, G. Brugerolle, R. A. Fensome, S. Fredericq, T. Y. James, S. Karpov, P. Kugrens, J. Krug, C. E. Lane, L. A. Lewis, J. Lodge, D. H. Lynn, D. G. Mann, R. M. McCourt, L. Mendoza, O. Moestrup, S. E. Mozley-Standridge,

T. Nerad, C. A. Shearer, A. V. Smirnov, F. W. Spiegel, and M. F. J. R. Taylor. 2005. The new higher level classification of eukaryotes with emphasis on the taxonomy of protists. *J. Euk. Microbiol.* **52:**399–451.

Amaral Zettler, L. A., M. A. Messerli, A. D. Laatsch, P. J. S. Smith, and M. L. Sogin. 2003. From genes to genomes: beyond biodiversity in Spain's Rio Tinto. *Biol. Bull.* **204:**205–209.

Andersen, O. K., J. C. Goldman, D. A. Caron, and M. R. Dennett. 1986. Nutrient cycling in a microflagellate food chain: III. Phosphorus dynamics. *Mar. Ecol. Prog. Ser.* **31:**47–55.

Andersen, R. A., G. W. Saunders, M. P. Paskind, and J. P. Sexton. 1993. Ultrastructure and 18S rRNA gene sequence for *Pelagomonas calceolata* gen. et sp. nov. and the description of a new algal class, the *Pelagophyceae* classis nov. *J. Phycol.* **29:**701–715.

Anderson, D. M., and J. S. Ramsdell. 2005. HARRNESS: a framework for HAB research and monitoring in the United States for the next decade. *Oceanography* **18:**238–245.

Anderson, O. R. 1980. Radiolaria, p. 1–42. *In* M. Levandowsky and S. H. Hutner (ed.), *Biochemistry and Physiology of Protozoa*, vol. 3. Academic Press, New York, NY.

Armbrust, E. V., J. A. Berges, C. Bowler, B. R. Green, D. Martinez, N. H. Putnam, S. G. Zhou, A. E. Allen, K. E. Apt, M. Bechner, M. A. Brzezinski, B. K. Chaal, A. Chiovitti, A. K. Davis, M. S. Demarest, J. C. Detter, T. Glavina, D. Goodstein, M. Z. Hadi, U. Hellsten, M. Hildebrand, B. D. Jenkins, J. Jurka, V. V. Kapitonov, N. Kroger, W. W. Y. Lau, T. W. Lane, F. W. Larimer, J. C. Lippmeier, S. Lucas, M. Medina, A. Montsant, M. Obornik, M. S. Parker, B. Palenik, G. J. Pazour, P. M. Richardson, T. A. Rynearson, M. A. Saito, D. C. Schwartz, K. Thamatrakoln, K. Valentin, A. Vardi, F. P. Wilkerson, and D. S. Rokhsar. 2004. The genome of the diatom *Thalassiosira pseudonana*: ecology, evolution, and metabolism. *Science* **306:**79–86.

Baldauf, S. L. 2003. The deep roots of eukaryotes. *Science* **300:**1703–1706.

Baldauf, S. L., A. J. Roger, I. Wenk-Siefert, and W. F. Doolittle. 2000. A kingdom-level phylogeny of eukaryotes based on combined protein data. *Science* **290:**972–977.

Bé, A. W. H., O. R. Anderson, W. W. Faber, Jr., and D. A. Caron. 1983. Sequence of morphological and cytoplasmic changes during gametogenesis in the planktonic foraminifer *Globigerinoides sacculifer* (Brady). *Micropaleontology* **29:**310–325.

Bé, A. W. H., H. J. Spero, and O. R. Anderson. 1982. Effects of symbiont elimination and reinfection on the life processes of the planktonic foraminifer *Globigerinoides sacculifer. Mar. Biol.* **70:**73–86.

Berney, C., J. Fahrni, and J. Pawlowski. 2004. How many novel eukaryotic 'kingdoms'? Pitfalls and limitations of environmental DNA surveys. *BMC Biol.* **4:**2–13.

Boenigk, J., S. Jost, T. Stoeck, and T. Garstecki. 2006. Differential thermal adaptation of clonal strains of a protist morphospecies originating from different climatic zones. *Environ. Microbiol.* **9:**593–602.

Boenigk, J., K. Pfandl, P. Stadler, and A. Chatzinotas. 2005. High diversity of the "Spumella-like" flagellates: an investigation based on the SSU rRNA gene sequences of isolates from habitats located in six different geographic regions. *Environ. Microbiol.* **7:**685–697.

Bolch, C. J. S. 2001. PCR protocols for genetic identification of dinoflagellates directly from single cysts and plankton cells. *Phycologia* **40:**162–167.

Buckley, M., and R. J. Roberts. 2007. Reconciling microbial systematics and genomics. American Academy of Microbiology, Washington, D.C.

Calbet, A., and M. R. Landry. 2004. Phytoplankton growth, microzooplankton grazing, and carbon cycling in marine systems. *Limnol. Oceanogr.* **49:**51–57.

Calkins, G. N. 1901. Marine protozoa from Woods Hole. *Bull. Bur. Fish.* **21:**413–468.

Caron, D. A., J. C. Goldman, O. K. Andersen, and M. R. Dennett. 1985. Nutrient cycling in a microflagellate food chain. II. Population dynamics and carbon cycling. *Mar. Ecol. Prog. Ser.* **24:**243–254.

Caron, D. A., E. L. Lim, H. Kunze, E. M. Cosper, and D. M. Anderson. 1989. Trophic interactions between nano- and microzooplankton and the "brown tide," p. 265–294. *In* E. M. Cosper, V. M. Bricelj, and E. J. Carpenter (ed.), *Novel Phytoplankton Blooms: Causes and Impacts of Recurrent Brown Tides and Other Unusual Blooms*, vol. 35. Springer-Verlag, Berlin, Germany.

Caron, D. A., A. F. Michaels, N. R. Swanberg, and F. A. Howse. 1995. Primary productivity by symbiont-bearing planktonic sarcodines (Acantharia, Radiolaria, Foraminifera) in surface waters near Bermuda. *J. Plankton Res.* **17:**103–129.

Caron, D. A., and N. R. Swanberg. 1990. The ecology of planktonic sarcodines. *Rev. Aquat. Sci.* **3:**147–180.

Carvalho, W. F., and E. Graneli. 2006. Acidotropic probes and flow cytometry: a powerful combination for detecting phagotrophy in mixotrophic and heterotrophic protists. *Aquat. Microb. Ecol.* **44:**85–96.

Cavalier-Smith, T. 1998. A revised six-kingdom system of life. *Biol. Rev.* **73:**203–266.

Choi, J. W., and F. Peters. 1992. Effects of temperature on two psychrophilic ecotypes of a heterotrophic nanoflagellate, *Paraphysomonas imperforata. Appl. Environ. Microbiol.* **58:**593–599.

Coats, D. W., E. J. Adam, C. L. Gallegos, and S. Hedrick. 1996. Parasitism of photosynthetic dinoflagellates in a shallow subestuary of Chesapeake Bay, USA. *Aquat. Microb. Ecol.* **11:**1–9.

Coats, D.W., and M. G. Park. 2002. *Parasitism of photosynthetic dinoflagellates by three strains of Amoebophrya* (Dinophyta): parasite survival, infectivity, generation time, and host specificity. *J. Phycol.* **38:**520–528.

Countway, P. D., R. J. Gast, M. R. Dennett, P. Savai, J. M. Rose, and D. A. Caron. 2007. Distinct protistan assemblages characterize the euphotic zone and deep sea (2500 m) of the western N. Atlantic (Sargasso Sea and Gulf Stream). *Environ. Microbiol.* **9:**1219–1232.

Courties, C., A. Vaquer, M. Troussellier, J. Lautier, M. Chretiennot-Dinet, J. Neveux, C. Machado, and H. Claustre. 1994. Smallest eukaryotic organism. *Nature* **370:**255.

Crawford, D.W. 1989. *Mesodinium rubrum*: the phytoplankter that wasn't. *Mar. Ecol. Prog. Ser.* **58:**161–174.

Croft, M. T., M. J. Warren, and A. G. Smith. 2006. Algae need their vitamins. *Eukaryotic Cell* **5:**1175–1183.

Daniels, R. B., T. L. Richardson, and H. W. Ducklow. 2006. Food web structure and biogeochemical processes during oceanic phytoplankton blooms: an inverse model analysis. *Deep-Sea Res.* **53:**532–554.

Dawson, S. C., and N. R. Pace. 2002. Novel kingdom-level eukaryotic diversity in anoxic environments. *Proc. Natl. Acad. Sci. USA* **99:**8324–8329.

DeLong, E. F. 2004. Microbial population genomics and ecology: the road ahead. *Environ. Microbiol.* **6:**875–878.

Dolven, J. K., C. Lindqvist, V. A. Albert, K. R. Bjørklund, T. Yuasa, O. Takahashi, and S. Mayama. 2007. Molecular diversity of alveolates associated with neritic North Atlantic radiolarians. *Protist* **158:**65–76.

Edgcomb, V. P., D. T. Kysela, A. Teske, A. D. Gomez, and M. L. Sogin. 2002. Benthic eukaryotic diversity in the Guaymas Basin hydrothermal vent environment. *Proc. Natl. Acad. Sci. USA* **99:**7658–7662.

Edvardsen, B., K. Shalchian-Tabrizi, K. S. Jakobsen, L. Medlin, K., E. Dahl, S. Brubak, and E. Paasche. 2003. Genetic variability and molecular phylogeny of *Dinophysis* species (*Dinophyceae*) from Norwegian waters inferred from single cell analyses of rDNA. *J. Phycol.* **39:**395–408.

Falkowski, P. G., and J. Raven. 1997. *Aquatic Photosynthesis*. Blackwell Scientific, Oxford, United Kingdom.

Fauré-Fremiet, E. 1924. Contribution à la connaissance des infusoires planktoniques. *Bull. Biol. France Belgique (Suppl)* **6:**1–171.

Fawley, M. J., K. P. Fawley, and M. A. Buchheim. 2004. Molecular diversity among communities of freshwater microchlorophytes. *Microb. Ecol.* **48:**489–499.

Fawley, M.W., K. P. Fawley, and H. A. Owen. 2005. Diversity and ecology of small coccoid green algae from Lake Itasca, Minnesota, USA, including *Meyerella planktonica*, gen. et sp nov. *Phycologia* **44:**35–48.

Fenchel, T. 1982a. Ecology of heterotrophic microflagellates. I. Some important forms and their functional morphology. *Mar. Ecol. Prog. Ser.* **8:**211–223.

Fenchel, T. 1982b. Ecology of heterotrophic microflagellates. III. Adaptations to heterogeneous environments. *Mar. Ecol. Prog. Ser.* **9:**25–33.

Finlay, B. J. 2002. Global dispersal of free-living microbial eukaryote species. *Science* **296:**1061–1063.

Finlay, B. J., and K. J. Clarke. 1999. Apparent global ubiquity of species in the protist genus *Paraphysomonas*. *Protist* **150:**419–430.

Finlay, B. J., and T. Fenchel. 1999. Divergent perspectives on protist species richness. *Protist* **150:**229–233.

Foissner, W. 1999. Protist diversity: estimates of the near-imponderable. *Protist* **72:**6578–6583.

Gast, R. J. 2006. Molecular phylogeny of a potentially parasitic dinoflagellate isolated from the solitary radiolarian, *Thalassicolla nucleata*. *J. Euk. Microbiol.* **53:**43–45.

Gast, R. J., M. R. Dennett, and D. A. Caron. 2004. Characterization of protistan assemblages in the Ross Sea, Antarctica by denaturing gradient gel electrophoresis. *Appl. Environ. Microbiol.* **70:**2028–2037.

Gast, R. J., D. M. Moran, M. R. Dennett, and D. A. Caron. 2007. Kleptoplasty in an Antarctic dinoflagellate: caught in evolutionary transition? *Environ. Microbiol.* **9:**39–45.

Girvan, M. S., C. D. Campbell, K. Killham, J. I. Prosser, and L. A. Glover. 2005. Bacterial diversity promotes community stability and functional resilience after perturbation. *Environ. Microbiol.* **7:**301–313.

Goldman, J. C., and D. A. Caron. 1985. Experimental studies on an omnivorous microflagellate: implications for grazing and nutrient regeneration in the marine microbial food chain. *Deep-Sea Res.* **32:**899–915.

Goldman, J. C., D. A. Caron, O. K. Andersen, and M. R. Dennett. 1985. Nutrient cycling in a microflagellate food chain: I. Nitrogen dynamics. *Mar. Ecol. Prog. Ser.* **24:**231–242.

Gordon, N., D. L. Angel, A. Neori, N. Kress, and B. Kimor. 1994. Heterotrophic dinoflagellates with symbiotic cyanobacteria and nitrogen limitation in the Gulf of Aqaba. *Mar. Ecol. Prog. Ser.* **107:**83–88.

Gran, H. H. 1912. Pelagic plant life, p. 307–386. *In* J. Murray and J. Hjort (ed.), *The Depths of the Ocean*. MacMillan, London, United Kingdom.

Griffiths, B. S., H. L. Kuan, K. Ritz, L. A. Glover, A. E. McCaig, and C. Fenwick. 2004. The relationship between microbial community structure and functional stability, tested experimentally in an upland pasture soil. *Microb. Ecol.* **47:**104–113.

Groisillier, A., R. Massana, K. Valentin, D. Vaulot, and L. Guillou. 2006. Genetic diversity and habitats of two enigmatic marine alveolate lineages. *Aquat. Microb. Ecol.* **42:**277–291.

Gruebl, T., M. E. Frischer, M. Sheppard, M. Neumann, A. N. Maurer, and R. F. Lee. 2002. Development of an 18S rRNA gene-targeted diagnostic for the blue crab parastie *Hematodinium* sp. *Dis. Aquat. Org.* **49:**61–70.

Guillou, L., M.-J. Chrétiennot-Dinet, L. K. Medlin, H. Claustre, S. Loiseaux-de Goer, and D. Vaulot. 1999a. *Bolidomonas*: a new genus with two species belonging to a new algal class, the *Bolidophyceae* (Heterokonta). *J. Phycol.* **35:**368–381.

Guillou, L., W. Eikrem, M. J. Chretiennot-Dinet, F. Le Gall, R. Massana, K. Romari, C. Pedros-Alio, and D. Vaulot. 2004. Diversity of picoplanktonic prasinophytes assessed by direct nuclear SSU rDNA sequencing of environmental samples and novel isolates retrieved from oceanic and coastal marine ecosystems. *Protist* **155:**193–214.

Guillou, L. R., M.-J. Chrétiennot-Dinet, S. Boulben, S. Y. Moon-van der Staay, and D. Vaulot. 1999b. *Symbiomonas scintillans* gen. et sp. nov. and *Picophagus flagellatus* gen. et sp. nov. (Heterokonta): two new heterotrophic flagellates of picoplanktonic size. *Protist* **150:**383–398.

Hackett, J. D., D. M. Anderson, D. L. Erdner, and D. Bhattacharya. 2004. Dinoflagellates: a remarkable evolutionary experiment. *Am. J. Bot.* **91:** 1523–1534.

Haeckel, E. 1887. Report on Radiolaria collected by H.M.S. Challenger during the 1873–1876, p. 1–1760. *In* C.W. Thompson and J. Murray (ed.), *The Voyage of the H.M.S. Challenger*, vol. 18. Her Majesty's Stationary Office, London, United Kingdom.

Hausmann, K., and N. Hülsmann. 1996. *Protozoology*. Georg Thieme Verlag, Stuttgart, Germany.

Hemleben, C., M. Spindler, and O. R. Anderson. 1988. *Modern Planktonic Foraminifera*. Springer-Verlag, New York, NY.

Hill, H. Z., H. T. Epstein, and J. A. Schiff. 1966. Studies of chloroplast development in euglena. XIV. Sequential interactions of ultraviolet light and photoreactivating light in green colony formation. *Biophys. J.* **6:**135–144.

Holzmann, M., A. Habura, H. Giles, S. S. Bowser, and J. Pawlowski. 2003. Freshwater foraminiferans revealed by analysis of environmental DNA samples. *J. Euk. Microbiol.* **50:**135–139.

Hoppenrath, M., and B. S. Leander. 2006. Ebriid phylogeny and the expansion of the Cercozoa. *Protist* **157:**279–290.

Jeon, S.-O., J. Bunge, T. Stoeck, K. J.-A. Barger, S.-H. Hong, and S. S. Epstein. 2006. Synthetic statistical approach reveals a high degree of richness of microbial eukaryotes in an anoxic water column. *Appl. Environ. Microbiol.* **72:**6578–6583.

Jeong, H. J. 1999. The ecological roles of heterotrophic dinoflagellates in marine planktonic community. *J. Euk. Microbiol.* **46:**390–396.

Jeong, H. J., Y. D. Yoo, J. Y. Park, J. Y. Song, S. T. Kim, S. H. Lee, K. Y. Kim, and W. H. Yih. 2005. Feeding by red-tide dinoflagellates: five species newly revealed and six species previously known to be mixotrophic. *Aquat. Microb. Ecol.* **40:**133–150.

Johnson, M. D., D. Oldach, D. F. Delwiche, and D. K. Stoecker. 2007. Retention of transcriptionally active cryptophyte nuclei by the ciliate *Myrionecta rubra*. *Nature* **445:**426–428.

Karpov, S. A., D. Bass, A. P. Mylnikov, and T. Cavalier-Smith. 2006. Molecular phylogeny of *Cercomonadidae* and kinetid patterns of *Cercomonas* and *Eocercomonas* gen. nov. (Cercomonadida, Cercozoa). *Protist* **157:**125–158.

Kawachi, M., M. Atsumi, H. Ikemoto, and S. Miyachi. 2002. *Pinguiochrysis pyriformis* gen. et sp. nov. (*Pinguiophyceae*), a new picoplanktonic alga isolated from the Pacific Ocean. *Phycol. Res.* **50:**49–56.

Ki, J.-S., G. Y. Jang, and M.-S. Han. 2004. Integrated method for single-cell DNA extraction, PCR amplification and sequencing of ribosomal DNA from harmful dinoflagellates *Cochlodinium polykrikoides* and *Alexandrium catenella*. *Mar. Biotechnol.* **6:**587–593.

Kofoid, C. A., and T. Skogsberg. 1928. The free-living unarmoured dinoflagellates. *Mem. Univ. California, Berkeley* **5:**563.

Kolodziej, K., and T. Stoeck. 2007. Cellular identification of a novel uncultured marine stramenopile (MAST-12 Clade) small-subunit rRNA gene sequence from a Norwegian estuary by use of fluorescence in situ hybridization-scanning electron microscopy. *Appl. Environ. Microbiol.* **73:**2718–2726.

Lee, J. J., and O. R. Anderson (ed.). 1991. *The Biology of Foraminifera*. Academic Press, London, United Kingdom.

Lee, J. J., S. H. Hutner, and E. C. Bovee (ed.). 1985. *An Illustrated Guide to the Protozoa*. Society of Protozoologists, Lawrence, KS.

Lee, J. J., G. F. Leedale, and P. Bradbury. 2000. *An Illustrated Guide to the Protozoa*. Allen Press, Inc., Lawrence, KS.

Lee, J. J., C. W. Reimer, and M. E. McEnery. 1980. The identification of diatoms isolated as endosymbionts from larger foraminifera from the Gulf of Eilat (Red Sea) and the description of 2 new species, *Fragilaria shiloi* sp. nov. and *Navicula reissii* sp. nov. *Botanica Marina* **23:**41–48.

Lee, J. J., K. Sang, B. ter Kuile, E. Strauss, P. J. Lee, and W. W. Faber, Jr. 1991. Nutritional and related experiments on laboratory maintenance of three species of symbiont-bearing, large foraminifera. *Mar. Biol.* **109:**417–425.

Leidy, J. 1879. Fresh-water rhizopods of North America. *U.S. Geological Survey of the Territories* **12:**1–234.

Lessard, E. J. 1991. The trophic role of heterotrophic dinoflagellates in diverse marine environments. *Mar. Microb. Food Webs* **5:**49–58.

Lim, E. E., L. A. Amaral, D. A. Caron, and E. F. DeLong. 1993. Application of rRNA-based probes for observing marine nanoplanktonic protists. *Appl. Environ. Microbiol.* **59:**1647–1655.

Lim, E. L., D. A. Caron, and M. R. Dennett. 1999. The ecology of *Paraphysomonas imperforata* based on studies employing oligonucleotide probe identification in coastal water samples and enrichment culture. *Limnol. Oceanogr.* **44:**37–51.

Lohmann, H. 1902. Die Coccolithophoridae, eine Monographie der coccolithen bildenden Flagellaten. *Arch. Prot.* **1:**89–165.

López-García, P., A. Lopez-Lopez, D. Moreira, and F. Rodríguez-Valera. 2001a. Diversity of free-living prokaryotes from a deep-sea site at the Antarctic Polar Front. *FEMS Microbiol. Ecol.* **36:**193–202.

López-García, P., F. Rodríguez-Valera, C. Pedrós-Alió, and D. Moreira. 2001b. Unexpected diversity of small eukaryotes in deep-sea Antarctic plankton. *Nature* **409:**603–607.

López-Garcia, P., A. Vereshchaka, and D. Moreira. 2007. Eukaryotic diversity associated with carbonates and fluid-seawater interface in Lost City hydrothermal field. *Environ. Microbiol.* **9:**546–554.

Lovejoy, C., R. Massana, and C. Pedros-Alio. 2006. Diversity and distribution of marine microbial eukaryotes in the Arctic Ocean and adjacent seas. *Appl. Environ. Microbiol.* **72:**3085–3095.

Lovejoy, C., W. F. Vincent, S. Bonilla, S. Roy, M.-J. Martineau, R. Terrado, M. Potvin, R. Massana, and C. Pedrós-Alió. 2007. Distribution, phylogeny and growth of cold-adapted picoprasinophytes in Arctic Seas. *J. Phycol.* **43:**78–89.

Luo, Q., L. R. Krumholz, F. Z. Najar, A. D. Peacock, B. A. Roe, D. C. White, and M. S. Elshahed. 2005. Diversity of the microeukaryotic community in sulfide-rich Zodletone Spring (Oklahoma). *Appl. Environ. Microbiol.* **71:**6175–6184.

Massana, R., V. Balagué, L. Guillou, and C. Pedrós-Alió. 2004. Picoeukaryotic diversity in an oligotrophic coastal site studied by molecular and culturing approaches. *FEMS Microbiol. Ecol.* **50:**231–243.

Massana, R., J. Castresana, V. Balagué, L. Guillou, K. Romari, A. Groisillier, K. Valentin, and C. Pedrós-Alió. 2004. Phylogenetic and ecological analysis of novel marine stramenopiles. *Appl. Environ. Microbiol.* **70:**3528–3534.

Massana, R., L. Guillou, B. Diez, and C. Pedros-Alio. 2002. Unveiling the organisms behind novel eukaryotic ribosomal DNA sequences from the ocean. *Appl. Environ. Microbiol.* **68:**4554–4558.

Massana, R., L. Guillou, R. Terrado, I. Forn, and C. Pedrós-Alió. 2006a. Growth of uncultured heterotrophic flagellates in unamended seawater incubations. *Aquat. Microb. Ecol.* **45:**171–180.

Massana, R., R. Terrado, I. Form, C. Lovejoy, and C. Pedrós-Alió. 2006b. Distribution and abundance of uncultured heterotrophic flagellates in the world oceans. *Environ. Microbiol.* **8:**1515–1522.

McGrady-Steed, J., P. M. Harris, and P. J. Morin. 1997. Biodiversity regulates ecosystem predictability. *Nature* **390:**162–165.

Michaels, A. F., D. A. Caron, N. R. Swanberg, F. A. Howse, and C. M. Michaels. 1995. Planktonic sarcodines (Acantharia, Radiolaria, Foraminifera) in surface waters near Bermuda: abundance, biomass and vertical flux. *J. Plankton Res.* **17:**131–163.

Modeo, L., G. Petroni, G. Rosati, and D. J. S. Montagnes. 2003. A multidisciplinary approach to describe protists: redescriptions of *Novistrombidium testaceum* and *Strombidium inclinatum* Montagnes, Taylor and Lynn 1990 (Ciliophora, Oligotrichia). *J. Euk. Microbiol.* **50:**175–189.

Moon-van der Staay, S. Y., R. De Wachter, and D. Vaulot. 2001. Oceanic 18S rDNA sequences from picoplankton reveal unsuspected eukaryotic diversity. *Nature* **409:**607–610.

Moreira, D., and P. Lopez-Garcia. 2003. Are hydrothermal vents oases for parasitic protists? *Trends Parasitol.* **19:**556–558.

Naeem, S., and L. Shibin. 1997. Biodiversity enhances ecosystem stability. *Nature* **390:**507–509.

Not, F., K. Valentin, K. Romari, C. Lovejoy, R. Massana, K. Töbe, D. Vaulot, and L. K. Medlin. 2007. Picobiliphytes: a marine picoplanktonic algal group with unknown affinities to other eukaryotes. *Science* **315:**253–255.

Okamoto, N., and I. Inouye. 2007. A secondary symbiosis in progress. *Science* **310:**287.

Park, M. G., S. K. Cooney, J. S. Kim, and D. W. Coats. 2002a. Effects of parasitism on diel vertical migration, phototaxis/geotaxis, and swimming speed of the bloom-forming dinoflagellate *Akashiwo sanguinea*. *Aquat. Microb. Ecol.* **29:**11–18.

Park, M. G., S. K. Cooney, W. Yih, and D. W. Coats. 2002b. Effects of two strains of the parastic dinoflagellate *Amoebophrys* on growth, photosynthesis, light absorption, and quantum yield of bloom-forming dinoflagellates. *Mar. Ecol. Prog. Ser.* **227:**281–292.

Park, M. G., S. Kim, H. S. Kim, G. Myung, Y. G. Kang, and W. Yih. 2006. First successful culture of the marine dinoflagellate *Dinophysis*. *Aquat. Microb. Ecol.* **45:**101–106.

Park, M. G., W. Yi, and D. W. Coats. 2004. Parasites and phytoplankton, with special emphasis on dinoflagellate infections. *J. Euk. Microbiol.* **51:**145–155.

Patterson, D. J. 1999. The diversity of eukaryotes. *Am. Nat.* **154** (Suppl)**:**S96–S124.

Patterson, D. J., and J. Larsen (ed.). 1991. *The Biology of Free-Living Heterotrophic Flagellates*. Clarendon Press, Oxford, United Kingdom.

Pedrós-Alió, C. 2006. Microbial diversity: can it be determined? *Trends Microbiol.* **14:**257–263.

Peters, F., J. W. Choi, and T. Gross. 1996. *Paraphysomonas imperforata* (Protista, Chrysomonadida) under different turbulence levels: feeding, physiology and energetics. *Mar. Ecol. Prog. Ser.* **134:** 235–245.

Podar, M., C. B. Abulencia, M. Walcher, D. Hutchinson, K. Zengler, J. A. Garcia, T. Holland, D. Cotton, L. Hauser, and M. Keller. 2007. Targeted access to the genomes of low-abundance organisms in complex microbial communities. *Appl. Environ. Microbiol.* **73:**3205–3214.

Potter, D., T. C. LaJeunesse, G. W. Saunders, and R. A. Andersen. 1997. Convergent evolution masks extensive biodiversity among marine coccoid picoplankton. *Biodivers. Conserv.* **9:**99–107.

Raghunathan, A., H. R. Ferguson, Jr., C. J. Bornarth, W. Song, M. Driscoll, and R. S. Lasken. 2005. Genomic DNA amplification from a single bacterium. *Appl. Environ. Microbiol.* **71:** 3342–3347.

Richardson, T. L., G. A. Jackson, H. W. Ducklow, and M. R. Roman. 2006. Spatial and seasonal patterns of carbon cycling through planktonic food webs of the Arabian Sea determined by inverse analysis. *Deep-Sea Res.* **534:**555–575.

Robbens, S., B. Khadaroo, A. Camasses, E. Derelle, C. Ferraz, D. Inze, Y. van de Peer, and H. Moreau. 2005. Genome-wide analysis of core cell cycle genes in the unicellular green alga *Ostreococcus tauri*. *Mol. Biol. Evol.* **22:**589–597.

Romari, K., and D. Vaulot. 2004. Composition and temporal variability of picoeukaryote communities at a coastal site of the English Channel from 18S rDNA sequences. *Limnol. Oceanogr.* **49:**784–798.

Rose, J. M., and D. A. Caron. 2007. Does low temperature constrain the growth rates of heterotrophic protists? Evidence and implications for algal blooms in cold water. *Limnol. Oceanogr.* **52:**886–895.

Ruiz Sebastian, C., and C. O'Ryan. 2001. Single-cell sequencing of dinoflagellate (*Dinophyceae*) nuclear ribosomal genes. *Molec. Ecol. Notes* **1:**329–331.

Sanders, R. W., D. A. Caron, and U.-G. Berninger. 1992. Relationships between bacteria and heterotrophic nanoplankton in marine and fresh water: an inter-ecosystem comparison. *Mar. Ecol. Prog. Ser.* **86:**1–14.

Sanders, R. W., and K. G. Porter. 1988. Phagotrophic phytoflagellates. *Adv. Microb. Ecol.* **10:**167–192.

Scala, S., N. Carels, A. Falciatore, M. L. Chiusano, and C. Bowler. 2002. Genome properties of the diatom *Phaeodactylum tricornutum*. *Plant Physiol.* **129:**993–1002.

Schewiakoff, W. 1926. Die Acantharia des Golfes von Neapel, p. 1–755. *In* G. Bardi (ed.), *Fauna e Flora del Golfo di Napoli*, vol. 37. R Friedlander und Sohn, Berlin, Germany.

Schlegel, M. 1994. Molecular phylogeny of eukaryotes. *Trends Ecol. Evol.* **9:**330–335.

Schlegel, M. 1991. Protist evolution and phylogeny as discerned from small subunit ribosomal RNA sequence comparisons. *Eur. J. Protistol.* **27:** 207–219.

Schloss, P. D., and J. Handelsman. 2004. Status of the microbial census. *Microbiol. Mol. Biol. Rev.* **68:**686–691.

Schnepf, E., and M. Elbrächter. 1992. Nutritional strategies in dinoflagellates. *Eur. J. Protistol.* **28:**3–24.

Shalchian-Tabrizi, K., W. Eikrem, D. Klaveness, D. Vaulot, M. A. Minge, F. LaGall, K. Romari, J. Throndsen, A. Botnen, R. Massana, H. A. Thomsen, and K. S. Jakobsen. 2006. Telonemia, a new protist phylym with affinity to chromist lineages. *Proc. R. Soc. London, Ser. B* **273:**1833–1842.

Shalchian-Tabrizi, K., H. Kauserud, R. Massana, D. Klaveness, and K. S. Jakobsen. 2007. Analysis of enviornmental 18S ribosomal RNA sequences reveals unknown diversity of the cosmopolitan phylum *Telonemia*. *Protist* **158:**173–180.

Sherr, B. F., E. B. Sherr, D. A. Caron, D. Vaulot, and A. Z. Worden. 2007. Oceanic protists. *Oceanography* **20:**102–106.

Sherr, E. B., and B. F. Sherr. 2002. Significance of predation by protists in aquatic microbial food webs. *Antonie Van Leeuwenhoek* **81:**293–308.

Simpson, A. G. B., and A. J. Roger. 2004. The real 'kingdoms' of eukaryotes. *Curr. Biol.* **14:** R693–696.

Slapeta, J., P. López-Garcia, and D. Moreíra. 2006a. Global dispersal and ancient cryptic species in the smallest marine eukaryotes. *Mol. Biol. Evol.* **23:**23–29.

Slapeta, J. R., D. Moreira, and P. López-García. 2006b. Present status of the molecular ecology of Kathablepharids. *Protist* **157:**7–11.

Sogin, M. L., H. G. Morrison, J. A. Huber, D. M. Welch, S. M. Huse, P. R. Neal, J. M. Arrieta, and G. J. Herndl. 2006. Microbial diversity in the deep sea and the underexplored "rare biosphere." *Proc. Natl. Acad. Sci. USA* **103:**12115–12120.

Stoeck, T., and S. Epstein. 2003. Novel eukaryotic lineages inferred from small-subunit rRNA analyses of oxygen-depleted marine environments. *Appl. Environ. Microbiol.* **69:**2657–2663.

Stoeck, T., B. Hayward, G. T. Taylor, R. Varela, and S. S. Epstein. 2006. A multiple PCR-primer approach to access the microeukaryotic diversity in environmental samples. *Protist* **157:**31–43.

Stoeck, T., G. T. Taylor, and S. S. Epstein. 2003. Novel eukaryotes from the permanently anoxic Cariaco Basin (Caribbean Sea). *Appl. Environ. Microbiol.* **69:**5656–5663.

Stoeck, T., A. Zuendorf, A. Behnke, and H.-W. Breiner. 2007. A molecular approach to identify active microbes in environmental eukaryote clone libraries. *Microb. Ecol.* **53:**328–339.

Stoecker, D. K. 1999. Mixotrophy among dinoflagellates. *J. Euk. Microbiol.* **46:**397–401.

Stoecker, D. K., M. Putt, L. H. Davis, and A. E. Michaels. 1991. Photosynthesis in *Mesodinium rubrum*: species:specific measurements and comparison to community rates. *Mar. Ecol. Prog. Ser.* **73:**245–252.

Stoecker, D. K., M. W. Silver, A. E. Michaels, and L. H. Davis. 1989. Enslavement of algal chloroplasts by four *Strombidium* spp. (Ciliophora, Oligotrichida). *Mar. Microb. Food Webs* **3:**79–100.

Stokes, A. C. 1878. *Der Organismus der Infusionthiere*, vol. 3. W. Englemann, Leipzig, Germany.

Swanberg, N. R., and G. R. Harbison. 1980. The ecology of *Collozoum longiforme*, sp. nov., a new colonial radiolarian from the equatorial Atlantic Ocean. *Deep-Sea Res.* **27A:**715–732.

Takano, Y., and T. Horiguchi. 2005. Acquiring scanning electron microscopical, light microscopical and multiple gene sequence data from a single dinoflagellate cell. *J. Phycol.* **42:**251–-256.

Takishita, K., H. Miyake, M. Kawato, and T. Maruyama. 2005. Genetic diversity of microbial eukaryotes in anoxic sediment around fumaroles on a submarine caldera floor based on the small-subunit rDNA phylogeny. *Extremophiles* **9:**185–196.

Yuasa, T., O. Takahashi, J. K. Dolven, S. Mayama, A. Matsuoka, D. Honda, and K. R. Bjorklund. 2006. Phylogenetic position of the small solitary phaeodarians (Radiolaria) based on 18S rDNA sequences by single cell PCR analysis. *Mar. Micropaleontol.* **59:**104–114.

Zaslavskaia, L. A., J. C. Lippmeier, P. G. Kroth, K. E. Apt, and A. R. Grossman. 2000. Transformation of the diatom *Phaeodactylum tricornutum* (*Bacillariophyceae*) with a variety of detectable marker and reporter genes. *J. Phycol.* **36:**379–386.

MICROBIAL BIOGEOGRAPHY: PATTERNS IN MICROBIAL DIVERSITY ACROSS SPACE AND TIME

Noah Fierer

6

Biogeography is a science that attempts to describe and explain spatial patterns of biological diversity and how these patterns change over time (Ganderton and Coker, 2005; Lomolino et al., 2006). In other words, biogeographers seek to answer the seemingly simple question: Why do organisms live where they do? While biogeography has traditionally focused on macroorganisms, i.e., plants and animals, microbiologists have studied biogeographical questions for many decades, and there has been a recent resurgence in interest in microbial biogeography (Green and Bohannan, 2006; Martiny et al., 2006; Ramette and Tiedje, 2007a). This resurgence has been led, in part, by advancements in molecular tools that allow us to survey uncultivated microbes in the environment and a growing recognition that microbial taxa are the most biologically diverse taxa on Earth.

At present, the study of microbial biogeography is in its infancy. Even the existence of microbial biogeography has been recently called into question ["There is no biogeography for anything smaller than 1 millimeter," Finlay quoted in Whitfield (2005)]. If this statement was correct, this chapter would be very brief. However, we know that a wide variety of microbial taxa exhibit biogeographical patterns; microbial communities are not homogeneous across habitat types, and within a given habitat, microbial diversity can vary between locations separated by millimeters to thousands of kilometers. If microbial biogeography did not exist, there would be no spatial or temporal heterogeneity in microbial communities, and global patterns in microbial diversity could be predicted by studying the microbial community in a single location at a single point in time. Unfortunately this is not the case; documenting and understanding patterns in microbial biogeography are not so simple.

Probably the best summary of microbial biogeography as a research field was provided by an unlikely source, D. Rumsfeld, former Secretary of Defense of the United States, who said "... there are things we know we know. We also know there are known unknowns; that is to say, we know there are some things we do not know. But there are also unknown unknowns—the ones we don't know we don't know" (Feb. 12, 2002, Department of Defense

Noah Fierer, Department of Ecology & Evolutionary Biology, Cooperative Institute for Research in Environmental Sciences, University of Colorado, Boulder, CO 80305.

Accessing Uncultivated Microorganisms: from the Environment to Organisms and Genomes and Back
Edited by Karsten Zengler

news briefing). Although he was not referring to biogeography when he uttered this phrase, it is a useful framework for thinking about microbial biogeography in that the "things we know we know" are relatively few, the "known unknowns" are abundant, and in coming decades we are likely to discover many phenomena that are currently "unknown unknowns." While reading this chapter, it will become readily apparent that the science of microbial biogeography is currently as mature as "macro"-bial (i.e., plant and animal) biogeography was in the 19th century. Just as early naturalists set out on voyages to document the diversity of plants and animals in uncharted lands, we are attempting to document "uncharted" microbial diversity, and we currently lack a comprehensive understanding of how (and why) microbial diversity changes across space and time.

The immaturity of the field of microbial biogeography is not due to lack of interest in the topic. The first paradigm in microbial biogeography, "everything is everywhere, but, the environment selects," was offered by Baas Becking (1934) more than 70 years ago, and his adage continues to be cited in nearly every recent publication on microbial biogeography (de Wit and Bouvier, 2006). Although the field of microbial biogeography is not new, we now have the methods available to survey a large portion of the microbial diversity on Earth and to quantify the biogeographic patterns exhibited by microbes living in a wide range of environments. As these techniques and methodologies continue to improve at a nearly exponential rate, the field of microbial biogeography is poised for significant advances.

In this chapter, I do not attempt to summarize everything that is known about microbial biogeography. The field of biogeography encompasses a wide breadth of research topics, and covering all topics related to microbial biogeography would be a Sisyphean task. In addition, microbes inhabit a wide range of habitats, from hot springs to the deep subsurface, and it is highly improbable that we would observe similar biogeographical patterns across the full range of possible microbial habitats. At the same time, it is also unlikely that all microbial taxa share similar biogeographical patterns, as the term "microbe" encompasses a broad array of taxa (e.g., bacteria, fungi, archaea, viruses, and protists) that are phylogenetically distinct and distinct with respect to their morphologies, physiologies, and life histories. For these reasons, this chapter should not be considered a comprehensive review of "microbial biogeography," as there is unlikely to be a common set of concepts and patterns unifying the field of microbial biogeography. Instead, I primarily focus on selected topics that are particularly relevant to researchers studying uncultivated microbes in natural environments in order to illustrate what we do, or do not, currently know about their biogeography. Most of the examples are drawn from research on bacteria, as bacterial biogeography has received far more attention than the biogeography of other microbial groups.

MICROBIAL DISPERSAL AND COLONIZATION

From work on plants and animals, we know that dispersal is likely to be one of the key processes shaping microbial biogeography and macroecological patterns (Hubbell, 2001; Lomolino et al., 2006). There is currently some debate regarding the extent of microbial dispersal. Finlay (2002) has argued that any organism less than 1 mm in size is likely to be ubiquitous due to an essentially unlimited capacity for long-distance dispersal. This speculation is primarily based on the assumption that the high local abundance of microbes (the large number of individuals per unit area) increases the probability that individual microbes may travel a long distance and successfully colonize a remote location simply by chance (Fenchel, 2003; Finlay, 2002; Martiny et al., 2006). If we combine a high probability of dispersal with the ability to survive the long-distance transport, we would expect few geographic constraints on microbial dispersal (Fig. 1). In contrast, Papke and Ward (2004) have argued that geographic barriers to microbial dispersal are relatively common and physical isolation is an important

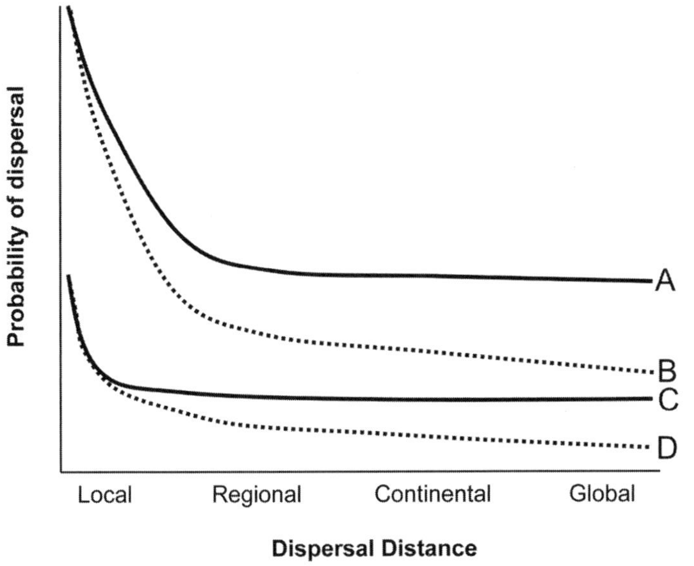

FIGURE 1 Hypothetical dispersal capabilities of microbes that differ in population densities and stress tolerances. (A) High population density, stress tolerant; (B) high population density, stress intolerant; (C) low population density, stress tolerant; (D) low population density, stress intolerant. Across larger spatial scales, microbial dispersal rates should be directly related to population densities in the source population and the ability to withstand biotic and abiotic stresses associated with dispersal. Figure based on Martiny et al. (2006).

driver of microbial evolution. They cite a handful of studies as evidence for the occurrence of microbial endemism, including work on hot spring microbes (Papke et al., 2003; Whitaker et al., 2003) and soil pseudomonads (Cho and Tiedje, 2000).

Unfortunately, the debate surrounding microbial dispersal is not likely to be resolved any time soon as there is limited information on actual rates of microbial dispersal. In a recent meta-analysis of published literature, Jenkins et al. (2007) concluded that "claims that microbes disperse widely cannot be tested by current data." However, they did find that the distance-mass relationship for passive dispersers was essentially random. In other words, the small size of microbes, in and of itself, does not necessarily mean that microbes have average dispersal distances that differ from those of larger plants or animals. In addition, both Jenkins et al. (2007) and Martiny et al. (2006) have speculated that the dispersal distances can vary considerably between microbial taxa. Such differences could arise from differences in the mode of transport, habitat characteristics, population densities, and the ability of the microbe to survive the transport process itself (Fig. 1).

At larger spatial scales, the active dispersal (self-propulsion) of microbes should be severely constrained (Jenkins et al., 2007; Martiny et al., 2006). However, passive dispersal may occur via a variety of mechanisms, including transport in the atmosphere, by water currents, or on or within larger plants and animals. Likewise, microbes that can go dormant for extended periods and survive harsh environmental conditions are more likely to be transported long distances (Fig. 1). This is particularly true for those microbes that are aerially dispersed, as the atmospheric environment poses a unique set of challenges due to the high levels of UV radiation, low moisture levels, and extremely oligotrophic conditions (Jones and Harrison, 2004; Lighthart, 1997; Madelin, 1994). Microbes inhabiting certain habitats, such as surface soils, plant leaf surfaces, or streams, are more likely to be dispersed longer distances than those in other habitats (such as subsurface soils and deep-sea sediments) where the potential for long-range transport is likely to be more limited. The same pattern should hold for microbes associated with plants or animals that can move (or be moved) long distances (such as whales, agricultural crops, and migratory birds) versus those microbes that are free-living or associated with organisms of more-limited dispersal abilities. Of course, this is a fundamental concept in epidemiology, illustrated most recently by the rapid intercontinental dispersal of avian flu by migratory birds (Rappole and Hubalek, 2006). All other factors being equal, those microbes that are more abundant in a given area are likely to be transported further as high densities may effectively broaden the dispersal distribution (Fig. 1).

Dispersal itself will not alter biogeographical patterns unless dispersal is accompanied by successful establishment (or colonization) of the new environment. If colonization rates are very low, we would expect to observe high levels of endemicity at the community level (Papke and Ward, 2004). A variety of biotic and abiotic processes may influence the frequency of successful colonization. Microbes that are generalists, i.e., those that are able to grow in a wide range of environments, are more likely to colonize new habitats than those microbes that can only grow under very specific conditions. Likewise, microbes that need to live in close association with other organisms (such as syntrophic microbes, specific pathogens, or species-specific mycorrhizae) are less likely to successfully colonize a "new" habitat than free-living microbes. We would expect habitats with more challenging environmental conditions to support lower colonization rates than those that are more hospitable. The amount of available niche space should also regulate the suitability of a habitat for colonization; if there is no "room" for an introduced microbe, it will not survive for long. Perhaps one example of this is the protective influence that healthy gut microflora can have against gastrointestinal pathogens. After prolonged antibiotic usage, the microbial community is disrupted and out of equilibrium, rendering the gastrointestinal system more susceptible to colonization by harmful pathogens (Guarner and Malagelada, 2003).

The processes associated with dispersal and colonization can be elucidated by research on community development in a previously lifeless environment, the process of primary succession. This has been elegantly demonstrated by work on plant community development on Indonesian islands sterilized by the eruption of Krakatau in 1883 (Whittaker et al., 1989). Patterns of primary succession have been documented in a variety of microbial systems including water pipes (Martiny et al., 2003), lake biofilms (Jackson et al., 2001), and recently deglaciated soils (Nemergut et al., 2007), but it is not clear if microbial succession follows similar patterns as those documented for plant communities (Jackson, 2003). With more studies on microbial community assembly during primary succession and careful analyses of the successional patterns, we can begin to estimate microbial dispersal/colonization rates and, possibly, determine how these rates are influenced by habitat type, phylogenetic characteristics, and environmental conditions.

WHY ARE MICROBIAL COMMUNITIES SO DIVERSE?

There is no question that microbial communities can be amazingly diverse. While some "extreme" environments, such as acid mine drainage (Baker and Banfield, 2003), harbor relatively few microbial taxa, the microbial diversity found in individual environmental samples is often very high. Small-subunit rRNA gene surveys, even those that are relatively large, rarely encompass the full extent of microbial diversity found in a sample (Fig. 2), making it difficult to accurately estimate the total taxonomic richness (Curtis and Sloan, 2005; Curtis et al., 2002; Hughes et al., 2001). This is particularly true for studies conducted in soil and sediment environments, where it has been estimated that individual samples are likely to harbor many tens of thousands of bacterial phylotypes (Gans et al., 2005; Hong et al., 2006; Torsvik et al., 2002; Tringe et al., 2005). Recent evidence suggests that bacteria are not unique in this regard, as other microbial groups (including protists, viruses, archaea, and fungi) may also exhibit very high levels of local phylogenetic diversity (Breitbart et al., 2002; Fierer et al., 2007a; O'Brien et al., 2005; Walsh et al., 2005) (Fig. 2). The statement by E. O. Wilson, "microbial diversity is beyond practical calculation" (Wilson, 1999), is likely to be accurate in many environments and for a variety of microbial taxa.

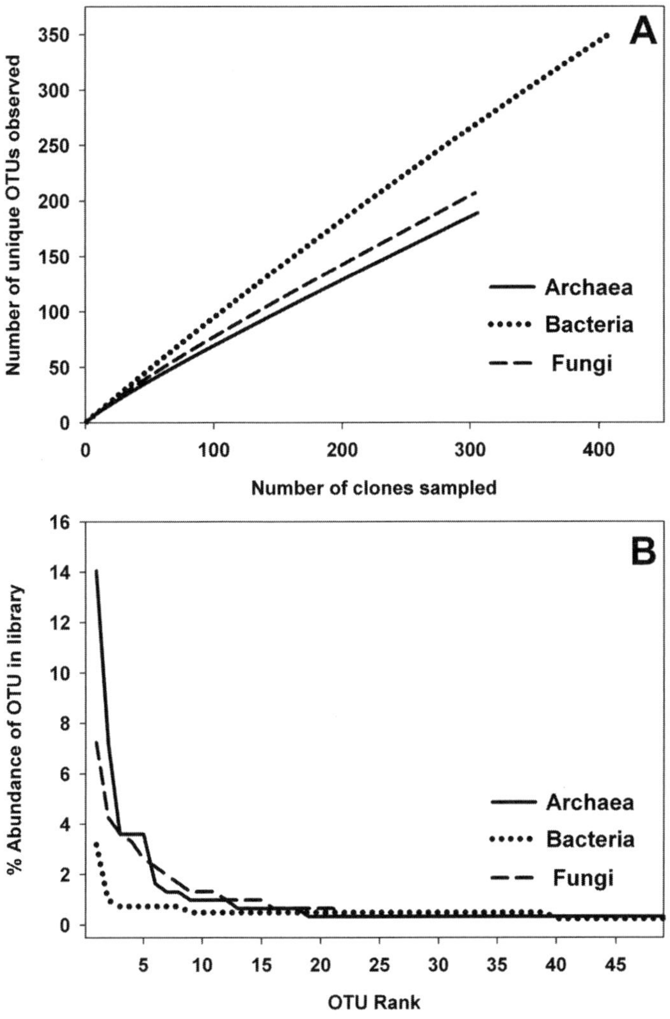

FIGURE 2 Comparison of rarefaction curves (A) and rank-abundance curves (B) for bacterial, archaeal, and fungal clone libraries targeting the small-subunit (16S, 18S) rRNA gene. Libraries constructed from a single desert soil sample collected in Joshua Tree, CA. OTUs are defined at the ≤97% sequence similarity level. For the rank-abundance curve (B), only the 50 most abundant OTUs are shown. All three rarefaction curves fail to asymptote, indicating that we have not surveyed the full extent of taxonomic richness in the sample. The differences in the slopes of the rarefaction curves (A) are a result of differences in community evenness (evident in B), not necessarily differences in overall richness. Data are from Fierer et al. (2007a).

The term "diversity" can be confusing in that it encompasses two very different components of community structure, richness and evenness. Richness is simply the number of unique operational taxonomic units (OTUs) in a given sample, area, or community. In contrast, evenness describes the distribution of individuals among the OTUs (the proportional abundances of OTUs), and evenness is maximized when all OTUs have the same number of individuals (Magurran, 2004). In most terrestrial and aquatic environments, microbial communities appear to have both high levels of richness and evenness. This is clearly evident if we examine taxon-accumulation curves, otherwise known as rarefaction curves, generated by plotting the cumulative number of unique OTUs against the size of the sampling effort (Fig. 2 and 3). The taxon-accumulation curves are often close to linear for soil and sediment microbial communities (Fig. 2 and 3), indicating that these communities are very even and any attempt to survey the full extent of microbial richness in a given sample would be a difficult (and expensive) effort given current sequencing technologies. For example, Schloss and Handelsman (2006) have estimated that a complete census of the unique bacteria (those with more than 3% divergence in their 16S rRNA gene sequences) in a single gram of

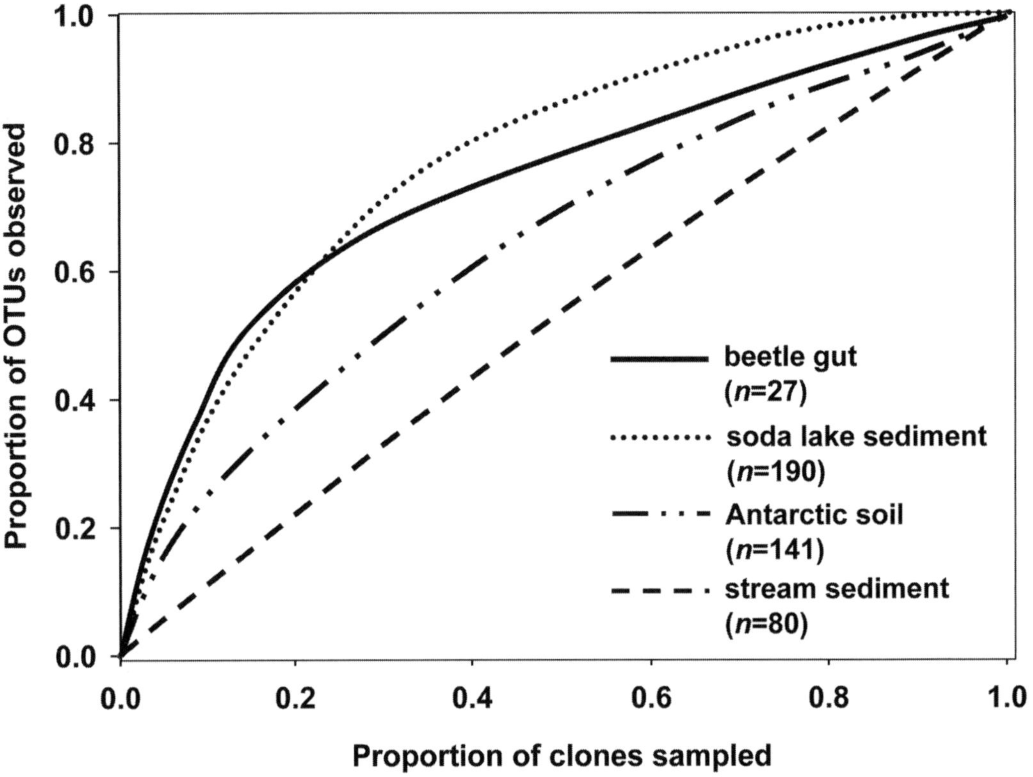

FIGURE 3 Comparison of rarefaction curves from bacterial communities found in different environments. All data are from bacterial clone libraries targeting the 16S rRNA gene with OTUs defined at the ≥97% sequence similarity. Data are from Vasanthakumar et al. (2006) for the beetle gut-associated bacteria, Wani et al. (2006) for the soda lake sediment, Lawley et al. (2004) for the Antarctic soil, and Fierer et al. (2007b) for the stream sediment. The total number of clones (*n*) in each library is indicated in the legend.

Alaskan soil would require sampling more than 480,000 sequences.

The near-linearity of many taxon-accumulation curves (Fig. 2 and 3) indicates that microbial communities commonly have a large number of rare OTUs, a so-called "long-tail" distribution (Fig. 2). We can model the taxon-abundance distribution of microbial communities by using a variety of mathematical functions, including types of lognormal, logarithmic, or power-law functions (Angly et al., 2005; Curtis et al., 2002; Dunbar et al., 2002; Fierer et al., 2007a; Hong et al., 2006; Schloss and Handelsman, 2006). The question of which mathematical function is most appropriate for describing bacterial community structure is subject to some debate. This debate is not likely to be resolved any time soon; plant and animal communities have been surveyed far more comprehensively than most microbial communities yet macro-ecologists have been arguing for decades over the choice of models used to describe species-abundance distributions (Hughes, 1986; Magurran, 2004). While we may not be able to accurately identify the specific taxon-abundance distribution in a given microbial community, we do know that bacterial communities are typically dominated by a few, more abundant taxa and many taxa that are relatively rare, a classic example of the "long-tail" phenomenon well studied by economists (Anderson, 2006).

Given that most environments harbor diverse microbial communities, the obvious question is: "Why are microbial communities so diverse?" Perhaps every researcher has his or her own hypothesis for the high diversity of microbial communities, and many of these hypotheses have not been tested or cannot be tested. What follows is a general overview of the categories of explanations that have been used to explain why many environments harbor highly diverse microbial communities. It is important to recognize that these hypotheses are not mutually exclusive, as some of the mechanisms and processes may act synergistically to affect levels of microbial diversity.

Environmental Complexity

At the scale at which microbes perceive their environment, most microbial habitats are spatially heterogeneous due to either biotic or abiotic factors (Kassen and Rainey, 2004), and a given sample can contain a large number of potential niches. For example, a 10-cm^2 sediment core may encompass a range of redox conditions with obligate aerobes, facultative anaerobes, and obligate anaerobes living in close proximity to one another. Likewise, a 1-g soil sample is likely to support microbes with a broad array of physiologies, including autotrophs (such as nitrifiers and methane oxidizers), aerobic heterotrophs that are either copiotrophic or oligotrophic, and anaerobes (such as denitrifiers and sulfate reducers). Even environments that appear to be relatively homogeneous can harbor a number of distinct microenvironments. For example, numerous *Pseudomonas* genotypes can arise from a single, ancestral genotype due to the availability of multiple ecological niches in different locations of an unshaken culture vessel (Kassen et al., 2000). Laboratory studies have demonstrated that there is a positive correlation between habitat heterogeneity ("patchiness") and the phylogenetic diversity of bacteria (Korona et al., 1994; Rainey et al., 2000), but such patterns have been more difficult to confirm in the field. Zhou et al. (2002) found that saturated subsurface soils contained less diverse bacterial communities than unsaturated soils, and they attributed this difference to the increased patchiness of the unsaturated soils. Their hypothesis has been supported by laboratory experiments (Treves et al., 2003), but it remains to be determined if there is a direct correlation between habitat complexity and microbial diversity, partly because it is so difficult to quantify environmental complexity given the large number of biotic and abiotic factors that interact to shape microbial habitats.

Body Size and Spatial Scaling

May (1988) and others (Azovsky, 2002; Morse et al., 1985; Ritchie and Olff, 1999) have hypothesized that smaller organisms should have a higher local diversity than larger organisms because of their ability to partition a given environment more finely. In other words, a decrease in body size increases the apparent number of habitats in a given environment as there is more of a fine-grained perception of environmental heterogeneity and a corresponding increase in the number of different ways the environment can be utilized by organisms. Of course, this hypothesis is similar to the "environmental complexity" hypothesis described above in that both hypotheses suggest that the high levels of microbial diversity are driven by the large number of potential niches in a given microbial habitat. However, it is important to recognize that the high levels of microbial diversity may be a direct result of our scale of inquiry; the samples analyzed by microbiologists are relatively small from our perspective, but incredibly large compared to the size of individual microbes living in that sample.

More specifically, surveying microbial diversity in individual environmental samples may be similar in magnitude to surveying the diversity of macroorganisms at continental scales. To illustrate this point, consider an environment that has 10^4 unique bacterial species in a 100-m^2 area, a reasonable estimate for soil and sediment samples (Gans et al., 2005; Hong et al., 2006; Torsvik et al., 2002). To directly compare the bacterial richness in this 100-m^2 area to bird species richness, we would have to survey bird species richness across the entire globe, which has approximately 10,000 bird species (Howard and Moore, 1991). This calculation is based on the assumption that species richness is correlated with the abundance of a taxon in a given area (Diamond, 1988; Siemann et al., 1996), which is largely a function of body size (May, 1988; Oindo et al., 2001), and birds (assume a body size of 10^{-3} m^3) are nearly 10^{15} times larger than an average bacterium. While this calculation is an obvious oversimplification, it does demonstrate that estimating microbial diversity in a relatively small area is analogous to estimating plant and animal diversity at much larger spatial scales. Whereas body size alone is not likely to account for the high diversity of soil microbes, once we reconcile differences in spatial scale, the local richness of microbes may be more comparable to the observed levels of plant and animal richness.

Speciation and Extinction Rates

High levels of microbial diversity could also be driven by high rates of speciation, low rates of extinction, or some combination of these two processes. Unfortunately, we do not have good estimates of microbial speciation and extinction rates in the field (Horner-Devine et al., 2004b; Ramette and Tiedje, 2007a), so we can only speculate on the importance of these competing processes. We would expect bacteria to have lower rates of extinction than most metazoans because bacteria are probably less likely to die of starvation or harsh environmental conditions, bacteria do not typically die of old age, and they are capable of rapid, asexual reproduction (Dykhuizen, 1998; Horner-Devine et al., 2004b; Ramette and Tiedje, 2007a). Likewise, the large number of individuals present in most bacterial populations and high rates of dispersal may effectively buffer microbial taxa from changes in the environment or other stochastic processes that could lead to extinction. One could also speculate that speciation rates should be higher for microbes than for plants and animals because bacteria often have short generation times, high rates of horizontal gene transfer, large population sizes, an ability to finely partition a given environment into distinct niches, and often engage in direct interspecies interactions (both positive and negative) that may contribute to ecological specialization (Dykhuizen, 1998; McArthur, 2006; Papke and Ward, 2004). We know from experimental studies that rates of speciation can be very rapid for laboratory strains of bacteria grown under controlled conditions (Elena and Lenski, 2003; Lenski et al., 1991; Rainey et al., 2000), but we do not know if speciation rates are also rapid for the majority of bacteria living in more natural conditions. If bacteria really do have high rates of speciation and low rates of extinction, then this combination of processes could contribute to the high levels of bacterial richness observed at both local and global scales.

The "Storage Effect"

Chesson and colleagues have outlined a hypothesis, termed the "storage effect," to explain the maintenance of species coexistence (Chesson, 1994; Chesson and Huntly, 1989; Chesson and Warner, 1981). Although the storage effect has not been explicitly applied to microbes, the hypothesis may provide an elegant explanation for the high levels of local microbial diversity. The storage effect hypothesizes that temporal fluctuations in recruitment rates among species can lead to the stable coexistence of competitors. More specifically, species will remain in a community and not become extinct as long as three conditions are met: (i) competition is an important factor regulating community structure; (ii) environmental conditions vary and there are species-specific responses in recruitment rates to this variability, giving species (even rare species) the capacity to increase in population size on occasion; and (iii) organisms have a long-lived life stage to survive periods of poor recruitment when environmental conditions are less favorable. All other factors being equal, the storage effect predicts that we would observe maximum levels of diversity in communities of long-lived, fecund organisms living in environments that experience high levels of temporal variability.

In all likelihood, most bacterial communities probably satisfy the conditions outlined above.

Microbial habitats are often temporally variable, bacteria are likely to compete for limited resources (or space), and many bacteria are capable of rapid growth rates given the appropriate environmental conditions. In addition, distinct phylogenetic groups of bacteria often have distinct environmental requirements (with respect to redox levels, substrate preferences, pH, and light availability, for example), and many bacteria can survive in a dormant or semidormant state for prolonged periods. Testing the applicability of the storage effect hypothesis to microbial communities would not be easy given our limited knowledge of microbial life histories. Nevertheless, the storage effect hypothesis may provide a comprehensive set of mechanisms to explain, and predict, levels of diversity in microbial systems.

ARE MICROBES GLOBALLY AS WELL AS LOCALLY DIVERSE?

We can assess microbial diversity at a variety of spatial scales, ranging from the diversity in an individual environmental sample to the diversity measured across large geographic regions. Typically local diversity is referred to as alpha diversity, whereas the total species richness over continents and biomes is referred to as gamma diversity (Lomolino et al., 2006; Magurran, 1988; Whittaker, 1975). As described above, we know that most microbial communities have a high local (alpha) diversity; however, there is currently some debate regarding the gamma diversity of microbes. In particular, Fenchel and Finlay have speculated in a series of papers (Fenchel, 1993; Fenchel et al., 1997; Finlay and Clarke, 1999; Finlay, 2002) that the global richness of microbes is not significantly higher than their local richness and a large percentage of the microbial taxa on Earth can be found in an individual sample collected from a single habitat. If correct, the global richness of a given microbial group could be calculated by surveying a handful of habitats, as distinct habitats would not necessarily harbor distinct species assemblages. The competing hypothesis is that gamma diversity far exceeds local diversity, there is minimal overlap in species assemblages between habitats, and the number of unique microbial taxa on Earth is enormous. The two opposing "sides" of this debate are graphically represented in Fig. 4.

Fenchel and Finlay's hypothesis that microscopic-sized organisms are locally very diverse, but globally species poor, is based on the idea that "smaller organisms tend to have wider or even cosmopolitan distribution, a higher efficiency of dispersal, a lower rate of allopatric speciation and lower rates of global extinction than do larger organisms" (Fenchel, 1993). Support for their hypothesis comes from the observation that a large proportion of the global richness of protozoa can be found in a given local area (Fenchel et al., 1997; Finlay, 2002). In particular, they found that 80% of the global species total in the genus *Paraphysomonas* was found in <0.1 cm^2 of sediment from a single freshwater pond (Finlay and Clarke, 1999). They conclude from this research that, at the global scale, microscopic-sized organisms should have lower levels of taxonomic richness than organisms that are of intermediate size (Fenchel, 1993). This conclusion is supported by species-area curves generated for different size classes of organisms living in Arctic marine sediments (Azovsky, 2002).

There have been a number of direct criticisms of Fenchel and Finlay's conclusions. Foissner (2006) has argued that their work is fundamentally flawed since they assume that the global richness of protozoa is a known quantity. Foissner speculates that the global diversity of protozoa is likely to be far higher than Fenchel and Finlay have estimated and therefore their estimate of the local:global diversity ratio is unreasonably high. He argues that Azovsky's work (2002) suffers from a similar flaw. There are a number of other specific criticisms of Fenchel and Finlay's work (and their conclusions), including their reliance on morphospecies definitions (Coleman, 2002), their undersampling of rare taxa (Foissner, 2006), and the unsupported extrapolation of their work on one taxonomic group to all microscopic organisms (Lachance, 2004). Perhaps more damning are the large number of

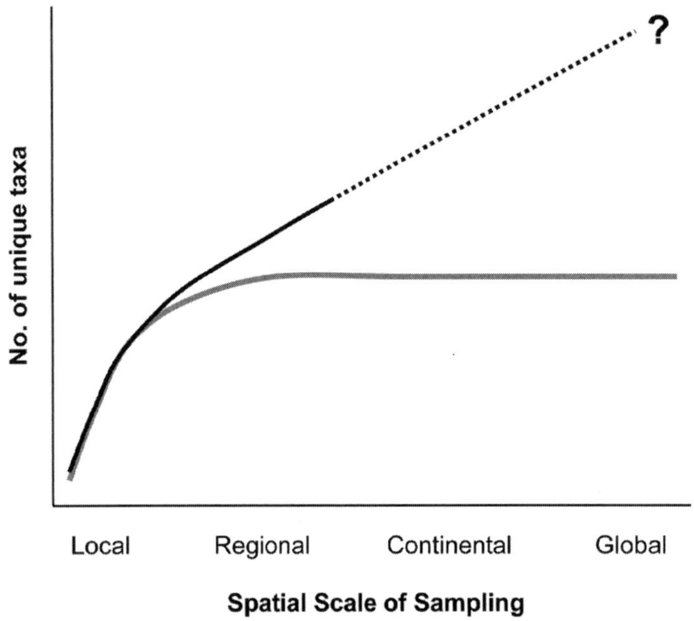

FIGURE 4 Hypothetical changes in the total number of unique microbial taxa identified from surveys of different spatial scales. The gray line represents the predictions of Fenchel and Finlay (Fenchel, 1993; Fenchel et al., 1997; Finlay, 2002); the black line represents the competing hypothesis that there is minimal overlap in species assemblages across habitats. The dashed line and the question mark indicate the high degree of uncertainty.

studies, reviewed here and elsewhere (Horner-Devine et al., 2004b; Martiny et al., 2006; Ramette and Tiedje, 2007b), that have documented considerable spatial heterogeneity in microbial community composition. If Fenchel and Finlay are correct, then microbes should have no biogeography, and the discovery of new microbial taxa should be far less common than has been observed.

Unfortunately, the debate surrounding the magnitude of global microbial richness has been fueled by speculation and a scarcity of hard data. Fortunately, sequencing efforts have been increasing at an exponential rate, and public databases (such as GenBank, http://www.ncbi.nlm.nih.gov/Genbank/index.html) are now filled with sequences of microbial small-subunit rRNA genes from a wide range of habitats and locations. We should be able to use these sequence data to quantify the degree of overlap in microbial assemblages between habitats and estimate (or roughly approximate) the lower bounds of microbial richness on Earth. Until this is done, the global richness of bacteria, fungi, and other microbial taxa will remain a question mark and the validity of these competing hypotheses cannot be assessed.

TAXON-AREA RELATIONSHIPS

One of the cornerstones in the field of modern biogeography is the equilibrium theory of island biogeography developed by MacArthur Wilson (1967). Put simply, this model represents the number of species inhabiting an island as an equilibrium between rates of immigration (colonization) and extinction. Although a number of criticisms have been leveled against MacArthur and Wilson's theory (Lomolino et al., 2006), their simple model is elegant in that it qualitatively predicts whether species numbers and species turnover rates will increase or decrease with changes in island (patch) size and the degree of isolation. In particular, MacArthur and Wilson's theory provides a conceptual explanation for the species-area relationship, one of the most-studied and best-documented patterns in plant and animal biogeography. The species-area relationship describes the pattern that species numbers tend to increase with increasing area and is generally expressed with the following equation:

$$S = cA^z$$

where S is species richness, c is a fitted constant, A is area, and z represents the slope when S and A are plotted on logarithmic scales (the slope of the species-area relationship). This equation, referred to as the Arrhenius equation or the power model, is commonly used to model species-area relationships with the steepness of the species-area relationship describing the rate at which communities differentiate in space (Fig. 5). The exponent z has been the focus of much inquiry, and there has been considerable debate surrounding the biological relevance of the z value (Lomolino et al., 2006). For plant and animal taxa, z values generally range from 0.1 to 0.2 in contiguous habitats and from 0.25 to 0.35 across discrete island habitats (Horner-Devine et al., 2004a; Rosenzweig, 1995). Recent studies, reviewed by Green and Bohannan (2006), Prosser et al. (2007), and Woodcock et al. (2006), have shown that microbes also demonstrate a positive species-area relationship with the taxonomic richness of microbes increasing with the amount of area surveyed (Fig. 5). However, it is unclear if microbial z values are comparable to those observed for plant and animal taxa. Studies of the taxa-area relationship for salt-marsh bacteria (Horner-Devine et al., 2004a), marine diatoms (Azovsky, 2002), soil fungi (Green et al., 2004), and soil bacteria (Fierer and Jackson, 2006) have found z values lower than those generally observed for comparable studies of plants and animals. However, studies of bacteria inhabiting sump tanks (van der Gast et al., 2005), tree holes (Bell et al., 2005), and forest soils (Noguez et al., 2005) have yielded z values that are similar (0.25 to 0.45) to those observed for plant and animal taxa. In a particularly elegant study of the microbial species-area relationship, Peay et al. (2007) examined the diversity of ectomycorrhizal fungi across "tree islands" and also found a species-area slope ($z \approx 0.2$) similar to that reported for macroorganisms (Fig. 5).

Although microbial taxon-area relationships have received considerable attention over the past few years, it is important to recognize that direct comparisons of z values from microbial taxon-area relationships must be considered carefully. One reason for this is that z values will vary with taxonomic resolution (Horner-Devine et al., 2004a), and the different methods used to assess microbial diversity (e.g., fingerprinting techniques, morphological analyses, and direct analyses of gene sequences) quantify community-level diversity at varying levels of taxonomic resolution. This issue becomes particularly problematic if we are trying to compare taxon-area relationships between micro- and macroorganisms since there is no uniform and consistent definition of what constitutes a microbial "species." Likewise, microbial communities are highly diverse, and it is often difficult to survey the full extent of microbial diversity in a given sample. Undersurveying a community will lead to an underestimation of z, and this may explain why the z values reported for microbial taxa are often lower than those reported for plant and animal taxa (Woodcock et al., 2006). It is also important to recognize that, since a number of different mechanisms may generate the apparent taxon-area curves, z values (in and of themselves) do not tell us what specific process, or processes, are driving the spatial differentiation in microbial communities.

IS MICROBIAL BIOGEOGRAPHY SHAPED BY ENVIRONMENTAL FACTORS OR HISTORY?

We know that microbial communities often vary across space, and we can confidently ignore Finlay's speculation that "there is no biogeography for anything smaller than 1 millimeter" (Whitfield, 2005). The question then becomes: What process or processes are responsible for generating the observed biogeographical patterns? From an oversimplified perspective, there are two general factors that may contribute to the formation of the biogeographical patterns: environmental heterogeneity and dispersal limitation. The idea that environmental heterogeneity drives biogeographic patterns is best summarized by the Baas Becking hypothesis "everything is everywhere, but the environment selects" (Baas Becking, 1934; de Wit and Bouvier, 2006). In other words, there is effectively no dispersal limitation, biogeographic patterns solely reflect contemporary environ-

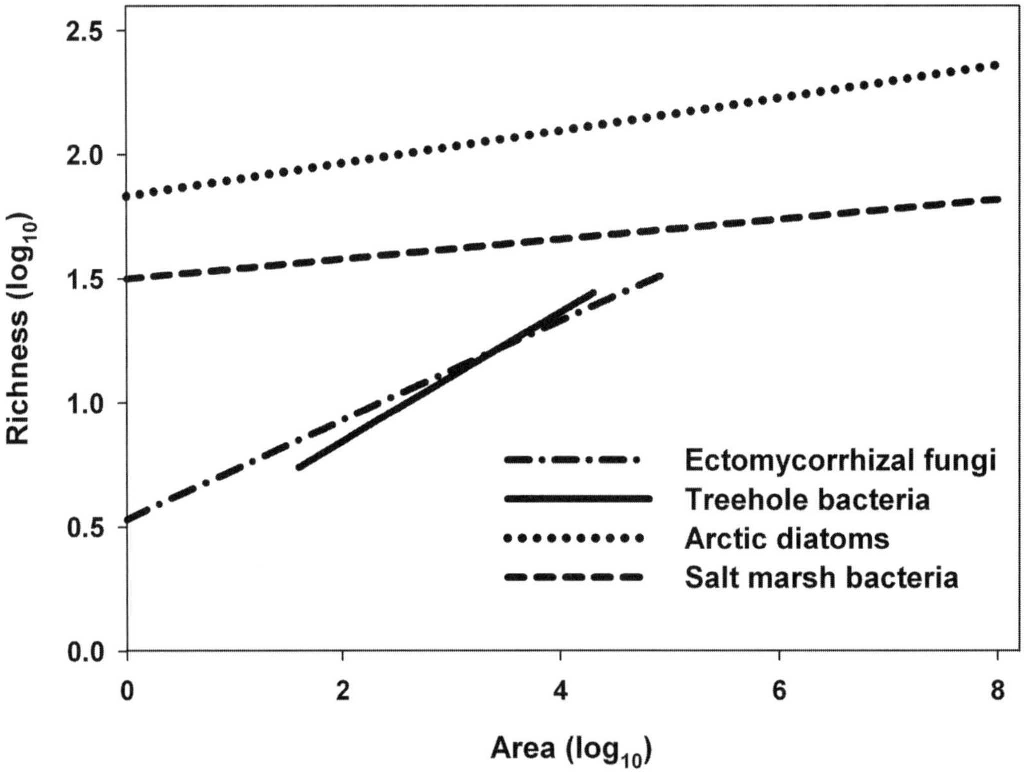

FIGURE 5 A comparison of published taxon-area relationships (TARs) from contiguous habitats (arctic diatoms and salt-marsh bacteria) and noncontiguous (island) habitats (treehole bacteria and ectomycorrhizal fungi). The TAR for arctic diatoms is from Azovsky (2002) and represents the number of diatom species in Arctic sediments versus area (m^2). The TAR for treehole bacteria is from Bell et al. (2005) and represents bacterial genetic diversity (determined by DGGE fingerprinting) versus the volume (ml) of water-filled treeholes. The TAR for salt-marsh bacteria is from Horner-Devine et al. (2004a) and represents the number of bacterial OTUs in a salt marsh (99% sequence similarity) versus area (cm^2). The TAR for ectomycorrhizal fungi is from Peay et al. (2007) and represents the number of ectomycorrhizal fungal species in "tree islands" of a given area (m^2).

mental conditions, and similar environments will harbor similar microbial taxa regardless of the geographic distance between the environments. The opposing hypothesis is that spatial variability in microbial communities is a product of historical events, namely dispersal limitation and (possibly) past environmental conditions (Martiny et al., 2006). If dispersal limitation is the primary driver of biogeographical patterns, then geographic distance should be the best predictor of genetic divergence between communities, and habitats in close proximity are more likely to share similar microbial taxa. Obviously these two hypotheses represent opposite ends of the spectrum, and microbial biogeography, like the biogeography of plants and animals (Lomolino et al., 2006), probably reflects some combination of both environmental heterogeneity and dispersal limitation (i.e., history). Nevertheless, it is worth considering these two processes independently and examining the limitations associated with using this strict dichotomy to understand the biogeographical patterns exhibited by microbes.

There is no shortage of evidence that environmental heterogeneity can, to some extent, directly influence the spatial heterogeneity in microbial communities. To cite just a few examples, pH has been found to be the best predictor of the continental-scale patterns exhibited by soil bacteria (Fierer and Jackson, 2006), estuarine bacterioplankton communities change along a salinity gradient (Crump et al., 2004), and shifts in hot-spring cyanobacterial communities correspond to temperature (Ward et al., 1998). These patterns can also be observed in experimental studies where changes in environmental conditions, e.g., substrate availability, redox potential, light intensity, and a wide range of other factors, can induce shifts in microbial community composition (Buckley and Schmidt, 2002; Horner-Devine et al., 2004b; McArthur, 2006). Of course, a correlation between one or more environmental factors and community composition does not, in and of itself, indicate that environmental heterogeneity is the sole factor influencing the observed biogeographical patterns.

Even in cases where environmental heterogeneity directly influences the spatial structure of microbial communities, it may still be difficult to determine the specific environmental factors that are directly responsible for generating the observed biogeographical patterns. First, the physical environment in most microbial habitats is highly variable, and it can be difficult to measure environmental characteristics at the fine levels of resolution that will be most relevant to microbes. Second, environmental influences are going to be highly dependent on the taxa in question and the scale of inquiry (Ganderton and Coker, 2005). Individual taxa may respond to different environmental factors, and those factors that correlate with the taxonomic structure of entire microbial communities may be unrelated to the taxonomic structure of subsets of the microbial community. For example, communities of marine cyanobacteria may shift with changes in light intensity, while nonphotosynthetic heterotrophs in the same environment may be more responsive to gradients of organic carbon bioavailability. Likewise, those factors that drive biogeographical patterns within a single habitat may not necessarily be important when we look across habitats. This is the so-called "paradox of scale" (Ganderton and Coker, 2005), the idea that different environmental factors will affect the same microbial community or population if we change our scale of inquiry. As an example, consider the spatial patterning of soil bacterial communities that may be strongly correlated with plant presence or absence within an individual plot (Kuske et al., 2002) but may appear to be correlated with soil pH at the continental scale (Fierer and Jackson, 2006). Third, microbes can exert a significant influence on their local environment, making it difficult to distinguish between environmental effects on the community and community impacts on the environment. This phenomenon is particularly evident in biofilm communities where microbes can effectively alter the environmental characteristics of their habitat.

There is some evidence the dispersal limitation may also influence biogeographical patterns, but oftentimes the effects of dispersal limitation can be difficult to distinguish from the effects of environmental heterogeneity. We know that microbial taxa can exhibit some degree of endemism (Cho and Tiedje, 2000; Papke and Ward, 2004), but only a handful of studies have specifically examined the effects of dispersal (i.e., geographic distance) versus environmental heterogeneity on microbial biogeography. Perhaps, the most widely cited study is that of Whitaker et al. (2003) where they examined *Sulfolobus* strains isolated from hot spring habitats and found that the geographic distance between hot springs, not the environmental characteristics of the hot springs, explained the biogeographic patterns. This study is not alone; other studies (Green et al., 2004; Papke et al., 2003; Reche et al., 2005) have also reported a negative correlation between geographic distance and the genetic similarity of microbial taxa with little to no influence of measured environmental heterogeneity on microbial community composition. We would expect that the influence of dispersal limitation on biogeographic patterns may be more apparent at finer levels of taxonomic resolution (Cho and Tiedje, 2000) where small phylogenetic differences between populations or communities can be more readily observed.

Likewise, the influence of dispersal limitation may be more apparent at continental or global scales than in studies that examine spatial structure over smaller scales (Martiny et al., 2006).

Although dispersal limitation is likely to have an important influence on microbial biogeography, designing studies to distinguish between the effects of environmental heterogeneity and dispersal limitation is difficult. Some of the studies that are frequently cited as evidence that dispersal limitation (i.e., geographic distance) exerts a major influence on microbial biogeography (see above) may be flawed in that they have not directly measured the environmental characteristics of the collected samples (e.g., Green et al., 2004) or they measured only a limited number of environmental characteristics (e.g., Reche et al., 2005). Without a thorough assessment of environmental characteristics at each sampled location and the temporal heterogeneity in the environmental characteristics, it cannot be assumed that dispersal limitation has a stronger influence than environmental heterogeneity on the observed biogeographical patterns. A correlation between geographic distance and genetic distance does not necessarily indicate that dispersal limitation drives biogeographical patterns. There is always a strong possibility that an unmeasured environmental characteristic may explain more of the variance in community structure than geographic distance, especially when we consider that habitats in close proximity are often similar with respect to their environmental characteristics. Even if we use statistical methods to separately quantify the influence of geographic distance and habitat heterogeneity (Martiny et al., 2006), we are still assuming that the degree of environmental similarity between samples has been sufficiently assessed.

Given the inherent difficulties associated with adequately characterizing microbial habitats, the most robust approach for quantifying the influence of dispersal limitation on microbial biogeography is to compare microbial communities across identical habitats in different geographic regions. Of course, finding identical habitats is nearly impossible as they would have to be the same with respect to their size, age (to allow establishment of microbial communities), and abiotic conditions (which may be influenced by the characteristics of the microbial assemblages). For this reason Foissner (2006) has argued that Baas Becking's hypothesis ("everything is everywhere, but the environment selects") is not a falsifiable hypothesis and is more valuable as a metaphor than as a scientific hypothesis.

How much of the spatial variation in microbial communities is driven by environmental heterogeneity versus dispersal limitation? It depends. It depends on a number of factors, including the taxonomic group in question, the scale of inquiry, habitat characteristics, and the level of taxonomic resolution. Microbial biogeography is likely to be driven by both environmental heterogeneity and dispersal limitation, but distinguishing between these two factors is not trivial and may be impossible given our inability to adequately assess microscale environmental characteristics and the difficulties associated with finding identical, but spatially separated, habitats. Even when we can assess the contributions of environmental heterogeneity and dispersal limitation to biogeographical patterns, it is highly likely that we will still find a large amount of unexplained variation due to unmeasured environmental heterogeneity, spatial structure, or ecologically neutral processes (Ramette and Tiedje, 2007a). Of course, these concerns are not unique to microbial biogeography. Even though we know far more about the natural history and spatial distribution of plants and animals, "macro"-bial biogeographers still struggle to predict (and explain) the spatial structure of macroorganisms.

FUTURE DIRECTIONS IN THE STUDY OF MICROBIAL BIOGEOGRAPHY

The field of microbial biogeography is on the cusp of rapid advancement. New tools and methods are emerging that will give us unprecedented abilities to survey individual microbial communities and document changes in microbial communities across space and time. There is also growing recognition that microbes exhibit biogeographical patterns and that, by studying these patterns, we may be able to develop biogeographical theories and hypotheses that apply across the entire tree of life, not just the small portion of the tree of life where we find macroorganisms. However, it is important to recognize that the "unknown unknowns" and "known unknowns" in microbial biogeography currently outnumber the "known knowns." For this reason, I conclude this chapter by highlighting some key topics where the gaps in our knowledge of microbial biogeography are particularly apparent. This list is neither unbiased nor exhaustive; I have simply highlighted a few research topics that may be ripe avenues for future research.

Taxon-Time Relationships

Although the field of biogeography principally focuses on the spatial distribution of organisms (Ganderton and Coker, 2005), the temporal aspects of microbial biogeography may be particularly important. Species turnover, the changes in species composition as new species arrive at a location and older species go extinct, is usually measured across years, decades, or centuries by plant and animal biogeographers (MacDonald, 2003). In microbial systems, turnover may be much more rapid, especially if we consider that the process of microbes leaving or entering a dormant state (thereby entering or leaving the active microbial community) may be akin to immigration/extinction processes. If the temporal turnover in microbial communities is rapid, there may be little consistency in the composition of the "active" microbial community across sampling dates. Likewise, if turnover rates are very low, there may be a weak correlation between microbial community composition and the environmental characteristics measured at the time of sampling. Unfortunately, turnover rates in the field are difficult to quantify as there are no robust methods for measuring generation times of individual microbial cells in situ, and it can be difficult to distinguish between "active" microbes and microbes that are in a dormant or semidormant state (even with RNA-based surveys of microbial communities). Turnover rates are likely to be highly variable within a given community as some populations may have generation times on par with *Escherichia coli* strains growing in the laboratory (<1 h) while other populations may remain viable, but nonreproductive, for months if not centuries (Kennedy et al., 1994; Kieft and Phelps, 1997). Likewise, we would expect microbial community turnover rates to vary across habitats. Those communities that experience relatively static environmental conditions, limited predation, and are resistant to disturbances should have particularly low turnover rates. For example, we would expect microbial communities residing in deep-sea sediments to have lower turnover rates (possibly weeks to months) than the planktonic communities found in surface waters (possibly hours to days). Just as taxon-area curves can provide a useful metric for comparing the spatial heterogeneity in microbial communities (see above), future research on taxon-time curves (Rosenzweig, 1995) may be useful for estimating microbial turnover rates and assessing those biotic and abiotic factors that influence the temporal heterogeneity in microbial community composition.

Viruses

Although viruses are ubiquitous and abundant in many environments, the biogeographical patterns exhibited by viruses have received little attention. Recent work in both terrestrial and aquatic environments suggests that the taxonomic diversity of viral communities is likely to be very high (Breitbart et al., 2004a; Breitbart et al., 2002; Williamson et al., 2005). It has been hypothesized that the composition of viral communities is relatively invariant across distinct locations and habitat types (Breitbart et al., 2004b; Breitbart and Rohwer, 2005) with viruses from one biome able to survive (and propagate) in other biomes (Sano et al., 2004). If these observations are confirmed with additional studies, they would suggest that viral biogeography is distinct from the biogeography of other microbial groups. Now that we are able to survey viral diversity in the environment using metagenomic tools (Edwards and Rohwer, 2005), we can begin to integrate viruses into the field of biogeography.

Incorporating Phylogenetics into Microbial Biogeography

Throughout this chapter I have emphasized how studies in microbial biogeography are more difficult to conduct than comparable studies of plant or animal biogeography, largely due to the problems associated with surveying microbial communities. However, because unculturable microbes are difficult to identify, microbial biogeographers (by necessity) often rely on nucleic acid sequence data to examine the spatial structure in microbial populations. This puts microbial biogeographers at a distinct advantage as their community surveys can directly incorporate information on evolutionary history to understand and explain observed biogeographical patterns. Although plant and animal biogeographers may also use phylogenetic approaches to examine biogeographical patterns, there is generally less of an incentive to conduct sequence-based surveys as species-level identification is (often) more straightforward.

Sequence-based surveys of microbial diversity (the most common being small-subunit rRNA-based surveys) yield a wealth of information, but this information is rarely mined to its full potential. Most studies in microbial biogeography compare diversity patterns by grouping sequences into one, or several, OTUs. The OTU-based approach is problematic because there is no consensus OTU definition. For example, a study that groups sequences at the 97% similarity level is not comparable to a study that groups sequences at the 99% similarity level. In addition, the OTU-based approach ignores evolutionary history, treating all OTUs equivalently even though some may be closely related and some distantly related. Phylogenetic approaches for analyzing the diversity of microbial communities are now available (Jones and Martin, 2006; Martin, 2002), and these methods can reveal patterns in the phylogenetic structure of microbial communities that would essentially be hidden with the standard OTU-based approach. For example, lineage per time plots can reveal shifts in number of divergent microbial lineages across an elevation gradient (Martin, 2002), shifts that would be difficult to discern if applying standard ecological statistics to OTU distributions. We can also use phylogenetic methods to quantify the pairwise distances between microbial communities (Lozupone et al., 2006; Lozupone and Knight, 2005). Since such methods incorporate information on the phylogenetic relationships between sequences, they are more sensitive than OTU-based approaches. For example, there may be zero overlap between two communities if we group sequences at the 97% sequence similarity level: however, these two communities could either be very similar (if all the sequences are from the same bacterial taxon) or markedly distinct (if the sequences from the two communities represent distinct phylogenetic lineages). The utility of applying phylogenetic-distance-based approaches to examine microbial biogeography is illustrated in two recent studies by Lozupone (Lozupone et al., 2007; Lozupone and Knight, 2007).

Latitude and Other Diversity Gradients

The latitudinal diversity gradient, whereby diversity tends to increase with decreases in latitude, is one of the most fundamental patterns in "macro"-bial biogeography. A wide range of plant and animal taxa exhibit an increase in diversity from the poles to the equator, and many competing hypotheses have been proposed to explain the pattern (Lomolino et al., 2006). Although the existence of these patterns in plant and animal taxa has been documented and studied for centuries, relatively few studies have explicitly tested whether microbial taxa also exhibit a latitudinal gradient in diversity. Hillebrand and Azovsky (2001) found that latitudinal gradients in richness are largely absent for diatoms, and they hypothesize that the strength of the latitudinal gradient is positively correlated with organism size. Other studies that have looked for latitudinal trends in microbial diversity have either found no relationship

(Fierer and Jackson, 2006) or a reverse pattern whereby richness increases with latitude (Buckley et al., 2003). These results are intriguing, but additional studies are needed before we can confidently conclude that microbes, unlike plants and animals, do not generally exhibit an increase in diversity with a decrease in latitude. More importantly, such studies can inform biogeographical theory by testing the universality of the latitudinal diversity pattern and the validity of the hypotheses that have been offered to explain the pattern.

Of course, the latitudinal diversity gradient is not the only diversity gradient to be frequently studied (and debated) by "macro"-bial biogeographers. Diversity-productivity and diversity-disturbance relationships have received considerable attention with many studies having observed maximum levels of plant and animal diversity in habitats that have intermediate frequencies of disturbance or intermediate productivity levels (Connell, 1978; Mittelbach et al. 2001; Rosenzweig, 1995). A number of experimental studies have observed that some (but not all) microbial taxa exhibit similar patterns (Buckling et al., 2000; Floder and Sommer, 1999; Horner-Devine et al., 2003; Li, 2002), highlighting the utility of using microbial taxa to test macroecological hypotheses.

Characterizing the Microbial Habitat at the Microbe Scale

The study of microbial biogeography is constrained by the enormous discrepancy between our scale of inquiry and the scale at which microbes live in their environment. We typically survey microbial communities in samples that are cubic centimeter to cubic meter in size, but these individual samples represent a wide array of distinct microbial habitats. As a result, we often lack detailed knowledge of the in situ characteristics of the microbial microenvironment and information on where specific microbial taxa are living in a given environment. Although some techniques are now employed to study microbes at the microbe scale (e.g., Crawford et al., 2005; Huang et al., 2007; Kuypers and Jørgensen, 2007; Teal et al., 2006), we often have to ignore the micron-scale complexity and use indirect methods to understand how microbes influence, and are influenced by, their environment. These constraints face all microbiologists studying microbes outside the laboratory (Madsen, 1998) and represent a set of conceptual and methodological barriers that are not typically encountered by biogeographers focusing on macroorganisms.

CONCLUSION

The study of microbial biogeography will help us move beyond anecdotal studies and observations to build a predictive understanding of microbial diversity and the factors influencing this diversity across space and time. Microbiologists may be able to use preexisting concepts in biogeography to understand microbial systems, or we may find that such concepts, which are largely derived from studies of plants and animals, are not directly applicable. Either way, the incorporation of microbiology into the field of biogeography promises to be a fruitful endeavor as many fundamental questions remain unanswered. Microbiologists can test biogeographical theories that are difficult, if not impossible, to test with plant and animal communities, and by studying microbial biogeography, we will move closer to understanding the full breadth of biological diversity on Earth.

REFERENCES

Anderson, C. 2006. *The Long Tail: Why the Future of Business Is Selling Less of More.* Hyperion, New York, NY.

Angly, F., B. Rodrigues-Brito, D. Bangor, P. McNairnie, M. Breitbart, P. Salamon, B. Felts, J. Nulton, J. Mahaffy, and F. Rohwer. 2005. PHACCS, an online tool for estimating the structure and diversity of uncultured viral communities using metagenomic information. *BMC Bioinformatics* **6:**41.

Azovsky, A. 2002. Size-dependent species-area relationships in benthos: is the world more diverse for microbes? *Ecography* **25:**273–282.

Baas Becking, L. 1934. *Geobiologie of inleiding tot de milieukunde.* Van Stockum & Zoon, The Hague, The Netherlands.

Baker, B. J., and J. F. Banfield. 2003. Microbial communities in acid mine drainage. *FEMS Microbiol. Ecol.* **44:**139–152.

Bell, T., D. Ager, J. I. Song, J. A. Newman, I. P. Thompson, A. K. Lilley, and C. J. van der Gast. 2005. Larger islands house more bacterial taxa. *Science* **308:**1884–1884.

Breitbart, M., B. Felts, S. Kelley, J. M. Mahaffy, J. Nulton, P. Salamon, and F. Rohwer. 2004a. Diversity and population structure of a near-shore marine-sediment viral community. *Proc. R. Soc. London, Ser. B* **271:**565–574.

Breitbart, M., J. H. Miyake, and F. Rohwer. 2004b. Global distribution of nearly identical phage-encoded DNA sequences. *FEMS Microbiol. Lett.* **236:**249–256.

Breitbart, M., and F. Rohwer. 2005. Here a virus, there a virus, everywhere the same virus? *Trends Microbiol.* **13:**278–284.

Breitbart, M., P. Salamon, B. Andresen, J. M. Mahaffy, A. M. Segall, D. Mead, F. Azam, and F. Rohwer. 2002. Genomic analysis of uncultured marine viral communities. *Proc. Natl. Acad. Sci. USA* **99:**14250–14255.

Buckley, D., and T. Schmidt. 2002. Exploring the biodiversity of soil—a microbial rain forest, p. 183–208. *In* J. Staley and A. Reysenbach (ed.), *Biodiversity of Microbial Life.* John Wiley & Sons, New York, NY.

Buckley, H. L., T. E. Miller, A. M. Ellison, and N. J. Gotelli. 2003. Reverse latitudinal trends in species richness of pitcher-plant food webs. *Ecol. Lett.* **6:**825–829.

Buckling, A., R. Kassen, G. Bell, and P. B. Rainey. 2000. Disturbance and diversity in experimental microcosms. *Nature* **408:**961–964.

Chesson, P. 1994. Multispecies competition in variable environments. *Theoretical Population Biol.* **45:**227–276.

Chesson, P., and N. Huntly. 1989. Short-term instabilities and long-term community dynamics. *Trends Ecol. Evol.* **4:**293–298.

Chesson, P. L., and R. R. Warner. 1981. Environmental variability promotes coexistence in lottery competitive-systems. *American Naturalist* **117:**923–943.

Cho, J. C., and J. M. Tiedje. 2000. Biogeography and degree of endemicity of fluorescent *Pseudomonas* strains in soil. *Appl. Environ. Microbiol.* **66:**5448–5456.

Coleman, A. 2002. Microbial eukaryote species. *Science* **297:**337.

Connell, J. H. 1978. Diversity in tropical rain forests and coral reefs—high diversity of trees and corals Is maintained only in a non-equilibrium state. *Science* **199:**1302–1310.

Crawford, J., J. Harris, K. Ritz, and I. Young. 2005. Towards an evolutionary ecology of life in soil. *Trends Ecol. Evol.* **20:**81–87.

Crump, B. C., C. S. Hopkinson, M. L. Sogin, and J. E. Hobbie. 2004. Microbial biogeography along an estuarine salinity gradient: combined influences of bacterial growth and residence time. *Appl. Environ. Microbiol.* **70:**1494–1505.

Curtis, T., and W. Sloan. 2005. Exploring microbial diversity—a vast below. *Science* **309:**1331–1333.

Curtis, T. P., W. T. Sloan, and J. W. Scannell. 2002. Estimating prokaryotic diversity and its limits. *Proc. Natl. Acad. Sci. USA* **99:**10494–10499.

de Wit, R., and T. Bouvier. 2006. 'Everything is everywhere, but, the environment selects'; what did Baas Becking and Beijerinck really say? *Environ. Microbiol.* **8:**755–758.

Diamond, J. 1988. Factors controlling species diversity: overview and synthesis. *Annals Missouri Botanical Garden* **75:**117–129.

Dunbar, J., S. M. Barns, L. O. Ticknor, and C. R. Kuske. 2002. Empirical and theoretical bacterial diversity in four Arizona soils. *Appl. Environ. Microbiol.* **68:**3035–3045.

Dykhuizen, D. E. 1998. Santa Rosalia revisited: why are there so many species of bacteria? *Antonie van Leeuwenhoek* **73:**25–33.

Edwards, R., and F. Rohwer. 2005. Viral metagenomics. *Nat. Rev. Microbiol.* **3:**504–510.

Elena, S., and R. Lenski. 2003. Evolution experiments with microorganisms: the dynamics and genetic bases of adaptation. *Nat. Rev. Genet.* **4:**457–469.

Fenchel, T. 2003. Biogeography for bacteria. *Science* **301:**925–926.

Fenchel, T. 1993. There are more small than large species? *Oikos* **68:**375–378.

Fenchel, T., G. Esteban, and B. Finlay. 1997. Local versus global diversity of microorganisms: cryptic diversity of ciliated protozoa. *Oikos* **80:**220–225.

Fierer, N., M. Breitbart, J. Nulton, P. Salamon, C. Lozupone, R. T. Jones, M. Robeson, R. Edwards, B. Felts, R. Knight, F. Rohwer, and R. B. Jackson. 2007a. Metagenomic and small-subunit RNA analyses reveal the genetic diversity of bacteria, archaea, fungi, and viruses in soil. *Appl. Environ. Microbiol.* **73:**7059–7066..

Fierer, N., and R. Jackson. 2006. The diversity and biogeography of soil bacterial communities. *Proc. Natl. Acad. Sci. USA* **103:**626–631.

Fierer, N., J. L. Morse, S. T. Berthrong, E. S. Bernhardt, and R. B. Jackson. 2007b. Environmental controls on the landscape-scale biogeography of stream bacterial communities. *Ecology* **88:**2162–2173.

Finlay, B., and K. Clarke. 1999. Ubiquitous dispersal of microbial species. *Nature* **400:**828.

Finlay, B. J. 2002. Global dispersal of free-living microbial eukaryote species. *Science* **296:**1061–1063.

Floder, S., and U. Sommer. 1999. Diversity in planktonic communities: an experimental test of the intermediate disturbance hypothesis. *Limnol. Oceanogr.* **44:**1114–1119.

Foissner, W. 2006. Biogeography and dispersal of micro-organisms: a review emphasizing protists. *Acta Protozoologica* **45:**111–136.

Ganderton, P., and P. Coker. 2005. *Environmental Biogeography*. Pearson Education, Essex, United Kingdom.

Gans, J., M. Wolinsky, and J. Dunbar. 2005. Computational improvements reveal great bacterial diversity and high metal toxicity in soil. *Science* **309:**1387–1390.

Green, J., and B. Bohannan. 2006. Spatial scaling of microbial biodiversity. *Trends Ecol. Evol.* **21:**501–507.

Green, J., A. Holmes, M. Westoby, I. Oliver, D. Briscoe, M. Dangerfield, M. Gillings, and A. Beattie. 2004. Spatial scaling of microbial eukaryote diversity. *Nature* **432:**747–750.

Guarner, F., and J.-R. Malagelada. 2003. Gut flora in health and disease. *Lancet* **361:**512–519.

Hillebrand, H., and A. I. Azovsky. 2001. Body size determines the strength of the latitudinal diversity gradient. *Ecography* **24:**251–256.

Hong, S. H., J. Bunge, S. O. Jeon, and S. S. Epstein. 2006. Predicting microbial species richness. *Proc. Natl. Acad. Sci. USA* **103:**117–122.

Horner-Devine, M., M. Lage, J. Hughes, and B. Bohannan. 2004a. A taxa-area relationship for bacteria. *Nature* **432:**750–753.

Horner-Devine, M. C., K. M. Carney, and B. J. M. Bohannan. 2004b. An ecological perspective on bacterial biodiversity. *Proc. R. Soc. London, Ser. B* **271:**113–122.

Horner-Devine, M. C., M. A. Leibold, V. H. Smith, and B. J. M. Bohannan. 2003. Bacterial diversity patterns along a gradient of primary productivity. *Ecol. Lett.* **6:**613–622.

Howard, R., and A. Moore. 1991. *A Complete Checklist of the Birds of the World*. Academic Press, London, United Kingdom

Huang, W. E., K. Stoecker, R. Griffiths, L. Newbold, H. Daims, A. S. Whiteley, and M. Wagner. 2007. Raman-FISH: combining stable-isotope Raman spectroscopy and fluorescence in situ hybridization for the single cell analysis of identity and function. *Environ. Microbiol* **9:**1878–1889.

Hubbell, S. 2001. *The Unified Neutral Theory of Biodiversity and Biogeography*. Princeton University Press, Princeton, NJ.

Hughes, J. B., J. J. Hellmann, T. H. Ricketts, and B. J. M. Bohannan. 2001. Counting the uncountable: statistical approaches to estimating microbial diversity. *Appl. Environ. Microbiol.* **67:**4399–4406.

Hughes, R. 1986. Theories and models of species abundance. *American Naturalist* **128:**897–899.

Jackson, C. R. 2003. Changes in community properties during microbial succession. *Oikos* **101:**444–448.

Jackson, C. R., P. F. Churchill, and E. E. Roden. 2001. Successional changes in bacterial assemblage structure during epilithic biofilm development. *Ecology* **82:**555–566.

Jenkins, D., C. Brescacin, C. Duxbury, J. Elliott, J. Evans, K. Grablow, M. Hillegass, B. Lyon, G. Metzger, M. Olandese, D. Pepe, G. Silvers, H. Suresch, T. Thompson, C. Trexler, G. Williams, N. Williams, and S. Williams. 2007. Does size matter for dispersal distance? *Global Ecol. Biogeogr.* **16:**415–425.

Jones, A., and R. Harrison. 2004. The effects of meteorological factors on atmospheric bioaerosol concentrations—a review. *Science Total Environ.* **326:**151–180.

Jones, R., and A. Martin. 2006. Testing for differentiation of microbial communities using phylogenetic methods: accounting for uncertainty of phylogenetic inference and character state mapping. *Microb. Ecol.* **52:**408–417.

Kassen, R., A. Buckling, G. Bell, and P. B. Rainey. 2000. Diversity peaks at intermediate productivity in a laboratory microcosm. *Nature* **406:**508–512.

Kassen, R., and P. Rainey. 2004. The ecology and genetics of microbial diversity. *Annu. Rev. Microbiol.* **58:**207–231.

Kennedy, M. J., S. L. Reader, and L. M. Swierczynski. 1994. Preservation records of micro-organisms: evidence of the tenacity of life. *Microbiology* **140:**2513–2529.

Kieft, T., and T. Phelps. 1997. Life in the slow lane: activities of microorganisms in the subsurface. *In* P. Amy and D. Haldeman (ed.), *The Microbiology of the Terrestrial Deep Subsurface*. CRC Press, Boca Raton, FL.

Korona, R., C. Nakatsu, L. Forney, and R. Lenski. 1994. Evidence for multiple adaptive peaks from populations of bacteria evolving in a structured habitat. *Proc. Natl. Acad. Sci. USA* **91:**9037–9041.

Kuske, C. R., L. O. Ticknor, M. E. Miller, J. M. Dunbar, J. A. Davis, S. M. Barns, and J. Belnap. 2002. Comparison of soil bacterial communities in rhizospheres of three plant species and the interspaces in an arid grassland. *Appl. Environ. Microbiol.* **68:**1854–1863.

Kuypers, M. M. M., and B. B. Jørgensen. 2007. The future of single-cell environmental microbiology. *Environ. Microbiol.* **9:**6–7.

Lachance, M. 2004. Here and there or everywhere? *Bioscience* **54:**884.

Lawley, B., S. Ripley, P. Bridge, and P. Convey. 2004. Molecular analysis of geographic patterns of eukaryotic diversity in Antarctic soils. *Appl. Environ. Microbiol.* **70:**5963–5972.

Lenski, R., M. Rose, S. Simpson, and S. Tadler. 1991. Long-term experimental evolution in *Escherichia coli.* I. Adaptation and divergence during 2,000 generations. *American Naturalist* **138:**1315–1341.

Li, W. K. W. 2002. Macroecological patterns of phytoplankton in the northwestern North Atlantic Ocean. *Nature* **419:**154–157.

Lighthart, B. 1997. The ecology of bacteria in the alfresco atmosphere. *FEMS Microbiol. Ecol.* **23:**263–274.

Lomolino, M., B. Riddle, and J. Brown. 2006. *Biogeography*, 3rd ed. Sinauer Assoc., Sunderland, MA.

Lozupone, C., M. Hamady, S. Kelley, and R. Knight. 2007. Quantitative and qualitative β diversity measures lead to different insights into factors that structure microbial communities. *Appl. Environ. Microbiol.* **73:**1576–1585.

Lozupone, C., M. Hamady, and R. Knight. 2006. UniFrac—an online tool for comparing microbial community diversity in a phylogenetic context. *BMC Bioinformatics* **7:**371.

Lozupone, C., and R. Knight. 2005. UniFrac: a new phylogenetic method for comparing microbial communities. *Appl. Environ. Microbiol.* **71:**8228–8235.

Lozupone, C. A., and R. Knight. 2007. Global patterns in bacterial diversity. *Proc. Natl. Acad. Sci. USA* **104:**11436–11440.

MacArthur, R., and E. Wilson. 1967. *The Theory of Island Biogeography.* Princeton University Press, Princeton, NJ.

MacDonald, G. 2003. *Biogeography: Space, Time, and Life.* John Wiley & Sons, New York, NY.

Madelin, T. 1994. Fungal aerosols: a review. *J. Aerosol Sci.* **25:**1405–1412.

Madsen, E. 1998. Epistemology of environmental microbiology. *Environ. Sci. Tech.* **32:**429–439.

Magurran, A. 1988. *Ecological Diversity and Its Measurement.* Princeton University Press, Princeton, NJ.

Magurran, A. 2004. *Measuring Biological Diversity.* Blackwell Publishing, Oxford, United Kingdom.

Martin, A. P. 2002. Phylogenetic approaches for describing and comparing the diversity of microbial communities. *Appl. Environ. Microbiol.* **68:**3673–3682.

Martiny, A. C., T. M. Jorgensen, H. J. Albrechtsen, E. Arvin, and S. Molin. 2003. Long-term succession of structure and diversity of a biofilm formed in a model drinking water distribution system. *Appl. Environ. Microbiol.* **69:**6899–6907.

Martiny, J. B., B. J. M. Bohannan, J. H. Brown, R. K. Colwell, J. A. Fuhrman, J. L. Green, M. C. Horner-Devine, M. Kane, J. A. Krumins, C. R. Kuske, P. J. Morin, S. Naeem, L. Ovreas, A. L. Reysenbach, V. H. Smith, and J. T. Staley. 2006. Microbial biogeography: putting microorganisms on the map. *Nat. Rev. Microbiol.* **4:**102–112.

May, R. 1988. How many species are there on Earth? *Science* **247:**1441–1449.

McArthur, J. V. 2006. *Microbial Ecology: an Evolutionary Approach.* Elsevier, Boston, MA.

Mittelbach, G. G., C. F. Steiner, S. M. Scheiner, K. L. Gross, H. L. Reynolds, R. B. Waide, M. R. Willig, S. I. Dodson, and L. Gough. 2001. What is the observed relationship between species richness and productivity? *Ecology* **82:**2381–2396.

Morse, D. R., J. H. Lawton, M. M. Dodson, and M. H. Williamson. 1985. Fractal dimension of vegetation and the distribution of arthropod body lengths. *Nature* **314:**731–733.

Nemergut, D. R., S. P. Anderson, C. C. Cleveland, A. P. Martin, A. E. Miller, A. Seimon, and S. K. Schmidt. 2007. Microbial community succession in unvegetated, recently-deglaciated soils. *Microbial Ecol.* **53:**110–122.

Noguez, A., H. Arita, A. Escalante, L. Forney, F. Garcia-Oliva, and V. Souza. 2005. Microbial macroecology: highly structured prokaryotic soil assemblages in a tropical deciduous forest. *Global Ecol. Biogeogr.* **14:**241–248.

O'Brien, H., J. Parrent, J. Jackson, J. Moncalvo, and R. Vilgalys. 2005. Fungal community analysis by large-scale sequencing of environmental samples. *Appl. Environ. Microbiol.* **71:**5544–5550.

Oindo, B., A. Skidmore, and H. Prins. 2001. Body size and abundance relationship: an index of diversity for herbivores. *Biodiversity Conservation* **10:**1923–1931.

Papke, R. T., N. B. Ramsing, M. M. Bateson, and D. M. Ward. 2003. Geographical isolation in hot spring cyanobacteria. *Environ. Microbiol.* **5:**650–659.

Papke, R. T., and D. M. Ward. 2004. The importance of physical isolation to microbial diversification. *FEMS Microbiol. Ecol.* **48:**293–303.

Peay, K., T. Bruns, P. Kennedy, S. Bergemann, and M. Garbelotto. 2007. A strong species-area relationship for eukaryotic soil microbes: island size matters for ectomycorrhizal fungi. *Ecol. Lett.* **10:**470–480.

Prosser, J., B. Bohannan, T. Curtis, R. Ellis, M. Firestone, R. Freckleton, J. Green, L. Green, K. Killham, J. Lennon, A. Osborn, M. Solan, C. van der Gast, and J. Young. 2007. The role of ecological theory in microbial ecology. *Nat. Rev. Microbiol.* **5:**384–392.

Rainey, P., A. Buckling, R. Kassen, and M. Travisano. 2000. The emergence and maintenance of diversity: insights from experimental bacterial populations. *Trends Ecol. Evol.* **15:**243–247.

Ramette, A., and J. Tiedje. 2007a. Biogeography: an emerging cornerstone for understanding prokaryotic diversity, ecology, and evolution. *Microb. Ecol.* **53:**197–207.

Ramette, A., and J. Tiedje. 2007b. Multiscale responses of microbial life to spatial distance and environmental heterogeneity in a patchy ecosystem. *Proc. Natl. Acad. Sci. USA* **104:**2761–2766.

Rappole, J., and Z. Hubalek. 2006. Birds and influenza H5N1 virus movement to and within North America. *Emerg. Infect. Dis.s* **12:**1486–1492.

Reche, I., E. Pulido-Villena, R. Morales-Baquero, and E. Casamayor. 2005. Does ecosystem size determine aquatic bacterial richness? *Ecology* **86:**1715–1722.

Ritchie, M., and H. Olff. 1999. Spatial scaling laws yield a synthetic theory of biodiversity. *Nature* **400:**557–560.

Rosenzweig, M. 1995. *Species Diversity in Space and Time.* Cambridge University Press, Cambridge, United Kingdom.

Sano, E., S. Carlson, L. Wegley, and F. Rohwer. 2004. Movement of viruses between biomes. *Appl. Environ. Microbiol.* **70:**5842–5846.

Schloss, P. D., and J. Handelsman. 2006. Toward a census of bacteria in soil. *PLoS Comput.l Biol.* **2:**e92.

Siemann, E., D. Tilman, and J. Haarstad. 1996. Insect species diversity, abundance and body size relationships. *Nature* **380:**704–706.

Teal, T. K., D. P. Lies, B. J. Wold, and D. K. Newman. 2006. Spatiometabolic stratification of *Shewanella oneidensis* biofilms. *Appl. Environ. Microbiol.* **72:**7324–7330.

Torsvik, V., L. øvreås, and T. F. Thingstad. 2002. Prokaryotic diversity: magnitude, dynamics, and controlling factors. *Science* **296:**1064–1066.

Treves, D. S., B. Xia, J. Zhou, and J. M. Tiedje. 2003. A two-species test of the hypothesis that spatial isolation influences microbial diversity in soil. *Microb. Ecol.* **45:**20–28.

Tringe, S., C. vonMering, A. Kobayashi, A. Salamov, K. Chen, H. Chang, M. Podar, J. Short, E. Mathur, J. Detter, P. Bork, P. Hugenholtz, and E. Rubin. 2005. Comparative metagenomics of microbial communities. *Science* **308:**554–557.

van der Gast, C., A. Lilley, D. Ager, and I. Thompson. 2005. Island size and bacterial diversity in an archipelago of engineering machines. *Environ. Microbiol.* **7:**1220–1226.

Vasanthakumar, A., I. Delalibera, J. Handelsman, K. D. Klepzig, P. D. Schloss, and K. F. Raffa. 2006. Characterization of gut-associated bacteria in larvae and adults of the southern pine beetle, *Dendroctonus frontalis* Zimmermann. *Environ. Entomol.* **35:**1710–1717.

Walsh, D. A., R. T. Papke, and W. F. Doolittle. 2005. Archaeal diversity along a soil salinity gradient prone to disturbance. *Environ. Microbiol.* **7:**1655–1666.

Wani, A. A., V. P. Surakasi, J. Siddharth, R. G. Raghavan, M. S. Patole, D. Ranade, and Y. S. Shouche. 2006. Molecular analyses of microbial diversity associated with the Lonar soda lake in India: an impact crater in a basalt area. *Res. Microbiol.* **157:**928–937.

Ward, D., M. Ferris, S. Nold, and M. Bateson. 1998. A natural view of microbial biodiversity within hot spring cyanobacterial mat communities. *Microbiol. Molec. Biol. Rev.* **62:** 1353–1370.

Whitaker, R. J., D. W. Grogan, and J. W. Taylor. 2003. Geographic barriers isolate endemic populations of hyperthermophilic archaea. *Science* **301:**976–978.

Whitfield, J. 2005. Biogeography: Is everything everywhere? *Science* **310:**960–961.

Whittaker, R. 1975. *Communities and Ecosystems*, 2nd ed. Macmillan, New York, NY.

Whittaker, R., M. Bush, and K. Richards. 1989. Plant recolonization and vegetation succession on the Krakatau Islands, Indonesia. *Ecol. Monogr.* **59:**59–123.

Williamson, K. E., M. Radosevich, and K. E. Wommack. 2005. Abundance and diversity of viruses in six Delaware soils. *Appl. Environ. Microbiol.* **71:**3119–3125.

Wilson, E. O. 1999. *The Diversity of Life.* Penguin, London, United Kingdom.

Woodcock, S., T. P. Curtis, I. M. Head, M. Lunn, and W. T. Sloan. 2006. Taxa-area relationships for microbes: the unsampled and the unseen. *Ecol Lett.* **9:**805–812.

Zhou, J., B. Xia, D. S. Treves, L. Y. Wu, T. L. Marsh, R. V. O'Neill, A. V. Palumbo, and J. M. Tiedje. 2002. Spatial and resource factors influencing high microbial diversity in soil. *Appl. Environ. Microbiol.* **68:**326–334.

THE LEAST COMMON DENOMINATOR: SPECIES OR OPERATIONAL TAXONOMIC UNITS?

Ramon Rosselló-Mora and Arantxa López-López

7

THE KINDS OF UNITS THAT WE OBSERVE IN THE ENVIRONMENT

Molecular techniques have given microbial ecologists an entirely new dimension for understanding natural ecosystems. The acceptance that the cultured fraction of microorganisms is almost negligible in comparison with the enormous expected diversity (Amann et al., 1995; Whitman et al., 1998) has led microbial ecologists to search for conceptual units to quantify such diversity. The straightforward idea that occurs to ecologists is "species." Actually, although species is the unit for understanding diversity, it is also used in macroecology as a main unit to qualitatively and quantitatively understand and compare ecosystems (e.g., Begon et al., 2006). The tendency of microbiology as a young science has been to adapt the concepts in use for macroeukaryotes to prokaryotic systematics, taxonomy, and ecology. However, there is a problem, since, after years of controversies, it seems that the term "species" encompasses different multiple conceptual formulations, each one conceived for the special purpose of each discipline (Rosselló-Mora, 2005). Nevertheless, it is simply a problem of homonymy and is a similar problem to the one that occurs in higher eukaryote taxonomies, where different disciplines use the same term to explain different things (Ereshefsky, 1998). In contrast to evolutionary microbiologists, microbial ecologists have understood that with the hitherto approaches of information retrieval one cannot ensure that the units observed can be identified as taxonomic species. Consequently, myriad numbers of terms are being used to describe diversity, and some of them, i.e., operational taxonomic unit (OTU), are recommended because they are neutral and distanced from philosophical discussions (Horner-Devine et al., 2006). In this chapter, we list and discuss the common terms used by microbial ecologists to describe diversity, as well as attempt to explain their virtues and pitfalls. In addition, we to some extent discuss the meaning of species.

Ramon Rosselló-Mora and Arantxa López-López, Marine Microbiology Group, Institut Mediterrani d'Estudis Avançats (IMEDEA, CSIC-UIB), E-07190 Esporles, Illes Balears, Spain.

SPECIES ETYMOLOGY, A HISTORICAL PERSPECTIVE

Species as a term is being used by many disciplines in and outside biology. All of them base its use on being able to explain the basic unit for

Accessing Uncultivated Microorganisms: from the Environment to Organisms and Genomes and Back
Edited by Karsten Zengler © 2008 ASM Press, Washington, DC

describing the kind of diversity with which they are dealing. However, although the conceptual basis in each discipline is unavoidably different, which is also due to the intrinsic differences between disciplines related to understanding and/or observing each particular system, the term is used equally. Actually, this is the basis of the so called "species problem." In addition, this "problem" is not only circumscribed to microbial systematics and related disciplines because there is also a similar ongoing and much older debate among "higher" eukaryote systematists (Rosselló-Mora, 2005). This, therefore, is a clear case of homonymy (Ereshefsky, 1998), and the use of the same term to explain fundamentally different kinds of things is driving an ongoing debate that does not seem easy to resolve. However, it is advocated here that the property of the use of the term "species" should be given primarily to taxonomy. In fact, other uses of the same term may be illegitimate, and, for these reasons, accepting pluralism (Ereshefsky, 1998) may solve an encysted problem that hampers conceptual developments.

Taxonomy is an essential discipline of the biological sciences because it provides a framework for the scientific community to facilitate understanding and knowledge exchange. It appears as a scientific need for classifying biological things in an ordered system. The first classification system was devised by Aristotle at least 2,400 years ago; he generated the first hierarchy based on creationist and essentialist tenets (Ereshefky, 1994). This system was based on only the two categories "species" and "genus," which were motivated by recurrent observations about the world (Hey, 2001). Subsequently, the hierarchical system has been qualitatively and quantitatively enlarged with many more categories, initially by Linnaeus, and later by Mayr and Simpson, as a response to the observation that life was much more complex than at first thought (Ereshefsky, 1994). It is remarkable that despite all the technological and conceptual developments achieved in the last and present centuries, the way we understand order in nature is still based on the taxonomic schema applied to higher eukaryotes devised by Linnaeus three centuries ago. Microbiologists adopted the system directly from the botanical and zoological taxonomies, and microbial classification was constructed following intuitive criteria of how a category could be circumscribed (Rosselló-Mora and Amann, 2001). Of course, this intuition was followed by more solid criteria developed in parallel to the techniques that periodically appeared as a result of technological improvements.

In any case, the category of "species" has been the property of taxonomy ever since, and has been the basis for any taxonomic system hitherto created. Prokaryote taxonomy has a short, but very dynamic, history of developments that occurred in parallel to technological progresses, and this is also true for the conception of what a species may be. Apart from the debates concerning what species may or should be, taxonomists agree that we have achieved a species concept that applies to all prokaryotic organisms, and that it is the best we can currently get (Stackebrandt et al., 2002). Although the latest version of the concept is a result of recurrent revisits and improvements of previous formulations, the essence of the concept has not changed at all. Prokaryote taxonomists describe species as "a category that circumscribes a monophyletic, and (preferably) genomically coherent group of individual isolates/strains sharing a high degree of similarity in (many) independent features, comparatively tested under highly standardized conditions" (Stackebrandt et al. 2002).

The current prokaryotic species concept fulfils the requisites of being universally applicable to the cellular nuclei-less microorganisms. However, the problem that is a recurrent discussion theme is in how we define species, i.e., that pragmatic aspect of establishing the parameters to circumscribe them. For some, the current definition is too conservative (e.g., Dykhuizen, 1998; Konstantinidis et al., 2006; Whitman et al., 1998), whereas for others it seems to be artificial (e.g., Staley, 2006; Whitaker, 2006). Indeed, we believe, and have already insisted, that the species concept is an artificial construct of the human mind and that this has been discussed not only by biologists but also by philosophers (reviewed by Ereshefsky, 1994; Ereshefsky, 1998; Hey, 2001; Rosselló-Mora and Amann, 2001; Rosselló-Mora, 2003; Rosselló-Mora, 2005). However, biologists insist on finding the existence of natural units that are universally comparable, especially between eukaryotes and prokaryotes. On the other hand, it seems that after many years of discussion and within the framework of the current knowledge of genetic and phenotypic microbial diversity, criticizers of the taxonomic species concept and definition have started to believe in pluralism, e.g., "given the diversity of speciation mechanisms described for macrobes, there is no reason for microbiologists to expect a single concept or definition to apply to all micro-organisms, which represent greater than two-thirds of the metabolic and genetic diversity of the planet" (Whitaker, 2006), or "the Linnaean binomial nomenclature followed in prokaryotic classification suggests analogy between prokaryotic and eukaryotic species which are completely different biological systems" (Gevers et al., 2006). Such affirmation reinforces the earlier statements made by taxonomists and philosophers acknowledging that different taxonomies may be constructed for different biological levels of complexity (Mishler and Donoghue, 1982).

Here, we advocate for the species concept and, hence, the definition is an artificial construct of the human mind. However, the idea of a species being a unit that is phylogenetically, genomically, and phenotypically coherent is in itself a universally applicable concept. The main and most controversial problem is the definition, since the parameters used to circumscribe units are being continuously evaluated and discussed within the framework of new technological developments. These problems have been very well exemplified when achievements such as small-subunit (SSU) rRNA gene reconstructions were implemented, and, currently, by the enormous flow of genomic information (Rosselló-Mora, 2005). The additional problem in circumscribing species within the framework of the taxonomic definition is that, due to the enormous heterogeneity and evolutive divergence of the prokaryotic world, most probably what is valid for certain phylogenetic groups may not be applicable for others. Thus, at least among taxonomists, we need to be (and generally are) flexible in the way we circumscribe taxa. Actually, the parameters that are mandatory for describing new taxa vary from one taxonomic group to another. The most important goal when classifying new taxa is guaranteeing identification, genetic and/or phenotypic property prediction, and stability of the system.

It is almost accepted that phylogenetic coherence is basically demonstrated by the reconstruction of phylogenies based on SSU rRNA gene sequences (Stackebrandt et al., 2002). It is also acknowledged that these reconstructions are generally corroborated by other single-gene (Ludwig and Schleifer, 2005) or concatenated gene reconstructions of a wide range of sets (e.g., Ciccarelli et al., 2006; Soria-Carrasco et al., 2006). Hitherto, the most parsimonious and cost-effective way to understand monophyly is by analyzing SSU rRNA gene sequences. However, it is important to note here that, despite the belief of some scientists, SSU rRNA gene analysis is not itself a suitable parameter for constructing the prokaryotic taxonomy. The SSU rRNA gene is indeed the primary backbone of the current taxonomic schema, but the different categories, ranging from domain to species, are circumscribed by additional phenotypic discriminative properties.

Nevertheless, the most remarkable pitfall of the SSU rRNA approach is the lack of resolution at the species level (Stackebrandt et al., 2002; Rosselló-Mora and Amann, 2001). The temptation has been to use this parameter additionally as the substitute parameter for DNA-DNA reassociation experiments (DDH) when trying to demonstrate genomic coherence (e.g., Stackebrandt and Goebel, 1994). Indeed, DDH is the method that, since the middle of the last century, has been used to show genomic coherence for a given group of strains. Despite its many pitfalls, DDH has been extensively used as the genomic parameter to embrace taxonomic species. Its influence on the taxonomic classification system is equivalent to the "interbreeding" basis for several eukaryotic taxonomies (Rosselló-Mora, 2006), and this influence cannot be underestimated. As happens with the interbreeding principle for macrobe taxonomies, DDH is heavily criticized by many microbiologists, and the search for substitute approaches has been emphasized (Stackebrandt et al., 2002). However, for decades, it has been the best way to genomically circumscribe species, although multilocus sequence analysis (MLSA) has now been proposed as the substitute for DDH (Stackebrandt et al., 2002). This approach is based on the amplification and sequence analyses of several housekeeping genes used to construct a dendrogram, as well as observing the patterns of genotype clustering. These analyses have been very useful to type pathogenic strains for which enough genetic information is available (Stackebrandt et al., 2002). Unfortunately, there are some limitations to this approach, since (i) the kinds of genes in use vary from species to species analyzed, (ii) the reduced set of genes represents a very small part of the organisms' genome, (iii) the design of the primers and successful amplification of the genes are achieved only if the genomic information is exhaustive enough for the species under study, and (iv), as stated by Hanage et al. (2006), "a major problem with this approach to the definition of species is deciding whether resolved clusters should be considered to be different lineages within a species or deserve to be assigned species status." In any case, the current experiences of genome comparisons indicate that finding a substitute for DDH may be difficult unless whole genomes are routinely sequenced (Konstantinidis et al., 2006).

The third step to fulfill when classifying new species is showing phenotypic coherence. In this case, finding a discriminative phenotypic property is mandatory for any new classification. Despite the fact that this principle has been emphasized as being of paramount importance in any description (e.g., Stackebrandt et al., 2002; Wayne et al., 1987), research into the phenotype of prokaryotes for taxonomic purposes has been heavily abandoned in favor of the molecular approaches. The difficulties encountered in standardizing tests and the time-consuming activities for phenotyping strains had led to some disinterest in the way to describe a phenotype. However, because a considerable part of the genome will never be expressed (Lerat and Ochman, 2005; Ochman and Davalos, 2006; Siew and Fischer, 2003), together with the fact that new technologies may increase the knowledge of the metabolic diversity of single organisms (Rosselló-Mora et al., 2008), a resurgence of the importance of the phenotype is expected.

THE NEED FOR PURE CULTURES

The main constraint to the classification of new species within the established taxonomic system is the need to have a pure culture of the organisms that will represent the new taxon. This principle is mainly responsible for the reduced number of currently classified species. This is especially dramatic because the cultured organisms negligibly represent the real environmental diversity (Whitman et al., 1998). Some scientists complain about this because, when compared with higher eukaryote taxonomies, prokaryotes are very much at a disadvantage (e.g., Pedrós-Alió, 2006; Staley, 2006). The need for pure cultures as the main principle for any new classification has the advantage that at least one strain, the named type strain, has to be deposited in at least two different international culture collections. These platforms are responsible for keeping the specimen alive and ready to be distributed among the scientific community. Making at least one strain of each classified species available to the scientific community guarantees, on the one hand, that any previous experiment can be repeated, and, on the other hand, the stability of the classification. In addition, making living specimens publicly available may also prevent potential misuse of experimental data.

Nevertheless, despite the previous reasons, the imperative of having a laboratory isolate as the study material has many disadvantages and they are the basis of many complaints among microbiologists. When observing natural ecosystems, scientists have access to a hidden diversity that deserves to be described. However, as will be argued below, the current techniques in use cannot guarantee that taxa descriptions are exhaustive enough to be placed within a stable classification system. On the other hand, there may be conspicuous organisms in environmental samples that cannot be brought to pure culture, but, due to their intrinsic characteristics and the possibility of retrieving enough of their genetic, phenotypic, and ecological information, the taxonomic system allows their description in provisional taxa named "*candidatus*" (Murray and Schleifer, 1994; Murray and Stackebrandt, 1995).

THE TAXONOMIC SPECIES RECOGNITION AMONG NONCULTURED ORGANISMS

As argued above, the species category is in any case a construct of the human mind. In the case of prokaryotes, a species is considered as a unit when a phylogenetic, genomic, and phenotypic coherence can be shown. The methods to define the boundaries of putative species may change because of scientific developments, but the description of any new taxon, nevertheless, has to be as exhaustive as possible.

As stated, the phylogenetic coherence of uncultured organisms in a given natural sample can be primarily based on retrieving the populations of SSU rRNA gene sequences and the subsequent separation and analysis of the mixed gene populations with different molecular approaches, such as, for example, cloning or denaturing gradient gel electrophoresis (DGGE) (as described thoroughly in other chapters of this book). For many reasons, SSU rRNA has been the selected candidate gene for organismal phylogenetic analyses (Ludwig and Schleifer, 2005), and this has led to the compilation of a database of over 500,000 entries of complete or partial sequences for cultured and uncultured organisms (Cole et al., 2007; or http://www.arb-silva.de). Despite this, however, there are many other genes that may be suitable for calculating organismal phylogenies, although these genes have to have previously shown an evolutionary coherence with SSU rRNA (Ludwig and Schleifer, 2005). This premise is especially important in the metagenomic approaches where the identity of the owner organism of cloned environmental genome pieces will be extrapolated from the presence and analysis of any such genes (e.g., DeLong et al., 2006). A parallel approach to the identification of the metagenome's ownership may be extrapolated after finding taxon-specific signatures within a given genome (Teeling et al., 2004; Wang et al., 2005). However, this latter approach is based on the hitherto sequenced genome database and, in the light of the expected diversity, a taxon assignment may be very unclear in the future.

More complex is the achievement of demonstrating that a given population of organisms shares enough genomic coherence to be considered as a species within the current taxonomic framework. As mentioned before, the demonstration of species genomic coherence is primarily carried out by DDH assays. First attempts to quantify the number of species present in a natural sample by culture-independent approaches had been done by the analysis of the reassociation kinetics of complex DNA mixtures (Torsvik et al., 1990; Torsvik et al., 2002). However, although this approach does not provide information on the identity of the putative species of the sample, it gives an excellent idea about the complexity of the genome mixture, which is directly related to the community structure of the sampled environment. In any case, the method relies on the assumption that each genome may represent a different species, and this fact is biased due to the important intraspecific diversity of the known species genomes (Konstantinidis et al., 2006). As will be argued below, such results are better not discussed in terms of species but rather in terms of genome diversity. The second important approach for understanding genomic coherence has traditionally been the analysis of the GC content of the genomes.

Although this is a vague parameter and not yet reliably discriminative between uncultured organisms, it has been used to reveal the putative content of the predominant prokaryotic population in environments of low species complexity (Øvreås et al., 2003). As stated above, MLSA has been called on to substitute for DDH when classifying new species (Stackebrandt et al., 2002). This method, based on sequencing and analyzing a set of housekeeping genes, has almost no applicability to uncultured organisms when trying to circumscribe species. Retrieving population mixtures of independent genes by PCR cannot give clear information of how these genes can be interlinked within a given genome.

The most modern approach for understanding the genomic diversity present in a given microbial population is the metagenomic clone library achievement. As stated, the putative identity of a cloned stretch of DNA can be achieved either if a gene coding for a SSU rRNA gene is found (e.g., Tyson et al., 2004; Legault et al., 2006) or by the extrapolation of the organismal identity after finding housekeeping genes with a phylogeny parallel to that of the rRNA (e.g., DeLong et al., 2006). In this case, MLSA would be of help in showing genomic coherence if the studied metagenome fragments encode for several genes that are common among all fragments. Another approach in showing coherence may result from the analysis of signature usage (Wang et al., 2005) or tetranucleotide usage frequency (Teeling et al., 2004). However, given the enormous diversity expected and the simplicity of the parameters measured, such approaches can only be used as exclusive, rather than inclusive, properties, i.e., two different usages or signatures might indicate divergence, but two similar or identical usages cannot guarantee genome relatedness.

Finally, the circumscription of a species in the taxonomy sense needs a phenotypic coherence or a characteristic phenotypic property to be found for the population under study. Although molecular tools are allowing information retrieval at the genomic level, the detection and interpretation of the expression of the organism's genotype are currently not very straightforward. Several culture-independent methods have been developed for understanding certain metabolic properties of uncultured organisms. Approaches such as microautoradiography combined with fluorescence in situ hybridization had often shed light on the ability of certain organisms to undergo substrate uptakes by visualizing single-cell radioactive material incorporation (Wagner et al., 2006). In addition, the use of stable isotopes had helped in revealing the fraction of a given community able to uptake and assimilate certain substrates to biomass in the form of DNA (Friedrich, 2006), RNA (Whiteley et al., 2006), or lipids (Evershed et al., 2006). Nevertheless, although these approaches alone cannot be of much help in circumscribing uncultured species, revealing certain phenotypic properties of conspicuous microorganisms that can be monitored by alternative methods may help in the description of novel taxa with the provisional taxonomic status of *candidatus* (Rosselló-Mora and Amann, 2001).

Altogether, it seems that the information that can be retrieved currently from uncultured microbial populations may not be sufficient to guarantee the basic premises of taxonomy, which are operationality and stability of the constructed classification system and predictability of the properties of the identified organisms. In any case, the potential cumulative information on single prokaryotic populations by current methods is still very sparse in comparison with that of organisms brought into pure culture. As a result, taxonomy cannot accept the classification of uncultured organisms because the system will tend to fail. Only the provisional status of *candidatus* can provide a place for some uncultured organisms within the prokaryote classification system.

As can be gathered from ecologists' affirmations, there is a need to use the concept of species as the basic unit for their ecosystem measurements (e.g, Horner-Devine et al., 2006; Pedrós-Alió, 2006; Whitaker, 2006). The two most important stepping stones for finding a solution to this problem are (i) the monistic view of the classification of living things, and (ii) that the species term is subject to homonymy. Both problems can be solved by accepting pluralism (Ereshefsky, 1998; Rosselló-Mora, 2005), which means accepting that different classification systems can be created and that each of them is legitimate if it accomplishes the purposes for which it was devised. Actually, in the case of molecular microbial ecology, a hidden and unofficial classification is being used where different lineages of uncultured organisms are being named, e.g., SAR lineages (Giovanonni et al., 1990) or OP lineages (Hugenholtz et al., 1998). In any case, the definitive solution to the problem of homonymy will be achieved by refusing the use of the term species to name the basic unit for describing the uncultured diversity and devising a term that is accepted by the scientific community that will make use of it. In this chapter, we discuss the different terms used by microbial ecologists to name the units that they observe, their suitability, and which of them may (from our point of view) have an optimal applicability.

TERMS AND MEANINGS

Species and Taxa

As we have discussed, the use of the term species to name the units that microbial ecologists measure may enter into conflict with that used within the framework of taxonomy and/or evolutionary microbiology. Nevertheless, it is important to note here that we understand that the use of the term species by microbial ecologists is always legitimate when they discuss their units in an abstract way. Indeed, from their theoretical point of view, or when discussing putative taxa, the use of species will never be mistaken if scientists are aware of the limitations of their knowledge. This means that they talk about units that are unable to be detected. However, when scientists discuss the units they retrieve as measurable data (i.e., clone sequences, DGGE bands, terminal restriction fragment length polymorphism (T-RFLP) patterns, fluorescence in situ hybridization and 4′,6′-diamidino-2-phenylindole (DAPI) counts, amplified intergenic spacer analysis bands, etc.), the use of the term species is incorrect and will lead to confusion simply because there is no guarantee that these units can be understood as species in the taxonomic way.

In our opinion, there is also confusion when using the term taxa. Actually, this term is often used as a synonym for species, as can be implicitly understood from the text (e.g., Curtis and Sloan, 2004; Coenye et al., 2005; Pedrós-Alió, 2006). In this regard, we would rather favor the use of the term species instead of taxa when talking abstractly. Species is a category, and each species is a taxon, but the same is true for each genus or family, or any category within the devised taxonomical hierarchy. Comparing the number of taxa between two ecosystems will always lead to confusion if one does not clarify which kinds of taxa are being compared. The term taxa is even more abstract than any of the hierarchical categories. For instance, if one is stating that a given ecosystem has more taxa than another one, the truth of the affirmation may depend on the taxa being considered, since an ecosystem richer in species could be poorer in terms of genera or families. Thus, we recommend avoiding the use of taxa unless it is previously clarified.

Alternative Terms

Microbial ecologists, in general, are aware that the units they are dealing with are far from being species. However, there is an implicit desire that, in the future, the concepts and terms among disciplines will be unified and universal units can be applied. While waiting for this to happen, and in the framework of the techniques currently in use, myriad numbers of terms are being used, depending on the information that can be retrieved.

PHYLOTYPES AND RIBOTYPES

In general, most of the techniques applied to microbial ecology for revealing prokaryotic community structures are based on retrieving and analyzing ribosomal RNA sequences, in particular that of the small subunit, for finally reconstructing a phylogenetic tree. Nearly complete sequences are often retrieved and cloned in libraries that are, in some cases, exhaustively sequenced. In general, there is the assumption that each single sequence represents a single organism, although scientists are aware that more than a single operon may be encoded within a single genome (Acinas et al., 2004). However, each different sequence can be considered to be a different phylogenetic type or phylotype, as used now for more than a decade (e.g., Polz and Cavanaugh, 1995). The use of the suffix -type is similar to that applied in prokaryote taxonomy. However, microbial taxonomists recommend avoiding the use of the epithet "-type" and substitute it by "-var" or "-form" to avoid confusion with the strict use meaning nomenclatural type (Sneath, 1992). However, here, we do not see the need to call each sequence a different phylovar or phyloform, since we do not see a conflict with prokaryotic taxonomy. Phylotype may be the most acceptable term to name a single unique sequence. Phylotype could also apply to those sets of sequences forming a tight clade that are considered to be below the variability threshold of a single population (Zaballos et al., 2005).

There are, however, other terms used to describe each unique phylogenetic branch. The term "ribotype" has been used in a variety of environmental studies in a similar way as phylotype (e.g., Acinas et al., 2004; Venter et al., 2004; Luna et al., 2006). Although the use of the prefix "ribo-" as relative to the ribosome may seem adequate, the term had been previously applied in clinical and epidemiological studies to discriminate strains or species by restriction fragment length polymorphism analysis of rRNA operons (Anisimov et al., 2004; Nightingale et al., 2005; MacCanell et al., 2006). Those techniques based on DNA extraction and 16S rDNA amplification/restriction complemented with Southern hybridizations had been used in microbial systematics for epidemiological typing or to classify species. As the technique is based on the use of restriction enzymes with few target sequences in the genome, there is no guarantee that identical ribotypes share identical gene sequences or vice versa. In addition, in eukaryotes the term ribotype is used to refer to the differences in the RNA pools generated through RNA processing events (alternative splicing and RNA editing) that can generate many different messages from a single gene (Herbert and Rich, 1999). Thus, here, we recommend avoiding the use of ribotype to name different gene sequences as it may enter into conflict with the true etymological origin of the term. To add to the confusion, in some cases, the wrong acceptance of the term ribotype can even be synonymized with "ribospecies" (e.g., Foissner, 2006; Zavarzin, 2006), despite both terms being essentially different from the etymological point of view. In summary, the rRNA sequence uniqueness may appear in the literature as being a phylotype, ribotype, or ribospecies. However, because of etymological reasons, the best recommendation is to use phylotype to name these unique sequences within a database.

Apart from clone libraries that normally generate high-quality sequence data, there are other techniques based on rRNA amplification and subsequent electrophoretic segregation of the individual fragments using a chemical denaturing gradient (Muyzer et al., 1993), a thermal denaturing gradient (Heuer et al., 1997), or restriction fragment sizes (T-RFLP) (Marsh, 1999) and internal transcribed spacer lengths (amplified intergenic spacer analysis) (Fisher and Triplett, 1999), among others. These methods are based on segregating PCR fragments, due to primary structure polymorphisms, through electrophoresis. All of them generate banding or peak patterns that migrate differently in the gel and are directly linked to sequence differences. Banding patterns can be compared, and in some cases, can even be used to generate a cumulative database (T-RFLP) (e.g., http://rdp8.cme.msu.edu/html/TAP-trflp.html), but they rarely lead to the fragment sequence. In any case, each different band or peak detected after electrophoresis, which corresponds to a different sequence, can be assumed to be a different phylogenetic type or phylotype. However, one has to assume that there is no guarantee that a single band or peak represents a single phylotype, which is especially true for those techniques that only amplify a small portion of the gene. It may happen that a single observed band contains up to three different sequences (e.g, Rosselló-Mora et al., 1999) or viceversa where an identical sequence may appear in different bands (Benlloch et al., 2002). As is discussed below, there is a growing group of scientists who prefer the use of OTU to name their units of comparison. As is argued later, both terms (phylotype and OTU) can be used equally, always considering that the first indicates phylogenetic uniqueness and the second an operational unit to perform numerical analyses.

ECOTYPE AND GEOTYPE

The prefix "eco-" seems very suitable for naming the relevant ecological units that ecologists observe in their systems. Actually, the term ecotype has been used to name the infraspecific subdivisions that show local adaptations (Begon et al., 2006). In addition, as often defined in dictionaries, it is "a population of a species that survives as a distinct group through environmental selection and isolation and that is comparable with a taxonomic subspecies" (e.g., Merriam-Webster's dictionary, www.m-w.com) or "a subspecific form within a true species, resulting from selection within a particular habitat and therefore adapted genetically to that habitat..." (Lawrence, 1996). In any case, the use of this term is becoming more popular among microbiologists to name those distinct populations inhabiting different niches for which an adaptation can be shown (Rocap et al., 2003; López-López et al., 2005; Hahn and Pöckl, 2005). However, again, the term is subjected to homonymy, at least within the framework of microbiology, and this could lead in the future to nomenclatural discussion similar to that of species. Molecular evolutionary researchers used this term to discriminate genetic clusters or "populations of cells in the same ecological niche, which would all be outcompeted by any adaptive mutant coming from the population" (Cohan, 2001). Although the definitions seem to be similar, they are implicitly different as the second refers to simultaneously occurring populations in the same niche, contrary to the former. In any case, the use of the term is legitimate when referring to subdivisions within a given species either because clear adaptive phenotypes can be determined (e.g., Rocap et al., 2003) or because differences exist at the genetic level (e.g., López-López et al., 2005). However, because genomes might harbor a significant portion that will never be expressed (Lerat and Ochman, 2005; Ochman and Davalos, 2006; Siew and Fischer, 2003), differences at the genetic level may not guarantee the identification of ecotypes in the macrobe ecologists' sense. In addition, it is, at least from the etymological point of view, a mistake to equate ecotypes with species. Furthermore, the term "geotype" has been used to differentiate between distinct geographic bacterial populations as a result of physical isolation (Papke and Ward, 2003). This term was originally created for discriminating geographically isolated genotypes of viruses (Lipskaya et al., 1995), and therefore a similar application to prokaryotes seems legitimate.

OPERATIONAL TAXONOMIC UNIT

The most generally applicable term is OTU, simply because it is the most abstract concept of a unit, which is even independent from the biological nature of the things to be counted. The OTU comes from the pheneticists' school that developed numerical taxonomy (Sokal and Sneath, 1973). Actually, any observed unit that is susceptible to numerical analysis could be named OTU. The important premise is that all OTUs represent identical kinds of things in the same way as discussed above regarding the comparison of different taxa among samples. DGGE bands or T-RFLP peaks can be named OTUs and analyzed independently. However, OTUs derived from DGGE bands cannot be equated to OTUs derived from T-RFLP peaks. This term is widely used among microbial ecologists (e.g., Massana et al., 2000; Zaballos et al., 2005; Daffonchio et al., 2006; Fuhrman, 2006; Sogin et al., 2006), and some of them favor its use because it helps in understanding different levels of diversity patterns (Horner-Devine et al., 2006). The only conflict that we may find is that OTU refers to a taxonomical unit, and ecologists analyze data for purposes other than classification. In addition, most, if not all, OTUs that are the subject of study by microbial ecologists are based on polymorphisms of a given homologous gene family. For this reason, and as we have discussed above, each band or peak can be equated to a different phylotype. As we understand, the current use of OTU in the framework of microbial ecology is legitimate, although if one desires to avoid nomenclatural confusions and unfruitful discussions, the term could be changed by substituting the word "taxonomic" for other perhaps more appropriate terms, such as phylogenetic (OPU, as suggested by Lucas Stal), ecological (OEU), diversity (ODU), or biodiversity (OBU).

CLADE, CLUSTER, OR LINEAGES

The numerical analysis of sequences produces trees that are calculated with different algorithms, such as maximum likelihood, maximum parsimony, or Bayesian inference, and they are generally based on models of evolution (Ludwig and Klenk, 2001). However, banding patterns are most commonly analyzed by unweighted pair group method with averages (UPGMA), which renders dendrograms that group OTUs in terms of overall similarity (Sokal and Sneath, 1973). The first approach in trying to reconstruct phylogeny is based on cladistic analysis, whether the second is based on phenetic or clustering analysis or not, as is well clarified by Sneath (1989). Then, being purist, those groups of units that can be seen within a given branch after a cladistic reconstruction may be called "clades." However, those groups that are obtained after, for instance, UPGMA analysis, may deserve the use of the term "cluster." In any case, it is easy to find the wrong use of the term in the literature, especially the use of cluster. On the other hand, lineage refers to those common descendants of a given ancestor. Thus, for pragmatic reasons, and because neighbor joining analyses cannot be clearly considered cladistic (and they are the most common analyses in the literature), the term lineage may be the best choice when referring to those branches enclosing monophyletic types.

GENOTYPE

The use of this term is not as common as the previously discussed terms. In the framework of microbial ecology, it has been used to name different rRNA operon sequences (e.g., Allewalt et al., 2006) or different profiles generated by the use of repetitive sequences as amplification primers (e.g., Wilson et al., 2005). The term genotype and its use can be subjected to a rather controversial philosophical discussion. In principle, the term can be synonymized with genetic information (Sneath, 1989), and, as stated earlier (Rosselló-Mora and Amann, 2001), we understand that the organism's genotype is the information hidden in the genome nucleotide sequence (genomic information) that codes for the synthesis of a protein with, for example, metabolic implications (phenotype). Actually, we further understand that genotype cannot be equated with gene or genome sequence, which could also even be considered as a phenotype. In any case, we agree with the use of the term "genomic" for large amounts of information in the genome (Sneath, 1989) or genome type for each distinct genome or metagenome fragment. In fact, it may be argued that behind each rRNA type there is a characteristic genotype (information). However, and due to most of the reasons expressed in previous sections, we still consider it more adequate to use phylotype for referring to distinct rRNA operon sequences.

FINAL CONSIDERATIONS

In summary, the techniques currently in use to reveal microbial community structures and dynamics are not resolutive enough to retrieve information that can be equated to the species, as circumscribed in prokaryote taxonomy. In this regard, ecologists are using myriad numbers of terms, some of which suffer from homonymy. Among all the terms discussed that may be appropriate for naming the minimally objective observable unit in ecology by hitherto established techniques, phylotype and OTU (or one of the variations obtained by substituting the word "taxonomic") are perhaps the most easily applicable. Actually, we favor the use of phylotype instead of ribotype, as the former can apply to any kind of sequence, whereas ribotype (independently of the homonymy problem) may apply only to ribosomic sequences. We recommend the use of the term species in any case when it is used in an abstract way, in other words, not referring to observed sequences or bands or peaks. We also recommend avoiding the use of the term taxa as a synonym of species, since it may lead to confusion.

ACKNOWLEDGMENTS

The authors acknowledge the research support obtained through the European Excellence Network Marbef (GOCE-CT-2003-505446) and the Spanish PN research project CLG2006-12714-C02-02. We acknowledge Pepa Antón, Antje Wichels, Peter Kämpfer, Rudolf Amann, and the entire Marbef "microbial community" for critical reading and improvements, as well as helpful discussions.

REFERENCES

Acinas, S. G., V. Klepac-Ceraj, D. E. Hunt, C. Pharino, I. Ceraj, D. L. Distel, and M. F. Polz. 2004. Fine-scale phylogenetic architecture of a complex bacterial community. *Nature* **430:**551–554.

Allewalt, J. P., M. M. Bateson, N. P. Revsbech, K. Slack, and D. M. Ward. 2006. Effect of temperature and light on growth of and photosynthesis by *Synechococcus* isolates typical of those predominating in the octopus spring microbial mat community of Yellowstone National Park. *Appl. Environ. Microbiol.* **72:**544–550.

Amann, R. I., W. Ludwig, and K.-H. Schleifer. 1995. Phylogenetic identification and in situ detection of individual cells without cultivation. *Microbiol. Rev.* **59:**143–169.

Anisimov, A. P., L. E. Lindler, and G. B. Pier. 2004. Intraspecific diversity of *Yersinia pestis*. *Clin. Microbiol. Rev.* **2:**434–464.

Begon, M., C. R. Townsed, and J. L. Harper. 2006. *Ecology, from Individuals to Ecosystems*, 4th ed. Blackwell Publishing Ltd., Oxford, United Kingdom.

Benlloch, S., A. López-López, E.O. Casamayor, L. øvreås, V. Goddard, F. L. Daae, G. Smerdon, R. Massana, I. Joint, F. Thingstad, C. Pedrós-Alió, and F. Rodríguez-Valera. 2002. Prokaryotic genetic diversity throughout the salinity gradient of a coastal solar saltern. *Environ. Microbiol.* **4:**349–360.

Ciccarelli, F. D., T. Doerks, C. von Mering, C. J. Creevey, B. Snel, and P. Bork. 2006. Toward automatic reconstruction of a highly resolved tree of life. *Science* **311:**1283–1287.

Coenye, T., D. Gevers, Y. Van de Peer, P. Bañadme, and J. Swings. 2005. Towards a prokaryotic genomic taxonomy. *FEMS Microbiol. Rev.* **29:**147––167.

Cohan, F. M. 2001. Bacterial species and speciation. *Syst. Biol.* **50:**513–524.

Cole, R., B. Chai, R. J. Farris, Q. Wang, A. S. Kulam-Syed-Mohideen, D. M. McGarrell, A. M. Bandela, E. Cardenas, G. M. Garrity, and J. M. Tiedje. 2007. The ribosomal database project (RDP-II): introducing *myRDP* space and quality controlled public data. *Nucleic Acids Res.* **35:** D169–D172. doi: 10.1093/nar/gkl889.

Curtis, T. P., and W. T. Sloan. 2004. Prokaryotic diversity and its limits: microbial community structure in nature and implications for microbial ecology. *Curr. Opin. Microbiol.* **7:**221–226.

Daffonchio, D., S. Borin, T. Brusa, L. Brusetti, P. W. van der Wielen, H. Bolhuis, M. M. Yakimov, G. D'Auria, L. Giuliano, D. Marty, C. Tamburini, T. J. McGenity, J. E. Hallsworth, A. M. Sass, K. N. Timmis, A. Tselepides, G. J. de Lange, A. Hubner, J. Thomson, S. P. Varnavas, F. Gasparoni, H. W. Gerber, E. Malinverno, C. Corselli, J. Garcin, B. McKew, P. N. Golyshin, N. Lampadariou, P. Polymenakou, D. Calore, S. Cenedese, F. Zanon, S. Hoog, and Biodeep Scientific Party. 2006. Stratified prokaryote network in the oxic-anoxic transition of a deep-sea halocline. *Science* **440:**203–207.

DeLong, E. F., C. M. Preston, T. Mincer, V. Rich, S. J. Hallam, N.-U. Frigaard, A. Martínez, M. B. Sullivan, R. Edwards, R. Rodríguez Brito, S. W. Chisholm, and D. M. Karl. 2006. Community genomics among stratified microbial assemblages in the ocean's interior. *Science* **311:**496–503.

Dykhuizen, D. E. 1998. Santa Rosalia revisited: why are there so many species of bacteria? *Antonie van Leeuwenhoek* **73:**25–33.

Ereshefsky, M. 1994. Some problems with the Linnaean hierarchy. *Philosophy of Science* **61:**186–205.

Ereshefsky, M. 1998. Species pluralism and anti-realism. *Philosophy of Science* **65:**103–120.

Evershed, R. P., Z. M. Crossman, I. D. Bull, H. Mottram, J. A. J. Dungait, P. J. Maxfield, and E. L. Brennand. 2006. ^{13}C-labelling of lipids to investigate microbial communities in the environment. *Curr. Opin. Biotechnol.* **17:**72–82.

Fisher, M. M., and E. W. Triplett. 1999. Automated approach for ribosomal intergenic spacer analysis of microbial diversity and its application to freshwater bacterial communities. *Appl. Environ. Microbiol.* **65:**4630–4636.

Foissner, W. 2006. Biogeography and dispersal of micro-organisms: a review emphasizing protists. *Acta Protozool.* **45:**111–136.

Friedrich, M. W. 2006. Stable-isotope probing of DNA: insights into the function of uncultivated microorganisms from isotopically labelled metagenomes. *Curr. Opin. Biotechnol.* **17:**59–66.

Fuhrman, J. A., I. Hewson, M. S. Schwalbach, J. A. Steele, M.V. Brown, and S. Naeem. 2006. Annually reoccurring bacterial communities are predictable from ocean conditions. *Proc. Natl. Acad. Sci. USA* **103:**13104–13109.

Gevers, D., P. Dawyndt, P. Vandamme, A. Willems, M. Vancanneyt, J. Swings, and P. DeVos. 2006. Stepping stones towards a new prokaryotic taxonomy. *Philos. Trans. R. Soc. London, Ser. B* **361:**1911–1916.

Giovannoni, S. J., T. B. Britschgi, C. L. Moyer, and K. G. Field. 1990. Genetic diversity in Sargasso Sea bacterioplankton. *Nature* **345:**60–63.

Hahn, M. W., and M. Pöckl. 2005. Ecotypes of planktonic actinobacteria with identical 16S rRNA genes adapted to thermal niches in temperate subtropical, and tropical freshwater habitats. *Appl. Environ. Microbiol.* **71:**766–773.

Hanage, W. P., C. Fraser, and B. G. Spratt. 2006. Sequences, sequence clusters and bacterial species. *Philos. Trans. R. Soc. London B* **361:** 1917–1927.

Herbert, A., and A. Rich. 1999. RNA processing in evolution. The logic of soft-wired genomes. *Ann. N. Y. Acad. Sci.* **870:**119–132.

Heuer, H., M. Krsek, P. Baker, K. Smalla, and E. M. Wellington. 1997. Analysis of actinomycete communities by specific amplification of genes encoding 16S rRNA and gel electrophoretic separation in denaturing gradients. *Appl. Environ. Microbiol.* **63:**3233–3241.

Hey, J. 2001. *Genes, Categories and Species.* Oxford University Press Inc., New York, NY.

Horner-Devine, M. C., K. M. Carney, and B. J. M. Bohannan. 2006. An ecological perspective on bacterial biodiversity. *Proc. R. Soc. London B.* **271:**113–122.

Hugenholtz, P., C. Pitulle, K. L. Hershberger, and N. R. Pace. 1998. Novel division level bacterial diversity in a Yellowstone hot spring. *J. Bacteriol.* **180:**366–376.

Konstantinidis, K. T., A. Ramette, and J. M. Tiedje. 2006. The bacterial definition in the genomic era. *Philos. Trans. R. Soc. London B* **361:**1929–1940.

Lawrence, E. 1996. *Henderson's Dictionary of Biological Terms*, 11th ed. Longman Singapore Publishers (Pte) Ltd., Singapore.

Legault, B. A., A. López-López, J. C. Alba-Casado, F. Doolittle, H. Bolhuis, F. Rodríguez-Valera, and T. Papke. 2006. Environmental genomics of "Haloquadratum walsbyi" in a saltern crystallizer indicates a large pool of accessory genes in an otherwise coherent species. *BCM Genomics* **7:**171.

Lerat, E., and H. Ochman. 2005. Recognizing the pseudogenes in bacterial genomes. *Nucleic Acids Res.* **33:**3125–3132.

Lipskaya, G. Y., E. A. Chervonskaya, G. I. Belova, S.V. Maslova, T. N. Kutateladze, S. G. Drozdov, M. Mulders, M. A. Pallansch, O. M. Kew, and V. I. Agol. 1995. Geographical genotypes (geotypes) of poliovirus case isolates from the former Soviet Union: relatedness to other known poliovirus genotypes. *J. Gen. Virol.* **76:**1687–1699.

López-López, A., S. G. Bartual, L. Stal, O. Onyshchenko, and F. Rodríguez-Valera. 2005. Genetic analysis of housekeeping genes reveals a deep-sea ecotype of *Alteromonas macleodii* in the Mediterranean Sea. *Environ. Microbiol.* **7:**649–659.

Ludwig, W., and H.-P. Klenk. 2001. Overview: a phylogenetic backbone and taxonomic framework for prokaryotic systematics, p. 49–65. *In* D. J. Brenner, N. R. Krieg, J. T Staley, and G.M. Garrity (ed.), *Bergey's Manual of Systematic Bacteriology*, 2nd ed. Springer SBM, New York, NY.

Ludwig, W., and K.-H. Schleifer. 2005. Molecular phylogeny of *Bacteria* based on comparative sequence analysis of conserved genes, p. 70–98. *In* J. Sapp (ed.), *Microbial Phylogeny and Evolution, Concepts and Controversies.* Oxford University Press, Oxford, United Kingdom.

Luna, G. M., A. Dell'Anno, and R. Danovaro. 2006. DNA extraction procedure: a critical issue for bacterial diversity assessment in marine sediments. *Environ. Microbiol.* **8:**308–320.

MacCannell, D. R., T. J. Louie, D. B. Gregson, M. Laverdiere, A. C. Labbe, F. Laing, and S. Henwick. 2006. Molecular analysis of *Clostridium difficile* PCR ribotype 027 isolates from eastern and western Canada. *J. Clin. Microbiol.* **44:**2147–2152.

Marsh, T. L. 1999. Terminal restriction fragment length polymorphism (T-RFLP): an emerging method for characterizing diversity and homologous populations of amplification products. *Curr. Opin. Microbiol.* **2:**323–327.

Massana, R, E. F. DeLong, and C. Pedrós-Alió. 2000. A few cosmopolitan phylotypes dominate planktonic archaeal assemblages in widely different oceanic provinces. *Appl. Environ. Microbiol.* **66:**1777–1787.

Mishler, B. D., and M. J. Donoghue. 1982. Species concepts: a case for pluralism. *Syst. Zool.* **31:**491–503.

Murray, R. G. E., and K.-H. Schleifer. 1994. Taxonomic note: a proposal for recording the properties of putative taxa of prokaryotes. *Int. J. Syst. Bacteriol.* **44:**174–176.

Murray, R. G. E., and E. Stackebrandt. 1995. Taxonomic note: implementation of the provisional status *Candidatus* for incompletely described prokaryotes. *Int. J. Syst. Bacteriol.* **45:**186–187.

Muyzer G., E. C. de Waal, and A. G. Uitterlinden. 1993. Profiling of complex microbial populations by denaturing gradient gel electrophoresis analysis of polymerase chain reaction-amplified genes coding for 16S rRNA. *Appl. Environ. Microbiol.* **59:**695–700.

Nightingale, K. K., K. Windham, and M. Wiedmann. 2005. Evolution and molecular phylogeny of *Listeria monocytogenes* isolated from human and animal listeriosis cases and foods. *J. Bacteriol.* **187:**5537–5551.

Ochman, H, and L. M. Davalos. 2006. The nature and dynamics of bacterial genomes. *Science* **311:**1730–1733.

Øvreås, L., F. L. Daae, V. Torsvik, and F. Rodríguez-Valera. 2003. Characterization of microbial diversity in hypersaline environments by melting profiles and reassociation kinetics in combination with terminal restriction fragment length polymorphism (T-RFLP). *Microb. Ecol.* **46:**291–301.

Papke, R. T., N. B. Ramsing, M. M. Bateson, and D. M. Ward. 2003 Geographical isolation in hot spring cyanobacteria. *Environ. Microbiol.* **5:**650–659.

Pedrós-Alió, C. 2006. Marine microbial diversity: can it be determined? *Trends Microbiol.* **14:**257–263.

Polz, M. F., and C. M. Cavanaugh. 1995. Dominance of one bacterial phylotype at a mid-Atlantic ridge hydrothermal vent site. *Proc. Natl. Acad. Sci. USA* **92:**7232–7236.

Rocap, G., F. W. Larimer, J. Lamerdin, S. Malfatti, P. Chain, N. A. Ahlgren, A. Arellano, M. Coleman, L. Hauser, W. R. Hess, Z. I. Johnson, M. Land, D. Lindell, A. F. Post, W. Regala, M. Shah, S. L. Shaw, C. Teglich, M. B. Sullivan, C. S. Ting, A. Tolonen, E. A. Webb, E. R. Zinser, and S. W. Chisholm. 2003. Genome divergence in two *Prochlorococcus* ecotypes reflects oceanic niche differentiation. *Nature* **424:**1042–1047.

Rosselló-Mora, R. 2003. The species problem: can we achieve a universal concept? *Syst. Appl. Microbiol.* **26:**323–326.

Rosselló-Mora, R. 2005. Updating prokaryotic taxonomy. *J. Bacteriol.* **187:**6255–6257.

Rosselló-Mora, R. 2006. DNA-DNA reassociation methods applied to microbial taxonomy and their critical evaluation, p. 23–50. *In* E. Stackebrandt (ed.), *Molecular Identification, Systematics, and Population Structure of Prokaryotes.* Springer, Berlin, Germany.

Rosselló-Mora, R., and R. Amann. 2001. The species concept for prokaryotes. *FEMS Microbiol. Rev.* **25:**39–67.

Rosselló-Móra, R., M. Lucio, A. Peña, J. Brito-Echeverria, A. López-López, M. Valens-Vadell, M. Frommberger, J. Antón, and P. Schmitt-Kopplin. 2008. Metabolic evidence for biogeographic isolation of the extremophilic bacterium *Salinibacter ruber. ISME J.* doi:10.1038/ismej.2007.93. (E-publish ahead of print.)

Rosselló-Mora, R., B. Thamdrup, H. Schäfer, R. Weller, and R. Amann. 1999. The response of the microbial community of marine sediments to organic carbon input under anaerobic conditions. *Syst. Appl. Microbiol.* **22:**237–248.

Siew, N., and D. Fischer. 2003. Analysis of singleton ORFans in fully sequenced microbial genomes. *Proteins* **53:**241–251.

Sneath, P. H. A. 1989. Analysis and interpretation of sequence data for bacterial systematics: the view of a numerical taxonomist. *Syst. Appl. Microbiol.* **12:**15–31.

Sneath, P. H. A. 1992. *International Code of Nomenclature of Bacteria* (1990 revision). American Society for Microbiology, Washington, DC.

Sogin, M. L., H. G. Morrison, J. A. Huber, D. M. Welch, S. M. Huse, P. R. Neal, J. M. Arrieta, and G. J. Herndl. 2006. Microbial diversity in the deep sea and the underexplored "rare biosphere." *Proc. Natl. Acad. Sci. USA* **103:**12115–12120.

Sokal, P. H., and R. R. Sneath. 1973. *Numerical Taxonomy.* W. H. Freeman & Company, San Francisco, CA.

Soria-Carrasco, V., M. Valens-Vadell, A. Peña, J. Antón, R. Amann, J. Castresana, and R. Rosselló-Mora. 2006. Phylogenetic position of *Salinibacter ruber* based on concatenated protein alignments. *Syst. Appl. Microbiol.* **30:**171–179.

Stackebrandt, E., W. Frederiksen, G Garrity, P. A. D. Grimont, P. Kämpfer, M. C. J. Maiden, X. Nesme, R. Rosselló-Mora, J. Swings, H. G. Trüper, L. Vauterin, A. C. Ward, and W. B. Whitman. 2002. Report of the ad hoc committee for the re-evaluation of the species definition in bacteriology. *Int. J. Syst. Evol. Microbiol.* **52:**1043–1047.

Stackebrandt, E., and B. M. Goebel. 1994. Taxonomic note: a place for DNA-DNA reassociation and 16S rRNA sequence analysis in the present species definition in bacteriology. *Int. J. Syst. Bacteriol.* **44:**846–849.

Staley, J. T. 2006. The bacterial species dilemma and the genomic-phylogenetic species concept. *Philos. Trans. R. Soc. London B* **361:**1899–1909.

Teeling, H., J. Waldmann, T. Lombardot, M. Bauer, and F. O. Glöckner. 2004. TETRA: a web-service and a stand-alone program for the analysis and comparison of tetranucleotide usage patterns in DNA sequences. *BMC Bioinfo.* **5:**163.

Torsvik, V., J. Goksoyr, and F. L. Daae. 1990. High diversity in DNA of soil bacteria. *Appl. Environ. Microbiol.* **56:**782–787.

Torsvik, V., L. Øvreås, and T. F. Thingstad. 2002. Prokaryotic diversity—magnitude, dynamics, and controlling factors. *Science* **296:**1064–1066.

Tyson, G. W., J. Chapman, P. Hugenholtz, E. E. Allen, R. J. Ram, P. M. Richardson, V. V. Solovyev, E. M. Rubin, D. S. Rokhsar, and J. F. Banfield. 2004. Community structure and metabolism through reconstruction of microbial genomes from the environment. *Nature* **428:**37–43.

Venter, J. C., K. Remington, J. F. Heidelberg, A. L. Halpern, D. Rusch, J. A. Eisen, D. Wu, I. Paulsen, K. E. Nelson, D. E. Fouts, S. Levy, A. H. Knap, M. W. Lomas, K. Nealson, O. White, J. Peterson, J. Hofmann, R. Parsons, H. Baden-Tillson, C. Pfannkoch, Y. H. Rogers, and H. O. Smith. 2004. Environmental genome shotgun sequencing of the Sargasso Sea. *Science.* **304:**66–74.

Wagner, M., P. H. Nielsen, A. Loy, J. P. Nielsen, and H. Daims. 2006. Linking microbial community structure with function: fluorescence in situ hybridization-microautoradiography and isotope arrays. *Curr. Opin. Biotechnol.* **17:**83–91.

Wang, Y., K., Hill, S. Singh, and L. Kari. 2005. The spectrum of genomic signatures: from dinucleotides to chaos game representation. *Gene* **346:**173–185.

Wayne, L G., D. J. Brenner, R. R. Colwell, P. A. D. Grimont, O. Kandler, M. I. Krichevsky, L. H. Moore, W. E. C. Moore, R. G. E. Murray, E. Stackebrandt, M. P. Starr, and H. G. Trüper. 1987. Report of the ad hoc committee on reconciliation of approaches to bacterial systematics. *Int. J. Syst. Bacteriol.* **37:**463–464.

Whitaker, R. J. 2006. Allopatric origins of microbial species. *Philos. Trans. R. Soc. London B* **361:**1975–1984.

Whiteley, A. S., M. Manefield, and T. Lueders. 2006. Unlocking the "microbial black box" using RNA-based stable isotope probing technologies. *Curr. Opin. Biotechnol.* **17:**67–71.

Whitman, W. B., D. C. Coleman, and W. J. Wiebe. 1998. Prokaryotes: the unseen majority. *Proc. Natl. Acad. Sci. USA* **9:**6578–6583.

Wilson, A. E., O. Sarnelle, B. A. Neilan, T. P. Salmon, M. M. Gehringer, and M. E. Hay. 2005. Genetic variation of the bloom-forming cyanobacterium *Microcystis aeruginosa* within and among lakes: implications for harmful algal blooms. *Appl. Environ. Microbiol.* **71:**6126–6133.

Zaballos, M., A. López-López, L. øvreås, S. Bartual, G. D'Auria, J. C. Alba, B. Legault, R. Pushker, F. L. Daae, and F. Rodríguez-Valera. 2005. Comparison of prokaryotic diversity at offshore oceanic locations reveals a different microbiota in the Mediterranean Sea. *FEMS Microbiol. Ecol.* **56:**359–370.

Zavarzin, G. A. 2006. Does evolution make the essence of biology? *H. Russ. Acad. Sci.* **76:**292–302.

MEASURING DIVERSITY

Jed A. Fuhrman

8

"Diversity" refers to the types of organisms present in an environment and the relative proportions of those types. Pedrós-Alió (2006) has pointed out that "the total biodiversity of an ecosystem is composed of two elements: first, a set of abundant taxa that carry out most ecosystem functions, grow actively and suffer intense losses through predation and viral lysis. These taxa are retrievable with molecular techniques but are difficult to grow in culture. Second, there is a seed bank of many rare taxa that are not growing or grow extremely slowly, do not experience viral lysis and predation is reduced. Such taxa are seldom retrieved by molecular techniques but many can be grown in culture." This chapter deals with the molecular, cultivation, and a few other techniques to evaluate microbial diversity in natural systems. While it does not include much discussion about which members of the community carry out most ecosystem functions, it is important to keep in mind that different approaches to measuring diversity can sometimes address components of the community with different contributions to the activity of the system.

Jed A. Fuhrman, Department of Biological Sciences and Wrigley Institute, University of Southern California, Los Angeles, CA 90089-0371.

The various ways to determine diversity could easily fill a book; therefore, this brief chapter basically provides an outline of the various approaches that are used to determine diversity of bacteria and archaea in natural habitats, and some guidelines regarding which methods may be most appropriate for specific environments and specific scientific questions. Also, it must be realized that this topic is undergoing massive revision as new molecular biological and other highly technical approaches are being brought to the fore. Therefore, this chapter describes some of these new methods in detail, even though they are under development, because they are likely to become particularly important in the future.

Diversity analysis might be performed for a number of purposes, and the most appropriate techniques to use will vary with the goals of the study. Certain techniques may allow study of subsets of the microbial community in great detail but at the same time entirely miss other community members. For example, one might be interested in the portion of the community that is performing a certain function, such as photosynthesis, and this might be studied by analyzing extracted photosynthetic pigments (Stolz, 1990) or possibly by flow cytometric

Accessing Uncultivated Microorganisms: from the Environment to Organisms and Genomes and Back
Edited by Karsten Zengler © 2008 ASM Press, Washington, DC

analysis of pigment contents of individual cells (Chisholm et al., 1988), or even genes for pigment synthesis (Beja et al., 2000; Schwalbach and Fuhrman, 2005). Several of the techniques described below rely on the ability to identify particular preselected components of the microbial community. However, in addition to learning about known types of microorganisms, a study of the general community structure can also potentially find previously unknown microbial types, and some of these may be the most interesting. As might be expected, when one finds novel organisms, it takes considerable time to characterize such organisms.

Methods are described in detail below, and most are listed with brief descriptive information in Table 1.

MICROSCOPY

The oldest method for obtaining microbial diversity information is to examine the sample with a microscope and characterize the microbes by their morphology. With some aquatic environments, one can in fact identify many types of organisms this way. An example is microbial mats; numerous types have readily recognizable morphologies, particularly by electron microscopy (Stolz, 1990). Many colonial microorganisms have distinctive colony morphologies that can be recognized by standard light microscopy or even with the naked eye. However, this approach is risky because different species, because of convergent evolution, may have similar morphotypes and may thus be misidentified. More to the point, in most aquatic environments, the vast majority of bacteria have nondistinct and/or variable morphologies, so this method is inappropriate for them (Sieburth, 1979). Related to identification by morphology is the identification of certain kinds of organisms by their natural fluorescence properties as examined by epifluorescence microscopy. For example, methanogens contain the F420 cofactor that fluoresces green when excited with light at 420-nm wavelength. More commonly, such identification may be based on photosynthetic pigments (often in conjunction with general size and morphology information) that can yield distinctive colors of fluorescence. For example, marine *Synechococcus* cells containing phycoerythrin have a distinctive golden color fluorescence when excited with blue light and a bright orange-red fluorescence when excited by green light. These pigment "signatures" can be more precisely determined by flow cytometry, which provides quantitative fluorescence (and other) information on many thousands of individual cells in a short time. This approach in fact was used to discover the presence of large quantities of tiny marine prochlorophytes in the ocean that had evaded detection by standard visual epifluorescence microscopy techniques (Chisholm et al., 1988).

CULTURING

The most common traditional method to analyze microbial diversity involves culturing the organisms from the habitat in question and identifying the cultures by standard techniques. If the organisms of interest are culturable, then this approach may be suitable. It goes without saying that the culturable organisms must be viable to be detected, and such viable counts are often thought to avoid the problem of counting inactive organisms that may be of less interest. However, there are several important caveats. First, even for culturable organisms, it has been suggested that many individuals may be viable but still not culturable (Roszak and Colwell, 1987). A second and bigger concern is that culture conditions that have typically been used to perform such counts usually recover on the order of 1% or less of the total number of organisms in aquatic habitats (Jannasch and Jones, 1959; Ferguson et al., 1984; Lee and Fuhrman, 1991), even though the majority of organisms in such habitats can be shown to be metabolically and synthetically active (Fuhrman and Azam, 1982). The reasons for this inability to culture the organisms are somewhat speculative and are generally thought to be the result of culture conditions that do not adequately mimic the natural growth conditions. Possibilities include wrong substrates, trace metal sensitivity, excessive storage product for-

TABLE 1 General summary of methods to assess diversity and community structure of natural microbial communities: features of the most common applications of the various approaches (Note that these are broad generalizations and do not apply in all instances [see text].)

Method	Coverage[a]	Sensitivity[b]	Phylogenetic resolution[c]	Cost	Notes
Microscopy	High	Moderate	Usually low, varies	Low	Epifluorescence has highest coverage
Culturing	Low	High	High	Low	New techniques increase coverage
Immunological	Low	Moderate-high	High	Moderate	Typically requires cultures first
Lipids	High	Moderate	Moderate	Moderate	Usually needs cultures to interpret
DNA-DNA Hybridization	High	Moderate	Moderate-high	Low-moderate	Community comparisons only, does not resolve taxa
16S cloning	High	Moderate-high	High-moderate	Moderate-low	Allows "ID" of previous-unknowns via placement on a phylogenetic tree
Fingerprinting, DGGE	High	Moderate	High	Low-moderate	Gel image-based interpretation
Fingerprinting, T-RFLP	High	Moderate	Moderate-high	Low-moderate	Some taxa missed if restriction site is very near primer; output is fragment list
Fingerprinting, ARISA	High	Moderate	Moderate-high	Low-moderate	Some taxa missed if they lack linked 16S-23S genes; output is fragment list
FISH	High	Moderate	Selectable, low-high	Low-moderate	ID and visualization
Metagenomics	High	Moderate-low	High	High	Needs large study to assess diversity via typical phylogenetic markers

[a]Coverage is the proportion of naturally occurring microorganisms detected by the method.

[b]Sensitivity is the ability to detect rare organisms.

[c]Phylogenetic resolution is the ability to distinguish relatives from each other (high resolution separates close relatives best; low resolution separates at the level of phyla or kingdoms).

mation (leading to cell damage), and viral infection, among others. Recent attempts to use dilution cultures (i.e., diluting a sample with filtered seawater such that only a few bacteria are in each tube) have greatly increased the percentage of bacteria that can be cultured, but for unknown reasons many of such cultures stop growth when abundances reach about 10^5 cells per milliliter (Button et al., 1993). At such abundances, the classical identification tests cannot be readily performed, but it is much better to have a pure culture than not. Linking identification to function is much easier with cultivated organisms, and recent attempts to first use molecular approaches to identify the dominant organisms in an environment and then culture them have been very valuable (Connon and Giovannoni, 2002; Rappé et al., 2002; Simu and Hagstrom, 2004). In any case, many aquatic microbiologists are concerned about accepting the results of culture-based approaches for determining diversity in a comprehensive manner, and one rarely sees such studies in the literature.

IMMUNOLOGICAL APPROACHES

Immunological approaches have been used primarily to characterize and count nitrifying bacteria (Ward, 1982) and cyanobacteria (Campbell et al., 1983). To prepare the antibodies for such a study, one must first culture the organisms and then use the culture to vaccinate an animal. The antibodies are purified from serum and then labeled with a fluorescent tag, such as fluorescein. The antibodies are mixed with a sample, unattached antibodies are rinsed away, and the cells are then observed by epifluorescence microscopy. Of course, this method presupposes that one must have cultures of the organisms in question, and so this approach would miss organisms that are resistant to cultivation. Also, one must know something about the cross-reactivity of the antibodies to other organisms. Often, antibodies can be specific for a particular species or strain. This approach has the potential to identify nonculturable members (viable or not) of a particular serotype, which is functionally defined as the type(s) to which the antibody binds. It has also been found to be useful for organisms that might be cultured, but grow slowly, such as nitrifying bacteria (Ward, 1982).

LIPID ANALYSIS

Different groups of microorganisms have different types of lipids, and this fact has been used extensively for microbial identification. For environmental work, lipid analysis has been used as an indicator of community structure and diversity, with certain classes of lipids (or particular lipids) being used as markers for certain groups. In general, the lipids are extracted in organic solvents and analyzed by gas chromatography. In aquatic systems, this work has been pioneered by White and colleagues (briefly reviewed in White, 1994). The approach provides information not only on cell type but also about nutritional status or stress. Because most such analytical methods require a substantial number of microbes of each type in each sample, this approach has been used primarily for sediments. However, with the appropriate instrumentation, it is now possible to extend such work to planktonic environments.

A particular application of lipid analysis in recent years has involved archaeal lipids. Archaea had traditionally been thought to be restricted to "extreme" environments, but it was then discovered by genetic techniques that archaea are common in marine plankton systems, especially the deep sea (Fuhrman et al., 1992). While genetics-based (fluorescent in situ hybridization [FISH]) techniques have been used to observe and count these archaea (Fuhrman and Ouverney, 1998; DeLong et al., 1999), an alternative approach has been lipid analysis. This permits not only detection of the archaeal lipids themselves as indicators (Hoefs et al., 1997; DeLong et al., 1998), but separation of the lipids permits isotopic analysis that can be used to examine sources of carbon, from natural isotopes (Hoefs et al., 1997; Pearson et al., 2001) or from added tracers (Wuchter et al., 2003; Blumenberg et al., 2005), permitting differentiation of processes between different groups of organisms.

LOW-MOLECULAR-WEIGHT RNA PROFILES

The molecular size distribution pattern of low-molecular-weight RNA (including tRNAs and 5S rRNA) is thought to be unique within narrow phylogenetic groups of microorganisms. This concept has been applied to natural aquatic planktonic communities by Höfle (Höfle, 1992; Höfle and Brettar, 1996): the whole community RNA profile is examined on an electrophoretic gel and compared with profiles of other communities and of standard cultures. As long as the bands on the gel are well resolved, one can see patterns of similarity and dissimilarity between natural communities. It is also possible to detect patterns that suggest the presence of specific organisms or groups, and bands may be excised and sequenced for partial identification.

NON-SEQUENCE-BASED DNA APPROACHES

DNA-DNA Hybridization

There are situations in which one wishes to know the community structure to learn whether two microbial communities contain the same or different organisms, yet the quantitative information on species compositions is not of particular interest. That is, one is asking simply whether two communities are the same or different. In such situations, it is possible to perform a DNA-DNA hybridization assay with total DNA extracted from the two communities (Lee and Fuhrman, 1990). In this approach, extracted prokaryotic DNA is labeled by nick translation, and samples are compared two at a time (reciprocal hybridizations). The results are scored by expressing the cross-hybrids as a percentage of the self-hybrids on the same filter. This percentage is expected to be the sum total of the shared common fractions between the two filters. For example, if sample 1 has 10% species A, 40% species B, and 50% species C, and sample 2 has 25% species A, 30% species C, and 45% species D, the shared common fraction is 10% + 30% = 40%. Tests with mixtures of pure culture DNAs have shown that the results are usually as expected, with 100% hybridization of identical or nearly identical samples ranging down to about 5 to 10% hybridization between samples sharing few or no species (the 5 to 10% represents well-conserved DNA sequences that cross-hybridize between distantly related organisms). This approach has been used to see broad differences between ocean basins, between depths at stratified locations, and over seasonal scales at the same location, and it has also shown similarities between some communities over time and space (Lee and Fuhrman, 1991).

Other DNA-Based Approaches

Another type of DNA-DNA hybridization is analysis of reassociation rates of single-sample DNA that is melted (strands separated). The rate is related to the complexity of the DNA and thus can be an index of the diversity of the sample. In other words, it indicates the number of species present but not what types they are. In practice, this method relies on changes in UV absorption between single- and double-stranded DNA, and one must make several assumptions in order to interpret results from complex natural communities. Torsvik et al. (1990) examined DNA extracted from soil microorganisms in this way, and the very slow reassociation of most of the DNA was interpreted mathematically as indicating the presence of about 4,000 completely different genomes in a 30-g sample from a deciduous forest. This was 200 times higher than the diversity from standard plate counts and graphically demonstrates the remarkably high potential overall diversity of microbial communities and its undersampling by conventional means. However, it should be pointed out that fingerprinting methods have indicated the large majority of some soil communities are made up of only perhaps a few dozen distinguishable taxa (Osborn et al., 2000) (see also terminal restriction fragment length polymorphism [T-RFLP] method below). This suggests that a very large part of the overall diversity in these samples is in rare organisms at the tail of the distribution.

A method to "fingerprint" a microbial community based on the quantitative distribution of genomes with different percent G+C contents was also described (Holben and Harris, 1995). In this approach, community DNA is centrifuged in a density gradient that separates the DNA based on percent G C content, yielding a profile. Although the percent G+C is not an unambiguous identifier for particular groups, the profile of DNA along the percent G+C gradient can strongly indicate the presence of certain groups (prompting probe analysis for verification) (see below). Furthermore, profiles from different samples can indicate differences between the microbial communities over time and space.

Another method that deserves mention here is PCR from repetitive sequences in DNA to yield electrophoretic banding patterns (genomic fingerprints) that can be distinct for particular groups or strains (Versalovic et al., 1994). Such methods may be applicable in the future for characterizing community structure, particularly for simple communities with little complexity.

16S rRNA-BASED APPROACHES

16S rRNA Sequences from Biomass

The difficulty of culturing bacteria and the ordinary requirement that a culture be available to identify a species have presented a major dilemma to microbial ecologists. About 20 years ago, Pace et al. (1986) and Olsen et al. (1986) presented the elegant idea that cultures are not necessary to identify the organisms present in a natural habitat. The idea came against the backdrop of the increasing use of molecular phylogeny to help define microbial systematics (Woese, 1987). It was becoming clear that the nucleotide sequences in molecules like 16S rRNA are very powerful tools in determining phylogenies and consequently in microbial systematics. Large databases of such sequences have been made available; e.g., the 16S rRNA database currently contains hundreds of thousands of organisms (http://rdp.cme.msu.edu/). The new idea was to use molecular biological techniques to obtain 16S rRNA sequences directly from the organisms freshly collected from the natural habitat without culturing them. These sequences could then be compared with those in the databases and with each other to learn how they fit into the microbial phylogenetic framework. Even if the sequence is unknown from previous work, it can be placed in relation to known organisms and other sequences from nature. A further major benefit is that the sequences can be used to make probes for quantitative composition analyses of microbial communities. This is proving to be a very powerful approach and is being augmented by inclusion of 23S rRNA data as well. A major aspect of the power of this approach is that the data are in the form of sequences that are universally understood and readily analyzed. Methods have focused on 16S rRNA genes because the molecule is universally distributed, has some extremely conserved segments (especially because it is not protein-encoding, so there are not "silent mutations") as well as variable regions, and yields an RNA product that has many copies per cell and can be used directly as a hybridization target. The 16S rRNA sequence analyses can be done in a few ways. If one is interested in learning what types of organisms are present in a sample, it is best to use the approaches that do not restrict the results to certain groups. The initial method proposed by Pace et al. (1986) called for extraction of DNA from all the microorganisms, fragmentation of that DNA with a restriction enzyme, and ligation to DNA from a bacteriophage (e.g., lambda) to make a library containing bits of DNA from all of the original organisms in the sample. Today this would be called a "metagenomic" library because the collective genomic content of the entire assemblage is all there, but that term was not in use in 1986 (and now metagenome research is becoming a "hot" topic—see later section). In the original analysis by Pace et al., that library is then screened for fragments coding for 16S rRNA by low-stringency hybridization to rRNA from a culture (or from cultures representing very broad groups like the domains *Bacteria*, *Archaea*, and *Eukarya*). The low stringency, combined with the moderately high level of conservation of this molecule, is expected to allow virtually any 16S rRNA to be detected. Screening with another rRNA, rather than with universally conserved oligonucleotide probes, is preferable because of the large number of false positives expected as a result of random matches between the oligonucleotide and the myriad genes from the natural community (in our laboratory, we found such false positives were very common). The positive clones are then sequenced, and the sequences are aligned to sequences from a database and analyzed for phylogenetic relationships by a computer. This library approach was used successfully by Schmidt et al. (1991) with marine plankton.

Ward et al. (1990) took a different approach to cloning 16S rRNA sequences of microorganisms from a well-studied hot spring at Yellowstone National Park. They extracted RNA from the natural sample and performed reverse transcription with universal primers to make cDNA that was then cloned and sequenced. The length of the cloned products was usually a few hundred bases, which is suitable for identification and some general phylogenetic analyses. The results yielded numerous clone types that were not the same as the organisms that had previously been cultured from that hot spring and had been thought to be the dominant types (by microscopical observation as well as culture work). This has been seen as good evidence that the culture-based approach finds only a subset of the natural diversity and that morphological identification of even distinctive organisms can be deceptive. More recent extension of the idea of working from rRNA directly has been comparison of clone libraries generated by amplification of DNA (by PCR—see below) to those from rRNA (by reverse transcription PCR) extracted directly from natural samples (Mills et al., 2005; Moeseneder et al., 2005). These take note of the point that rRNA tends to reflect growth of cells (Kemp et al., 1993), so the rRNA-based libraries are expected to be biased toward the more active organisms.

PCR-Based Cloning and Sequencing

One major alternative-related approach to direct cloning of 16S rRNA genes has been used with great success and has in fact dominated the field: the PCR, first applied by Giovannoni et al. (1990). With PCR methods, DNA is extracted from freshly collected organisms, PCR is performed, and with the primers of choice (see below), the PCR products are ligated into a phage or plasmid vector; these are cloned by standard techniques, and the clones are sequenced and analyzed phylogenetically. This approach is particularly suited to planktonic communities because there is usually very little DNA to work with. For example, with typical bacterial abundances of 10^6 bacteria per milliliter and typical bacterial DNA content of a few femtograms per cell, there is on the order of a few micrograms of bacterial DNA per liter. The geometric amplification inherent in PCR means that one can begin with 1 ng of total genomic DNA (representing roughly 1 million bacteria) and end up with micrograms of amplified 16S rRNA genes, sufficient for a variety of subsequent assays.

The existence of regions of the 16S (18S in eukaryotes) rRNA molecule that are essentially invariant among all known organisms means that universal (for all three domains) primers can be used in PCR; the longest distance between such primers is about 860 bases, which is adequate (but not ideal) for most phylogenetic analyses (Fuhrman et al., 1992; Fuhrman et al., 1993). The universal (or nearly universal) nature of the primers allows the broadest coverage, but it can also be a problem. For example, the high copy number of nucleus-encoded 18S rRNA genes in some eukaryotes may cause eukaryotic PCR products to swamp those of prokaryotes if they are together in the sample. One solution is to try to remove eukaryotes by filtration, as can be done in marine plankton; Fuhrman et al. (1992; 1993) used glass fiber filtration that removed essentially all of the eukaryotes but only about 10% of the prokaryotes. Alternatively, one can choose more specific primers to target specific groups. An example of this might be to use bacteria-specific primers to avoid amplifying eukaryotic genes when eukaryotes and prokaryotes may not be easily separated; however, one must remember that chloroplasts and mitochondria contain "bacterial" 16S rRNA sequences, so these primers do not completely avoid interference from eukaryotes (in practice, the plastids but not mitochondria tend to amplify with bacteria). On the positive side, there are nearly universal bacterial primers that allow cloning of almost the whole 16S rRNA gene (nearly 1,500 bases), maximizing the available phylogenetic information (Amann et al., 1995). However, the so-called universal primers should be used with care, in part because they are not exactly universal. Even with some ambiguous bases, such primers have mismatches with certain known groups, and one should check updated databases before embarking on this work. For example, the nearly universal bacterial primer set does not match many *Planctomyces* sequences well, and therefore one may suspect that other unknown groups may be missed. The commonly used primer at *Escherichia coli* position 1492, often treated as if it is universal, is not really so. Primer and probe matches to the current rRNA database can be examined at http://rdp.cme.msu.edu/ (Probe Match program) or with programs like ARB (Ludwig et al., 2004; http://www.arb-home.de). While the PCR annealing conditions can be adjusted to allow priming in the presence of some mismatches, there still may be some rRNA genes that do not amplify well, and this possibility must be considered in interpreting the results.

Among results from PCR-based cloning and sequencing of 16S rRNA genes from marine plankton, Giovannoni et al. (1990) used moderately specific PCR primers designed primarily for cyanobacteria, yet still found a novel proteobacterial group (SAR 11) in addition to a cyanobacterial group. Fuhrman et al. (Fuhrman et al., 1992, 1993; Fuhrman and Davis, 1997) used universal primers with microbial DNA collected from the deep sea and the euphotic zone and found novel groups of both archaea and bacteria. DeLong et al. (1993) examined "marine snow" with bacterial primers and found that the clones were distinctly different from those collected from surrounding free-living bacteria. As was found by Schmidt et al. (1991), who used a phage library, very few of these marine clones (with the exception of cyanobacteria) were closely related to previously known cultures. The SAR11 group was found at several marine locations and depths. Some of the clones were so distant as likely to be considered separate phyla from those previously studied. DeLong and coworkers (DeLong, 1992; DeLong et al., 1994) used archaeal primers and found two distinct archaeal groups in coastal temperate and polar waters; these were in the same groups as found by Fuhrman et al. (1992, 1993) at different locations and with universal primers. When specific primers are used, even a relatively minor component can be amplified, detected, and studied. As an example, ammonium-oxidizing bacterial 16S rRNA genes were specifically amplified from plankton by Voytek and Ward (1995).

The PCR methods have been used with samples from sediments and microbial mats (Barns et al., 1994; Devereux and Mundfrom, 1994; Reysenbach et al., 1994; Moyer et al., 1995), as well as deep-sea holothurian guts (McInerney et al., 1995), often showing unexpected bacterial and archaeal diversity. Although such material is far more concentrated than plankton (more organisms per unit volume), making it easier to obtain enough material for analysis, there are many substances that can interfere with molecular analyses, so the DNA may need extensive purification. Also, such organisms may be difficult to extract. A further consideration with at least some of such samples is that they should be frozen or extracted immediately upon sampling; this is because it has been found that storage of sediments (especially if the sediments were initially anaerobic but stored in aerobic conditions, or vice versa) can lead to significant and rapid shifts in species compositions (Rochelle et al., 1994). Such rapid potential changes seem most likely in rich material with rapid potential growth.

High-Throughput Sequence Analysis

New approaches are being developed rapidly to generate large quantities of sequences from environmental samples for diversity analysis. An example is pyrosequencing, a bead-based approach that can generate tens of thousands or more short sequences in a single run. Currently the length of each sequence is about 100 bases, but it is reportedly increasing to about 200 or more in the near future with various improvements. A recent study that applied this approach was that of Sogin et al. (2006), who amplified a hypervariable region of the 16S rRNA genes by PCR and used pyrosequencing to characterize thousands of unique "tag" sequences (fragments suitable for partial phylogenetic analysis) from deep-ocean samples. Thus the observed diversity of these samples was about 2,000 to 6,000 operational taxonomic units per sample at the 0.03 sequence difference level, calculated to extrapolate to about 5,000 to 20,000 different taxa mathematically (see "Extrapolations from Observed Diversity," below). These authors reported that the extent of diversity was real and not an artifact of errors in sequencing, which has been a concern with this new approach. This method and others now in the development pipeline have the potential to generate enormous amounts of data on diversity, including learning much about the long "tail" of the taxon-abundance curves, but care will be needed to interpret such results.

Fragment-Based Analyses

Some studies have used information about the 16S rRNA clones short of partial or full sequences. These include restriction fragment length polymorphism analysis (Britschgi and Giovannoni, 1991; DeLong et al., 1993; Moyer et al., 1994), which can be useful in grouping clones together. However, such analysis is based on only a few base positions in the sequence, and so it can lack resolution between closely related groups, particularly when universal primers are used for the initial amplification. On the other hand, it is possible to do the PCR with more specific primers, followed by restriction analysis, to indicate rather rapidly the presence of particular groups or types of interest and to compare different samples with respect to these groups. One could envision some fairly specific analyses by this approach with judicious selection of primers and restriction enzymes. Even with universal primers, the results can have resolution adequate for many types of studies.

Assemblage Fingerprinting

A related and powerful application has been the use of fingerprinting methods that give a snapshot of the entire microbial community at once, with the ability to tentatively identify different components. These methods include T-RFLP (Avaniss-Aghajani et al., 1994; Liu et al., 1997; Osborn et al., 2000), length heterogeneity PCR (LHPCR) (Suzuki et al., 1998), and amplified ribosomal intergenic spacer analysis (ARISA) (Fisher and Triplett, 1999; Brown et al., 2005). In T-RFLP, PCR is performed with one fluorescent and one nonfluorescent primer, and the products are cut with a restriction enzyme and analyzed for the size of the fluorescent end products. The presence or absence of restriction sites, as well as positions and lengths of insertions and deletions, leads to different length products. The result looks like a chromatogram, with peaks representing different taxa. In LHPCR, the length variations of the entire PCR product (due to insertions and deletions) permit separations. In ARISA, it is variations in the spacer length between the 16S and 23S rRNA genes. In these methods, it is possible for multiple taxa to have the same length of detected product, so "identification" of a particular peak on the basis of database information is not definitive (and some very rapidly growing organisms, rare in seawater, may have more than one length among multiple rRNA operons). Nevertheless, clone libraries from the environment in question can be used to find the most likely identification of peaks in the fingerprints (Gonzalez et al., 2000; Brown et al., 2005). In general, ARISA has the highest potential phylogenetic resolution because this spacer varies greatly and is far less conserved than the rRNA gene itself; T-RFLP has the potential for greater phylogenetic resolution than LHPCR, as the presence or absence of restriction sites, plus the ability to choose different enzymes, adds variety. An important feature of the fingerprinting methods described above (T-RFLP, LHPCR, ARISA) is that the results are in the form of data on the amount of different PCR products of particular fragment lengths. In other words, these methods result in data that are discrete numbers such as the proportion of total fluorescence in a given fragment length. Such discrete data can be put into a database in tabular form and compared statistically, permitting exact comparisons between different gel runs and even different laboratories. This allows statistically rigorous examination of sampling strategies and the data themselves, within and between studies. This is not usually true of data from the fingerprinting methods described in the next paragraph, where the identity of an organism relates to the position of a band on a gel that can vary between different runs and particularly between different laboratories.

A different type of fingerprint analysis is denaturing gradient gel electrophoresis (DGGE) and thermal gradient gel electrophoresis (TGGE), which are ways to separate similar-length nucleic acid molecules on the basis of small differences in the sequences. Another conceptually related approach is single-stranded conformational polymorphism (SSCP). Muyzer and colleagues (Muyzer et al., 1993; Muyzer and Smalla, 1998) have described the use of gradient gel electrophoresis analysis of PCR products to separate different components of the PCR mixtures from natural samples, and Schwieger and Tebbe (1998) described SCCP. Typically, extracted DNA from an environmental sample is amplified with bacterial primers, and the products are separated by one of these approaches. One can get an idea of the broad diversity (number of different bacterial types) of a sample by examining the number of different bands in such an analysis. It is also possible to use probes to characterize individual bands, or the bands may be excised, cloned, and sequenced (or sometimes directly sequenced) for a detailed phylogenetic analysis. This latter ability to cut out the bands for characterization is an advantage of DGGE and TGGE over the current versions of T-RFLP, LHPCR, or ARISA in that the identity of individual components can be verified. However, the T-RFLP, LHPCR, and ARISA methods yield specific size information on the products, suitable for database analysis and comparison to known sequences (for presumptive identification), whereas DGGE or TGGE banding patterns are much harder to standardize. Also, in a comparison with marine communities, T-RFLP was found to detect more taxa than DGGE (Moeseneder et al., 1999), although this may be specific to the particular conditions of the study and the gels or fragment analyzers used. The resolution of methods that rely on examining a gel directly (like DGGE) is typically not as fine as resolution of typical fragment analyzers used for T-RFLP or ARISA; in other words, one can typically distinguish up to about 50 to 100 different band locations on a standard-length DGGE gel, but there are several hundred distinct fragment lengths detectable by methods like ARISA. Nevertheless, when comparing fragment analysis results from different samples, gels, and laboratories, it is important to consider the precision of fragment size calling so that comparisons are statistically sound (Hewson and Fuhrman, 2006). Finally, it should be noted that fragment analysis-based fingerprinting has been reported to detect very minor components of the community, with taxa whose individuals represent about 0.1% of the community being detectable above background (Hewson and Fuhrman, 2004).

Choosing Cloning versus Fingerprints

Cloning and sequencing provide a great deal of detailed phylogenetic information, permitting placement of the organisms on a tree. Fingerprinting, even with matching clones to help identify individual taxa, provides a presumptive identification at best, and because there are far fewer characters that determine the identity, there is necessarily far less resolution and often some ambiguity. On the other hand, fingerprinting has the potential to show the entire bacterial assemblage in a single rapid assay, including minor components that would take hundreds or even thousands of clones to find. So it is a tradeoff between resolution, cost, time for analysis, and coverage. Coverage of clone libraries can be estimated with programs like DOTUR (http://www.plantpath.wisc.edu/fac/joh/DOTUR.html) or EstimateS (http://viceroy.eeb.uconn.edu/EstimateS) (see also "Extrapolations from Observed Diversity" below). Fingerprints are also more readily compared statistically, due to their relative simplicity. Therefore, in studies where composition of many samples is to be compared, fingerprinting may have several advantages, but if one wishes an exhaustive analysis of a particular sample or a few samples with high resolution, cloning and sequencing hundreds or thousands of clones are preferred.

POTENTIAL BIASES

There is still a question about possible biases and errors in the molecular methods: Have we replaced culturing biases with unknown biases? Potential biases could arise at a few stages. In the extraction stage, one can check microscopically to see that substantially all the cells have lysed, and this has been done with some of the extraction techniques with aquatic samples (Fuhrman et al., 1993). However, this is often not done, and this step deserves close attention. The PCR may introduce biases due to variations in primer binding or extension efficiency. There are also possible biases in the cloning step, as it is known that some sequences clone more readily than others (and some not at all). These possible problems indicate that caution is still in order when one is interpreting data on the relative amounts of different clones in libraries, and one cannot assume that clones are found in proportion to their natural abundance (but see below). There is also serious concern about possible chimeras or heteroduplexes being formed during PCR amplification (Acinas et al., 2004; Ashelford et al., 2005), and one needs to check clone libraries for the chimeras at least with programs like Chimera Check at rdp (http://rdp.cme.msu.edu/), bellerophon (http://foo.maths.uq.edu.au/~huber/bellerophon.pl), mglobalCHI (http://www-hto.usc.edu/software/mglobalCHI/index.html), Pintail, or Mallard (http://www.cf.ac.uk/biosi/research/biosoft/index.html) (Komatsoulis and Waterman, 1997; Huber et al., 2004; Ashelford et al., 2005), and also consider measures to reduce chances of heteroduplex formation (Acinas et al., 2004). However, even if there are biases, they are probably quite unrelated to culturing biases, and so these approaches are yielding much new information on what organisms are present. Can we yet say if cultures are representative of "typical" aquatic prokaryotes? Direct comparisons between cultivation and cloning results have been inconclusive. Suzuki et al. (1997) compared 127 cultivated organisms and 58 16S rRNA clones from the same Oregon coastal sample, and reported little overlap, also noting that even the culturable marine organisms are poorly represented in sequence databases. Pinhassi et al. (1997) took a different approach, using whole-genome hybridization (from 48 culturable organisms) toward community DNA from the northern Baltic Sea. They reported significant genomic overlap, suggesting the cultures are representative of a substantial part of the native community. One group readily found in 16S rRNA clone libraries and also cultures the so-called marine alpha-proteobacteria (Gonzalez and Moran, 1997). On the other hand, some readily cloned sequences, such as SAR86, have not yet been cultured at the time of this writing. Therefore, it appears that cultures may represent important components of native communities but may also miss important ones as well.

There is also the question of quantitative biases. As more molecular analysis is done, there is a temptation to compare results from libraries or fingerprints quantitatively. Can this be done at all, with the several presumed PCR biases operating in concert? There are a few studies that have examined quantitative fingerprint data, with encouraging results. With laboratory-created mixtures of organisms, Lueders and Friedrich (2003) found that T-RFLP peak quantity was proportional to original template amounts. In an extensive field study, Brown et al. (2005) found that quantitative ARISA estimates of the proportion of *Prochlorococcus* in monthly oceanic field samples (over 4 years) closely matched independent estimates of that organism by flow cytometry ($r^2 = 0.86$). While these two examples obviously cannot be applied universally to all organisms, it is interesting that these fingerprinting approaches seem to have the potential to yield accurate proportions of at least some taxa despite presumed biases. With the potential for biases, note that it is generally better to make comparisons in the relative proportion of a particular taxon between samples, rather than comparing the relative proportions of different taxa in an absolute sense.

QUANTITATIVE PROBES AND PCR

In the quantitative analysis of community structure, oligonucleotide probes are powerful tools that avoid possible biases in cloning and yield a more direct measure of the target groups of interest. One way this can be done is with oligonucleotide probes hybridized to bulk nucleic acid extracted from the aquatic habitat in question. Probe sequences are determined from sequence databases and can be universal or specific to certain domains, groups, or even some species (Amann et al., 1995). It is generally preferred to use RNA as the target instead of DNA because DNA is likely to yield far more false positives (unintentional hits) because mixed genomic DNA from innumerable species will have an immense variety of genes. RNA also has the benefit of being present in ribosomes and thus much more abundant as a target and also related to the cellular growth rate (although the exact relationship in natural communities is still uncertain). Quantitative bulk hybridizations to extracted RNA from aquatic habitats have been reported by Giovannoni et al. (1990) and DeLong et al. (1994), showing the relative abundances of the SAR11 cluster and archaea, respectively. It should be noted that to best standardize such probe binding, it is ideal to have a culture of the organism or group in question to determine the relative binding of different probes empirically.

A more recent quantitative measure of template quantity that has been applied to community structure studies is a process called real-time quantitative PCR (sometimes abbreviated QPCR or RT-QPCR), in which amplification from field samples is compared to side-by-side amplification of standards. There are a few different versions of this procedure, all of which are based on determining which cycle during the PCR a threshold amount of specific amplification occurs. Some versions rely on fluorescent internal probes (e.g., "TaqMan" assay), and some simply monitor the amount of total DNA amplified (Heid et al., 1996; Kim, 2001; Wilhelm and Pingoud, 2003). The dynamic range of the measurements often exceeds 4 orders of magnitude (sometimes reaching 10 orders of magnitude). An early application in seawater was by Suzuki and colleagues (2000, 2001), using 16S rRNA-based primers and probes. The approach can also be applied to a functional group, e.g., to examine the abundance in seawater of *pufM* genes that code for bacteriochlorophyll synthesis (Schwalbach and Fuhrman, 2005). It is important to realize that although this method avoids direct PCR bias in measurement of template amounts, interpretation of results quantitatively for the organisms themselves requires knowledge or assumptions about sampling and DNA extraction efficiency (very difficult to verify), gene copy numbers, and genome sizes. All these factors go into calculating organism abundance from template abundance. The approach used by Schwalbach and Fuhrman (2005) avoided the need to know DNA extraction efficiency. Instead, they relied on first physically separating bacteria from eukaryotes by size fractionation before DNA extraction, and then using an average bacterial genome size to estimate the number of bacteria represented per nanogram of DNA in each assay extract. Had eukaryotic DNA been mixed in with the extract, the authors could not have known how much bacterial DNA was used in each assay.

FISH

Although quantitative hybridization can be readily used with natural samples, one is not usually interested in knowing the fraction of RNA coming from particular groups but instead wants to know the proportions of individuals in those groups. For such work, it is ideal to tag each cell type with specific probes that allow visual identification. The preferred mode of observation has been epifluorescence microscopy with fluorescent oligonucleotide probes to rRNA (reviewed by Amann et al., 1995). Flow cytometry has also been used to automate the analysis with success (Wallner et al., 1995). This area has blossomed recently. However, it has been found to work most easily with relatively rich environments, probably because slowly growing cells in relatively oligotrophic environments have few ribosomes and thus fluoresce dimly. Nevertheless, some approaches have yielded useful data with difficult samples like marine plankton. Lee et al. (1993) found that multiple probes yield enough fluorescence for standard visual observation of about 75% of the bacteria in marine plankton. Fuhrman et al. (1994) found that about 75% of the 4′,6′-diamidino-2-phenylindole (DAPI)-countable cells from coastal marine plankton may be seen even with single fluorescent probes when video image intensification is used to boost the brightness of the images, and about 8% fluoresced with an archaea-specific probe. Ouverney and Fuhrman (1997) found that typically 90 to 95% of the total DAPI-countable cells can be tagged with a "universal" 16S rRNA probe after a marine plankton sample is treated for an hour with the antibiotic chloramphenicol. This ostensibly stops protein synthesis (and prevents changes in community composition) while allowing rRNA synthesis to continue, apparently permitting particularly good binding of probes to the RNA. Bright fluorescence can also be obtained with long probes (hundreds of bases or more) containing multiple fluorochromes, such as polyribonucleotides (Ludwig et al., 1994; Trebesius et al., 1994; DeLong et al., 1999); however, with relatively conserved targets such as16S rRNA, such long probes are best for tagging broad rather than narrow phylogenetic groups. Enzymatic amplification is another way to produce a bright signal with relatively few probe molecules, by using an enzyme-linked probe that reacts with a substrate to deposit brightly fluorescent product near the site of the probe, a procedure recently called CARD FISH and applicable to aquatic field samples (Schönhuber et al., 1999; Pernthaler et al., 2002). There is also the possibility of using in situ PCR to produce fluorescent products within particular target cells (Hodson et al., 1995; Chen et al., 2000), although the ability to use this as a general method in natural communities is still under development. Laser confocal microscopy (Amann et al., 1995) is another promising approach with excellent sensitivity that allows visualization of dim fluorescence and also permits examination of microbes in complex matrices. This is clearly an area that will progress rapidly.

PROBE/MICROAUTORADIOGRAPHY COMBINATION

A useful development is the ability to combine microautoradiographic characterization of the activity of single cells with 16S rRNA probes to identify those cells (Lee et al., 1999; Ouverney and Fuhrman, 1999; Cottrell and Kirchman, 2000). With such methods, one can start to measure the activities of individual cells within natural mixed communities while at the same time learning which taxa are responsible for each measured property. This is a natural extension of community structure analysis. In one of the most recent and sensitive applications, Teira et al. (2004) combined CARD FISH with microautoradiography to examine archaeal abundance and activity in the deep sea. This approach thus greatly reduces the need to separate and/or cultivate an organism to learn what it is doing in its natural habitat. Such an approach is a particularly powerful means to dissect the "black box" of natural microbial communities, going far beyond simply learning the taxonomic breakdown of community composition. There is little doubt that in the future, such single-cell probe approaches will be common tools in studies of microbial community structure.

COMMUNITY METAGENOMICS

In what may be considered the ultimate expression of diversity analysis, metagenomic studies aim to characterize the genes from the entire community, ideally showing how these genes are ordered within the genomes of the constituent organisms. The original phage library concept described by Pace et al. and Olsen et al. in 1986 (Olsen et al., 1986; Pace et al., 1986) is in fact a metagenomic study, but the more recent versions aim to sequence all the DNA (or long continuous stretches at least), rather than just looking for 16S rDNA or other specific genes. This whole-genome focus has been facilitated by high-throughput automated cloning and sequencing systems, and most recently, clone-free sequencing. Of course, there are practical limits on the extent of resolution of such studies (i.e., fraction of the community that is well characterized), depending on the diversity of the community, with simple communities easier to analyze for a given amount of sequencing effort. There are two basic ways that metagenomes are investigated: shotgun approaches and large insert libraries (DeLong, 2005). Both start with extracted community DNA. In the shotgun approach, the DNA is randomly sheared into relatively small fragments of a few thousand bases each, and the pieces are sequenced from the ends (typically around 600 to 800 bases sequenced at a time). With a sufficient number of pieces and sufficient amount of redundant sequencing, it should be possible to have enough overlap to permit computerized assembly of the fragments into continuous scaffolds, and even intact genomes. Complex diverse systems with many close relatives present particular challenges for this approach, as it becomes extremely difficult and expensive to assemble long stretches of sequence when most of the fragments are "singletons" (no overlap with others) and similarities in conserved regions among close relatives may make true overlaps hard to confirm (DeLong, 2005). In contrast, the large insert library approach does not have problems with assembling sequences within each insert, and the inserts may each represent 10% of a genome (e.g., 300-kb insert size for a 3-Mb genome), but it is still difficult to find overlapping inserts when the sample is highly diverse. For "traditional" diversity studies, where the idea is to make lists of how many of each "kind" of organism is present in a given sample, both these types of studies have the potential to generate significant amounts of data, but it should be clear that if this is the *only* objective, it makes more sense to focus on particularly phylogenetically informative genes such as 16S rDNA rather than having most of the sequencing effort on other genes (a given gene typically represents 0.1% of a genome). Metagenomic studies are typically limited by low genome coverage of diverse communities due to the sheer enormity of the task. On the other hand, if one wishes to link community structure and function and does not have representative cultures with which to work, the metagenomic approach is potentially very powerful (Beja et al., 2000).

There are relatively few metagenomic studies suitable for diversity analysis at the time of this writing (Venter et al., 2004; DeLong, 2005; Schleper et al., 2005; Tringe et al., 2005; DeLong et al., 2006; Rusch et al., 2007). In one of the earliest of such studies, Tyson et al. (2004) examined an acid-mine drainage community with very low diversity and were able to assemble the genomes of a few dominant organisms, and provide evidence about a few other community members. In contrast, Venter et al. (2004) examined a diverse marine plankton community from the Sargasso Sea (actually from multiple nearby samples) and reported there were about 1,800 different kinds of bacteria present, as indicated by the number of distinct phylogenetically informative genes. The most abundant kinds of organisms in most samples were too diverse to have their genomes assembled (even with >1 billion bases sequenced), yet the data still provided extensive information on diversity and community structure based on 16S rDNA and several other well-conserved genes (like *recA*) that provide phylogenetic information. The two genomes in one sample that were assembled by the Venter study were challenged by DeLong (2005) as likely contaminants because they appeared to be clonal on the basis of rare polymorphisms, and unlike anything seen in many previous or subsequent similar marine studies, while also strongly resembling terrestrial bacteria. It is brought up here as a reminder that contamination control in sample collection and processing is critical and not always easy. A multiple habitat (marine and terrestrial) shotgun library comparison by Tringe et al. (2005) found habitat-specific patterns of many genes. More recently, a large insert library study of a depth profile from 10 m to 4,000 m in the Pacific Ocean near Hawaii was reported by DeLong et al. (2006), and showed strong stratification in the distribution of genes, many of which matched sequences recently found via whole-genome sequence analysis of marine bacterial cultures. Unexpectedly, there were no reported 16S rRNA genes from the abundant SAR11 cluster in the five samples from 10-m to 500-m depths, despite the presence of many other genes related to those from the one known culture of that group whose genome has been sequenced (Giovannoni et al., 2005). This raises questions about unknown biases in this approach. Rusch et al. (2007) extended the Venter et al. (2004) study to include 6.3 billion base pairs of data from ocean surface samples along a several thousand kilometer transect from the North Atlantic off Canada through Panama into the South Pacific. They reported 85% of the assembled sequence and 57% of the unassembled data were unique at a 98% sequence identity cutoff, indicating a very high degree of diversity within the bacterial and archaeal communities. These authors developed tools to "recruit" environmental sequences to known microbial genomes, showing how various relatives of these previously sequenced organisms are distributed along the transect, based on all genes and not just phylogenetic markers like 16S rRNA. They also developed "extreme assembly" of similar but not identical genomic fragments that allowed construction of multichimeric assemblies to investigate genomic features of organisms in abundant clusters of close relatives.

Pyrosequencing

Pyrosequencing can also be used for metagenomic analysis, yielding large quantities of relatively short sequences, and operating without PCR or cloning biases (yet potentially with other biases, as all methods seem to have to some extent). Edwards et al. (2006) used this method to characterize microbial assemblages from two locations in a deep mine. The sequence information was sufficient to identify significant differences between the samples in terms of many physiological and biochemical functions or subsystems, such as carbon utilization, iron acquisition mechanisms, nitrogen assimilation, and respiratory pathways. At this time, some of the limitations of interpreting pyrosequencing data involve the difficulty in handling and comparing hundreds of thousands of sequences at once. Pyrosequencing and other related approaches are improving and yielding longer or more sequences, and these productive approaches will need to be matched with more powerful bioinformatics tools (and bigger computers) to be able to fully utilize the massive amounts of data generated.

Community Analysis from Protein-Encoding Genes

A recent analysis by von Mering et al. (2007) used 31 different protein-encoding genes from metagenomic datasets as a more robust way to analyze phylogenetic composition quantitatively, compared to a single marker like 16S rRNA. These authors compared shotgun metagenomic data from a soil sample, open ocean, acid-mine drainage, and ocean whalefall, and they reported that certain communities evolve faster than others. The method also enabled determination of preferred habitats for entire microbial clades and provided evidence that such habitat preferences can be remarkably stable over time.

With regard to biases, data quality, and overall coverage of the community, genome library methods that do not use PCR avoid the potential for PCR biases (although to date, the extent of such biases has not yet been shown to be a major obstacle). But it is important to remember that any method relying on DNA still may have sampling or extraction biases, and with metagenomic methods that include a cloning step, certain genes may be lethal to cells harboring the vectors (an example is viral genes, but there are many other possibilities) or otherwise be excluded.

EXTRAPOLATIONS FROM OBSERVED DIVERSITY

It is not possible to exhaustively sample all the microorganisms in natural environments, and in practice, only an extremely tiny fraction of the total diversity is ever observed directly (i.e., if the total diversity is considered to include every taxon that is present in a single individual). One might legitimately debate the extent to which this total diversity is meaningful biologically or philosophically. This was addressed in the quote from Pedrós-Alió (2006) in the opening paragraph of this chapter. So if one wants to know the diversity of bacteria and archaea that are relatively active and turning over, molecular techniques appear to be the most suitable, but they might miss the rare individuals that still contribute to the overall diversity of the system and might be important in adapting to new conditions.

So while identifying all the bacteria and archaea in a typical natural system is not currently practical for most environments (except perhaps for extreme low-diversity environments like acid-mine drainage), some extrapolations can be made based on what we can observe. If one wishes to estimate the total number of different species of bacteria (setting aside the thorny species concept problems for now), there are mathematical extrapolation approaches that attempt to estimate the overall species richness on the basis of a subsample or series of subsamples (Colwell and Coddington, 1994; Hughes et al., 2001; Bohannan and Hughes, 2003; Schloss and Handelsman, 2005; Chao et al., 2006).

Some calculations are based on the assumption that the taxon-abundance curve has a particular shape, such as a lognormal distribution. Thus, from the number of species (or whatever taxonomic unit) observed in given individual samples, Curtis et al. (2002) used such an assumption and calculated that a milliliter of seawater typically has about 160 different taxa and the whole ocean up to 2 million.

A few estimators that have developed some popularity, in part due to their simplicity, distill the calculation into a simple equation, such as the Chao1 estimator, which estimates the total number of taxa as

$$S_{\mathrm{Chao1}} = S_{\mathrm{obs}} + (n_1)^2/(2n_2)$$

where S_{obs} is the observed richness (total number of taxa), n_1 is the number of singletons (taxon appearing only once), and n_2 is the number of doubletons (taxon appearing exactly twice). Note that this has a maximum possible value of $(S^2_{\mathrm{obs}} - 1)/2$ when one species in the sample is a doubleton and all others are singletons. Therefore, S_{Chao1} will underestimate richness and strongly correlates with sample size until S_{obs} reaches at least the square root of twice the total richness (Colwell and Coddington, 1994). The coverage of samples is also often examined with "collectors' curves" that plot the total observed number of taxa versus the sampling effort; the incremental increase in new taxa is determined as more individuals are sampled, and such curves tend to flatten out as the total diversity is approached. There are several alternate approaches to estimating the number of taxa, with programs such as EstimateS (http://purl.oclc.org/estimates) or DOTUR (Schloss and Handelsman, 2005).

MICRODIVERSITY

A major question about community diversity relates to the extent of phylogenetic resolution we wish to have. Recent studies indicate that many or most natural aquatic microbial communities consist of thousands or more genetically distinguishable different types of bacteria and archaea, typically consisting in large part of tight clusters of many close relatives in several broad groups (Garcia-Martinez and Rodriguez-Valera, 2000; Acinas et al., 2004; Brown and Fuhrman., 2005). With so much sequencing power, we can distinguish very fine differences to the point of eventually identifying each single individual if desired. The question then becomes: what extent of diversity information is necessary for each question at hand? Do we want to know about bacteria or archaea with certain broad or specific functions, or those that might be considered a given "species," by whatever definition one uses, or population genetics of individuals? This depends on each study and goes beyond the scope of this chapter, but it ultimately helps determine which approaches one wishes to use and how the results are interpreted.

ACKNOWLEDGMENTS

I thank N. Pace, S. Giovannoni, and G. Olsen for helping me get involved in this work many years ago; S. Lee and R. Amann for helpful discussions; and R. Hicks and R. Christian for reviewing an earlier version of the manuscript.

This work was supported by NSF Microbial Observatory grant MCB0084231 and NSF grant OCE0527034.

REFERENCES

Acinas, S. G., V. Klepac-Ceraj, D. E. Hunt, C. Pharino, I. Ceraj, D. L. Distel, and M. F. Polz. 2004. Fine-scale phylogenetic architecture of a complex bacterial community. *Nature* **430:**551–554.

Amann, R. I., W. Ludwig, and K. H. Schleifer. 1995. Phylogenetic identification and in situ detection of individual microbial cells without cultivation. *Microbiol. Rev.* **59:**143–169.

Ashelford, K. E., N. A. Chuzhanova, J. C. Fry, A. J. Jones, and A. J. Weightman. 2005. At least 1 in 20 16S rRNA sequence records currently held in public repositories is estimated to contain substantial anomalies. *Appl. Environ. Microbiol.* **71:**7724–7736.

Avaniss-Aghajani, E., K. Jones, D. Chapman, and C. Brunk. 1994. A molecular technique for identification of bacteria using small subunit ribosomal RNA sequences. *Biotechniques* **17:**144–149.

Barns, S. M., R. E. Fundyga, M. W. Jeffries, and N. R. Pace. 1994. Remarkable archaeal diversity detected in a Yellowstone National Park hot spring environment. *Proc. Natl. Acad. Sci. USA* **91:**1609–1613.

Beja, O., L. Aravind, E. V. Koonin, M. T. Suzuki, A. Hadd, L. P. Nguyen, S. B. Jovanovich, C. M. Gates, R. A. Feldman, J. L. Spudich, E. N. Spudich, and E. F. DeLong. 2000. Bacterial rhodopsin: evidence for a new type of phototrophy in the sea. *Science* **289:**1902–1906.

Blumenberg, M., R. Seifert, K. Nauhaus, T. Pape, and W. Michaelis. 2005. In vitro study of lipid biosynthesis in an anaerobically methane-oxidizing microbial mat. *Appl. Environ. Microbiol.* **71:**4345–4351.

Bohannan, B. J., and J. Hughes. 2003. New approaches to analyzing microbial biodiversity data. *Curr. Opin. Microbiol.* **6:**282–287.

Britschgi, T., and S. J. Giovannoni. 1991. Phylogenetic analysis of a natural marine bacterioplankton population by rRNA gene cloning and sequencing. *Appl. Environ. Microbiol.* **57:**1707–1713.

Brown, M. V., and J. A. Fuhrman. 2005. Marine bacterial microdiversity as revealed by internal transcribed spacer analysis. *Aquat. Microb. Ecol.* **41:**15–23.

Brown, M. V., M. S. Schwalbach, I. Hewson, and J. A. Fuhrman. 2005. Coupling 16S-ITS rDNA clone libraries and ARISA to show marine microbial diversity:development and application to a time series. *Environ. Microbiol.* **7:**1466–1479.

Button, D. K., F. Schuts, P. Quang, R. Martin, and B. R. Robertson. 1993. Viability and isolation of marine bacteria by dilution culture:theory, procedures, and initial results. *Appl. Environ. Microbiol.* **59:**881–891.

Campbell, L., E. J. Carpenter, and V. J. Iacono. 1983. Identification and enumeration of marine chroococcoid cyanobacteria by immunofluorescence. *Appl. Environ. Microbiol.* **46:**553–559.

Chao, A., R. L. Chazdon, R. K. Colwell, and T. J. Shen. 2006. Abundance-based similarity indices and their estimation when there are unseen species in samples. *Biometrics* **62:**361–371.

Chen, F., B. Binder, and R. E. Hodson. 2000. Flow cytometric detection of specific gene expression in prokaryotic cells using in situ RT-PCR. *FEMS Microbiol. Lett.* **184:**291–296.

Chisholm, S. W., R. J. Olson, E. R. Zettler, J. Waterbury, R. Goericke, and N. Welschmeyer. 1988. A novel free-living prochlorophyte abundant in the oceanic euphotic zone. *Nature* **334:**340–343.

Colwell, R. K., and J. A. Coddington. 1994. Estimating terrestrial biodiversity through extrapolation. *Philos. Trans. R. Soc. London B* **345:**101–118.

Connon, S. A., and S. J. Giovannoni. 2002. High-throughput methods for culturing microorganisms in very-low-nutrient media yield diverse new marine isolates. *Appl. Environ. Microbiol.* **68:**3878–3885.

Cottrell, M. T., and D. L. Kirchman. 2000. Natural assemblages of marine Proteobacteria and members of the Cytophaga-Flavobacter cluster consuming low- and high-molecular-weight dissolved organic matter. *Appl. Environ. Microbiol.* **66:**1692–1697.

Curtis, T. P., W. T. Sloan, and J. W. Scannell. 2002. Estimating prokaryotic diversity and its limits. *Proc. Natl. Acad. Sci. USA* **99:**10494–10499.

DeLong, E. F. 1992. Archaea in coastal marine environments. *Proc. Natl. Acad. Sci. USA* **89:**5685–5689.

DeLong, E. F. 2005. Microbial community genomics in the ocean. *Nat. Rev. Microbiol.* **3:**459–469.

DeLong, E. F., D. G. Franks, and A. A. Alldredge. 1993. Phylogenetic diversity of aggregate-attached vs. free-living marine bacterial assemblages. *Limnol. Oceanogr.* **38:**924–934.

DeLong, E. F., L. L. King, R. Massana, H. Cittone, A. Murray, C. Schleper, and S. G. Wakeham. 1998. Dibiphytanyl ether lipids in nonthermophilic crenarchaeotes. *Appl. Environ. Microbiol.* **64:**1133–1138.

DeLong, E. F., C. M. Preston, T. Mincer, V. Rich, S. J. Hallam, N. U. Frigaard, A. Martinez, M. B. Sullivan, R. Edwards, B. R. Brito, S. W. Chisholm, and D. M. Karl. 2006. Community genomics among stratified microbial assemblages in the ocean's interior. *Science* **311:**496–503.

DeLong, E. F., L. T. Taylor, T. L. Marsh, and C. M. Preston. 1999. Visualization and enumeration of marine planktonic archaea and bacteria by using polyribonucleotide probes and fluorescent in situ hybridization. *Appl. Environ. Microbiol* **65:**5554–5563.

DeLong, E. F., K. Y. Wu, B. B. Prezelin, and R. V. M. Jovine. 1994. High abundance of archaea in antarctic marine picoplankton. *Nature* **371:**695–697.

Devereux, R., and G. W. Mundfrom. 1994. A phylogenetic tree of 16S rRNA sequences from sulfate-reducing bacteria in a sandy sediment. *Appl. Environ. Microbiol.* **60:**3437–3439.

Edwards, R. A., B. Rodriguez-Brito, L. Wegley, M. Haynes, M. Breitbart, D. M. Peterson, M. O. Saar, S. Alexander, E. C. Alexander, and F. Rohwer. 2006. Using pyrosequencing to shed light on deep mine microbial ecology. *BMC Genomics* **7:**57.

Ferguson, R. L., E. N. Buckley, and A. V. Palumbo. 1984. Response of marine bacterioplankton to differential filtration and confinement. *Appl. Environ. Microbiol.* **47:**49–55.

Fisher, M. M., and E. W. Triplett. 1999. Automated approach for ribosomal intergenic spacer analysis of microbial diversity and its application to freshwater bacterial communities. *Appl. Environ. Microbiol.* **65:**4630–4636.

Fuhrman, J. A., and F. Azam. 1982. Thymidine incorporation as a measure of heterotrophic bacterioplankton production in marine surface waters: evaluation, and field results. *Mar. Biol.* **66:**109–120.

Fuhrman, J. A., and A. A. Davis. 1997. Widespread archaea and novel bacteria from the deep sea as shown by 16S rRNA gene sequences. *Mar. Ecol. Prog. Ser.* **150:**275–285.

Fuhrman, J. A., S. H. Lee, Y. Masuchi, A. A. Davis, and R. M. Wilcox. 1994. Characterization of marine prokaryotic communities via DNA and RNA. *Microb. Ecol.* **28:**133–145.

Fuhrman, J. A., K. McCallum, and A. A. Davis. 1992. Novel major archaebacterial group from marine plankton. *Nature* **356:**148–149.

Fuhrman, J. A., K. McCallum, and A. A. Davis. 1993. Phylogenetic diversity of subsurface marine microbial communities from the Atlantic and Pacific Oceans. *Appl. Environ. Microbiol.* **59:**1294–1302.

Fuhrman, J. A., and C. C. Ouverney. 1998. Marine microbial diversity studied via 16S rRNA sequences: cloning results from coastal waters and counting of native archaea with fluorescent single cell probes. *Aquat. Ecol.* **32:**3–15.

Garcia-Martinez, J., and F. Rodriguez-Valera. 2000. Microdiversity of uncultured marine prokaryotes:the SAR11 cluster and the marine archaea of Group I. *Mol. Ecol.* **9:**935–948.

Giovannoni, S. J., T. B. Britschgi, C. L. Moyer, and K. G. Field. 1990. Genetic diversity in Sargasso Sea bacterioplankton. *Nature* **345:**60–63.

Giovannoni, S. J., H. J. Tripp, S. Givan, M. Podar, K. L. Vergin, D. Baptista, L. Bibbs, J. Eads, T. H. Richardson, M. Noordewier, M. S. Rappe, J. M. Short, J. C. Carrington, and E. J. Mathur. 2005. Genome streamlining in a cosmopolitan oceanic bacterium. *Science* **309:**1242–1245.

Gonzalez, J. M., and M. A. Moran. 1997. Numerical dominance of a group of marine bacteria in the alpha-subclass of the class Proteobacteria in coastal seawater. *Appl. Environ. Microbiol.* **63:**4237–4242.

Gonzalez, J. M., R. Simo, R. Massana, J. S. Covert, E. O. Casamayor, C. Pedrós-Alió, and M. A. Moran. 2000. Bacterial community structure associated with a dimethylsulfoniopropionate-producing North Atlantic algal bloom. *Appl. Environ. Microbiol.* **66:**4237–4246.

Heid, C. A., J. Stevens, K. J. Livak, and P. M. Williams. 1996. Real time quantitative PCR. *Genome Res.* **6:**986–994.

Hewson, I., and J. A. Fuhrman. 2004. Richness, and diversity of bacterioplankton species along an estuarine gradient in Moreton Bay, Australia. *Appl. Environ. Microbiol.* **70:**3425–3433.

Hewson, I., and J. A. Fuhrman. 2006. Improved strategy for comparing community fingerprints. *Microb. Ecol.* **51:**147–153.

Hodson, R. E., W. A. Dustman, R. P. Garg, and M. A. Moran. 1995. In situ PCR for visualization of microscale distribution of specific genes and gene products in prokaryotic communities. *Appl. Environ. Microbiol.* **61:**4074–4082.

Hoefs, M. J. L., S. Schouten, J. W. deLeeuw, L. L. King, S. G. Wakeham, and J. S. Sinninghe Damsté. 1997. Ether lipids of planktonic archaea in the marine water column. *Appl. Environ. Microbiol.* **63:**3090–3095.

Höfle, M. G. 1992. Bacterioplankton community structure and dynamics after large-scale release of nonindigenous bacteria as revealed by low-molecular-weight-RNA analysis. *Appl. Environ. Microbiol.* **58:**3387–3394.

Höfle, M. G., and I. Brettar. 1996. Genotyping of heterotrophic bacteria from the central Baltic Sea by use of low-molecular-weight RNA profiles. *Appl. Environ. Microbiol.* **62:**1383–1390.

Holben, W. E., and D. Harris. 1995. DNA-based monitoring of total bacterial community structure in environmental samples. *Mol. Ecol.* **4:**627–631.

Huber, T., G. Faulkner, and P. Hugenholtz. 2004. Bellerophon: a program to detect chimeric sequences in multiple sequence alignments. *Bioinformatics* **20:**2317–2319.

Hughes, J. B., J. J. Hellmann, T. H. Rocketts, and B. J. M. Bohannan. 2001. Counting the uncountable: statistical approaches to estimating microbial diversity. *Appl. Environ. Microbiol.* **67:**4399–4406.

Jannasch, H. W., and G. E. Jones. 1959. Bacterial populations in sea water as determined by different methods of enumeration. *Limnol. Oceanogr.* **4:**128–139.

Kemp, P. F., S. Lee, and J. Laroche. 1993. Estimating the growth-rate of slowly growing marine-bacteria from RNA-content. *Appl. Environ. Microbiol.* **59:**2594–2601.

Kim, D. W. 2001. Real time quantitative PCR. *Exp. Mol. Med.* **33:**101–109.

Komatsoulis, G. A., and M. S. Waterman. 1997. A new computational method for detection of chimeric 16S rRNA artifacts generated by PCR amplification from mixed bacterial populations. *Appl. Environ. Microbiol.* **63:**2338–2346.

Lee, N., P. H. Nielsen, K. H., Andreasen, S. Juretschko, J. L. Nielsen, K. H. Schleifer, and M. Wagner. 1999. Combination of fluorescent in situ hybridization and microautoradiography —a new tool for structure-function analyses in microbial ecology. *Appl. Environ. Microbiol.* **65:**1289–1297.

Lee, S., and J. A. Fuhrman. 1990. DNA hybridization to compare species compositions of natural bacterioplankton assemblages. *Appl. Environ. Microbiol.* **56:**739–746.

Lee, S., and J. A. Fuhrman. 1991. Spatial and temporal variation of natural bacterioplankton assemblages studied by total genomic DNA cross-hybridization. *Limnol. Oceanogr.* **36:**1277–1287.

Lee, S. H., and J. A. Fuhrman. 1991. Species composition shift of confined bacterioplankton studied at the level of community DNA. *Mar. Ecol. Prog. Ser.* **79:**195–201.

Lee, S. H., C. Malone, and P. F. Kemp. 1993. Use of multiple 16S ribosomal-RNA-targeted fluorescent-probes to increase signal strength and measure cellular RNA from natural planktonic bacteria. *Mar. Ecol. Prog. Ser.* **101:**193–201.

Liu, W. T., T. L. Marsh, H. Cheng, and L. J. Forney. 1997. Characterization of microbial diversity by determining terminal restriction fragment length polymorphisms of genes encoding 16S rRNA. *Appl. Environ. Microbiol.* **63:**4516–4522.

Ludwig, W., S. Dorn, N. Springer, G. Kirchhof, and K. H. Schleifer. 1994. PCR-based preparation of 23S rRNA-targeted group-specific polynucleotide probes. *Appl. Environ. Microbiol.* **60:**3234–3244.

Ludwig, W., O. Strunk, R. Westram, L. Richter, H. Meier, Yadhukumar, A. Buchner, T. Lai, S. Steppi, G. Jobb, W. Forster, I. Brettske, S. Gerber, A. W. Ginhart, O. Gross, S. Grumann, S. Hermann, R. Jost, A. Konig, T. Liss, R. Lussmann, M. May, B. Nonhoff, B. Reichel, R. Strehlow, A. Stamatakis, N. Stuckmann, A. Vilbig, M. Lenke, T. Ludwig, A. Bode, and K. H. Schleifer. 2004. ARB: a software environment for sequence data. *Nucleic Acids Res.* **32:**1363–1371.

Lueders, T., and M. W. Friedrich. 2003. Evaluation of PCR amplification bias by terminal restriction fragment length polymorphism analysis of small-subunit rRNA and mcrA genes by using defined template mixtures of methanogenic pure cultures, and soil DNA extracts. *Appl. Environ. Microbiol.* **69:**320–326.

McInerney, J. O., M. Wilkinson, J. W. Patching, T. M. Embley, and R. Powell. 1995. Recovery and phylogenetic analysis of novel archaeal rRNA sequences from a deep-sea deposit feeder. *Appl. Environ. Microbiol.* **61:**1646–1648.

Mills, H. J., R. J. Martinez, S. Story, and H. J. Mills. 2005. Characterization of microbial community structure in Gulf of Mexico gas hydrates: comparative analysis of DNA- and RNA-derived clone libraries. *Appl. Environ. Microbiol.* **71:**3235–3247.

Moeseneder, M. M., J. M. Arrieta, and G. J. Herndl. 2005. A comparison of DNA- and RNA-based clone libraries from the same marine bacterioplankton community. *FEMS Microbiol. Ecol.* **51:**341–352.

Moeseneder, M. M., J. M. Arrieta, G. Muyzer, C. Winter, and G. J. Herndl. 1999. Optimization of terminal-restriction fragment length polymorphism analysis for complex marine bacterioplankton communities and comparison with denaturing gradient gel electrophoresis. *Appl. Environ. Microbiol.* **65:**3518–3525.

Moyer, C. L., F. C. Dobbs, and D. M. Karl. 1994. Estimation of diversity and community structure through restriction fragment length polymorphism distribution analysis of bacterial 16S rRNA genes from a microbial mat at an active, hydrothermal vent system, Loihi Seamount, Hawaii. *Appl. Environ. Microbiol.* **60:**871–879.

Moyer, C. L., F. C. Dobbs, and D. M. Karl. 1995. Phylogenetic diversity of the bacterial community from a microbial mat at an active, hydrothermal vent system, Loihi Seamount, Hawaii. *Appl. Environ. Microbiol.* **61:**1555–1562.

Muyzer, G., and K. Smalla. 1998. Application of denaturing gradient gel electrophoresis DGGE, and temperature gradient gel electrophoresis TGGE in microbial ecology. *Antonie van Leeuwenhoek* **73:**127–141.

Muyzer, G., E. C. de Waal, and A. G. Uitterlinden. 1993. Profiling of complex microbial populations by denaturing gradient gel electrophoresis analysis of polymerase chain reaction-amplified genes coding for 16S rRNA. *Appl. Environ. Microbiol.* **59:**695–700.

Olsen, G. J., D. L. Lane, S. J. Giovannoni, and N. R. Pace. 1986. Microbial ecology and evolution: a ribosomal RNA approach. *Annu. Rev. Microbiol.* **40:**337–365.

Osborn, A. M., E. R. B. Moore, and K. N. Timmis. 2000. An evaluation of terminal-restriction fragment length polymorphism T-RFLP analysis for the study of microbial community structure, and dynamics. *Environ. Microbiol.* **2:**39–50.

Ouverney, C. C., and J. A. Fuhrman. 1997. Increase in fluorescence intensity of 16S rRNA in situ hybridization in natural samples treated with chloramphenicol. *Appl. Environ. Microbiol.* **63:**2735–2740.

Ouverney, C. C., and J. A. Fuhrman. 1999. Combined microautoradiography-16S rRNA probe technique for determination of radioisotope uptake by specific microbial cell types in situ. *Appl. Environ. Microbiol.* **65:**1746–1752.

Pace, N. R., D. A. Stahl, D. L. Lane, and G. J. Olsen. 1986. The analysis of natural microbial populations by rRNA sequences. *Adv. Microbiol. Ecol.* **9:**1–55.

Pearson, A., A. P. McNichol, B. C. Benitez-Nelson, J. M. Hayes, and T. I. Eglinton. 2001. Origins of lipid biomarkers in Santa Monica Basin surface sediment: a case study using compound-specific Delta C-14 analysis. *Geochim. Cosmochim. Acta* **65:**3123–3137.

Pedrós-Alió, C. 2006. Marine microbial diversity: can it be determined? *Trends Microbiol.* **14:**257–263.

Pernthaler, A., J. Pernthaler, and R. Amann. 2002. Fluorescence in situ hybridization and catalyzed reporter deposition for the identification of marine bacteria. *Appl. Environ. Microbiol.* **68:**3094–3101.

Pinhassi, J., U. L. Zweifel, and A. Hagstrom. 1997. Dominant marine bacterioplankton species found among colony-forming bacteria. *Appl. Environ. Microbiol.* **63:**3359–3366.

Rappé, M. S., S. A. Connon, K. L. Vergin, and S. J. Giovannoni. 2002. Cultivation of the ubiquitous SAR11 marine bacterioplankton clade. *Nature* **418:**630–633.

Reysenbach, A.-L., G. W. Wickham, and N. R. Pace. 1994. Phylogenetic analysis of the hyperthermophilic pink filament community in Octopus Spring, Yellowstone National Park. *Appl. Environ. Microbiol.* **60:**2113–2119.

Rochelle, P. A., B. A. Cragg, J. C. Fry, R. J. Parkes, and A. J. Weightman. 1994. Effect of sample handling on estimation of bacterial diversity in marine sediments by 16S rRNA gene sequence analysis. *FEMS Microbiol. Ecol.* **15:**215–226.

Roszak, D. B., and R. R. Colwell. 1987. Survival strategies of bacteria in the natural environment. *Microbiol. Rev.* **51:**365–379.

Rusch, D. B., A. L. Halpern, G. Sutton, K. B. Heidelberg, S. Williamson, S. Yooseph, D. Wu, J. A. Eisen, J. M. Hoffman, K. Remington, K. Beeson, B. Tran, H. Smith, H. Baden-Tillson, C. Stewart, J. Thorpe, J. Freeman, C., Andrews-Pfannkoch, J. E. Venter, K. Li, S. Kravitz, J. F. Heidelberg, T. Utterback, Y. H. Rogers, L. I. Falcon, V. Souza, G. Bonilla-Rosso, L. E. Eguiarte, D. M. Karl, S. Sathyendranath, T. Platt, E. Bermingham, V. Gallardo, G. Tamayo-Castillo, M. R. Ferrari, R. L. Strausberg, K. Nealson, R. Friedman, M. Frazier, and J. C. Venter. 2007. The Sorcerer II Global Ocean Sampling Expedition: Northwest Atlantic through Eastern Tropical Pacific. *PLoS Biol* **5:**e77.

Schleper, C., G. Jurgens, and M. Jonuscheit. 2005. Genomic studies of uncultivated archaea. *Nat. Rev. Microbiol.* **3:**479–488.

Schloss, P. D., and J. Handelsman. 2005. Introducing DOTUR, a computer program for defining operational taxonomic units and estimating species richness. *Appl. Environ. Microbiol.* **71:**1501–1506.

Schmidt, T. M., E. F. DeLong, and N. R. Pace. 1991. Analysis of a marine picoplankton community by 16S rRNA gene cloning and sequencing. *J. Bacteriol.* **173:**4371–4378.

Schönhuber, W., B. Zarda, S. Eix, R. Rippka, M. Herdman, W. Ludwig, and R. Amann. 1999. In situ identification of cyanobacteria with horseradish peroxidase-labeled, rRNA-targeted oligonucleotide probes. *Appl. Environ. Microbiol.* **65:**1259–1267.

Schwalbach, M. S., and J. A. Fuhrman. 2005. Wide-ranging abundances of aerobic anoxygenic phototrophic bacteria in the world ocean revealed by epifluorescence microscopy and quantitative PCR. *Limnol. Oceanogr.* **50:**620–628.

Schwieger, F., and C. C. Tebbe. 1998. A new approach to utilize PCR-single-strand-conformation polymorphism for 16S rRNA gene-based microbial community analysis. *Appl. Environ. Microbiol.* **64:**4870–4876.

Sieburth, J. M. 1979. *Sea Microbes*. Oxford University Press, New York, NY.

Simu, K., and A. Hagstrom. 2004. Oligotrophic bacterioplankton with a novel single-cell life strategy. *Appl. Environ. Microbiol.* **70:**2445–2451.

Sogin, M. L., H. G. Morrison, J. A. Huber, D. M. Welch, S. M. Huse, P. R. Neal, J. M. Arrieta, and G. J. Herndl. 2006. Microbial diversity in the deep sea and the underexplored rare biosphere. *Proc. Natl. Acad. Sci. USA* **103:**12115–12120.

Stolz, J. F. 1990. Distribution of phototrophic microbes in the flat laminated microbial mat at Laguna Figueroa, Baja California, Mexico. *BioSystems* **23:**345–357.

Suzuki, M., M. S. Rappé, and S. J. Giovannoni. 1998. Kinetic bias in estimates of coastal picoplankton community structure obtained by measurements of small-subunit rRNA gene PCR amplicon length heterogeneity. *Appl. Environ. Microbiol.* **64:**4522–4529.

Suzuki, M. T., C. M. Preston, F. P. Chavez, and E. F. DeLong. 2001. Quantitative mapping of bacterioplankton populations in seawater: field tests across an upwelling plume in Monterey Bay. *Aquat. Microb. Ecol.* **24:**117–127.

Suzuki, M. T., M. S. Rappé, Z. W. Haimberger, H. Winfield, N. Adair, J. Strobel, and S. J. Giovannoni. 1997. Bacterial diversity among small-subunit rRNA gene clones, and cellular isolates from the same seawater sample. *Appl. Environ. Microbiol.* **63:**983–989.

Suzuki, M. T., L. T. Taylor, and E. F. DeLong. 2000. Quantitative analysis of small-subunit rRNA genes in mixed microbial populations via 5′-nuclease assays. *Appl. Environ. Microbiol.* **66:**4605–4614.

Teira, E., T. Reinthaler, A. Pernthaler, J. Pernthaler, and G. J. Herndl. 2004. Combining catalyzed reporter deposition-fluorescence in situ hybridization and microautoradiography to detect substrate utilization by bacteria, and archaea in the deep ocean. *Appl. Environ. Microbiol.* **70:**4411–4414.

Torsvik, V., J. Goksøyr, and F. L. Daae. 1990. High diversity of DNA of soil bacteria. *Appl. Environ. Microbiol.* **56:**782–787.

Trebesius, K., R. Amann, W. Ludwig, K. Muhlegger, and K. H. Schleifer. 1994. Identification of whole fixed bacterial cells with nonradioactive 23S rRNA-targeted polynucleotide probes. *Appl. Environ. Microbiol.* **60:**3228–3235.

Tringe, S. G., C. von Mering, A. Kobayashi, A. A. Salamov, K. Chen, H. W. Chang, M. Podar, J. M. Short, E. J. Mathur, J. C. Detter, P. Bork, P. Hugenholtz, and E. M. Rubin. 2005. Comparative metagenomics of microbial communities. *Science* **308:**554–557.

Tyson, G. W., J. Chapman, P. Hugenholtz, E. E. Allen, R. J. Ram, P. M. Richardson, V. V. Solovyev, E. M. Rubin, D. S. Rokhsar, and J. F. Banfield. 2004. Community structure and metabolism through reconstruction of microbial genomes from the environment. *Nature* **428:**37–43.

Venter, J. C., K. Remington, J. F. Heidelberg, A. L. Halpern, D. Rusch, J. A. Eisen, D. Y. Wu, I. Paulsen, K. E. Nelson, W. Nelson, D. E. Fouts, S. Levy, A. H. Knap, M. W. Lomas, K. Nealson, O. White, J. Peterson, J. Hoffman, R. Parsons, H. Baden-Tillson, C. Pfannkoch, Y. H. Rogers, and H. O. Smith. 2004. Environmental genome shotgun sequencing of the Sargasso Sea. *Science* **304:**66–74.

Versalovic, J., M. Schnieder, F. J. de Bruijn, and J. R. Lupski. 1994. Genomic fingerprinting of bacteria using repetitive sequence-based polymerase chain reaction. *Methods Mol. Cell Biol.* **5:**25–40.

von Mering, C., P. Hugenholtz, J. Raes, S. G. Tringe, T. Doerks, L. J. Jensen, N. Ward, and P. Bork. 2007. Quantitative phylogenetic assessment of microbial communities in diverse environments. *Science* **315:**1126–1130.

Voytek, M. A., and B. B. Ward. 1995. Detection of ammonium-oxidizing bacteria of the beta subclass of the class Proteobacteria in aquatic samples with PCR. *Appl. Environ. Microbiol.* **61:**1444–1450.

Wallner, G., R. Erhart, and R. Amann. 1995. Flow cytometric analysis of activated sludge with rRNA-targeted probes. *Appl. Environ. Microbiol.* **61:**1859–1866.

Ward, B. B. 1982. Oceanic distribution of ammonium-oxidizing bacteria determined by immunofluorescent assay. *J. Mar. Res.* **40:**1155–1172.

Ward, D. M., R. Weller, and M. M. Bateson. 1990. 16S rRNA sequences reveal numerous uncultured microorganisms in a natural community. *Nature* **345:**63–65.

White, D. C. 1994. Is there anything else you need to understand about the microbiota that cannot be derived from analysis of nucleic acids? *Microb. Ecol.* **28:**163–166.

Wilhelm, J., and A. Pingoud. 2003. Real-time polymerase chain reaction. *Chembiochem.* **4:**1120–1128.

Woese, C. R. 1987. Bacterial evolution. *Microbiol. Rev.* **51:**221–271.

Wuchter, C., S. Schouten, H. T. S. Boschker, and J. S. Sinninghe Damsté. 2003. Bicarbonate uptake by marine crenarchaeota. *FEMS Microbiol. Lett.* **219:**203–207.

METAGENOMICS AS A TOOL TO STUDY BIODIVERSITY

Karen E. Nelson

9

Metagenomics has been defined as the genomic analysis of microorganisms by direct extraction and cloning of DNA from an assemblage of microorganisms (Handelsman, 2004). Although by virtue of new sequencing technologies such as 454 pyrosequencing, cloning is no longer essential for metagenomic projects (http://www.454.com/). The field of metagenomics has been the natural evolution from sequencing the genomes of single species. The genomics explosion has occurred at an accelerated rate since the mid-1990s with the release of the complete genome of *Haemophilus influenzae* (Fleischmann et al., 1995). The intrigue associated with metagenomics is that it allows us to study the genetic material of many individual species simultaneously, without having to first do a culturing step. This is particularly relevant given that estimates are that we can only culture 1% of the microbial species in nature (Amann et al., 1995).

HISTORY OF GENOMICS

In the mid-1990s the real launch of the genomic era began with the availability of the complete genome sequence of *H. influenzae* (Fleischmann et al., 1995). Since that major success, there have been numerous examples of organisms that represent all domains of life, including the completion of hundreds of microbial genomes (with an additional estimated hundreds of partial genomes) (Mongodin et al., 2005a). The traditional microbial and eukaryotic genome sequencing projects have been based on the Sanger sequencing method (Fleischmann et al., 1995; see below), which has been responsible for the successful completion of hundreds of genomes including that of humans. For the majority of these projects, sequence reads have been generated from gene libraries that are created in different types of cloning vectors (Frangeul et al., 1999).

The shotgun sequencing approach for generating completely random sequences has been the traditional strategy used in genome sequencing projects. The sequencing of ordered-clone libraries for overlapping clones has also been used although to a limited extent (Frangeul et al., 1999). For the shotgun sequencing approach, more than one library size is normally used so that complete coverage of the organisms' genome is obtained, with the smaller-sized insert libraries giving more depth of coverage and the larger-sized insert library

Karen E. Nelson, The J. Craig Venter Institute, 9704 Medical Center Drive, Rockville, MD 20850.

Accessing Uncultivated Microorganisms: from the Environment to Organisms and Genomes and Back
Edited by Karsten Zengler © 2008 ASM Press, Washington, DC

providing scaffolding to the sequence (Frangeul et al., 1999). Many of the earlier microbial sequencing projects such as that for *Escherichia coli* (Blattner et al., 1997), as well as the larger eukaryotic projects, have taken advantage of previously generated genetic and physical maps as well as ordered libraries that were constructed using cosmids, lambda clones, yeast artificial chromosome or bacterial artificial chromosome (BAC) clones (Frangeul et al., 1999) for the sequencing projects.

What Did Some of the First Genomes Give Us Insight Into?

The genomes that were initially chosen for whole-genome sequencing primarily represented the major pathogens as well as model organisms such as *E. coli* and *Bacillus subtilis*. A few unusual types, such as the archaea *Methanococcus jannaschii* and *Archaeoglobus fulgidus*, were also sequenced in the early days. This focus later changed as sequencing costs went down and technologies improved to include a wider range of favorite microbial species and strains, as well as organisms of agricultural and environmental significance (reviewed by Nelson et al., 2000). From the bioinformatics analyses of these genomes, it soon became obvious that we had a limited appreciation of the extent of diversity in the microbial world. This was particularly evident in the range of lateral gene transfer and metabolic diversity that was encountered as these genomes were investigated more closely (Nelson et al., 1999; Nierman et al., 2001).

One of the major discoveries is the extent of gene transfer, genome rearrangements, and gene shuffling that can occur in many microbial species (DeBoy et al., 2006; Mongodin et al., 2005b). Clustered regularly interspaced short palindromic repeats (CRISPRs) (Jansen et al., 2002) that consist of a 30-base pair repeat element interspersed with a unique sequence of approximately the same length have been identified in numerous microbial species, and were initially thought to be involved in the mobilization of DNA (Bolotin et al., 2005; Haft et al., 2005; Jansen et al., 2002). More recently, the intervening spacer sequences in CRISPRs have been shown to have a possible origin from preexisting chromosomal sequences and sometimes from transmissible elements such as bacteriophage and conjugative plasmids (Bolotin et al., 2005). It was also found that viruses do not reside in cells that carry virus-specific CRISPR spacer sequences but could be found within closely related strains that did not carry these sequences. Thus, a role for CRISPRs in immunity to foreign DNA was proposed (Bolotin et al., 2005; Mojica et al., 2005; Pourcel et al., 2005). In their groundbreaking publication, Barrangou and colleagues (Barrangou et al., 2007) have since clearly demonstrated that CRISPRs provide acquired resistance against viruses in prokaryotes. They found that bacteria integrated new spacers that were originally from phage genomic sequences following viral challenge and that the removal or addition of spacers modified the phenotype of the cell with respect to phage-resistance (Barrangou et al., 2007).

As a result of these initial efforts, we have come to realize that in many cases the genome of a single species cannot give insight into the extent of diversity within a species. Many projects have been initiated to sequence multiple strains of the same species (for example, the efforts by NIAID, http://www.niaid.nih.gov/dmid/genomes/mscs/default.htm), always with interesting findings (Fouts et al., 2005; Nelson et al., 2004), and today, many genomic efforts are focused on sequencing multiple isolates and strains and providing new insights into species diversity and the dynamic nature of the prokaryotic genome.

History of Studying Microbial Communities

The use of the small-subunit (SSU) ribosomal RNA gene as a phylogenetic marker to study bacterial and archaeal diversity, as well as the composition of various environments and natural communities, has clearly spawned the entire field of microbial ecology. The work that was started by Norman Pace and colleagues involved the sequencing and comparative analysis of 5S and 16S rRNA genes to estimate the diversity of microorganisms in a sample, thereby avoiding the culturing step.

Although the studies of SSU rRNA have undoubtedly resulted in tremendous quantities of information about microbial communities, it is clearly limited to the amount of information that can be inferred, even in concert with other molecular approaches such as restriction fragment length polymorphism and fluorescent in situ hybridization (Hugenholtz et al., 1998; Spiegelman et al., 2005). PCR of 16S rDNA is inherently biased because of the "universal primers," which do not amplify all 16S sequences effectively, particularly in mixed population samples. Nondominant community members (in terms of total numbers of cells) in environmental samples can readily go undetected (Gill et al., 2006). These techniques in particular have revealed limited to no information on the genetic diversity of the microbial inhabitants in these environments. For example, the physiological role that is being played by individual species that have been identified by 16S rDNA sequencing cannot be defined. Some level of analogy may be drawn by extrapolation to available genetic information from related species, but it is apparent from whole-genome sequencing and microarray analysis that species that appear to be closely related from 16S rDNA sequences may have tremendous differences in genome composition (DeBoy et al., 2006; Mongodin et al., 2005b).

16S rRNA gene surveys have continued to expand, and as of summer 2007 the current version of the Ribosomal Database Project (http://rdp.cme.msu.edu/) holds an estimated 368,406 aligned and annotated 16S rRNA gene sequences, clearly demonstrating the extent of microbial diversity in the environment and hinting at what remains to be discovered.

New techniques are constantly being developed to address diversity questions with respect to microbial communities, by focusing on the regions of the ribosomal operons that have the highest information content (for example, Kysela et al., 2005; Neufeld et al., 2004; Neufeld and Mohn, 2005; Sogin et al., 2006). Sequencing of these high-information-content regions can allow deeper surveys of particular environments. Serial analysis of ribosomal DNA (SARD) is a new technique that has been developed by Ashby and colleagues at Taxon Biosciences in California (http://www.taxon.com/) for the detection and quantitation of ribosomal sequences in microbial communities. SARD also allows the measurement of "rare" sequences or minority members in environmental samples (Ashby et al., 2007). In a recent publication of this group (Ashby et al., 2007), short tags from the V5 region were ligated together. From four agricultural soils, 37,008 SARD tags comprising 3,127 unique sequences were identified (Ashby et al., 2007), and the method was found to be highly reproducible. Most of the tags that were observed belonged to abundance classes present at less than 1% of the community (Ashby et al., 2007), meaning that over 99% of the unique tags made up less than 1% of the community. SARD was proposed as a good means "to explore the ecological role of these rare members of microbial communities in qualitative and quantitative terms" (Ashby et al., 2007).

Sogin's group described the abundance of microbial diversity in the oceans using the 454 technology (http://www.454.com/) to sequence 118,000 PCR amplicons that span the V6 region. Their results again showed that "bacterial diversity estimates for the diffuse flow vents of Axial Seamount and the deep water masses of the North Atlantic are much greater than any published description of marine microbial diversity" (Sogin et al., 2006).

A significant amount of work has gone into studying the unculturables in the environment in attempts to generate additional information on their biology beyond just 16S rRNA inference. As mentioned above, species can have identical 16S rDNA sequences but have very significant differences in their biology. For example, from the genomes of the first two *E. coli* strains to be sequenced (strains O157:H7 by comparison with the genome of the nonpathogenic laboratory strain *E. coli* K-12) lateral gene transfer was found to be extensive, with 1,387 new genes including virulence factors and prophages encoded in strain-specific clusters in strain O157:H7 (Blattner et al., 1997; Perna et al., 2001).

Connon and Giovannoni (2002) have employed microtiter plates and cell arrays for cell enumeration from small aliquots of cultures with very low densities, and through the use of extinction cultures they were able to successfully culture up to 14% of the cells collected from coastal seawater (Connon and Giovannoni, 2002), many of which had not been previously cultured. Zengler and colleagues described a method for high-throughput cultivation that is based on single-cell encapsulation and flow cytometry and that allows cells to grow in conditions that simulate their own environments (Zengler et al., 2002).

The Process of Doing Metagenomic Studies

Metagenomic studies have evolved from approaches to sequence and close the genomes of individual species and most studies to date have relied on (i) isolating DNA from an environmental sample,(ii) cloning or not cloning depending on the sequencing technology to be used in the study, (iii) transforming the clones, and(iv) screening and sequencing the resulting transformants (Fig. 1) (Handelsman, 2004).

Construction of representative libraries for sequencing was one of the main challenges that

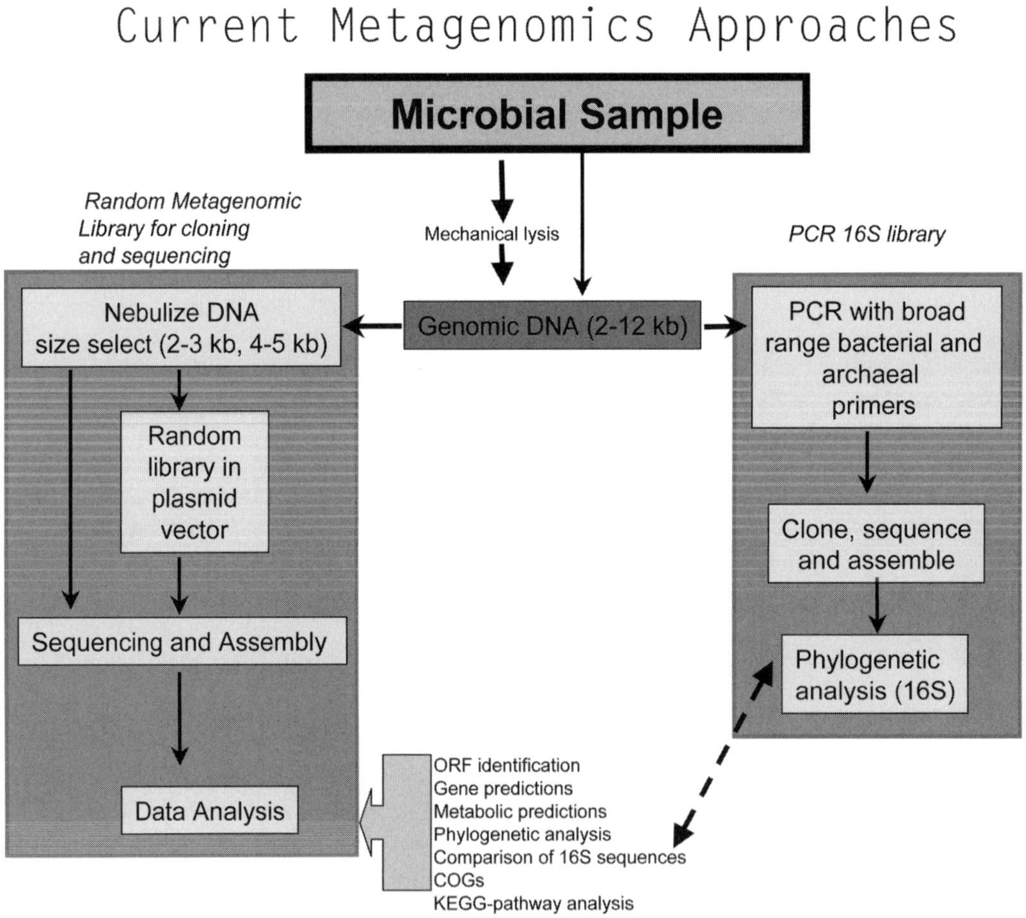

FIGURE 1 Present methods for performing metagenomic studies. (Adapted with permission from Steve Gill, University of Buffalo.)

had to be addressed in the launch of the metagenomics era, particularly as very often the quantities of available DNA are limited. This challenge has been heavily addressed through the group of DeLong, initially through the construction of large insert libraries (Béjà et al., 2000a) from environmental samples. This work, along with that of other groups (for example, Schleper et al., 1998; Stein et al., 1996), enabled deciphering the genetic information of uncultured species which can currently be achieved by sequencing genomic libraries that are created directly from environmental DNA (Béjà et al., 2000b; Béjà et al., 2002; Liles et al., 2003; Quaiser et al., 2002). DeLong and coworkers pioneered the sequencing of large fragments of DNA to characterize genes from uncultivated species (Béjà et al., 2000b; DeLong, 1992). In the first of a series of reports, they analyzed a 40-kb DNA fragment from a marine archaeon that had been cloned into a fosmid (DeLong, 1992). DeLong and coworkers subsequently constructed BAC libraries with an average insert size of 80 kb and a maximum insert size of 155 kb (Béjà et al., 2000b). That group demonstrated the value and utility of BAC libraries for providing genetic information on uncultivated species and laid the foundation for subsequent metagenomic studies in other environments. Béjà and colleagues used large genomic fragments (~80 kb) from a diverse population of largely uncultivated microorganisms to assign functional properties to archaeal clones (Béjà et al., 2000a).

TECHNOLOGIES FOR SEQUENCING COMMUNITIES

Currently, the most widely used method for genomic and metagenomic sequencing is the Sanger dideoxynucleotide chain termination method, which works on the incorporation of 2′, 3′ dideoxyribonucleoside triphosphates (ddNTPs) onto a template DNA strand (Sanger et al., 1977). Pyrosequencing is the base for a promising new technology developed by 454 Life Sciences (http://www.454.com/) (Hyman, 1988; Margulies et al., 2005; Ronaghi et al., 1996; Ronaghi et al., 1998). The 454 technology presents a major advantage over traditional sequencing technologies by eliminating the need for cloning vectors, which are known to present biases in terms of the clonability of certain DNA fragments (http://www.454.com/). This sequencing technology also reads through secondary structures readily and has a capacity to produce very large amounts of sequence; current estimates are of sequencing 20 megabases per 4.5-h run (http://www.454.com/enabling-technology/index.asp).

White and colleagues at The University of Illinois and The Institute for Genomic Research (now the J. Craig Venter Institute) in Rockville, MD, have recently used 454 technology to analyze the complex microbiome of the rumen (estimated to hold a dense and complex 500 to 1,000 microbial species) (J. M. Brulc, D. A. Antonopoulos, M. E. Berg, M. Wilson, R. E. Edwards, E. D. Frank, J. B. Emerson, K. E. Nelson, and B. A. White, unpublished data). The 454 technology and the subsystems-based annotations available in the SEED database (see below) were used to gain a better understanding of the metabolic potential of these metagenomes. The results from this study showed that for the three animals included in this study that were on identical diets, the community structures were markedly different with respect to nutrient utilization. From 178,713, 264,849, and 345,317 reads from the different animals (35,115,534 bases total) from the fiber-associated rumen metagenomes, the authors were able to detail the different metabolic profiles that are associated with the different microbial populations in the three different animals (Brulc et al., unpublished).

As mentioned above, Sogin and colleagues (Sogin et al., 2006) have also used pyrosequencing technology to develop a global, in-depth description of the diversity of deep-sea microbes. Here they generated approximately 118,000 amplicons that spanned a hypervariable region of the 16S rRNA gene, and they also investigated the composition of paired samples at different depths within the water column in the North Atlantic. Researchers have also sequenced more than 1 million base pairs of Neanderthal DNA and demonstrated sequence changes between modern and ancient humans (Noonan et al., 2006). This work was particularly significant as previous understanding of human and Neanderthal evolution was based on artifacts from excavation sites (Noonan et al., 2006). From the metagenomic sequence, human-Neanderthal divergence time could be calculated based on random loci, and the results suggested that both groups shared a common ancestor 706,000 years ago, with the ancestral populations splitting approximately 370,000 years ago (Noonan et al., 2006). The Neanderthal and human genomes were also found to be at least 99.5% identical (Noonan et al., 2006).

In the study of Poinar's group that focused on paleogenomics, they screened eight of what appeared to be the best-preserved mammoth remains in the collections of the Mammoth Museum (Poinar et al., 2006). Using 454 sequencing technology, the authors obtained a total of 28 million base pairs of DNA, 13 million base pairs of which were identified as mammoth DNA. The results showed that the mammoth DNA was 98.55% identical to that of the African elephant, with an estimated divergence date of 5 to 6 million years (Poinar et al., 2006).

Edwards and colleagues (Edwards et al., 2006) used the pyrosequencing technique of 454 Life Sciences to generate environmental genome sequences from two adjacent sites in the Soudan Mine, MN (Edwards et al., 2006). Although the average read lengths were shorter than those generated with traditional sequencing methods, the authors were able to highlight significant differences in metabolic potential in each environment, particularly in terms of the distinct biochemistries. They concluded that pyrosequencing would be widely used to sequence environmental samples because of the speed, cost, and technical advantages (Edwards et al., 2006).

Venter's group at the J. Craig Venter Institute in Rockville, MD, analyzed both the usability and costliness of a merged sequencing approach that combines both 3730xl Sanger sequencing where longer reads are traditionally obtained and the shorter reads from 454 sequencing to give microbial genome assemblies (Goldberg et al., 2006). These assemblies were compared to genomes that were obtained by using Sanger sequencing strategies in isolation (Goldberg et al., 2006). The authors found that the absence of a cloning bias and an ability to read through highly variable regions inclusive of hard stops make 454 technology a great improvement over Sanger methodology. However, the shorter reads could not reliably span repetitive areas, and an absence of paired ends limited scaffolding information. The 454 technology was, however, capable of sequencing regions of the genome where Sanger sequencing was not effective. The merging of the two techniques also resulted in an 86% gap reduction in most of the species included in the study (Goldberg et al., 2006). The authors found that the combination of the two sequencing technologies resulted in a higher-quality genome assembly and that 454 technology significantly reduced work and manual labor necessary for traditional closing/finishing of genomes (Goldberg et al., 2006).

COMMUNITIES THAT HAVE BEEN SUBJECT TO METAGENOMIC SEQUENCING

In addition to those mentioned above, a number of environments have been subject to detailed large-scale metagenomic analysis using high-throughput sequencing and data analysis applications. Among the communities that have been the subject of comprehensive metagenomic studies are soils (Tringe et al., 2005; Voget et al., 2003), the human gastrointestinal tract (Gill et al., 2006), the human oral cavity, the oceans (Venter et al., 2004), whale fall (Tringe et al., 2005), the rumen (Brulc et al., unpublished), and acid-mine drains (Tyson et al., 2004).

In one of the earliest metagenomic studies, Voget and coworkers (Voget et al., 2003) examined a soil metagenomic library to identify more than 15 different genes encoding novel biocatalysts. Cosmid libraries were screened for functionality to putative agarases, a putative stereoselective amidase, two cellulases, an alpha-amylase, a 1,4-alpha-glucan branching enzyme, and two pectate lyases, as well as two lipases (Voget et al., 2003). Goodman's group also conducted one of the earlier studies but used BAC soil libraries to generate more than 1 Gbp of cloned DNA (Rondon et al., 2000). Their initial expression screens in *E. coli* revealed a range of activities including lipases, amylases, and hemolytic activities (Rondon et al., 2000).

Sebat's group (Sebat et al., 2003) screened a metagenomic library that was constructed from a microcosm of groundwater microorganisms and identified genes of ecological importance, including some with hydrogen oxidation, nitrate reduction, and transposition activities (Sebat et al., 2003). Drinking water biofilms (Schmeisser et al., 2003) have been studied with a metagenomic approach, whereby a study was initiated to characterize drinking water biofilms grown on rubber-coated valves by employing three different strategies. Sequence analysis of 650 16S rDNA clones indicated a high diversity within the biofilm communities, with the majority of the microbes being closely related to the *Proteobacteria* (Schmeisser et al., 2003). The sequence information was used to set up a database containing the phylogenetic and genomic information on this model microbial community (Schmeisser et al., 2003).

Banfield's group at Berkeley has published a series of papers on a community approach to analyzing the mixed populations associated with acid-mine drainage fields. They used a metagenomic approach to study the microbial community structure of an acid-mine biofilm (Tyson et al., 2004). The relative simplicity of this system allowed the recovery of near-complete genomes of *Leptospirillum* and *Ferroplasma* species, as well as the partial recovery of three other species from 75 Mbp of sequence data (Tyson et al., 2004). This analysis was followed by a proteomic study that investigated gene expression and key activities again in a natural acid-mine drainage microbial biofilm community (Ram et al., 2005). Here the authors were able to identify 2,033 proteins from the five most abundant species in the biofilm (Ram et al., 2005).

In the recently concluded metagenomic analysis of the human gastrointestinal tract (Gill et al., 2006), a total of 139,521 sequences were generated by Sanger sequencing technology (approximately 78 Mbp of unique DNA sequence), and a total of 6,000 16S rDNA clones were also sequenced. Assembly of these sequences resulted in 33,753,108 base pairs of unique DNA sequence that represented diverse microbial species from both the archaeal and bacterial domains. From the comparative genome analyses, the authors were able to identify novel virulence factors and re-create a metabolic reconstruction of some of the ongoing processes in the gastrointestinal tract (Gill et al., 2006). At about the same time, Dore's group at INRA in France published a study that used a metagenomic approach to investigate the diversity of the fecal microbiota in patients with Crohn's disease (Manichanh et al., 2006). Here they used a fosmid vector to construct two genomic libraries, each composed of 25, 000 clones. Approximately 125 nonredundant ribotypes were identified that demonstrated a reduced complexity of these samples from patients with Crohn's disease.

Edwards and colleagues (Edwards et al., 2006) were the first to use pyrosequencing for metagenomic applications. Here they used the pyrosequencing technique to generate environmental genome sequences from two adjacent sites in the Soudan Mine, MN. Through data analysis with the SEED database (http://theseed.uchicago.edu/FIG/index.cgi), the authors were able to highlight significant differences in metabolic potential in each environment, particularly in terms of the distinct biochemistries. At the time of this data analysis in 2006, the SEED database comprised 351 subsystems, the majority of which were present in either of the metagenomic samples. For example, the results demonstrated that the metabolism of phosphorus was more dominant in oceanic surfaces than terrestrial environments (Edwards et al., 2006). The investigators concluded that pyrosequencing would be widely used to sequence environmental samples because of the speed, cost, and technical advantages (Edwards et al., 2006).

From the metagenomic analysis of the rumen (Brulc et al., unpublished) differences in community structure as related to diet were clearly identifiable. Again, the 454 technology-generated sequences were analyzed primarily with subsystems-based annotations from the SEED database (http://theseed.uchicago.edu/FIG/index.cgi) and revealed the metabolic potential of these metagenomes. The authors highlight the vast amount of previously unexplored DNA to which they now have access and the potential for accessing new enzymes for industrial and agricultural use (Brulc et al., unpublished).

Whole-genome shotgun sequencing of microbial populations collected on tangential flow and impact filters from seawater samples from the Sargasso Sea generated a total of 1.045 billion base pairs of nonredundant sequence (Venter et al., 2004). This sequence was annotated and analyzed to elucidate the gene content, diversity, and relative abundance of the organisms within these samples, estimated to derive from at least 1,800 genomic species based on sequence relatedness, including 148 previously unknown bacterial phylotypes. The 1.6-Gbp dataset also included more than 1.2 million new genes. Variation in species present and stoichiometry suggest substantial oceanic microbial diversity (Venter et al., 2004).

Venter and colleagues followed this study with a massive expedition, data collection, and analysis study that was recently published in *PLoS Biology* (http://collections.plos. org/plosbiology/gos-2007.php). For this study, "gigabasepairs" of data represent genes from microbes collected in waters off Canada, the Panama Canal, the Galapagos Islands, and the Pacific and the South Pacific Gyre (Parthasarathy et al., 2007). The assemblies and associated annotated peptides are available (http://www.ncbi.nlm.nih.gov/Genbank), and the data are also freely available through Cyberinfrastructure for Advanced Marine Microbial Ecology Research and Analysis (AMMERA) (http://camera.calit2.net; Parthasarathy et al., 2007).

The metagenomic dataset that was generated from this global expedition included the above-mentioned Sargasso Sea dataset and represents the largest dataset collected to date, comprising close to 7.7 million sequencing reads (Rusch et al., 2007).

VIRAL METAGENOMICS

Viruses are commonly found in marine environments. A significant amount of interest revolves around oceanic phage populations because they can account for differences in bacterial strains and can mobilize genetic information through lateral gene transfer (Angly et al., 2006). Rohwer's group has conducted a series of studies to investigate viral diversity using metagenomic approaches (Breitbart et al., 2002; Breitbart et al., 2003; Paul et al., 2002). In one of the earlier studies, shotgun libraries were made from two near-shore viral communities, and most sequences had no significant hits to previously deposited sequences (Breitbart et al., 2002). Angly and colleagues from Rohwer's group (Angly et al., 2006) subsequently conducted a metagenomic analysis of 184 viral assemblages through pyrosequencing of the uncultured viral populations. These samples had been collected over a 10-year period, and were derived from 68 sites in four major oceanic regions: the Arctic Ocean, the coast of British Columbia, the Gulf of Mexico, and the Sargasso Sea (Angly et al., 2006). As with the rumen metagenome study of White and colleagues at the University of Illinois (Brulc et al., unpublished), the sequences that were generated in this study were compared to the SEED nonredundant database. Most of the viral sequences that were identified from this study were novel, with cyanophages and a newly discovered clade of single-stranded DNA phages dominating the samples that were obtained from the Sargasso Sea sample, but the authors did find that prophage-like sequences were most common in the Arctic (Angly et al., 2006). The authors presented data that showed variation in viral assemblages based of different geographic locations and that indicated regional diversity of phages can be almost as high as global diversity, possibly as a result of viral migration (Angly et al., 2006).

The group of Culley (Culley et al., 2006) used metagenomic approaches to analyze coastal RNA virus communities. In their study, reverse-transcribed whole-genome shotgun sequencing was applied to uncultivated marine RNA virus assemblages and resulted in a range of RNA viruses, including distant relatives of viruses infecting arthropods and higher plants (Culley et al., 2006), demonstrating that the ocean acts as a reservoir for many unknown RNA viruses.

ISSUES WITH ASSEMBLY AND PROCESSING METAGENOMIC DATA

The majority of metagenomic projects to date have relied on available assembly tools to stitch the massive numbers of sequences together. The recent global ocean sampling expedition, for example, that involved traversing several thousand kilometers from the North Atlantic to the South Pacific yielded more than 7.7 million sequencing reads (~6.3 billion bp). The dataset was assembled using the Celera assembler, which requires very stringent criteria (reviewed by Rusch et al., 2007). This dataset was subsequently used for annotation. Interestingly, only a very small fraction of the assemblies (9%) resulted in scaffolds that were longer than 10 kb in length, and a majority of the sequences (53%) did not go into assemblies (Rusch et al., 2007). Assembling these sequences with varying levels of stringency (98%, 94%, 90%, 85%, and 80%) resulted in increased lengths of the assemblies as the cutoff decreased.

These authors also used "extreme assembly" to resolve any potential conflicts. Here, mate-pairing information is not used, and only contigs are produced rather than scaffolds. The authors used any large contigs from these alternate assemblies to look at variations in population structure. Seed alignments with a particular phylogenetic marker present (primarily those associated with the ribosomal operon) were also used to generate contigs greater than 100 kb in size for many species (Rusch et al., 2007).

To address this question, a group led by Kyrpides at the Department of Energy Joint Genome Institute recently published a study of simulated datasets that varied in complexity (Mavromatis et al., 2007). The design of the datasets was such that they would mimic metagenomic data. The sequence reads were assembled using Phrap, Arachne, and JAZZ, and fgenesb and CRITICA/GLIMMER were used as the gene-finding pipelines. The authors highlight that depending on the algorithm used, chimeric contigs and scaffolds that are composed of different species are always possible (Mavromatis et al., 2007). From their analysis, they found that Phrap resulted in a lower number of singletons and gave the largest number of contigs in all the simulated datasets. Arachne produced the least number of chimeras (Mavromatis et al., 2007).

For gene prediction, Fgenesb identified 10 to 30% more reference genes, and the Arachne assemblies, which were predicted to be of higher quality, resulted in more accurate gene predictions than the other two assemblers.

What this all means is that for the high quantities of data that have been generated to date in metagenomic studies, there is still a need to standardize the assembly and data analysis protocols.

TOOLS FOR GENOMIC AND METAGENOMIC COMPARISONS

The sequencing of individual genomes has allowed the development of various tools and databases for addressing complex biological questions. These tools have to a certain extent allowed accurate predictions on genome organization, gene content, novel genes, gene families, biochemical pathways, transport and metabolism, gene transfer events, etc., and functional genomic studies. Over the years, more integrated platforms have been developed that allow the researcher to access all necessary information related to a single species, along with comparative data from one central location. The Comprehensive Microbial Resource (http://cmr.tigr.org/tigr-scripts/CMR/

CmrHomePage.cgi), for example, freely displays data associated with all the publicly available, completely sequenced genomes. All the species are available in a single location, and cross-genome comparisons are possible. A tutorial is also available for first-time users. Among the analyses possible via the Comprehensive Microbial Resource, one can compare multiple genomes for sequence homologies, single nucleotide polymorphisms, and atypical genome composition.

The U.S. Department of Energy Joint Genome Institute has the Integrated Microbial Genomes data management system that is also freely available (http://img.jgi.doe.gov/). The latest version of this system contains 2,815 genomes including 687 bacteria and 41 archaea along with multiple eukaryotic genomes. Both draft and complete genomes are included.

Many of the tools that have been developed for the computational analysis of single genomes have, out of necessity, been applied directly to the analysis of environmental metagenomic data. In many cases, however, the analysis has proved certain challenges that have been related to the assembly of the data, phylogenetic analysis, sequence polymorphisms, comparisons between multiple communities (Chen and Pachter, 2005), and overall interpretation of individual species from mixed incomplete genomes. In addition, challenges include the gene identification and prediction process.

In traditional single-genome sequencing projects, the genome data from closely related species can serve as "scaffolds" for the newly assembled data. In metagenomic projects where novel species will be encountered, there is no backbone of scaffold for the data to be aligned to, and very often the sequences become unassembled fragments, with no real matches, and to which conventional approaches for the identification of the sequences cannot be applied (for example, Gill et al., 2006; Noguchi et al., 2006). Unassembled reads make up approximately 50% of the assembled sequences in the Sargasso Sea dataset (Chen and Pachter, 2005; Venter et al., 2004). Less than 2% of the Sargasso Sea matched at 90% identity or higher with sequences in other databases (Tress et al., 2006). Tress and colleagues propose that the high proportion of sequence fragments, as encountered in the Sargasso Sea study, limits the ability to make conclusions with respect to distribution of sequences and that these sequence fragments may bias any subsequent studies (Tress et al., 2006).

Binning has been used in some metagenomic projects to sort the data based on GC composition (Venter et al., 2004). Genomic features that include phylogenetic markers and/or codon biases can also be used to separate out "similar" sequences. However, gene transfer events limit the interpretation of these sequences, as very distant species may harbor similar pieces of genetic information (Chen and Pachter, 2005).

Predicted genes can be blast searched against a number of databases including the Clusters of Orthologous Genes (COGs) and the Kyoto Encyclopedia of Genes and Genomes (KEGG) databases. In the metagenomic study of the human gastrointestinal tract by Gill and colleagues (Gill et al., 2006) COG- and KEGG-based analyses revealed an enrichment for glycan, amino acid, and xenobiotic metabolism, methanogenesis, as well as 2-methyl-D-erythritol 4-phosphate pathway-mediated biosynthesis of vitamins and isoprenoids.

Overbeek and colleagues developed the SEED database (http://theseed.uchicago.edu/FIG/index.cgi), which is a curated database of microbial genomes with an interface of subsystems (Overbeek et al., 2005). The subsystems include annotations that are based on complete and partial biochemical pathways as well as clusters of genes that function together or are considered to be related (Brulc et al., unpublished). By comparing within and across various samples, one is able to identify major functions that can be correlated with the individual metagenome sample (Edwards et al., 2006). As mentioned above, the SEED database has been used successfully to analyze the assembled metagenomic sequences that were generated using pyrosequencing techniques (Edwards et al., 2006). Major features of the SEED database include access to KEGG maps associated with the data being analyzed, metabolic overviews, the generation of spreadsheets that represent the occurrence of particular functional roles, and tools to support phylogenetic analysis (Overbeek et al., 2005; http://theseed.uchicago.edu/FIG/index.cgi).

Schuster's group has described MEGAN for the analysis of metagenomic data (Huson et al., 2007). The data to be analyzed are imported as raw DNA sequences and subsequently compared to other datasets with BLAST; MEGAN then analyzes the taxonomical content of the data, which is assigned with a simple algorithm (Huson et al., 2007). The result is that the data can be put through a preliminary analysis without an initial assembly, and a statistical output can be obtained (Huson et al., 2007).

Noguchi and colleagues have recently described "MetaGene: prokaryotic gene finding from environmental genome shotgun sequences" (Noguchi et al., 2006). This program is designed to predict genes from fragmented genomic sequences (Noguchi et al., 2006), and primarily uses di-codon frequencies in its analyses. The program extracts all open reading frames (ORFs) from a dataset and gives a score from the base composition and length of the ORF. A high score for ORFs is derived from orientations and distances of neighboring ORFs in addition to the scores for the ORFs themselves (Noguchi et al., 2006). The program is estimated to have a sensitivity of 95% and a specificity of 90% for artificial shotgun sequences (Noguchi et al., 2006). MetaGene runs in a very short time, and the authors state that analysis of the sequences derived from the Sargasso Sea dataset took only 15 minutes. The software is currently available at http://metagene.cb.k.u-tokyo.ac.jp/.

WET LAB TOOLS FOR THE ANALYSIS OF METAGENOMIC DATA

Suppression subtractive hybridization (SSH) was originally developed to amplify rare mRNA transcripts in mRNA libraries from eukaryotic tissues (Bonaldo et al., 1996). SSH introduced the use of specially designed adaptors. When these adaptors are ligated to themselves or to small DNA fragments, the secondary structure introduced by their sequence content effectively inhibits the PCR amplification in the final preparation. This decreases the detection of background fragments and allows the identification of larger DNA of sufficient "informational" length. An amended version of SSH was used successfully to compare the genomes of two strains of *Helicobacter pylori* (Akopyants et al., 1998). The patented kit, which is produced by BD Biosciences, is based on this publication and provides a convenient collection of the adaptors and enzymes to successfully construct SSH libraries (http://www.bd.com/aboutbd/).

White and colleagues (Galbraith et al., 2004) based at The University of Illinois were the first to apply the SSH technique to a mixed microbial community when they compared the metagenomes of microbial consortia from rumen samples. The SSH study was conducted with samples of total genomic DNA obtained from rumen fluid samples of two hay-fed steers. Unique fragments were found in almost all (95 of 96) of the randomly sequenced clones. In this study, the authors clearly demonstrated that SSH can allow a clear discerning and identification of DNA fragments that are unique to one complex community relative to another.

Meyer and colleagues (Meyer et al., 2007) proposed a new method for the recovery of reading frames from metagenomic samples. Their method, which is based on subtractive hybridization magnetic bead capture technology, is an adaptation of the method of Jacobsen such that a target gene can be retrieved. Multiple target genes can be amplified and bound to streptavidin beads, which are then used as hybridization probes to recover full-length sequence.

METAPROTEOMICS: THE COLLECTIVE PROTEINS IN A SAMPLE

The value gained from the sequencing and data analysis of microbial communities cannot be denied, but it is evident that the sequence information alone does not give any insight into roles of individual species and protein activity in these environments (Kan et al., 2005). Metaproteomics, which represents the collective analysis of all proteins in a sample, is being developed as one avenue to gain further insight into these environments. Here, the quantity of the proteins extracted is important, and two-dimensional polyacrylamide gel electrophoresis coupled with mass spectrometry (MS)-based protein identification relying on mass-based (matrix-assisted laser desorption ionization–time of flight MS) or sequence-based (liquid chromatography electrospray ionization tandem MS) methods are thought to be the most appropriate for metaproteomic studies (Kan et al., 2005). Kan and colleagues applied metaproteomic approaches to evaluate a constructed community and a Chesapeake Bay microbial community. Their study demonstrates the potential "to link metagenomic data, taxonomic diversity, functional diversity and biological processes in natural environments" (Kan et al., 2005).

USING THE METAGENOME TO EVALUATE THE TRANSCRIPTOME

Metagenomics is also being used to describe the transcriptome of various environmental samples. In the case of eukaryotic organisms that contain introns, the transcriptome is usually accessed by reverse transcription (Grant et al., 2006). Grant and colleagues constructed cDNA libraries from RNA in environmental samples (Grant et al., 2006). Here, the transcriptome could be accessed without any prior knowledge of the specific organisms that are in the sample.

FUTURE DIRECTIONS IN SEQUENCING

It is clear that newer and less expensive methods for sequencing will continue to develop, one example being "Polony" sequencing using a single cell (Podar et al., 2007; Shendure et al., 2005; Zhang et al., 2006).

For environments where there is a low abundance of DNA, multiple displacement amplification (MDA) can be successfully used to amplify the limited DNA quantities. Keller's group used MDA on low-biomass samples to amplify whole genomes from environmental sediments from nitrate- and heavy-metal-contaminated soils (Abulencia et al., 2006). Clone libraries were constructed from the amplified gDNA, and analysis revealed a high level of similarity to known proteins (Abulencia et al., 2006). This polymerase has been used for rolling-circle amplification (Detter et al., 2002), but there appears to be a bias in genome amplification of mixed populations (Abulencia et al., 2006). The MDA kit is commercially available as the GenomiPhi DNA amplification kit (Amersham Biosciences, Piscataway, NJ).

Yokouchi and colleagues (Yokouchi et al., 2006) also addressed the issues associated with obtaining enough DNA from samples with MDA using phi29 polymerase. Here they optimized their reaction conditions with *Synechocystis* sp. strain PCC6803. The bacterial metagenome from corals was amplified and analyzed, again demonstrating the reliability of using this technique in samples that have a low number of microbial cells (Yokouchi et al., 2006).

Finally, Lasken's group (Raghunathan et al., 2005) successfully amplified single-cell DNA 5 billion-fold with the MDA technique. This clearly demonstrated that techniques such as MDA can allow access to the genetic information of any microbial species, particularly when coupled with the use of a flow sorter or laser tweezers (Podar et al., 2007).

THE POTENTIAL OF THE METAGENOMICS REVOLUTION: REAL AND PERCEIVED POTENTIAL

The field of genomics has rapidly expanded to the point where we are now able to sequence complex communities without the need for first culturing the individual species that are present in these environments. Most recently, Venter and colleagues have described an unprecedented level of microbial diversity in samples collected from a global ocean sampling expedition (Rusch et al., 2007). Metagenomics provides the capacity to characterize genetic diversity present in samples irrespective and regardless of the availability of laboratory culturing techniques. Markers that are identified from these mixed populations can be used to monitor individual species or provide insight into metabolic capabilities within these environments (for example, the study on the human gastrointestinal tract).

A recent recommendation was made by the National Research Council for a Global Metagenomics Initiative (http://www.genomeweb.com/issues/news/139188-1.html; http://www8.nationalacademies.org/onpinews/newsitem.aspx?RecordID=11902). The report states that metagenomics "presents the greatest opportunity to revolutionize understanding of the microbial world." The report highlights the importance of organisms in performing functions including balancing the atmosphere's composition, fighting disease, and supporting plant growth. The report proposes detailed analysis of carefully chosen microbial communities worldwide. Novel techniques and community databases would be obvious benefits of such a system as well as groups of scientists working together to disseminate advances, agree on standards, and develop guidelines for best practices, the committee said. The National Science Foundation, the National Institutes of Health, and the U.S. Department of Energy sponsored this report.

In summary, the true diversity that exists in nature is nowhere close to being fully sampled or studied, even for those environments that have already been subject to metagenomics. Hopefully, the current cost limitations will soon be eliminated and further progress can be made in this field.

REFERENCES

Abulencia, C. B., D. L. Wyborski, J. A. Garcia, M. Podar, W. Chen, S. H. Chang, H. W. Chang, D. Watson, E. L. Brodie, T. C. Hazen, and M. Keller. 2006. Environmental whole-genome amplification to access microbial populations in contaminated sediments. *Appl. Environ. Microbiol.* **72:**3291–3301.

Akopyants, N. S., A. Fradkov, L. Diatchenko, J. E. Hill, P. D. Siebert, S. A. Lukyanov, E. D. Sverdlov, and D. E. Berg. 1998. PCR-based subtractive hybridization and differences in gene content among strains of *Helicobacter pylori*. *Proc. Natl. Acad. Sci. USA* **95:**13108–13113.

Amann, R. I., W. Ludwig, and K. H. Schleifer. 1995. Phylogenetic identification and in situ detection of individual microbial cells without cultivation. *Microbiol. Rev.* **59:**143–169.

Angly, F. E., B. Felts, M. Breitbart, P. Salamon, R. A. Edwards, C. Carlson, A. M. Chan, M. Haynes, S. Kelley, H. Liu, J. M. Mahaffy, J. E. Mueller, J. Nulton, R. Olson, R. Parsons, S. Rayhawk, C. A. Suttle, and F. Rohwer. 2006. The marine viromes of four oceanic regions. *PLoS Biol.* **4:**e368.

Ashby, M. N., J. Rine, E. F. Mongodin, K. E. Nelson, and D. Dimster-Denk. 2007. Serial analysis of ribosomal DNA and the unexpected dominance of rare members of microbial communities. *Appl. Environ. Microbiol.* **73:**4532–4542.

Barrangou, R., C. Fremaux, H. Deveau, M. Richards, P. Boyaval, S. Moineau, D. A. Romero, and P. Horvath. 2007. CRISPR provides acquired resistance against viruses in prokaryotes. *Science* **315:**1709–1712.

Béjà, O., L. Aravind, E. V. Koonin, M. T. Suzuki, A. Hadd, L. P. Nguyen, S. B. Jovanovich, C. M. Gates, R. A. Feldman, J. L. Spudich, E. N. Spudich, and E. F. DeLong. 2000a. Bacterial rhodopsin: evidence for a new type of phototrophy in the sea. *Science* **289:**1902–1906.

Béjà, O., E. V. Koonin, L. Aravind, L. T. Taylor, H. Seitz, J. L. Stein, D. C. Bensen, R. A. Feldman, R. V. Swanson, and E. F. DeLong. 2002b. Comparative genomic analysis of archaeal genotypic variants in a single population and in two different oceanic provinces. *Appl. Environ. Microbiol.* **68:**335–345.

Béjà, O., M. T. Suzuki, E. V. Koonin, L. Aravind, A. Hadd, L. P. Nguyen, R. Villacorta, M. Amjadi, C. Garrigues, S. B. Jovanovich, R. A. Feldman, and E. F. DeLong. 2000. Construction and analysis of bacterial artificial chromosome libraries from a marine microbial assemblage. *Environ. Microbiol.* **2:**516–529.

Blattner, F. R., G. Plunkett, 3rd, C. A. Bloch, N. T. Perna, V. Burland, M. Riley, J. Collado-Vides, J. D. Glasner, C. K. Rode, G. F. Mayhew, J. Gregor, N. W. Davis, H. A. Kirkpatrick, M. A. Goeden, D. J. Rose, B. Mau, and Y. Shao. 1997. The complete genome sequence of *Escherichia coli* K-12. *Science* **277:**1453–1474.

Bolotin, A., B. Quinquis, A. Sorokin, and S. D. Ehrlich. 2005. Clustered regularly interspaced short palindrome repeats (CRISPRs) have spacers of extrachromosomal origin. *Microbiology* **151:**2551–2561.

Bonaldo, M. F., G. Lennon, and M. B. Soares. 1996. Normalization and subtraction: two approaches to facilitate gene discovery. *Genome Res.* **6:**791–806.

Breitbart, M., I. Hewson, B. Felts, J. M. Mahaffy, J. Nulton, P. Salamon, and F. Rohwer. 2003. Metagenomic analyses of an uncultured viral community from human feces. *J. Bacteriol.* **185:**6220–6223.

Breitbart, M., P. Salamon, B. Andresen, J. M. Mahaffy, A. M. Segall, D. Mead, F. Azam, and F. Rohwer. 2002. Genomic analysis of uncultured marine viral communities. *Proc. Natl. Acad. Sci. USA* **99:**14250–14255.

Chen, K., and L. Pachter. 2005. Bioinformatics for whole-genome shotgun sequencing of microbial communities. *PLoS Comput. Biol.* **1:**106–112.

Connon, S. A., and S. J. Giovannoni. 2002. High-throughput methods for culturing microorganisms in very-low-nutrient media yield diverse new marine isolates. *Appl. Environ. Microbiol.* **68:**3878–3885.

Culley, A. I., A. S. Lang, and C. A. Suttle. 2006. Metagenomic analysis of coastal RNA virus communities. *Science* **312:**1795–1798.

DeBoy, R. T., E. F. Mongodin, J. B. Emerson, and K. E. Nelson. 2006. Chromosome evolution in the *Thermotogales*: large-scale inversions and strain diversification of CRISPR sequences. *J. Bacteriol.* **188:**2364–2374.

DeLong, E. F. 1992. Archaea in coastal marine environments. *Proc. Natl. Acad. Sci. USA* **89:**5685–5689.

Detter, J. C., J. M. Jett, S. M. Lucas, E. Dalin, A. R. Arellano, M. Wang, J. R. Nelson, J. Chapman, Y. Lou, D. Rokhsar, T. L. Hawkins, and P. M. Richardson. 2002. Isothermal strand-displacement amplification applications for high-throughput genomics. *Genomics* **80:**691–698.

Edwards, R. A., B. Rodriguez-Brito, L. Wegley, M. Haynes, M. Breitbart, D. M. Peterson, M. O. Saar, S. Alexander, E. C. Alexander, Jr., and F. Rohwer. 2006. Using pyrosequencing to shed light on deep mine microbial ecology under extreme hydrogeologic conditions. *BMC Genomics* **7:**57.

Fleischmann, R. D., M. D. Adams, O. White, R. A. Clayton, E. F. Kirkness, A. R. Kerlavage, C. J. Bult, J.-F. Tomb, B. A. Dougherty, J. M. Merrick, et al. 1995. Whole-genome shotgun sequencing and assembly of *Haemophilus influenzae* Rd genome. *Science* **269:**496–512.

Fouts, D. E., E. F. Mongodin, R. E. Mandrell, W. G. Miller, D. A. Rasko, J. Ravel, L. M. Brinkac, R. T. DeBoy, C. T. Parker, S. C. Daugherty, R. J. Dodson, A. S. Durkin, R. Madupu, S. A. Sullivan, J. U. Shetty, M. A. Ayodeji, A. Shvartsbeyn, M. C. Schatz, J. H. Badger, C. M. Fraser, and K. E. Nelson. 2005. Major structural differences and novel potential virulence mechanisms from the genomes of multiple campylobacter species. *PLoS Biol.* **3:**e15.

Frangeul, L., K. E. Nelson, C. Buchrieser, A. Danchin, P. Glaser, and F. Kunst. 1999. Cloning and assembly strategies in microbial genome projects. *Microbiology* **145:**2625–2634.

Galbraith, E. A., D. A. Antonopoulos, and B. A. White. 2004. Suppressive subtractive hybridization as a tool for identifying genetic diversity in an environmental metagenome: the rumen as a model. *Environ. Microbiol.* **6:**928–937.

Gill, S. R., M. Pop, R. T. Deboy, P. B. Eckburg, P. J. Turnbaugh, B. S. Samuel, J. I. Gordon, D. A. Relman, C. M. Fraser-Liggett, and K. E. Nelson. 2006. Metagenomic analysis of the human distal gut microbiome. *Science* **312:**1355–1359.

Goldberg, S. M., J. Johnson, D. Busam, T. Feldblyum, S. Ferriera, R. Friedman, A. Halpern, H. Khouri, S. A. Kravitz, F. M. Lauro, K. Li, Y. H. Rogers, R. Strausberg, G. Sutton, L. Tallon, T. Thomas, E. Venter, M. Frazier, and J. C. Venter. 2006. A Sanger/pyrosequencing hybrid approach for the generation of high-quality draft assemblies of marine microbial genomes. *Proc. Natl. Acad. Sci. USA* **103:**11240–11245.

Grant, S., W. D. Grant, D. A. Cowan, B. E. Jones, Y. Ma, A. Ventosa, and S. Heaphy. 2006. Identification of eukaryotic open reading frames in metagenomic cDNA libraries made from environmental samples. *Appl. Environ. Microbiol.* **72:**135–143.

Haft, D. H., J. Selengut, E. F. Mongodin, and K. E. Nelson. 2005. A guild of 45 CRISPR-associated (Cas) protein families and multiple CRISPR/Cas subtypes exist in prokaryotic genomes. *PLoS Comput. Biol.* **1:**e60.

Handelsman, J. 2004. Metagenomics: application of genomics to uncultured microorganisms. *Microbiol. Mol. Biol. Rev.* **68:**669–685.

Hugenholtz, P., B. M. Goebel, and N. R. Pace. 1998. Impact of culture-independent studies on the emerging phylogenetic view of bacterial diversity. *J. Bacteriol.* **180:**4765–4774.

Huson, D. H., A. F. Auch, J. Qi, and S. C. Schuster. 2007. MEGAN analysis of metagenomic data. *Genome Res.* **17:**377–386.

Hyman, E. D. 1988. A new method of sequencing DNA. *Anal. Biochem.* **174:**423–436.

Jansen, R., J. D. Embden, W. Gaastra, and L. M. Schouls. 2002. Identification of genes that are associated with DNA repeats in prokaryotes. *Mol. Microbiol.* **43:**1565–1575.

Kan, J., T. E. Hanson, J. M. Ginter, K. Wang, and F. Chen. 2005. Metaproteomic analysis of Chesapeake Bay microbial communities. *Saline Systems* **1:**7.

Kysela, D. T., C. Palacios, and M. L. Sogin. 2005. Serial analysis of V6 ribosomal sequence tags (SARST-V6): a method for efficient, high-throughput analysis of microbial community composition. *Environ. Microbiol.* **7:**356–364.

Liles, M. R., B. F. Manske, S. B. Bintrim, J. Handelsman, and R. M. Goodman. 2003. A census of rRNA genes and linked genomic sequences within a soil metagenomic library. *Appl. Environ. Microbiol.* **69:**2684–2691.

Manichanh, C., L. Rigottier-Gois, E. Bonnaud, K. Gloux, E. Pelletier, L. Frangeul, R. Nalin, C. Jarrin, P. Chardon, P. Marteau, J. Roca, and J. Dore. 2006. Reduced diversity of faecal microbiota in Crohn's disease revealed by a metagenomic approach. *Gut* **55:**205–211.

Margulies, M., M. Egholm, W. E. Altman, S. Attiya, J. S. Bader, L. A. Bemben, J. Berka, M. S. Braverman, Y. J. Chen, Z. Chen, S. B. Dewell, L. Du, J. M. Fierro, X. V. Gomes, B. C. Godwin, W. He, S. Helgesen, C. H. Ho, G. P. Irzyk, S. C. Jando, M. L. Alenquer, T. P. Jarvie, K. B. Jirage, J. B. Kim, J. R. Knight, J. R. Lanza, J. H. Leamon, S. M. Lefkowitz, M. Lei, J. Li, K. L. Lohman, H. Lu, V. B. Makhijani, K. E. McDade, M. P. McKenna, E. W. Myers, E. Nickerson, J. R. Nobile, R. Plant, B. P. Puc, M. T. Ronan, G. T. Roth, G. J. Sarkis, J. F. Simons, J. W. Simpson, M. Srinivasan, K. R. Tartaro, A. Tomasz, K. A. Vogt, G. A. Volkmer, S. H. Wang, Y. Wang, M. P. Weiner, P. Yu, R. F. Begley, and J. M. Rothberg. 2005. Genome sequencing in microfabricated high-density picolitre reactors. *Nature* **437:**376–380.

Mavromatis, K., N. Ivanova, K. Barry, H. Shapiro, E. Goltsman, A. C. McHardy, I. Rigoutsos, A. Salamov, F. Korzeniewski, M. Land, A. Lapidus, I. Grigoriev, P. Richardson, P. Hugenholtz, and N. C. Kyrpides. 2007. Use of simulated data sets to evaluate the fidelity of metagenomic processing methods. *Nat. Methods* **4:**495–500.

Meyer, Q. C., S. G. Burton, and D. A. Cowan. 2007. Subtractive hybridization magnetic bead capture: a new technique for the recovery of full-length ORFs from the metagenome. *Biotechnol. J.* **2:**36–40.

Mojica, F. J., C. Diez-Villasenor, J. Garcia-Martinez, and E. Soria. 2005. Intervening sequences of regularly spaced prokaryotic repeats derive from foreign genetic elements. *J. Mol. Evol.* **60:**174–182.

Mongodin, E. F., J. B. Emerson, and K. E. Nelson. 2005a. Microbial metagenomics. *Genome Biol.* **6:**347.

Mongodin, E. F., I. R. Hance, R. T. Deboy, S. R. Gill, S. Daugherty, R. Huber, C. M. Fraser, K. Stetter, and K. E. Nelson. 2005b. Gene transfer and genome plasticity in *Thermotoga maritima*, a model hyperthermophilic species. *J. Bacteriol.* **187:**4935–4944.

Nelson, K. E., R. A. Clayton, S. R. Gill, M. L. Gwinn, R. J. Dodson, D. H. Haft, E. K. Hickey, J. D. Peterson, W. C. Nelson, K. A. Ketchum, L. McDonald, T. R. Utterback, J. A. Malek, K. D. Linher, M. M. Garrett, A. M. Stewart, M. D. Cotton, M. S. Pratt, C. A. Phillips, D. Richardson, J. Heidelberg, G. G. Sutton, R. D. Fleischmann, J. A. Eisen, O. White, S. L. Salzberg, H. O. Smith, J. C. Venter, and C. M. Fraser. 1999. Evidence for lateral gene transfer between archaea and bacteria from genome sequence of *Thermotoga maritima*. *Nature* **399:**323–329.

Nelson, K. E., D. E. Fouts, E. F. Mongodin, J. Ravel, R. T. DeBoy, J. F. Kolonay, D. A. Rasko, S. V. Angiuoli, S. R. Gill, I. T. Paulsen, J. Peterson, O. White, W. C. Nelson, W. Nierman, M. J. Beanan, L. M. Brinkac, S. C. Daugherty, R. J. Dodson, A. S. Durkin, R. Madupu, D. H. Haft, J. Selengut, S. Van Aken, H. Khouri, N. Fedorova, H. Forberger, B. Tran, S. Kathariou, L. D. Wonderling, G. A. Uhlich, D. O. Bayles, J. B. Luchansky, and C. M. Fraser. 2004. Whole genome comparisons of serotype 4b and 1/2a strains of the food-borne pathogen *Listeria monocytogenes* reveal new insights into the core genome components of this species. *Nucleic Acids Res.* **32:**2386–2395.

Nelson, K. E., I. T. Paulsen, J. F. Heidelberg, and C. M. Fraser. 2000. Status of genome projects for nonpathogenic bacteria and archaea. *Nat. Biotechnol.* **18:**1049–1054.

Neufeld, J. D., and W. W. Mohn. 2005. Unexpectedly high bacterial diversity in arctic tundra relative to boreal forest soils, revealed by serial analysis of ribosomal sequence tags. *Appl. Environ. Microbiol.* **71:**5710–5718.

Neufeld, J. D., Z. Yu, W. Lam, and W. W. Mohn. 2004. Serial analysis of ribosomal sequence tags (SARST): a high-throughput method for profiling complex microbial communities. *Environ. Microbiol.* **6:**131–144.

Nierman, W. C., T. V. Feldblyum, M. T. Laub, I. T. Paulsen, K. E. Nelson, J. A. Eisen, J. F. Heidelberg, M. R. Alley, N. Ohta, J. R. Maddock, I. Potocka, W. C. Nelson, A. Newton, C. Stephens, N. D. Phadke, B. Ely, R. T. DeBoy, R. J. Dodson, A. S. Durkin, M. L. Gwinn, D. H. Haft, J. F. Kolonay, J. Smit, M. B. Craven, H. Khouri, J. Shetty, K. Berry, T. Utterback, K. Tran, A. Wolf, J. Vamathevan, M. Ermolaeva, O. White, S. L. Salzberg, J. C. Venter, L. Shapiro, and C. M. Fraser. 2001. Complete genome sequence of *Caulobacter crescentus*. *Proc. Natl. Acad. Sci. USA* **98:**4136–4141.

Noguchi, H., J. Park, and T. Takagi. 2006. MetaGene: prokaryotic gene finding from environmental genome shotgun sequences. *Nucleic Acids Res.* **34:**5623–5630.

Noonan, J. P., G. Coop, S. Kudaravalli, D. Smith, J. Krause, J. Alessi, F. Chen, D. Platt, S. Paabo, J. K. Pritchard, and E. M. Rubin. 2006. Sequencing and analysis of Neanderthal genomic DNA. *Science* **314:**1113–1118.

Overbeek, R., T. Begley, R. M. Butler, J. V. Choudhuri, H. Y. Chuang, M. Cohoon, V. de Crecy-Lagard, N. Diaz, T. Disz, R. Edwards, M. Fonstein, E. D. Frank, S. Gerdes, E. M. Glass, A. Goesmann, A. Hanson, D. Iwata-Reuyl, R. Jensen, N. Jamshidi, L. Krause, M. Kubal, N. Larsen, B. Linke, A. C. McHardy, F. Meyer, H. Neuweger, G. Olsen, R. Olson, A. Osterman, V. Portnoy, G. D. Pusch, D. A. Rodionov, C. Ruckert, J. Steiner, R. Stevens, I. Thiele, O. Vassieva, Y. Ye, O. Zagnitko, and V. Vonstein. 2005. The subsystems approach to genome annotation and its use in the project to annotate 1000 genomes. *Nucleic Acids Res.* **33:**5691–5702.

Parthasarathy, H., E. Hill, and C. Maccallum. 2007. Global Ocean Sampling Collection. *PLoS Biol.* **5:**e83.

Paul, J. H., M. B. Sullivan, A. M. Segall, and F. Rohwer. 2002. Marine phage genomics. *Comp. Biochem. Physiol. B Biochem. Mol. Biol.* **133:**463–476.

Perna, N. T., G. Plunkett III, V. Burland, B. Mau, J. D. Glasner, D. J. Rose, G. F. Mayhew, P. S. Evans, J. Gregor, H. A. Kirkpatrick, G. Posfai, J. Hackett, S. Klink, A. Boutin, Y. Shao, L. Miller, E. J. Grotbeck, N. W. Davis, A. Lim, E. T. Dimalanta, K. D. Potamousis, J. Apodaca, T. S. Anantharaman, J. Lin, G. Yen, D. C. Schwartz, R. A. Welch, and F. R. Blattner. 2001. Genome sequence of enterohaemorrhagic *Escherichia coli* O157:H7. *Nature* **409:**529–533.

Podar, M., C. B. Abulencia, M. Walcher, D. Hutchison, K. Zengler, J. A. Garcia, T. Holland, D. Cotton, L. Hauser, and M. Keller. 2007. Targeted access to the genomes of low-abundance organisms in complex microbial communities. *Appl. Environ. Microbiol.* **73:**3205–3214.

Poinar, H. N., C. Schwarz, J. Qi, B. Shapiro, R. D. Macphee, B. Buigues, A. Tikhonov, D. H. Huson, L. P. Tomsho, A. Auch, M. Rampp, W. Miller, and S. C. Schuster. 2006. Metagenomics to paleogenomics: large-scale sequencing of mammoth DNA. *Science* **311:**392–394.

Pourcel, C., G. Salvignol, and G. Vergnaud. 2005. CRISPR elements in *Yersinia pestis* acquire new repeats by preferential uptake of bacteriophage DNA, and provide additional tools for evolutionary studies. *Microbiology* **151:**653–663.

Quaiser, A., T. Ochsenreiter, H. P. Klenk, A. Kletzin, A. H. Treusch, G. Meurer, J. Eck, C. W. Sensen, and C. Schleper. 2002. First insight into the genome of an uncultivated crenarchaeote from soil. *Environ. Microbiol.* **4:**603–611.

Raghunathan, A., H. R. Ferguson, Jr., C. J. Bornarth, W. Song, M. Driscoll, and R. S. Lasken. 2005. Genomic DNA amplification from a single bacterium. *Appl. Environ. Microbiol.* **71:**3342–3347.

Ram, R. J., N. C. Verberkmoes, M. P. Thelen, G. W. Tyson, B. J. Baker, R. C. Blake II, M. Shah, R. L. Hettich, and J. F. Banfield. 2005. Community proteomics of a natural microbial biofilm. *Science* **308:**1915–1920.

Ronaghi, M., S. Karamohamed, B. Pettersson, M. Uhlen, and P. Nyren. 1996. Real-time DNA sequencing using detection of pyrophosphate release. *Anal. Biochem.* **242:**84–89.

Ronaghi, M., M. Uhlen, and P. Nyren. 1998. A sequencing method based on real-time pyrophosphate. *Science* **281:**363–365.

Rondon, M. R., P. R. August, A. D. Bettermann, S. F. Brady, T. H. Grossman, M. R. Liles, K. A. Loiacono, B. A. Lynch, I. A. MacNeil, C. Minor, C. L. Tiong, M. Gilman, M. S. Osburne, J. Clardy, J. Handelsman, and R. M. Goodman. 2000. Cloning the soil metagenome: a strategy for accessing the genetic and functional diversity of uncultured microorganisms. *Appl. Environ. Microbiol.* **66:**2541–2547.

Rusch, D. B., A. L. Halpern, G. Sutton, K. B. Heidelberg, S. Williamson, S. Yooseph, D. Wu, J. A. Eisen, J. M. Hoffman, K. Remington, K. Beeson, B. Tran, H. Smith, H. Baden-Tillson, C. Stewart, J. Thorpe, J. Freeman, C. Andrews-Pfannkoch, J. E. Venter, K. Li, S. Kravitz, J. F. Heidelberg, T. Utterback, Y. H. Rogers, L. I. Falcon, V. Souza, G. Bonilla-Rosso, L. E. Eguiarte, D. M. Karl, S. Sathyendranath, T. Platt, E. Bermingham, V. Gallardo, G. Tamayo-Castillo, M. R. Ferrari, R. L. Strausberg, K. Nealson, R. Friedman, M. Frazier, and J. C. Venter. 2007. The Sorcerer II global ocean sampling expedition: Northwest Atlantic through eastern tropical Pacific. *PLoS Biol.* **5:**e77.

Sanger, F., S. Nicklen, and A. R. Coulson. 1977. DNA sequencing with chain-terminating inhibitors. *Proc. Natl. Acad. Sci. USA* **74:**5463–5467.

Schleper, C., E. F. DeLong, C. M. Preston, R. A. Feldman, K. Y. Wu, and R. V. Swanson. 1998. Genomic analysis reveals chromosomal variation in natural populations of the uncultured psychrophilic archaeon *Cenarchaeum symbiosum*. *J. Bacteriol.* **180:**5003–5009.

Schmeisser, C., C. Stockigt, C. Raasch, J. Wingender, K. N. Timmis, D. F. Wenderoth, H. C. Flemming, H. Liesegang, R. A. Schmitz, K. E. Jaeger, and W. R. Streit. 2003. Metagenome survey of biofilms in drinking-water networks. *Appl. Environ. Microbiol.* **69:**7298–7309.

Sebat, J. L., F. S. Colwell, and R. L. Crawford. 2003. Metagenomic profiling: microarray analysis of an environmental genomic library. *Appl. Environ. Microbiol.* **69:**4927–4934.

Shendure, J., G. J. Porreca, N. B. Reppas, X. Lin, J. P. McCutcheon, A. M. Rosenbaum, M. D. Wang, K. Zhang, R. D. Mitra, and G. M. Church. 2005. Accurate multiplex polony sequencing of an evolved bacterial genome. *Science* **309:**1728–1732. Epub 2005 Aug 4.

Sogin, M. L., H. G. Morrison, J. A. Huber, D. M. Welch, S. M. Huse, P. R. Neal, J. M. Arrieta, and G. J. Herndl. 2006. Microbial diversity in the deep sea and the underexplored "rare biosphere." *Proc. Natl. Acad. Sci. USA* **103:**12115–12120.

Spiegelman, D., G. Whissell, and C. W. Greer. 2005. A survey of the methods for the characterization of microbial consortia and communities. *Can. J. Microbiol.* **51:**355–386.

Stein, J. L., T. L. Marsh, K. Y. Wu, H. Shizuya, and E. F. DeLong. 1996. Characterization of uncultivated prokaryotes: isolation and analysis of a 40-kilobase-pair genome fragment from a planktonic marine archaeon. *J. Bacteriol.* **178:**591–599.

Tress, M. L., D. Cozzetto, A. Tramontano, and A. Valencia. 2006. An analysis of the Sargasso Sea resource and the consequences for database composition. *BMC Bioinformatics* **7:**213.

Tringe, S. G., C. von Mering, A. Kobayashi, A. A. Salamov, K. Chen, H. W. Chang, M. Podar, J. M. Short, E. J. Mathur, J. C. Detter, P. Bork, P. Hugenholtz, and E. M. Rubin. 2005. Comparative metagenomics of microbial communities. *Science* **308:**554–557.

Tyson, G. W., J. Chapman, P. Hugenholtz, E. E. Allen, R. J. Ram, P. M. Richardson, V. V. Solovyev, E. M. Rubin, D. S. Rokhsar, and J. F. Banfield. 2004. Community structure and metabolism through reconstruction of microbial genomes from the environment. *Nature* **428:**37–43.

Venter, J. C., K. Remington, J. F. Heidelberg, A. L. Halpern, D. Rusch, J. A. Eisen, D. Wu, I. Paulsen, K. E. Nelson, W. Nelson, D. E. Fouts, S. Levy, A. H. Knap, M. W. Lomas, K. Nealson, O. White, J. Peterson, J. Hoffman, R. Parsons, H. Baden-Tillson, C. Pfannkoch, Y. H. Rogers, and H. O. Smith. 2004. Environmental genome shotgun sequencing of the Sargasso Sea. *Science* **304:**66–74.

Voget, S., C. Leggewie, A. Uesbeck, C. Raasch, K. E. Jaeger, and W. R. Streit. 2003. Prospecting for novel biocatalysts in a soil metagenome. *Appl. Environ. Microbiol.* **69:**6235–6242.

Yokouchi, H., Y. Fukuoka, D. Mukoyama, R. Calugay, H. Takeyama, and T. Matsunaga. 2006. Whole-metagenome amplification of a microbial community associated with scleractinian coral by multiple displacement amplification using phi29 polymerase. *Environ. Microbiol.* **8:**1155–1163.

Zengler, K., G. Toledo, M. Rappe, J. Elkins, E. J. Mathur, J. M. Short, and M. Keller. 2002. Cultivating the uncultured. *Proc. Natl. Acad. Sci. USA* **99:**15681–15686.

Zhang, K., A. Martiny, N. B. Reppas, K. W. Barry, J. Malek, S. W. Chisholm, and G. M. Church. 2006. Sequencing genomes from single cells by polymerase cloning. *Nat. Biotechnol.* **24:**680–686.

ARE MICROORGANISMS NONCULTURABLE OR NOT YET CULTURABLE?

III

NEW CULTIVATION STRATEGIES FOR TERRESTRIAL MICROORGANISMS

Peter H. Janssen

10

Soils are populated by microbial cells from all three domains of life: *Archaea*, *Bacteria*, and *Eukarya*. The general unculturability of soil bacteria has been recognized for 90 years, and is the focus of this review. The word "bacteria" is used here to include members of *Archaea* and *Bacteria*. If the relatively few culturable cells of soil bacteria were phylogenetically representative of the total community, or if the unculturable cells were dead, moribund, or inactive, this unculturability problem would not be of as much consequence as it is, now that we know that these explanations are not valid. The unculturable bacteria belong largely to poorly studied phyla, classes, and orders, and isolating them in laboratory culture often has been described as a challenge for microbiologists.

The majority of the more than 10^9 bacterial cells per gram of soil seem to be alive and therefore are potentially culturable. Collectively, these cells contain sufficient DNA for all to have genomes (Bakken and Olsen, 1989), and they have phospholipid contents comparable to those of viable cells (Lindahl et al., 1997). The mean ATP content of bacteria in soils (Contin et al., 2001) is similar to that of actively growing cells (Knowles, 1977), and the adenylate energy charge (Brookes et al., 1983) is typical of exponentially growing cells (Dawes, 1976). Bacterial cells in soil appear to have high levels of ribosomes as judged by their ability to bind fluorescently labeled DNA probes targeting their rRNA (Christensen et al., 1999). Up to 46% of microscopically detectable cells reduced the artificial electron acceptor iodonitrotetrazolium using a mixture of NADH and NADPH as electron donors, which cannot be expected to be transported by all species (MacDonald, 1980). With the Molecular Probes *Bac*Light Live/Dead stain, 91 to 97% of soil bacteria were judged to have intact cell envelopes with energized cytoplasmic membranes (Janssen et al., 2002; S. Cairnduff and P. H. Janssen, unpublished data).

SOIL BACTERIA ARE POORLY CULTURABLE

Culturing soil bacteria on a synthetic (though not necessarily defined) medium solidified with a gel such as agar, gelatin, or gellan gum has been the traditional method of choice since Koch (1881) introduced the technique. The bacteria are allowed to proliferate to form

Peter H. Janssen, AgResearch Limited, Palmerston North 4442, New Zealand.

Accessing Uncultivated Microorganisms: from the Environment to Organisms and Genomes and Back
Edited by Karsten Zengler © 2008 ASM Press, Washington, DC

visible colonies, and from these colonies, isolation of pure cultures and subsequent physiological and genetic investigations can be carried out. However, less than 40 years after Koch's landmark publication, Conn (1918) reported that the total microscopic count of soil bacteria was 10 to 70 times greater than plate (or viable) counts from the same soil samples. He argued that this discrepancy in culturability (equal to the viable count divided by the total microscopic count) could not be ascribed to insufficient dispersion of the cells or due to the presence of dead cells, and concluded that the difference could only be explained if the majority of soil bacteria did not grow on plates. Conn (1927) was of the opinion that the plate count was of little value in soil microbiology long before this became widely, although not universally, accepted. Subsequent investigators, using fluorescent stains, showed that total microscopic counts were up to 1,000 times greater than plate counts (Russell, 1950; Skinner et al., 1952; Olsen and Bakken, 1987). There are likely to be many reasons why only a small part of the bacterial community of soil can form detectable colonies, and these have been discussed many times (e.g., Barer and Harwood, 1999; Mukamolova et al., 2003; Overmann, 2006). These explanations have been described as "a standard litany of excuses that microbiologists give for being able to cultivate so few of the microbes in natural environments" (Kieft, 2000).

It is likely that some species, ones belonging to groups that generally are rare in soil, have opportunistic growth strategies and are able to respond rapidly to new opportunities such as nutrient flushes. Transferring a bacterium to a normal synthetic growth medium can be considered to be a transition to such a nutrient flush. Species that grow rapidly under such conditions may be termed *r*-strategists (Andrews and Harris, 1986) and seem preadapted to growth on plates of standard agar-solidified laboratory media. They are able to survive the transition from soil to plates and readily form colonies. These species may be overrepresented as a proportion of the total colonies formed, and so appear relatively more abundant than they actually are in the community that served as the inoculum (Janssen, 2006).

Other species, belonging to taxa that are poorly represented among cultured isolates, must be poorly preadapted to growth on routinely used laboratory media. They seem to grow slowly and, importantly, only seem to form detectable colonies rarely. Since members of these taxa are abundant in soil (Hugenholtz et al., 1998; Janssen, 2006), it has been inferred that members of these poorly culturable clades are largely *K*-strategists (Watve et al., 2000), well adapted to survival and growth in their resource-limited environment (Andrews and Harris, 1986). It is likely that this is due to a trade-off that results in low growth rates and possibly an inability to respond rapidly to sudden changes in their environment. At present there is little physiological evidence for these speculations, since the soil bacteria that have been studied in any detail are all capable of rapid growth, and are probably all *r*-strategists. However, cultivation of soil bacteria that may display the characteristics of true soil *K*-strategists can now be carried out routinely, albeit with some effort, and future comparisons with the better-known *r*-strategists will reveal any underlying physiological and genetic bases for these ecological strategies in soil bacteria.

There are higher-level taxa (families and orders) of bacteria in soils that are poorly represented by pure culture isolates, even within the well-studied phyla *Actinobacteria*, *Bacteroidetes*, *Firmicutes*, and *Proteobacteria* (Janssen, 2006). Within other phyla, like *Acidobacteria*, *Chloroflexus*, *Gemmatimonadetes*, *Planctomycetes*, and *Verrucomicrobia*, most lineages are poorly, or not at all, represented by pure cultures (Hugenholtz et al., 1998; Janssen, 2006). In addition, there are many other phyla that make up minor parts of the soil bacterial community and that are poorly represented by pure cultures (Janssen, 2006).

Consideration of the points discussed above leads to the conclusion that the plate count is not an appropriate tool for enumerating soil bacteria. The large discrepancies between the inoculum and the resultant cultures, both quantitatively (low culturability) and qualitatively (poor phylogenetic representation), have caused soil microbiologists to move away from employing this tool, except to estimate the number of *r*-strategists (e.g., Ellis et al., 2003). However, cultivation on solid media in plates has been found to be a powerful tool for isolating members of poorly studied phylogenetic groups, when the likely characteristics of *K*-strategists and the limitations of the approach are considered and the methodology modified appropriately. This chapter reviews some of the factors that allow successful isolation of a wider phylogenetic representation of soil bacteria than has traditionally been thought to be possible.

WHAT DO WE WANT TO CULTURE, AND WHY?

Having ruled out the use of the plate count as a method for enumerating soil bacteria, it becomes important to decide what the use of cultivation of soil bacteria on plates is. The diversity of soil bacteria species is very large, with estimates of thousands to millions of distinct genomes in samples of just a few grams of soil (Torsvik et al., 1990; Dykhuizen, 1998; Gans et al., 2005), and perhaps millions of different species in a ton of soil (Curtis et al., 2002). It seems that a comprehensive collection of all species is, at present, impossible. The aim of isolating bacteria from soil must then be to obtain a synoptic collection within a group of interest. This might consist of isolating representatives of the higher taxon diversity present in the environment, with an attempt to cover the phylogenetic diversity. Another aim might be to obtain representatives of groups with a known activity or function, and to understand their phylogenetic, phenotypic, and genetic diversity. Alternatively, the intention may be to gather as much genetic diversity as possible within a particular taxon, such as in the large collections of members of the subclass *Actinobacteridae* held by organizations engaged in the discovery of new bioactive compounds (e.g., Lazzarini et al., 2000).

Recently, a number of advances have been made that allow cultivation of a wider phylogenetic range of soil bacteria. The aim of these studies has been to bring a better representation of bacterial diversity into laboratory culture, rather than improve the plate count method as a tool for enumeration. The approach has to be one of compromises. Many isolates of soil bacteria, especially most of those of the poorly represented phyla such as *Acidobacteria*, *Chloroflexus*, *Gemmatimonadetes*, *Planctomycetes*, and *Verrucomicrobia*, are extremely difficult to maintain in pure culture. Isolates that are more easily handled in the laboratory are obviously preferable for subsequent genetic and biochemical studies, although these may not be good representatives of their group. For example, close relatives of *Chthoniobacter flavus*, a relatively easily isolated member of the class *Spartobacteria* of the phylum *Verrucomicrobia* (Sangwan et al., 2004; Sangwan et al., 2005), are only rarely detected in soils, while members of the more abundant clades of the class *Spartobacteria* have so far eluded cultivation (P. Sangwan and P. H. Janssen, unpublished data).

The routine cultivation of the marine ultramicrobacterium *Sphingopyxis alaskensis* strain RB2256 on a solid medium is thought to have become possible because of a physiological change that occurred during long-term storage of liquid cultures (Schut, 1994). The low incidence of colony formation by members of some groups of bacteria leads therefore to the caution that the few isolates that do form detectable colonies may be mutants capable of what a microbiologist considers reasonable growth under laboratory conditions. These are, of course, still useful laboratory models of the groups to which they belong, but care should be taken in interpreting aspects of their physiology, particularly related to *r*- versus *K*-selected growth strategies.

WHAT ARE THE CHARACTERISTICS OF "UNCULTURABLE" SOIL BACTERIA?

In recent systematic studies into improving the culturability of soil bacteria, researchers have often been content to see just what turns up. If there are many uncultured groups, even moderate success will initially appear to be significant (e.g., Janssen et al., 2002), given that culturing any so-called "unculturables" represents a breakthrough. In the process, much can be learned about which approaches are useful. Targeting more defined groups, such as *Verrucomicrobia* (Sangwan et al., 2005), *Planctomycetes* (Wang et al., 2002), and subdivision 6 of the phylum *Acidobacteria* (K. E. R. Davis and P. H. Janssen, unpublished data), has proven to be more challenging, although successful. Obtaining pure cultures of some specific groups, for example, soil archaea or members of the bacterial phylum TM-7 (Simon et al., 2005; Ferrari et al., 2005), has proven to be very difficult.

It has been suggested that most soil bacteria are aerobic heterotrophs (Bakken, 1997), and this is the assumption made in nearly all studies to isolate so-called "unculturable" soil bacteria. Targeting aerobic heterotrophs will obviously result in isolation of bacteria with that physiology, and will not tell us if this assumption is generally valid, even with culturabilities in the region of 24% (Davis et al., 2005). It has also been assumed that many of the so-called "unculturables" are *K*-strategists and will be slow growing. This does seem to be borne out by experimental evidence, because the few cultured isolates of so-called "unculturable" groups are slow growing, relative to members of better-studied groups (Davis et al., 2005). Studies of the characteristics of putative *K*-strategists, and the identification of underlying genetic characteristics, followed by a measure of the significance of these characteristics in partial genomes recovered using metagenomic approaches, would help confirm the validity of this assumption.

It has been postulated that metagenomic data will provide information on the physiology of unculturable bacteria (Streit and Schmitz, 2004). Extracting large fragments of genomic DNA directly from soil potentially allows genome sequence data from all soil organisms to be gathered. However, given the current difficulties in assembling even partial genomes because of the enormous genetic diversity (Tringe et al., 2005; Schloss and Handelsman, 2006), and the generally uninformative nature of genes associated with reliable phylogenetic markers (Liles et al., 2003), this is still a development for the future. Metagenomic approaches should eventually identify those groups that are not aerobic heterotrophs and are unable to grow under the conditions normally employed to isolate soil bacteria. Using phylogenetic information to enrich cells of target groups without cultivation may accelerate this approach significantly (Podar et al., 2007). Currently, more directed techniques can be used to determine the physiology of certain taxa. For example, the application of a combination of molecular ecological methods has suggested that some uncultured soil archaea are ammonia oxidizers (Leininger et al., 2006). A range of methods has been used to deduce which microorganisms have particular activities (Gray and Head, 2001; Overmann, 2006), including stable isotope probing techniques that have allowed identification of methylotrophs (Radajewski et al., 2000) and bacterial predators of other bacteria (Lueders et al., 2006).

MEDIUM CHOICE

Medium choice has a significant effect on the outcomes of attempts to obtain isolates of rarely cultured groups of soil bacteria. The purpose of the isolation medium should be to bring the bacteria from the soil into laboratory culture. Some bacteria belonging to rarely cultured groups do grow on commonly employed media once in pure culture but are not able to be cultured directly from soil using these media (Davis et al., 2005). The best isolation media therefore may not be the ones that support the best growth of the isolates once they are in pure culture (Leadbetter, 2003). In fact, such media may be poorly suited to the task for reasons that are discussed below.

Nonselective agar-based media that support growth of a wide range of heterotrophic bacteria are generally selected for cultivation studies. Some of the more widely used are trypticase soy agar (TSA) at one-tenth of the normal concentration but with the agar content adjusted to normal (0.1×TSA), different variants of

Winogradsky's salt solution agar (Winogradsky, 1949), and cold and hot extracted soil extract agar (Fischer, 1909; Olsen and Bakken, 1987), but hundreds of others have been devised and used. Although it has been pointed out that these commonly used media do not mimic the soil solution (Angle et al., 1991), these traditionally used media remain popular, probably because of their ease of preparation and the long history of their use. Soil extract, either autoclaved or filter sterilized, has long been used as a source of growth factors and as a base for media for culturing soil bacteria (Olsen and Bakken, 1987). The soil used to make the soil extract does not have to come from the site from which cultivation is attempted. In fact, it seems that any fresh fertile soil will suffice for the preparation of a good quality soil extract (James, 1958; Egdell et al., 1960; Jensen, 1968). The precise nature of the stimulatory factors in soil extract is unknown, but it has been argued that, for some bacteria, it can be satisfactorily replaced with calcium (Taylor, 1951) or with vitamins, amino acids, and other defined supplements (Lochhead and Chase, 1943; Lochhead and Thexton, 1952). The addition of vitamins and trace elements at low concentrations is recommended, and very low levels of nutrient broth can supply a range of potential growth factors.

The traditionally used media do not seem to be particularly useful for isolating so-called "unculturable" soil bacteria, although occasional notable isolates have been obtained (McCaig et al., 2001; Furlong et al., 2002; Rösch et al., 2002). Plate counts obtained on 0.1×TSA, on Winogradsky's salt solution agar, and on soil extract agar were much lower than those obtained on diluted nutrient broth and on medium VL55 supplemented with xylan (Davis et al., 2005). The VL55 and VL70 medium bases were devised to mimic the ratios of the inorganic components of the soil solution (P. H. Janssen, unpublished data). In addition, 0.1×TSA, Winogradsky's salt solution agar, and soil extract agar were vastly inferior to plates of solidified dilute nutrient broth (0.01 × normal concentration) and VL55/xylan in their ability to support colony formation by members of rarely isolated bacterial groups (Davis et al., 2005). Media based on VL55 and VL70 were used to culture many members of the phyla *Acidobacteria*, *Chloroflexi*, *Gemmatimonadetes*, *Planctomycetes*, and *Verrucomicrobia* and of the subclasses *Acidimicrobidae* and *Rubrobacteridae* of the phylum *Actinobacteria* (Janssen et al., 2002; Sait et al., 2002; Joseph et al., 2003; Sangwan et al., 2005; Sait et al., 2006). These media have low concentrations of all components. Only small amounts of carbon, nitrogen, phosphorus, and sulfur are actually required to produce good growth of colonies on plates (Table 1), and most microbiological media contain these elements in great excess.

Reactive components of the medium should be sterilized separately, by filtration if necessary, and then combined, which avoids unwanted reactions between medium com-

TABLE 1 Amounts of carbon, nitrogen, phosphorus, and sulfur required for colony formation by bacteria on plates[a]

Cell component	Percentage of cell dry weight	Amt required per colony (mmol)	Concn needed for 100 colonies (mmol liter^{-1})
C	50	7.2×10^{-4}	2.4[b]
N	14	4.3×10^{-5}	0.14
P	3	4.2×10^{-6}	0.014
S	1	1.4×10^{-6}	0.005

[a]These calculations are based on the following assumptions: A colony with a diameter of 1 mm and a height of 50 μm contains 1.5×10^7 cells, each 2 μm long, 1 μm in diameter, and with a dry weight of 2.8×10^{-13} g. The colony grows using an energy source that also serves as the carbon source, with a low efficiency of 1 mole of C assimilated from every 4 moles used for growth, to allow a high maintenance energy component during slow growth. Each plate contains 30 ml of medium. The cell size, weight, and composition data are from Neidhardt et al. (1990).

[b]Note that this is the concentration of C; 2.4 mmol of C liter^{-1} is equivalent to 0.4 mmol of glucose liter^{-1}.

ponents during autoclaving. Examples of such reactions are the well-known Maillard reaction between reducing sugars and amino acids, and the potential degradation of the buffers 2-(*N*-morpholino)-ethanesulfonic acid and 3-(*N*-morpholino)-propanesulfonic acid when autoclaved in the presence of glucose and so potentially other compounds (Good and Izawa, 1972). Short autoclaving times, often achieved more easily using small autoclaves, help reduce the effects of heat on medium components. Care should also be taken to avoid using new glassware, which often has a surface coating and can inhibit bacterial growth (P. H. Janssen, unpublished data), and to minimize possible contamination of media with plasticizers and detergents. A good-quality water source is required; the author prefers double-distilled water prepared in a glass still.

Not all soil bacteria can be expected to grow on a few generalized media. Different bacteria can be expected to have different requirements or be sensitive to different components of growth media. Increasing the range of media has been shown to increase the diversity of species cultured (Balestra and Misaghi, 1997; Sørheim et al., 1989; Palumbo et al., 1996). Stevenson et al. (2004) used a range of media and growth conditions to find those most suitable for supporting growth of members of the phyla *Acidobacteria* and *Verrucomicrobia* by screening using a PCR-based assay, and in this way identified media that supported growth of members of subdivision 1 of the phylum *Acidobacteria* and subdivision 4 of the phylum *Verrucomicrobia*.

CHOICE OF CARBON AND ENERGY SOURCE

The choice of carbon and energy source seems an obvious significant factor in medium design. Some carbon and energy sources result in higher plate counts than others, and polymeric substrates seem particularly useful in achieving good culturability (Chin et al., 1999; Schoenborn et al., 2004; Davis et al., 2005). In particular, heteropolymeric substrates have been shown to be useful for the isolation of a wide range of different soil bacteria (Sait et al., 2002; Joseph et al., 2003; Davis et al., 2005; K. E. R. Davis and P. H. Janssen, unpublished data). Humic acids also support growth of large number of colonies (K. E. R. Davis and P. H. Janssen, unpublished data), and a medium containing humic acids enabled the isolation of a member of the subdivision *Rubrobacteridae* of the phylum *Actinobacteria* (Monciardini et al., 2003).

Substrates need to be added at levels high enough to allow detection of growth, and so mimicking the concentrations found in bulk soil will not suffice. However, in practical terms such low concentrations are not generally achievable anyway. Agar- and gellan-solidified media with no added substrate also support good colony formation (Olsen and Bakken, 1987; Winding et al, 1994; Joseph et al., 2003), which suggests that achieving low substrate concentrations in media solidified with agar or gellan gum is technically difficult. Silica gel can be used to solidify media that are very low in background organic compounds, but preparation is technically more demanding than agar or gellan gum (Casida, 1968).

There is some evidence that high substrate concentrations inhibit colony formation, and recently there has been a trend to describe some media as "low substrate" media. Even so-called "low substrate" media have substrate concentrations in the region of hundreds of micromolar or greater, and so are well above the typical K_s values for aerobic bacteria (Button, 1985). High (and possibly so-called "low") substrate concentrations may induce substrate-accelerated death when starving cells are transferred to the high substrate concentrations required for laboratory media. Exposure of bacteria starved of certain nutrients to those limiting nutrients can result in substrate-accelerated death (Postgate and Hunter, 1964; Calcott and Postgate, 1972). The effect is not limited to carbon and energy sources, and so death may occur when cells are transferred from soil to any microbiological medium. Addition of cAMP to starved pure cultures prevents substrate-accelerated death (Calcott and Calvert, 1981). Addition of cAMP to media has also been shown to result in significant increases in the cultivation efficiency of marine and freshwater bacteria (Bruns et al., 2002; Bruns et al., 2003) but did not have an effect on the overall culturability of soil bacteria (Sangwan et al., 2005). A better understanding of the phenomenon of substrate-accelerated death is still required to determine its significance (Barer and Harwood, 1999). At present it is not clear if many of the bacteria in soil are sensitive to substrate-accelerated death, because they have not been studied in any detail. Those species that have been studied could be regarded as *r*-strategists, and it is not clear if *K*-strategists are more susceptible or less susceptible to this effect.

Increasing nutrient concentrations in the medium generally result in decreasing culturability of soil bacteria (Hattori, 1976; Ohta and Hattori, 1980; Olsen and Bakken, 1987), and richer media yield lower culturabilities than do media with lower nutrient concentrations (Watve et al., 2000; Aagot et al., 2001; Davis et al., 2005). It appears that some specific substrates have inhibitory or toxic effects at high concentrations, whereas others do not (Olsen and Bakken, 1987). Although the use of polymeric growth substrates resulted in higher culturabilities of soil bacteria than did monomeric growth substrates (Chin et al., 1999; Schoenborn et al., 2004; Davis et al., 2005), the isolates obtained using the polymers all grew with the monomers, suggesting that the monomers were detrimental to initiating growth on laboratory media, but not inhibitory once active cultures were obtained (Chin et al., 1999; P.H. Janssen, unpublished data). It is likely that employing polymeric growth substrates results in initially low concentrations of transportable substrates and that, as the cells initiate growth and begin to express the enzymes required for polymer hydrolysis, the concentration of small substrate molecules increases slowly, allowing expression of pathways to cope with the increased concentrations and so avoiding substrate-accelerated death or other substrate-induced inhibitory effects.

Some soil bacteria display an inability to grow with high substrate concentrations or on rich media even once active cultures are obtained (Hattori and Hattori, 1980; Davis et al., 2005), indicating that factors other than substrate-accelerated death also play a role. They can, however, be trained to grow at higher substrate concentrations, suggesting that only a few differences in gene regulation are involved in inhibition. This training may be the result of selecting for mutants, and a better understanding of this may help design better cultivation strategies. A more subtle training may be possible by continuous feeding on very low nutrient media after encapsulating cells in gel microdroplets or capturing the cells on membranes, which has been used successfully to isolate new freshwater, marine, and soil bacteria (Zengler et al., 2002; Hahn et al., 2004). With this technology, cells can be slowly adapted to increasing nutrient concentrations before plating and isolating pure cultures.

OTHER MEDIUM COMPONENTS

Components of the growth medium other than the carbon and energy source may inhibit growth. Some bacteria have been shown to be inhibited by phosphate at concentrations used as a pH buffer (Bartscht et al., 1999) or by copper at levels common in trace element mixes (Waterbury, 1991; Widdel, 1987). Lower concentrations of ammonia and phosphate have been shown to result in increased culturability of marine bacteria (Eilers et al., 2001). Even subtle differences in medium composition can have an impact on the survival of cells exposed to oxidative stress (De Spiegeleer et al., 2004). The relative concentrations of trace elements can affect phenotype (Leadbetter, 2003), and so could influence culturability. A wide range of other compounds, including amino acids, vitamins, and common cations such as K^+, Ca^{2+}, and Mg^{2+}, even at low concentrations, have been shown to inhibit some soil bacteria (Hattori and Hattori, 1980). Even low concentrations of NaCl, as low as 0.3 g per liter, can increase the lag period (Hattori, 1976), and isolates of subdivision 3 of the phylum *Verrucomicrobia* are inhibited by 2 g of NaCl per liter of medium (P. Sangwan and P. H. Janssen, unpublished data). All of these observations may explain why such a wide range of soil bacteria was isolated using diluted nutrient broth and media based on VL55 and VL70 (Janssen et al., 2002; Sait et al., 2002; Joseph et al., 2003; Davis et al., 2005). These media contain low concentrations of medium components other than the growth substrates and avoid the use of phosphate as a pH buffer, either having no buffer or using sulfonate buffers like 2-(*N*-morpholino)-ethanesulfonic acid and 3-(*N*-morpholino)-propanesulfonic acid. However, these, and related buffers, cannot be assumed to be innocuous (Good and Izawa, 1972).

GELLING AGENTS

Soil bacteria typically grow in water films but interact with and attach to surfaces (Gray et al., 1968; Hattori and Hattori, 1976). This may be why the culturability of soil bacteria on solidified media was higher than on the same media without a gelling agent (Olsen and Bakken, 1987; Janssen et al., 2002). Some soil bacteria isolated in liquid media grow as colonies on the surfaces of glass tubes in which they are cultured, and do not disperse into liquid culture (X. Chen and P. H. Janssen, unpublished data), and soil myxobacteria often grow on the solid surfaces of culture vessels when grown in liquid culture (Reichenbach and Dworkin, 1981). Such a preference for growth on surfaces may be another reason why some groups of soil bacteria are not isolated in liquid culture but can be isolated on solid media in plates (Schoenborn et al., 2004). Some of these can eventually be grown in liquid culture, but this may be due to selection of mutants.

Hattori (1980) reported that different viable counts were obtained using the same medium solidified with different agars and that more highly purified agars yielded higher counts when used in conjunction with low-nutrient media. This suggests that inhibitory substances in the agar prevent visible colony formation by some soil bacteria. Even high-grade agar should be washed before use to improve purity (Widdel and Bak, 1992). Similarly, the use of gellan gum as a solidifying agent resulted in increased culturability compared with agar (Janssen et al., 2002). Gellan plates are much clearer than agar plates, and very small colonies can be seen more easily. Agar and gellan should be autoclaved for only 15 min, and the use of small autoclaves reduces the heating and cooling times and so minimizes the effects of heat. The gelling agents should also be autoclaved separately. This is particularly important for media with pH <6, since acidified agar loses gelling strength when autoclaved.

Eliminating the solidifying agent completely and using serial liquid dilution in a most-probable-number format to isolate bacteria resulted in lower culturability (Janssen et al., 2002) and resulted in a smaller part of the higher taxon diversity being isolated (Schoenborn et al., 2004). This is probably due at least in part to the inherently low growth rates of many soil bacteria (Davis et al., 2005). When multiple cells of different species are transferred onto a solid surface, all have the possibility of being detected when they eventually form colonies. However, when multiple cells of different species are coinoculated into a liquid medium, the resultant culture will be dominated by the fastest-growing organisms present in the inoculum.

INOCULUM AND INTERACTIONS

Marine sediment communities change rapidly (<24 h) after sampling (Rochelle et al., 1994) and, like soils, are made up largely of active but nongrowing cells (Novitsky, 1987). It is therefore preferable to minimize the time between sampling and plating to <4 h if possible and to transport intact soil cores to the laboratory at ambient temperature. Sieving and storage could lead to redistribution of occluded substrates and result in nutrient flushes that may increase the number of *r*-strategists, diluting the target bacteria, and potentially producing antimicrobial compounds.

Soil particles are colonized by multiple cells that often occur in microcolonies (Gray et al., 1968; Hattori and Hattori, 1976). If these microcolonies or particles contain members of multiple bacterial groups and include rapidly growing species, the experimenter's aim of obtaining a representative collection of strains will be spoiled. Rapidly growing organisms will dominate the resultant colonies or liquid cultures. Since it is suspected that many of the poorly studied groups of soil bacteria consist mainly of slowly growing species, while well-studied groups contain many fast-growing species (Davis et al., 2005; Schoenborn et al., 2004), this will skew a collection of isolates toward members of the already well-studied groups. Different groups of bacteria are released from soil by different treatments (Aakra et al., 2000), and so the choice of cell dispersion or extraction method and of the diluent affects the outcome (Jensen, 1962; Hill and Gray, 1966; Lindahl, 1996; Dabek-Szreniawska and Hattori, 1981; Janssen et al., 2002). A compromise must be made between cell dispersal and cell damage (Ramsay, 1984; Katayama et al., 1998), and this will have to be determined empirically for each soil. A certain amount of cell damage and stress can be expected when soil suspensions are prepared, due to changes in osmotic potential, ionic composition, and oxygen tension, as well as through shearing and collision. Sulfonate pH buffers may even help cells survive osmotic stress by acting as osmoprotectants (Cayley et al., 1989; Mason and Blunden, 1989). Simple dispersal of the soil suspension after a step to release as many of the adherent bacteria as possible, e.g., using a mild sonication treatment (Janssen et al., 2002), seems preferable to cell extraction procedures, as a more representative fraction of the community is likely to grow, since species that strongly adhere to particles will not be excluded. A low-salt, low-phosphate diluent with the same general composition as the medium is recommended.

The size of the inoculum is an important consideration when attempting to isolate soil bacteria. Microorganisms can interact in a positive, neutral, or negative manner, and denser inocula will increase the likelihood of these interactions, and so potentially increase or decrease culturability. Davis et al. (2005) concluded that negative interactions have a greater impact than positive ones for general culturability, but it should be borne in mind that members of certain taxa may rely on interactions with other microorganisms or even with plants or soil mesofauna. For example, members of some lineages of the phylum *Verrucomicrobia* are symbionts of nematodes (Vanderkerkhove et al., 2000) or ciliates (Petroni et al., 2000). The use of liquid media will enhance the possibility of positive interactions with other microorganisms but will also increase the impacts of negative interactions. Again, the failure of liquid culture techniques to isolate so-called "unculturables" suggests that, in general, negative interactions have more impact than positive ones (Schoenborn et al., 2004).

Smaller inocula usually result in higher culturabilities (James and Sutherland, 1940; Jensen, 1962; Casida, 1968; Jensen, 1968), with a 10-fold dilution of the inoculum resulting in a twofold increase in culturability (Olsen and Bakken, 1987; Davis et al., 2005). This is probably due to slowed development of some nascent colonies due to inhibition by others when the colonies are too close together, perhaps by antibiotics or by the depletion of nutrients by faster-growing colonies so that slower-growing ones do not reach a detectable size. Members of the class "*Spartobacteria*" (subdivision 2) of the phylum *Verrucomicrobia* were never detected growing on plates that developed a mean of 29 colonies per plate from a mean inoculum of about 250 cells, and were only detected on plates that received a 10-fold smaller inoculum and on which a mean of 5 colonies per plate developed (Sangwan et al., 2005), suggesting that colony development was inhibited by the presence of other bacterial colonies on those plates receiving a denser inoculum. On the basis of the data of Davis et al. (2005) and Sangwan et al. (2005), the aim should be to inoculate plates with <50 cells and achieve <10 colonies per standard 9-cm diameter plate.

Even on plates where cells in the inoculum are well separated from one another, overgrowth can still be a problem. Spreading colony types were encountered more frequently on some media, and isolates displayed a spreading phenotype more often on those media (Davis et al., 2005). Diluted nutrient broth and media based on VL55 supported a low incidence of spreading phenotypes. Fungi also cause problems. Adding antifungal agents can decrease the number of bacterial colonies that form (P. H. Janssen, unpublished data), so choosing soils with small fungal populations, using very dilute inocula, and increasing the number of replicates seem to be the best solutions to this problem. A certain number of cultures will be lost to spreading colonies, fungal contamination, and also agar- and gellan gum-hydrolyzing microorganisms, which will make some plates unusable.

Soils also contain viruslike particles at levels that are easily detectable by electron microscopy (Ashelford et al., 2003), and bacteriophage titers for specific isolates are measurable (Marsh and Wellington, 1994). The stress of the transition to growth on laboratory media may result in the induction of lysogenic bacteriophage (Barer and Harwood, 1999), so minimizing these stresses could be important in this respect. Lytic phage could also result in low culturability, not only by contact of phage with host cells during cultivation but also by pseudolysogeny when the sudden change to a growing state allows infecting phage to complete their lytic cycle (Romig and Brodetsky, 1961; Ripp and Miller, 1997). It is probably difficult to prevent the death of already infected cells, but good physical separation of cells and phage particles may help reduce some cell death, and this may be part of the reason why culturability on plates is higher than in liquid culture.

The size of the soil sample can also be important. Ellingsøe and Johnsen (2002) found the culturability of bacteria was higher when 0.1-g or 1.0-g samples were used compared to 10-g samples. In line with this, Grundmann and Gourbière (1999) suggested that working with larger soil samples, in the region of grams, results in the mixing of bacteria from different microhabitats and can lead to competition and negative interactions during cultivation. Working from milligram quantities of soil means that bacteria that have already been selected by competition are cultured together. Culturing from multiple microfragmented samples should thus increase the diversity of soil bacteria obtained.

One way of avoiding negative interactions is to separate individual cells onto individual plates or into individual liquid cultures so that each gives rise to a pure culture, should it grow. Single cells can be isolated by micromanipulation (Casida, 1962), by optical or laser tweezers (Mitchell et al., 1993; Fröhlich and König, 2000), by gel encapsulation (Zengler et al., 2002), or by a range of other techniques (Brehm-Stecher and Johnson, 2004). However, separation of the cells by dilution to low densities, as recommended above, followed by plating on solidified media, is still a very effective and low-cost means of achieving almost the same effect.

GROWTH CONDITIONS: pH, TEMPERATURE, ATMOSPHERE

In addition to the obvious matching of the incubation temperature to the growth environment, lower temperatures may also have the advantage of lowering the rate of metabolism, which may reduce the rate of production of inhibitory compounds (discussed below). Incubation at 25°C yields more colonies than at 30°C (Thornton, 1922; Jensen, 1962), and temperatures of 20 to 25°C are recommended. Obviously, if samples are from colder climates, then the incubation temperature should be lowered.

It is also important to match the medium pH closely to that of the environment. Culturability was significantly higher when a medium with a pH of 5.5 was used to isolate bacteria from a soil with that pH than when medium at pH 7.0 was used (K. E. R. Davis and P. H. Janssen, unpublished data). Sait et al. (2006) identified moderately acidic pH values as an important factor in their success in isolating many members of subdivision 1 of the phylum *Acidobacteria*. This fits with the abundance of members of this group in different soils, in which they form relatively larger parts of the total community with decreasing soil pH (Sait et al., 2006). Pure cultures also display optima for growth at moderately acidic pH values of 4.5 to 6 (Eichorst et al., 2007; M. Sait and P. H. Janssen, unpublished data).

Stevenson et al. (2004) found that increased CO_2 partial pressures, mimicking those found in soils (Russell, 1950), increased the culturability of subdivision 1 acidobacteria, but Sait et al. (2006) and Eichorst et al. (2007) later showed that this was an effect on the medium pH. There was no difference in the general culturability of soil bacteria incubated under air enriched with 5% (vol/vol) CO_2 compared with air alone (K. E. R. Davis and P. H. Janssen, unpublished data), but different subsets of the soil community may have been cultured. This was not investigated. There is still a strong possibility that members of certain groups will display a requirement for an elevated CO_2 partial pressure, and so this should be borne in mind. This could be particularly significant for any potentially autotrophic groups, but there are also some types of catabolism in which CO_2 is a reactant (e.g., Ensign et al., 1998). However, heterotrophic CO_2 assimilation contributes only about 1% of microbial cell carbon in soil (Šantrůčková et al., 2005).

In addition to elevated CO_2 levels, soil air also can have lower than atmospheric O_2 partial pressures, and some groups may preferentially inhabit microoxic parts of the soil. Larger soil crumbs display strong gradients of O_2 (Sexstone et al., 1985), and of course, metabolism by microorganisms at sites with high substrate availabilities or affected by waterlogging will result in lower O_2 levels. The possibility of O_2 sensitivity of some groups of soil bacteria, not necessarily as anaerobes but as microaerophiles, should also be kept in mind. However, Stevenson et al. (2004) found no significant difference between the number of colonies formed when plates were incubated under 2% (vol/vol) oxygen compared with air, although, again, these experiments may have sampled two different subsets of the soil community.

TRANSITION TO GROWTH

Some soil bacteria can be expected to be in nongrowing long-term survival states, like endo- and exospores, cysts, or other physiologically or morphologically differentiated cell forms (Kolter et al., 1993). The existence of signals that trigger germination of spores in *Bacillus* spp. (Moir and Smith, 1990) may be a hint that other survival forms also respond to triggers that facilitate the transition to a growing state. These may be more significant for *r*-strategists. Spore germinants are typically common small molecules that might be released when fresh organic matter becomes available, and are components of many microbiological media.

It has been shown that adding specific signal compounds, for example from culture supernatants (Mukamolova et al., 1998a; Sun and Zhang, 1999; Shleeva et al., 2002; Freeman et al., 2002) or as pure compounds (Batchelor et al., 1997; Bussmann et al., 2001; Bruns et al., 2002), increases the cultivation efficiency of certain bacterial groups. A mix of acyl-homoserine lactones had no effect on the overall culturability of soil bacteria (Stevenson et al., 2004), although they did improve culturability of marine (Bruns et al., 2002) and freshwater (Bussmann et al., 2001) bacteria. Among the relatively small part of the total diversity of bacteria that has been investigated, different members of the phylum *Proteobacteria* use different acyl-homoserine lactones, while different members of the phylum *Firmicutes* use different posttranslationally modified peptides of 15 to 22 amino acids as signal molecules (Miller and Bassler, 2001). A mix of acyl-homoserine lactones did, however, appear to improve the incidence of colony formation by members of the phylum *Acidobacteria* from soil (Stevenson et al., 2004), but it remains to be shown that this was a direct effect on members of this group of bacteria.

A 16- to 17-kDa protein produced by *Micrococcus luteus* promotes resuscitation and growth of dormant cells of *M. luteus* and other members of the phylum *Actinobacteria* (Mukamolova et al., 1998b). Homologs of the gene coding for this protein have been detected in different members of this phylum, suggesting yet another type of cell-to-cell signal molecule. It is likely that some of the vast array of secondary metabolites produced by soil bacteria act as yet-unrecognized signal compounds (Kell et al., 1995). Given that the phylogenetic diversity of soil bacteria is very high, any one signal molecule is unlikely to have any significant impact on overall culturability but may affect culturability of a particular subset. The signal molecules for as-yet uncultured bacteria will not be known unless they can be deduced from metagenomic studies or found by trial and error.

It has been suggested that starved cells may accumulate inhibitory substances that prevent growth once suitable growth conditions are imposed (Kuznetzov et al., 1979; Mukamolova et al., 1995). The effects of inhibitory substances or metabolic products may be minimized by growing cells on membranes over an active microbial community in a simulated natural environment (Kaeberlein et al., 2002; Svenning et al., 2003; Bollmann et al., 2007), which will act as a sink for those compounds. These systems may also allow metabolic interaction between target cells and the rest of the community.

Electron transport processes in the presence of oxygen generate highly reactive oxygen species that can damage cell components and cause cell death (Bloomfield et al., 1998). Aerobic bacteria have mechanisms that can neutralize these radicals, but at low rates of growth these may not be expressed at high levels, and so a sudden transfer to a high substrate concentration may result in the rapid generation of large amounts of these radicals. However, Stevenson et al. (2004) did not find a significant increase in colony numbers when catalase was added to plates, and adding ascorbate or thioglycolate similarly did not result in increased culturability (P. H. Janssen, unpublished data). Perhaps these methods were ineffective, or reactive oxygen species were not a significant cause of cell death in these studies. Other methods to avoid generating reactive oxygen species include slowing down metabolism by using polymers as growth substrates and by incubating cultures at low temperatures.

INCUBATION TIME

If the majority of soil bacteria are *K*-selected species, they could be expected to have low growth rates even under laboratory conditions (Andrews and Harris, 1986). The mean growth rate of soil bacteria in situ is very low, with doubling times of greater than 100 days (Harris and Paul, 1994), which may favor the dominance of *K*-selected species. It has been noted that the time taken for bacteria to reach visible colony size can be long for soil bacteria (Jensen, 1968), and incubation times of 2 to 4 weeks have generally been used in the past (Jensen, 1968; Olsen and Bakken, 1987). With traditionally used media, this is sufficient for the maximum number of colonies to form (Davis et al., 2005), and on rich media, colony numbers reached a maximum after only a few days (Jensen, 1962). However, on more nutrient-poor media, colonies continue to form with extended incubation periods (Skinner et al., 1952; Davis et al., 2005).

Reasoning that *K*-selected autochthonous species may form visible colonies only after extended incubation, Janssen et al. (2002) incubated plates for 3 months and then identified members of rarely isolated bacterial groups among the colonies that had formed. This observation was extended in further studies (Sait et al., 2002; Joseph et al., 2003; Davis et al., 2005), resulting in the isolation of multiple strains belonging to the phyla *Acidobacteria*, *Chloroflexus*, *Gemmatimonadetes*, *Planctomycetes*, and *Verrucomicrobia* and to subclasses *Acidimicrobidae* and *Rubrobacteridae* of the phylum *Actinobacteria*. Davis et al. (2005) showed that members of these rarely isolated groups were found only among the later appearing colonies and that subcultured isolates of these groups consistently took longer to form visible colonies than did members of groups that are better represented by cultured isolates. From these experiments, it was concluded that well-studied bacterial groups contained many (but not only) rapidly-growing species, while the poorly studied groups contained predominantly slow-growing species. Increased lag periods may also contribute to these differences in time for visible colony appearance. In an independent study, Cavaletti et al. (2006) isolated 11 aerobic strains of the phylum *Chloroflexi* on plates that had been incubated for 8 weeks. Members of this phylum are commonly detected in soils (Janssen, 2006), but only very few soil isolates have been obtained. The incubation periods in all of these experiments were significantly longer than those used to isolate other soil bacteria designated as slow-growing, which appear within 1 week (Saito et al., 1998).

It is still not clear if the failure to isolate many members of the groups poorly represented by cultured isolates on traditionally used media is due to these media lacking essential components, due to inhibitory components in the medium, or due to lethal effects of the medium. However, it is clear that the perceived unculturability of at least some members of these groups was not due to them having any unusual growth requirements. They turned out to be slowly growing aerobic heterotrophs able to grow at pH values similar to that of the bulk soil from which they were isolated (Janssen et al., 2002; Sait et al., 2002; Joseph et al., 2003; Davis et al., 2005). It can be expected that a large part of the soil bacterial community will be similarly slow-growing species.

In addition to low growth rates, adaptation to laboratory media may be required. Soil bacteria (Davis et al., 2005) and bacteria in ice cores (Christner et al., 2000) took longer to form colonies when first plated than when they were subsequently transferred. Extended incubation and acclimatization may allow development of colonies from slow-growing or injured cells, or cells that require a long transition from a survival to a growing state (Christner et al., 2000; Zengler et al., 2002; Hahn et al., 2004; Bollmann et al., 2007).

Drying of plates during long incubations can be minimized if 5 or 6 plates are incubated together in a thin polyethylene zip-lock bag, with an empty plate at the bottom to prevent condensed water from flooding the plates. This also effectively prevents fungal contamination of large numbers of plates should a few plates be overgrown by fungi. Polyethylene films are permeable to oxygen and carbon dioxide but retain water (Bremner and Douglas, 1971).

DETECTING COLONIES

Obtaining visible colonies is another factor that may limit apparent culturability. All colonies on a plate do not develop at the same rate, even in pure cultures (Ishikuri and Hattori, 1985; Mochizuki and Hattori, 1986). Some cells, after extraction from soil, develop into microcolonies of just a few cells (Winding et al., 1994). Some of the microcolonies will proceed to form larger colonies. Even after 6 months of incubation, microscopic examination of plates revealed that there were numerous minicolonies that had diameters of between 50 and 200 μm (K. E. R. Davis and P. H. Janssen, unpublished data). Some of these minicolonies were formed by bacteria that belong to lineages that lack cultured representatives (K. E. R. Davis and P. H. Janssen, unpublished data). Watve et al. (2000) similarly found that after 1 month of incubation, many minicolonies were detectable. It is not clear if micro- or minicolonies would form visible colonies if incubated even longer than 6 months, or if they have a self-limiting growth phenotype.

Bacteria with a dispersive phenotype would also not be easily detected. This phenotype has been described among marine bacteria (Simu and Hagström, 2004). Given that gliding and spreading phenotypes are common among cultured soil bacteria (Reichenbach and Dworkin, 1981), it seems possible that bacteria with a more extreme dispersive phenotype have eluded detection. Similarly, bacteria that form flat sheets, only one or a few cells thick on plates, rather than more familiar heaped colonies, would also not be easily detected. Cells that display this phenotype have been observed on plates inoculated with soil (Ferrari et al., 2005; K. E. R. Davis and P. H. Janssen, unpublished data).

Detection of slow-growing bacteria belonging to specific groups by the use of targeted identification systems allows their presence in culture to be recognized more easily (Leadbetter, 2003). Group-specific oligonucleotide probes have been used to screen colonies that have particular DNA sequences. Screening large number of colonies is important, especially for bacteria that form visible colonies only rarely. By screening more than 1,200 bacterial colonies that developed on plates receiving small inocula, Sangwan et al.

(2005) cultured 14 isolates from two different classes (subdivisions) of the phylum *Verrucomicrobia*. The incidence of colonies of this phylum was therefore only 1.2%, although they were estimated to make up 6% of the bacterial community in the soil under investigation. Using fluorescently labeled oligonucleotide probes, Ferrari et al. (2005) observed cultured microcolonies of soil bacteria affiliated with the TM-7 phylum, which suggests that isolation of members of this group may be possible by modifying the conditions they employed. Sait et al. (2006) used oligonucleotide probes to detect members of subdivision 1 acidobacteria on solid media and to culture and isolate many members of this group that closely matched the diversity detected by cultivation-independent methods. 16S rRNA genes were amplified from a small amount of each colony, and these products were probed with oligonucleotides specific for the target group to identify colonies of interest. Similar methods could be easily adapted to detect bacteria that possess genes coding for specific functions.

CONCLUSIONS

Many useful guidelines for the cultivation of soil bacteria have been published by Parkinson et al. (1971) and Jensen (1962, 1968), and an excellent general introduction to isolating bacteria has been compiled by Overmann (2006). On the basis of recent insights, a number of general recommendations can be added to assist researchers attempting to isolate soil bacteria that belong to difficult-to-culture groups.

1. The medium should be chosen to allow transition of the target organisms from very slow growth in the soil to more rapid growth in the laboratory. If plates are used, growth does not have to be good at first. Only very small amounts of substrate, nitrogen, and phosphate are required to form a visible colony. Use whatever information is available from molecular ecological and other studies to devise the medium.

2. The inoculum should be dilute, with <50 cells per plate. Liquid cultures should aim for one cell per culture. This will lessen the impact of potentially negative interactions among the cells that do grow. Diluents should be based on the growth medium.

3. The growth conditions (especially pH and salt concentrations) should match the environment, and the incubation temperature should be low. The cultures should be incubated for a long time. The target group may be slow growing, have a long lag time, or both. The key ingredient is patience.

4. A very large number of colonies should be screened to find the few target colonies that have arisen. The abundance of colonies is not correlated with the abundance of the individual groups in the environment. A suitable detection system should be employed to find the target colonies among the many that will form.

ACKNOWLEDGMENTS

I sincerely thank the people who worked with me in my laboratory at the University of Melbourne for their enthusiasm, insights, and diligence; for sharing my belief that soil bacteria are not generally unculturable; and for sharing the joy of the successful isolation of some of these bacteria.

REFERENCES

Aagot, N., O. Nybroe, P. Nielsen, and K. Johnsen. 2001. An altered *Pseudomonas* diversity is recovered from soil by using nutrient-poor *Pseudomonas*-selective soil extract media. *Appl. Environ. Microbiol.* **67:**5233–5239.

Aakra, Å, M. Hesseløe, and L. R. Bakken. 2000. Surface attachment of ammonia-oxidizing bacteria in soil. *Microb. Ecol.* **39:**222–235.

Andrews, J. H., and R. F. Harris. 1986. *r*- and *K*-selection and microbial ecology. *Adv. Microb. Ecol.* **9:**99–147.

Angle, J. S., S. P. McGrath, and R. L. Chaney. 1991. New culture medium containing ionic concentrations of nutrients similar to that found in the soil solution. *Appl. Environ. Microbiol.* **57:**3674–3676.

Ashelford, K. E., M. J. Day, and J. C. Fry. 2003. Elevated abundance of bacteriophage infecting bacteria in soil. *Appl. Environ. Microbiol.* **69:**285–289.

Bakken, L. R. 1997. Culturable and nonculturable bacteria in soil, p. 47–61. *In* J. D. van Elsas, J. T. Trevors, and E. M. H. Wellington (ed.), *Modern Soil Microbiology*. Marcel Dekker, New York, NY.

Bakken, L. R., and R. A. Olsen. 1989. DNA-content of soil bacteria of different cell size. *Soil Biol. Biochem.* **21:**789–793.

Balestra, G. M., and I. J. Misaghi. 1997. Increasing the efficiency of the plate count method for estimating bacterial diversity. *J. Microbiol. Methods* **30:**111–117.

Barer, M. R., and C. R. Harwood. 1999. Bacterial viability and culturability. *Adv. Microb. Physiol.* **41:**93–137.

Bartscht, K., H. Cypionka, and J. Overmann. 1999. Evaluation of cell activity and of methods for the cultivation of bacteria from a natural lake community. *FEMS Microbiol. Ecol.* **28:**249–259.

Batchelor, S. E., M. Cooper, S. R. Chhabra, L. A. Glover, G. S. Stewart, P. Williams, and J. I. Prosser. 1997. Cell density-regulated recovery of starved biofilm populations of ammonia-oxidizing bacteria. *Appl. Environ. Microbiol.* **63:**2281–2286.

Bloomfield, S. F., G. S. A. B. Stewart, C. E. R. Dodd, I. R. Booth, and E. G. M. Power. 1998. The viable but non-culturable phenomenon explained? *Microbiology* **144:**1–3.

Bollmann, A., K. Lewis, and S. S. Epstein. 2007. Growth of environmental samples in a diffusion chamber increases the diversity of recovered isolates. *Appl. Environ. Microbiol.* **73:**6386–6390.

Brehm-Stecher, B. F., and E. A. Johnson. 2004. Single-cell microbiology: tools, technologies, and applications. *Microbiol. Mol. Biol. Rev.* **68:**538–559.

Bremner, J. M., and L. A. Douglas. 1971. Use of plastic films for aeration in soil incubation experiments. *Soil Biol. Biochem.* **3:**289–296.

Brookes, P.C., K. R. Tate, and D. S. Jenkinson. 1983. The adenylate energy charge of the soil microbial biomass. *Soil Biol. Biochem.* **15:**9–16.

Bruns, A., H. Cypionka, and J. Overmann. 2002. Cyclic AMP and acyl homoserine lactones increase the cultivation efficiency of heterotrophic bacteria from the central Baltic Sea. *Appl. Environ. Microbiol.* **68:**3978–3987.

Bruns, A., U. Nübel, H. Cypionka, and J. Overmann. 2003. Effect of signal compounds and incubation conditions on the culturability of freshwater bacterioplankton. *Appl. Environ. Microbiol.* **69:**1980–1989.

Bussmann, I., B. Philipp, and B. Schink. 2001. Factors influencing the cultivability of lake water bacteria. *J. Microbiol. Methods* **47:**41–50.

Button, D. K. 1985. Kinetics of nutrient-limited transport and microbial growth. *Microbiol. Rev.* **49:**270–297.

Calcott, P., and T. Calvert. 1981. Characterization of 3′,5′-cyclic AMP phosphodiesterase in *Klebsiella aerogenes* and its role in substrate-accelerated death. *J. Gen. Microbiol.* **122:**313–321.

Calcott, P. H., and J. R. Postgate. 1972. On substrate-accelerated death in *Klebsiella aerogenes*. *J. Gen. Microbiol.* **70:**115–122.

Casida, L. E. 1962. On the isolation and growth of individual cells from soil. *Can. J. Microbiol.* **8:**115–119.

Casida, L. E. 1968. Methods for the isolation and estimation of activity of soil bacteria, p. 97–122. *In* T. R. G. Gray and D. Parkinson (ed.), *The Ecology of Soil Bacteria*. Liverpool University Press, Liverpool, United Kingdom.

Cavaletti, L., P. Monciardini, R. Bamonte, P. Schumann, M. Rohde, M. Sosio, and S. Donadio. 2006. New lineage of filamentous, spore-forming, gram-positive bacteria from soil. *Appl. Environ. Microbiol.* **72:**4360–4369.

Cayley, S., M. T. Record, and B. A. Lewis. 1989. Accumulation of 3-(*N*-morpholino)-propanesulfonate by osmotically stressed *Escherichia coli* K-12. *J. Bacteriol.* **171:**3597–3602.

Chin, K.-J., D. Hahn, U. Hengstmann, W. Liesack, and P. H. Janssen. 1999. Characterization and identification of numerically abundant culturable bacteria from the anoxic bulk soil of rice paddy microcosms. *Appl. Environ. Microbiol.* **65:** 5042–5049.

Christensen, H., M. Hansen, and J. Sørensen. 1999. Counting and size classification of active soil bacteria by fluorescence in situ hybridization with an rRNA oligonucleotide probe. *Appl. Eviron. Microbiol.* **65:**1753–1761.

Christner, B. C., E. Mosley-Thompson, L. G. Thompson, V. Zagorodnov, K. Sandman, and J. N. Reeve. 2000. Recovery and identification of viable bacteria immured in glacial ice. *Icarus* **144:**479–485.

Conn, H. J. 1918. The microscopic study of bacteria and fungi in soil. *N. Y. Agr. Exp. Sta. Tech. Bull.* **64:**3–20.

Conn, H. J. 1927. The general soil flora. *N.Y. Agr. Exp. Sta. Tech. Bull.* **129:**3–10.

Contin, M., A. Todd, and P. C. Brookes. 2001. The ATP concentration in the soil microbial biomass. *Soil Biol. Biochem.* **33:**701–704.

Curtis, T. P., W. T. Sloan, and J. C. Scannell. 2002. Estimating prokaryotic diversity and its limits. *Proc. Natl. Acad. Sci. USA* **99:**10494–10499.

Dabek-Szreniawska, M., and T. Hattori. 1981. Winogradsky's salts solution as a diluting medium for plate count of oligotrophic bacteria in soil. *J. Gen. Appl. Microbiol.* **27:**517–518.

Davis, K. E. R., S. J. Joseph, and P. H. Janssen. 2005. Effects of growth medium, inoculum size, and incubation time on the culturability and isolation of soil bacteria. *Appl. Environ. Microbiol.* **71:**826–834.

Dawes, E. A. 1976. Endogenous metabolism and the survival of starved procaryotes. *Symp. Soc. Gen. Microbiol.* **26:**19–53.

De Spiegeleer, P., J. Sermon, A. Lietaert, A. Aertsen, and C. W. Michiels. 2004. Source of tryptone in growth medium affects oxidative stress resistance in *Escherichia coli*. *J. Appl. Microbiol.* **97:**124–133.

Dykhuizen, D. E. 1998. Santa Rosalia revisited: why are there so many species of bacteria? *Antonie van Leeuwenhoek* **73:**25–33.

Egdell, J. W., W. A. Cuthbert, C. A. Scarlett, and S. B. Thomas. 1960. Some studies of the colony count technique for soil bacteria. *J. Appl. Bacteriol.* **23:**69–86.

Eichorst, S. A., J. A. Breznak, and T. M. Schmidt. 2007. Isolation and characterization of soil bacteria that define *Terriglobus* gen. nov., in the phylum *Acidobacteria*. *Appl. Environ. Microbiol.* **73:**2708–2717.

Eilers, H., J. Pernthaler, J. Peplies, F. O. Glöckner, G. Gerdts, and R. Amann. 2001. Isolation of novel pelagic bacteria from the German Bight and their seasonal contributions to surface picoplankton. *Appl. Environ. Microbiol.* **67:**5134–5142.

Ellingsøe, P., and K. Johnsen. 2002. Influence of soil sample sizes on the assessment of bacterial community structure. *Soil Biol. Biochem.* **34:**1701–1707.

Ellis, R. J., P. Morgan, A. J. Weightman, and J. C. Fry. 2003. Cultivation-dependent and -independent approaches for determining bacterial diversity in heavy-metal-contaminated soil. *Appl. Environ. Microbiol.* **69:**3223–3230.

Ensign, S. A., F. J. Small, J. R. Allen, and M. K. Sluis. 1998. New roles for CO_2 in the microbial metabolism of aliphatic epoxides and ketones. *Arch. Microbiol.* **169:**179–187.

Ferrari, B. C., S. J. Binnerup, and M. Gillings. 2005. Microcolony cultivation on a soil substrate membrane system selects for previously uncultured soil bacteria. *Appl. Environ. Microbiol.* **71:**8714–8720.

Fischer, H. 1909. Bakteriologisch-chemishe Untersuchungen. Bakteriologischer Teil. *Landw. Jahrb.* **38:**355–364.

Freeman, R., J. Dunn, J. Magee, and A. Barrett. 2002. The enhancement of isolation of mycobacteria from a rapid liquid culture system by broth culture supernate of *Micrococcus luteus*. *J. Med. Microbiol.* **51:**92–93.

Fröhlich, J., and H. König. 2000. New techniques for the isolation of single prokaryotic cells. *FEMS Microbiol. Rev.* **24:**567–572.

Furlong, M. A., D. R. Singleton, D. C. Coleman, and W. B. Whitman. 2002. Molecular and culture-based analyses of prokaryotic communities from an agricultural soil and the burrows and casts of the earthworm *Lumbricus rubellus*. *Appl. Environ. Microbiol.* **68:**1265–1279.

Gans, J., M. Wolinsky, and J. Dunbar. 2005. Computational improvements reveal great bacterial diversity and high metal toxicity in soil. *Science* **309:**1387–1390.

Good, N. E., and S. Izawa. 1972. Hydrogen ion buffers. *Methods Enzymol.* **24:**53–68.

Gray, N. D., and I. M. Head. 2001. Linking genetic identity and function in communities of uncultured bacteria. *Environ. Microbiol.* **3:**481–492.

Gray, T. R. G., P. Baxby, I. R. Hill, and M. Goodfellow. 1968. Direct observation of bacteria in soil, p. 171–197. *In* T. R. G. Gray and D. Parkinson (ed.), *The Ecology of Soil Bacteria*. Liverpool University Press, Liverpool, United Kingdom.

Grundmann, G. L., and F. Gourbière. 1999. A micro-sampling approach to improve the inventory of bacterial diversity in soil. *Appl. Soil Ecol.* **13:**123–126.

Hahn, M. W., P. Stadler, Q. L. Wu, and M. Pöckl. 2004. The filtration-acclimatization-method for isolation of an important fraction of the not readily cultivable bacteria. *J. Microbiol. Methods* **57:**379–390.

Harris, D., and E. A. Paul. 1994. Measurements of bacterial growth rates in soil. *Appl. Soil Ecol.* **1:**277–290.

Hattori, T. 1976. Plate count of bacteria in soil on a diluted nutrient broth as a culture medium. *Rep. Inst. Agric. Res. Tohoku Univ.* **27:**23–30.

Hattori, T. 1980. A note on the effect of different types of agar on plate count of oligotrophic bacteria in soil. *J. Gen. Appl. Microbiol.* **26:**373–374.

Hattori, T., and R. Hattori. 1976. The physical environment in soil microbiology: an attempt to extend principles of microbiology to soil microorganisms. *CRC Crit. Rev. Microbiol.* **4:**423–461.

Hattori, R., and T. Hattori. 1980. Sensitivity to salts and organic compounds of soil bacteria isolated on diluted media. *J. Gen. Appl. Microbiol.* **26:**1–14.

Hill, I. R., and T. R. G. Gray. 1966. Magnetic stirring as a method of dispersing soil bacteria in diluents. *Soil Biol.* **5:**12–14.

Hugenholtz, P., B. M. Goebel, and N. R. Pace. 1998. Impact of culture-independent studies on the emerging phylogenetic view of bacterial diversity. *J. Bacteriol.* **180:**4765–-4774.

Ishikuri, S., and T. Hattori. 1985. Formation of bacterial colonies in successive time intervals. *Appl. Environ. Microbiol.* **49:**870–873.

James, N. 1958. Soil extract in soil microbiology. *Can. J. Microbiol.* **4:**363–370.

James, N., and M. Sutherland. 1940. Effect of numbers of colonies per plate on the estimate of the bacterial population in soil. *Can. J. Res. Section C* **18:**347–356.

Janssen, P. H. 2006. Identifying the dominant soil bacterial taxa in libraries of 16S rRNA and 16S rRNA genes. *Appl. Environ. Microbiol.* **72:**1719–1728.

Janssen, P. H., P. S. Yates, B. E. Grinton, P. M. Taylor, and M. Sait. 2002. Improved culturability of soil bacteria and isolation in pure culture of novel members of the divisions *Acidobacteria, Actinobacteria, Proteobacteria*, and *Verrucomicrobia*. *Appl. Environ. Microbiol.* **68:**2391–2396.

Jensen, V. 1962. Studies on the microflora of Danish beech forest soils. I. The dilution plate count technique for the enumeration of bacteria and fungi in soil. *Zentralbl. Bakteriol. Parasitenkd. Abt. 2* **116:**13–32.

Jensen, V. 1968. The plate count technique, p. 158–170. *In* T. R. G. Gray and D. Parkinson (ed.), *The Ecology of Soil Bacteria*. Liverpool University Press, Liverpool, United Kingdom.

Joseph, S. J., P. Hugenholtz, P. Sangwan, C. A. Osborne, and P. H. Janssen. 2003. Laboratory cultivation of widespread and previously uncultured soil bacteria. *Appl. Environ. Microbiol.* **69:**7210–7215.

Kaeberlein, T., K. Lewis, and S. Epstein. 2002. Isolating "uncultivable" microorganisms in pure culture in a simulated natural environment. *Science* **296:**1127–1129.

Katayama, A., K. Kai, and K. Fujie. 1998. Extraction efficiency, size distribution, colony formation and [^{3}H]-thymidine incorporation of bacteria directly extracted from soil. *Soil Sci. Plant Nutr.* **44:**245–252.

Kell, D. B., A. S. Kaprelyants, and A. Grafen. 1995. On pheromones, social behaviour and the functions of secondary metabolism in bacteria. *Trends Ecol. Evolution* **10:**126–129.

Kieft, T. L. 2000. Size matters: dwarf cells in soil and subsurface terrestrial environments, p. 19–46. *In* R. R. Colwell and D. J. Grimes (ed.), *Non-culturable Microorganisms in the Environment.* American Society for Microbiology, Washington, DC.

Knowles, C. J. 1977. Microbial metabolic regulation by adenine pools. *Symp. Soc. Gen. Microbiol.* **27:**241–283.

Koch, R. 1881. Zur Untersuchung von pathogenen Organismen. *Mitth. a. d. Kaiserl. Gesundheitsampte* **1:**1–48.

Kolter, R., D. A. Siegele, and A. Tormo. 1993. The stationary phase of the bacterial life cycle. *Annu. Rev. Microbiol.* **47:**855–874.

Kuznetsov, S. I., G. A. Dubinina, and N. A. Lapteva. 1979. Biology of oligotrophic bacteria. *Annu. Rev. Microbiol.* **33:**377–387.

Lazzarini, A., L. Cavaletti, G. Toppo, and F. Marinella. 2000. Rare genera of actinomycetes as potential producers of new antibiotics. *Antonie van Leeuwenhoek* **78:**399–405.

Leadbetter, J. R. 2003. Cultivation of recalcitrant microbes: cells are alive, well and revealing their secrets in the 21st century laboratory. *Curr. Opin. Microbiol.* **6:**274–281.

Leininger, S., T. Urich, M. Schloter, L. Schwark, J. Qi, G. W. Nicol, J. I. Prosser, S. C. Schuster, and C. Schleper. 2006. Archaea predominate among ammonia-oxidizing prokaryotes in soils. *Nature* **442:**806–809.

Liles, M. R., B. F. Manske, S. B. Bintrim, J. Handelsman, and R. M. Goodman. 2003. A census of rRNA genes and linked genomic sequences within a soil metagenomic library. *Appl. Environ. Microbiol.* **69:**2684–2691.

Lindahl, A., Å. Frostegård, L. Bakken, and E. Bååth. 1997. Phospholipid fatty acid composition of size fractionated indigenous soil bacteria. *Soil Biol. Biochem.* **29:**1565–1569.

Lindahl, V. 1996. Improved soil dispersion procedures for total bacterial counts, extraction of indigenous bacteria and cell survival. *J. Microbiol. Methods* **25:**279–286.

Lochhead, A. G., and F. E. Chase. 1943. Qualitative studies of soil microorganisms: V. Nutritional requirements of the predominant bacterial flora. *Soil Sci.* **55:**185–195.

Lochhead, A. G., and R. H. Thexton. 1952. Qualitative studies of soil bacteria. X: Bacteria requiring vitamin B_{12} as growth factor. *J. Bacteriol.* **63:**219–226.

Lueders, T., R. Kindler, A. Miltner, M. W. Friedrich, and M. Kaestner. 2006. Identification of bacterial micropredators distinctively active in a soil microbial food web. *Appl. Environ. Microbiol.* **72:**5342–5348.

MacDonald, R. M. 1980. Cytochemical demonstration of catabolism in soil micro-organisms. *Soil Biol. Biochem.* **12:**419–424.

Marsh, P., and E. M. H. Wellington. 1994. Phage-host interactions in soil. *FEMS Microbiol. Ecol.* **15:**99–108.

Mason, T. G., and G. Blunden. 1989. Quaternary ammonium and tertiary sulfonium compounds of algal origin as alleviators of osmotic stress. *Bot. Mar.* **32:**313–316.

McCaig A. E., S. J. Grayston, J. I. Prosser, and L. A. Glover. 2001. Impact of cultivation on characterisation of species composition of soil bacterial communities. *FEMS Microbiol. Ecol.* **35:**37–48.

Miller, M. B., and B. L. Bassler. 2001. Quorum sensing in bacteria. *Annu. Rev. Microbiol.* **55:**165–199.

Mitchell, J. G., R. Weller, M. Beconi, J. Sell, and J. Holland. 1993. A practical optical trap for manipulating and isolating bacteria from complex microbial communities. *Microb. Ecol.* **25:**113–119.

Mochizuki, M., and T. Hattori. 1986. Kinetics of microcolony formation of a soil oligotrophic bacterium, *Agromonas* sp. *FEMS Microbiol. Ecol.* **38:**51–55.

Moir, A., and D. A. Smith. 1990. The genetics of bacterial spore germination. *Annu. Rev. Microbiol.* **44:**531–553.

Monciardini, P., L. Cavaletti, P. Schumann, M. Rohde, and S. Donadio. 2003. *Conexibacter woesii* gen. nov., sp. nov., a novel representative of a deep evolutionary line of descent within the class *Actinobacteria*. *Int. J. Syst. Evol. Microbiol.* **53:**569–576.

Mukamolova, G. V., A. S. Kaprelyants, and D. B. Kell. 1995. Secretion of an antibacterial factor during resuscitation of dormant cells in *Micrococcus luteus* cultures held in an extended stationary phase. *Antonie van Leeuwenhoek* **67:**289–295.

Mukamolova, G. V., N. D. Yanopolskaya, D. B. Kell, and A. S. Kaprelyants. 1998a. On resuscitation from the dormant state of *Micrococcus luteus*. *Antonie van Leeuwenhoek* **73:**237–243.

Mukamolova, G. V., A. S. Kaprelyants, D. I. Young, M. Young, and D. B. Kell. 1998*b*. A bacterial cytokine. *Proc. Natl. Acad. Sci. USA* **95:**8916–8921.

Mukamolova, G. V., A. S. Kaprelyants, D. B. Kell, and M. Young. 2003. Adoption of the transiently non-culturable state—a bacterial survival strategy? *Adv. Microb. Physiol.* **47:**65–129.

Neidhardt, N. C., J. L. Ingraham, and M. Schaechter. 1990. *Physiology of the Bacterial Cell. A Molecular Approach*. Sinauer Associates, Inc., Sunderland, MA.

Novitsky, J. A. 1987. Microbial growth rates and biomass production in a marine sediment: evidence for a very active but mostly nongrowing community. *Appl. Environ. Microbiol.* **53:**2368–2372.

Ohta, H., and T. Hattori. 1980. Bacteria sensitive to nutrient broth medium in terrestrial environments. *Soil Sci. Plant Nutr.* **26:**99–107.

Olsen, R. A., and L. R. Bakken. 1987. Viability of soil bacteria: optimization of plate-counting techniques and comparisons between total counts and plate counts within different size groups. *Microb. Ecol.* **13:**59–74.

Overmann, J. 2006. Principles of enrichment, isolation, cultivation and preservation of prokaryotes, p. 80–136. *In* M. Dworkin, S. Falkow, E. Rosenberg, K.-H. Schleifer, and E. Stackebrandt (ed.), *The Prokaryotes,* 3rd ed. Vol. 1*: Symbiotic Associations, Biotechnology, Applied Microbiology*. Springer, New York, NY.

Palumbo, A. V., C. Zhang, S. Liu, S. P. Scarborough, S. M. Pfiffner, and T. J. Phelps. 1996. Influence of media on measurement of bacterial populations in the subsurface. *Appl. Biochem. Biotech.* **57/58:**905–914.

Parkinson, D., T. R. G. Gray, J. Holding, and H. M. Nagel-de-Boois. 1971. Heterotrophic microflora, p. 34–50. *In* J. Phillipson (ed.), *Methods of Study in Quantitative Soil Ecology: Population, Production and Energy Flow*. Blackwell Scientific Publications, Oxford, United Kingdom.

Petroni, G., S. Spring, K.-H. Schleifer, F. Verni, and G. Rosati. 2000. Defensive extrusive ectosymbionts of *Euplotidium* (Ciliophora) that contain microtubule-like structures are bacteria related to *Verrucomicrobia*. *Proc. Natl. Acad. Sci. USA* **97:**1813–1817.

Podar, M., C. B. Abulencia, M. Walcher, D. Hutchison, K. Zengler, J. A. Garcia, T. Holland, D. Cotton, L. Hauser, and M. Keller. 2007. Targeted access to the genomes of low-abundance organisms in complex microbial communities. *Appl. Environ. Microbiol.* **73:**3205–3214.

Postgate, J. R., and J. R. Hunter. 1964. Accelerated death of *Aerobacter aerogenes* starved in the presence of growth limiting substrates. *J. Gen. Microbiol.* **34:**459–473.

Radajewski, S., P. Ineson, N. R. Parekh, and J. C. Murrell. 2000. Stable-isotope probing as a tool in microbial ecology. *Nature* **403:**646–649.

Ramsay, A. J. 1984. Extraction of bacteria from soil: efficiency of shaking or ultrasonification as indicated by direct counts and autoradiography. *Soil Biol. Biochem.* **16:**475–481.

Reichenbach, H., and M. Dworkin. 1981. Introduction to the gliding bacteria, p. 315–327. *In* M. P. Starr, H. Stolp, H. G. Trüper, A. Balows, and H. G. Schlegel (ed.), *The Prokaryotes. A Handbook on Habitats, Isolation, and Identification of Bacteria,* vol. 1. Springer-Verlag, Heidelberg, Germany.

Ripp, S., and R. V. Miller. 1997. The role of pseudolysogeny in bacteriophage-host interactions in a natural freshwater environment. *Microbiology* **143:**2065–2070.

Rochelle, P. A., B. A. Cragg, J. C. Fry, R. J. Parkes, and A. J. Weightman. 1994. Effect of sample handling on estimation of bacterial diversity in marine sediments by 16S rRNA gene sequence analysis. *FEMS Microbiol. Ecol.* **15:**215–225.

Romig, W. R., and A. M. Brodetsky. 1961. Isolation and preliminary characterization of bacteriophages for *Bacillus subtilis*. *J. Bacteriol.* **82:**135–141.

Rösch, C., A. Mergel, and H. Bothe. 2002. Biodiversity of denitrifying and dinitrogen-fixing bacteria in an acid forest soil. *Appl. Environ. Microbiol.* **68:**3818–3829.

Russell, E. J. 1950. *Soil Conditions and Plant Growth,* 8th ed. Longmans, Green and Co., London, United Kingdom.

Sait, M., P. Hugenholtz, and P. H. Janssen. 2002. Cultivation of globally-distributed soil bacteria from phylogenetic lineages previously only detected in cultivation-independent surveys. *Environ. Microbiol.* **4:**654–666.

Sait, M., K. E. R. Davis, and P. H. Janssen. 2006. Effect of pH on the isolation and distribution of members of subdivision 1 of the phylum *Acidobacteria* occurring in soil. *Appl. Environ. Microbiol.* **72**:1852–1857.

Saito, A., H. Mitsui, R. Hattori, K. Minamisawa, and T. Hattori. 1998. Slow-growing and oligotrophic soil bacteria phylogenetically close to *Bradyrhizobium japonicum*. *FEMS Microbiol. Ecol.* **25**:277–286.

Sangwan, P., X. Chen, P. Hugenholtz, and P. H. Janssen. 2004. *Chthoniobacter flavus* gen. nov., sp. nov., the first pure-culture representative of subdivision two, *Spartobacteria* classis nov., of the phylum *Verrucomicrobia*. *Appl. Environ. Microbiol.* **70**:5875–5881.

Sangwan, P., S. Kovac, K. E. R. Davis, M. Sait, and P. H. Janssen. 2005. Detection and cultivation of soil verrucomicrobia. *Appl. Environ. Microbiol.* **71**:8402–8410.

Šantrůčková, H., M. I. Bird, D. Elhottová, J. Novák, T. Picek, M. Šimek, and R. Tykva. 2005. Heterotrophic fixation of CO_2 in soil. *Microb. Ecol.* **49**:218–225.

Schloss, P. D., and J. Handelsman. 2006. Toward a census of bacteria in soil. *PloS Comp. Biol.* **2**:e92.

Schoenborn, L., P. S. Yates, B. E. Grinton, P. Hugenholtz, and P. H. Janssen. 2004. Liquid serial dilution is inferior to solid media for isolation of cultures representing the phylum level diversity of soil bacteria. *Appl. Environ. Microbiol.* **70**:4363–4366.

Schut, F. 1994. *Ecophysiology of a Marine Ultramicrobacterium*. Ph.D. thesis, University of Groningen, The Netherlands.

Sexstone, A. J., N. P. Revsbech, T. P. Parkin, and J. M. Tiedje. 1985. Direct measurement of oxygen profiles and denitrification rates in soil aggregates. *Soil Sci. Soc. Amer. J.* **49**:645–651.

Shleeva, M. O., K. Bagramyan, M. V. Telkov, G. V. Mukamolova, M. Young, D. B. Kell, and A. S. Kaprelyants. 2002. Formation and resuscitation of "non-culturable" cells of *Rhodococcus rhodochrous* and *Mycobacterium tuberculosis* in prolonged stationary phase. *Microbiology* **148**:1581–1591.

Simon, H. M., C. E. Jahn, L. T. Bergerud, M. K. Sliwinski, P. J. Weimer, D. K. Willis, and R. M. Goodman. 2005. Cultivation of mesophilic soil crenarchaeotes in enrichment cultures from plant roots. *Appl. Environ. Microbiol.* **71**:4751–4760.

Simu, K., and Å. Hagström. 2004. Oligotrophic bacterioplankton with a novel single-cell life strategy. *Appl. Environ. Microbiol.* **70**:2445–2451.

Skinner, F. A., P. C. T. Jones, and J. E. Mollison. 1952. A comparison of a direct- and a plate-counting technique for the quantitative estimation of soil micro-organisms. *J. Gen. Microbiol.* **6**:261–271.

Sørheim, R., V. L. Torsvik, and J. Goksøyr. 1989. Phenotypic divergences between populations of soil bacteria isolated on different media. *Microb. Ecol.* **17**:181–192.

Stevenson, B. S., S. A. Eichorst, J. T. Wertz, T. M. Schmidt, and J. A. Breznak. 2004. New strategies for cultivation and detection of previously uncultured microbes. *Appl. Environ. Microbiol.* **70**:4748–4755.

Streit, W. R., and R. A. Schmitz. 2004. Metagenomics—the key to the uncultured microbes. *Curr. Opin. Microbiol.* **7**:492–498.

Sun, Z., and Y. Zhang. 1999. Spent culture supernatant of *Mycobacterium tuberculosis* H37Ra improves viability of aged cultures of this strain and allows small inocula to initiate growth. *J. Bacteriol.* **181**:7626–7628.

Svenning, M. M., I. Wartiainen, A. G. Hestnes and S. J. Binnerup. 2003. Isolation of methane oxidising bacteria from soil by use of a soil substrate membrane system. *FEMS Microbiol. Ecol.* **44**:347–354.

Taylor, C. B. 1951. Nature of the factor in soil-extract responsible for bacterial growth-stimulation. *Nature* **168**:115–116.

Thornton, H. G. 1922. On the development of a standardized agar medium for counting soil bacteria, with especial regard to the repression of spreading colonies. *Ann. Appl. Biol.* **9**:241–274.

Torsvik, V., J. Goksøyr, and F. L. Daae. 1990. High diversity in DNA of soil bacteria. *Appl. Environ. Microbiol.* **56**:782–787.

Tringe, S. G., C. von Mering, A. Kobayashi, A. A. Salamov, K. Chen, H. W. Chang, M. Podar, J. M. Short, E. J. Mathur, J. C. Detter, P. Bork, P. Hugenholtz, and E. M. Rubin. 2005. Comparative metagenomics of microbial communities. *Science* **308**:554–557.

Vandekerkhove, T. T. M., A. Willems, M. Gillis, and A. Coomans. 2000. Occurrence of novel verrucomicrobial species, endosymbiotic and associated with parthenogenesis in *Xiphinema americanum*-group species (Nematoda, Longidoridae). *Int. J. Syst. Evol. Microbiol.* **50**:2197–2205.

Wang, J., C. Jenkins, R. I. Webb, and J. A. Fuerst. 2002. Isolation of *Gemmata*-like and *Isosphaera*-like planctomycete bacteria from soil and freshwater. *Appl. Environ. Microbiol.* **68**:417–422.

Waterbury, J. B. 1991. The cyanobacteria—isolation, purification, and identification, p. 149–196. *In* M. P. Starr, H. Stolp, H. G. Trüper, A. Balows, and H. G. Schlegel (ed.), *The Prokaryotes. A Handbook on Habitats, Isolation, and Identification of Bacteria,* vol. 1. Springer-Verlag, Heidelberg, Germany.

Watve, M., V. Shejval, C. Sonawane, M. Rahalkar, A. Matapurkar, Y. Shouche, M. Patole, N. Phadnis, A. Champhenkar, K. Damle, S.

Karandikar, V. Kshirsagar, and M. Jog. 2000. The "K" selected oligophilic bacteria: a key to uncultured diversity? *Curr. Sci.* **78:**1535–1542.

Widdel, F. 1987. New types of acetate-oxidizing, sulfate-reducing *Desulfobacter* species, *D. hydrogenophilus* sp. nov., *D. latus* sp. nov., and *D. curvatus* sp. nov. *Arch. Microbiol.* **148:**286–291.

Widdel, F., and F. Bak. 1992. Gram-negative mesophilic sulfate-reducing bacteria, p. 3352–3378. *In* A. Balows, H. G. Trüper, M. Dworkin, W. Harder, and K. H. Schleifer (ed.), *The Prokaryotes. A Handbook on the Biology of Bacteria: Ecophysiology, Isolation, Identification, Applications*, 2nd ed., vol. 4. Springer-Verlag, New York, NY.

Winding, A., S. J. Binnerup, and J. Sørensen. 1994: Viability of indigenous soil bacteria assayed by respiratory activity and growth. *Appl. Environ. Microbiol.* **60:**2869–2875.

Winogradsky, S. 1949. *Microbiologie du Sol. Problèmes et Méthodes.* Masson, Paris, France.

Zengler, K., G. Toledo, M. Rappé, J. Elkins, E. J. Mathur, J. M. Short, and M. Keller. 2002. Cultivating the uncultured. *Proc. Natl. Acad. Sci. USA* **99:**15681–15686.

CULTIVATION OF MARINE SYMBIOTIC MICROORGANISMS

Todd A. Ciche

11

Symbioses, or the living together of dissimilar organisms, are prevalent in nature, and marine organisms contain a plethora of diverse symbiotic associations (McFall-Ngai and Ruby, 2000). The definition for symbiosis used in this chapter is that of de Bary, as being the nontransient interaction between dissimilar organisms, which range from parasitic to mutualistic (de Bary, 1879). Symbiosis has dramatically affected life on Earth. For example, eukaryotic cells in our own bodies have resulted from symbiotic mergers between microorganisms millions of years ago (Margulis, 1970). Today, entire biomes, including coral reefs and hydrothermal vent communities, rely on symbiosis for stability and efficiency, for example, the sulfide-oxidizing autotrophic symbionts of *Riftia pachyptila* (Table 1). Symbiosis, in sum, is a powerful evolutionary mechanism that has resulted in novel characteristics and allowed the exploitation of diverse environmental niches.

DEFINITION OF SYMBIOSIS

Unlike parasitic and pathogenic interactions that cause disease, symbiotic associations are usually operating incognito in healthy hosts and only become apparent if they are disrupted or upon close examination by the researcher. Although different in effect on host fitness, mutualism and pathogenesis share many similar characteristics that have recently been revealed by genomics, genetics, and molecular biology (Hentschel and Steinert, 2001). I use the general definition of symbiosis by de Bary, acknowledging similarities between pathogenic and mutual host-bacterial interactions, as long as they are nontransient, which result in a great range of fitness to the host (de Bary, 1879). I also use similarities between pathogenesis and symbiosis to develop a suite of criteria, described below, that one can use to infer symbiotic relationships involving bacteria, archaea, and even viruses.

A common conception of symbiosis is of two organisms cooperating for their common benefit. However, the change in fitness of each organism resulting from symbiosis is difficult to quantify and is usually dynamic with influences from environmental and genetic variables. Fitness is defined as the evolutionary success of an organism mostly influenced by the number of progeny produced. Thus, de Bary defined symbiosis as a gradient of interactions ranging

Todd A. Ciche, Department of Microbiology and Molecular Genetics, Michigan State University, 2215 Biomedical Physical Sciences Building, East Lansing, MI 48824-4310.

Accessing Uncultivated Microorganisms: from the Environment to Organisms and Genomes and Back
Edited by Karsten Zengler © 2008 ASM Press, Washington, DC

TABLE 1 Examples of marine microbial symbioses

Symbiosis	Type	Cultivated symbiont?	Reference(s)
Vibrio cholerae, CTXϕ phage	Broadened host range	NA	Walder and Mekalanos, 1996
Anaerobic, methane-oxidizing, sulfate-reducing consortium	Syntrophy	No	Boetius et al., 2000; Orphan et al., 2001
Nanoarchaeum equitans, *Ignicoccus hospitalis*	Mutualism, parasitism?	No	Waters et al., 2003; Paper et al., 2007
Sponges, diverse bacteria/archaea	Chemical defense?	Some	Enticknap et al., 2006; Fieseler et al., 2007; Sharp et al., 2007
Palaemon macrodactylus (shrimp), *Alteromonas* spp.	Chemical defense	Yes	Gil-Turnes et al., 1989
Theonella swinhoei, *Pseudomonas* spp.	Chemical defense	No	Piel et al., 2004
Bugula neritina, *Candidatus* Endobugula sertula	Chemical defense	No	Haygood and Davidson, 1997; Davidson et al., 2001; Lopanik et al., 2004
Shipworms (Bivalvia: *Teredinidae*), *Teredinibacter turnerae*	Cellulose degradation, nitrogen fixation	Yes	Waterbury et al., 1983; Sipe et al., 2000; Distel et al., 2002
Euprymna scolopes, *Vibrio fischeri*	Bioluminescence	Yes	Nyholm and McFall-Ngai, 2004
Kryptophanaron alfredi (flashlight fish), *Vibrio* spp.	Bioluminescence	No	Haygood and Cohn, 1986; Haygood and Distel, 1993
Squid accessory nidamental gland, α-, γ-Proteobacteria	Chemical defense?	Some	Kaufman et al., 1998; Pichon et al., 2005
Riftia pachyptila, sulfide-oxidizing autotrophic symbionts	Nutrition	No	Cavanaugh et al., 1981
Nematodes, polychaetes, shrimp-sulfide-oxidizing ϵ-Proteobacteria	Nutrition, epibionts	No	Polz et al., 1994; Haddad et al., 1995; Polz and Cavanaugh, 1995

[a]NA, not available.

from parasitism, commensalism, and mutualism where the host fitness is decreased, unaffected, or increased as a result of the interaction, respectively. It is also often assumed that the host is a eukaryotic organism, but the definition from de Bary does not exclude microorganism-microorganism or microorganism-phage interactions as being symbiotic; examples of these are shown in Table 1.

This chapter describes the concept of symbiosis in a broad sense, presents criteria and methods to demonstrate that an organism is symbiotic, and presents general strategies for the cultivation of marine symbiotic microorganisms.

KOCH'S POSTULATES APPLIED TO SYMBIOSIS

If we want to identify a potential symbiosis involving two or more organisms, we must first demonstrate a nontransient interaction, distinguish the residents (i.e., symbiont) from the transients (i.e., free-living "contaminants"), and assess the effect of the symbiosis on the fitness of the host. Since the broad definition of symbiosis includes parasitism and pathogenesis (i.e., synonymous with parasitism, except involving microorganism or virus parasites), criteria for determining the causative agent of disease (Koch's postulates) can be applied to symbiosis where the characteristics of the symbiosis (e.g., colonization, tissue structure, fitness) can be substituted for the symptoms of disease.

The contributions of many scientists, including Koch and Pasteur in the late 19th century, on the ability of bacteria to cause disease are useful for our understanding of host-bacterial interactions in general. Robert Koch presented four postulates (listed below) that, when fulfilled, establish causal rather than correlative evidence for the microbial nature of disease although failure to meet any of these criteria does not prove otherwise.

Koch's Postulates

1. Identify pathogen from a diseased host.
2. Add pathogen to a healthy host.
3. Observe disease in new host.
4. Identify pathogen from new diseased host.

Symbiotic microorganisms do not fit easily into Koch's postulates because the resulting observable fitness of the host is usually less obvious than acute disease. Moreover, isolation and usually cultivation of the pathogen are employed to fulfill Koch's postulates. Because symbiosis contains a wide range of interactions, the effect of the association on host fitness (observation of "disease" in postulate no. 3) will vary from unobservable (commensalisms) to obvious or even essential to host and/or symbiont fitness (mutualism). However, symbiosis is often accompanied by novel characteristics not observed in apo-(or non-) symbiotic organisms. Even for commensal relationships between symbiont and host, a nontransient presence (i.e., colonization) in host tissues can be reliably observed. These characteristics can be substituted for disease to adapt Koch's postulates to symbiosis (see below).

Koch's Postulates Adapted for Symbiosis

1. Identify symbiont from host.
2. Eliminate symbiont from host (e.g., via removal of symbiont-free host and propagation under symbiont-free conditions and/or antibiotic treatment).
3. Observe difference between symbiotic and aposymbiotic host.
4. Add symbiont to new aposymbiotic host and observe symbiotic state.
5. Identify symbiont from new host.

An example of the fulfillment of these modified postulates is the light organ symbiosis between the Hawaiian bobtail squid *Euprymna scolopes* and the bioluminescent bacterium *Vibrio fischeri*: symbiotic bacteria are readily isolated and identified from the light organs of symbiotic hosts (postulate no. 1), aposymbiotic hosts are not bioluminescent (postulate no. 3), addition of symbiotic bacteria to aposymbiotic hosts reestablishes the symbiotic function (i.e., bioluminescence of the light organ) (postulate no. 4), and symbiotic *V. fischeri* can be reisolated from symbiotic hosts (postulate no. 5). Isolation and cultivation of *V. fischeri* were essential to fulfilling Koch's postulates applied to symbiosis and illustrate the value of cultivation of symbiotic microorganisms (see below) (Ruby, 1996).

Many symbioses, however, present additional challenges to fulfilling Koch's postulates. These challenges include symbioses for which there is an (i) inability to isolate and cultivate symbiotic bacteria and (ii) inability to selectively eliminate the symbiont or generate an aposymbiotic host. These challenges are similar to those encountered when investigating pathogenesis caused by uncultured bacteria, viruses, or chronic infections. In 1996, Relman adapted Koch's postulates for criteria to determine the causative agent of disease by using sequence-based detection methods for pathogens (Fredericks and Relman, 1996). Similar sequence-based criteria can be adapted to symbiosis to provide evidence that an organism is symbiotic (listed below). These criteria are generally listed from strongest (i.e., by definition, a nontransient association between symbiont nucleic acid and host is required for symbiosis) to weakest (the specific nature of the association may be unknown). Even though many symbioses only partially meet these criteria, they are presented here to aid the researcher when examining symbiosis from a molecular perspective.

Biochemical and Molecular Criteria for Symbiosis

1. A nucleic acid sequence (or phylotype) belonging to the putative symbiont should be present in most cases of symbiosis and found preferentially in organs, tissues, or cells known to be involved in symbiosis (based on other anatomic, histologic, or chemical evidence of symbiosis).

2. Fewer or no copy numbers of symbiont-associated nucleic acid sequences should occur in aposymbiotic hosts or tissues not involved in housing symbionts.

3. Efforts should be made to demonstrate the microbial phylotype in situ (e.g., by fluorescence in situ hybridization on tissues or cellular level).

4. These sequence-based forms of evidence for microbial causation should be reproducible.

5. After elimination of symbionts (e.g., by antibiotic treatment), the symbiont-associated phylotype should decrease or become nondetectable, manifested in loss of symbiont-derived characteristics.

6. When nucleic acid sequence detection predates symbiosis (e.g., during ontogenetic development), or sequence copy number correlates with symbiosis (e.g., health of the host for mutualism), the sequence-symbiosis association is more likely to be a causal relationship.

7. The nature of the microorganism inferred from the available sequence should be consistent with the known characteristics for that group of organisms. When phenotype (e.g., microbial morphology and physiology) is predicted by sequence-based phylogenetic relationships, the meaningfulness of the sequence is enhanced but should be supported by other evidence (e.g., molecular or physiological).

These criteria have been met for many symbioses, including the chemoautotrophic symbiosis involving the giant tube worm *Riftia pachyptila* that inhabits hydrothermal vents (Cavanaugh et al., 1981). For instance, the symbiont nucleic acid (or phylotype) and bacteria-specific enzymes are always found in symbiotic tubeworms (criterion no. 1), the nonsymbiotic tissues of *R. pachyptila* are devoid of symbionts (criterion no. 2), the direct identification and localization of the symbiont have been observed via fluorescence in situ microscopy and immunohistochemistry of intracellular bacteria-specific enzymes (criterion no. 3), these results are reproducible from worm to worm (criterion no. 4), and sulfide oxidation and CO_2 fixation are metabolic properties consistent with certain members of the γ-*Proteobacteria* with which the symbiont is phylogenetically related (criterion no. 7) (Cary et al., 1993; Cavanaugh, 1994; Cavanaugh et al., 2006). These criteria are important for the study of not-yet-cultured microorganisms and also useful in validating the success of symbiont isolation and cultivation (see below).

DIVERSITY AND SIGNIFICANCE OF MARINE SYMBIOSES

Symbiosis is prevalent in almost all types of organisms, from bacteria to protists, plants, and animals. In fact, among plants and animals, symbioses with microorganisms are the norm, rather than the exception. Historically, the important and often essential role of symbiotic organisms to the lives of their hosts has been overlooked. For example, biologists studied for decades the symbiosis between leaf-cutting Attine ants and their fungal symbionts without realizing that a conspicuous patch of "waxy bloom" was an actinomycete symbiont (Currie et al., 1999). The symbiotic actinomycete produces antibiotics specific to the parasitic fungus *Escovopsis* that can infect the symbiotic fungal garden and is preferentially found in ant castes that groom the primary fungal symbiont. The source for some medically important secondary metabolites present in marine organisms has recently been determined to be of symbiont origin, suggesting that these pharmaceuticals might be produced in a sustainable manner using fermentative technology (Gil-Turnes et al., 1989; Piel et al., 2004; Sudek et al., 2007).

There is enormous diversity in the number of organisms that function as hosts, as well as the types of microbes that become symbionts. Examples of marine symbioses are listed in Table 1. I mention only a few examples here to illustrate the breadth and diversity of symbioses, both where the symbiont has been cultivated and is not yet cultivated, focusing on bacterial symbioses despite the abundance of many other host-eukaryotic associations (i.e., alga, protists, and fungi).

Bacteria-Phage

Viruses and phages, which by de Bary's definition can be considered symbiotic, affect their hosts in a variety of ways, from acute virulence, nonvirulence, or even benefit. Microbial genomes are littered with genes that confer added capabilities to the host organism and in many cases have been shown to be transmitted by phage. This is most evident in pathogens where toxins and/or virulence factors are encoded on a lysogenic phage (Brussow et al., 2004), e.g., the cholera toxin genes, *ctxAB,* are encoded and transmitted by CTXϕ, which itself utilizes the colonization factor, toxin coregulated pilus (TCP) encoded on another phagelike element, for its infection (Faruque and Mekalanos, 2003).

Microbe-Microbe

Microbe-microbe syntrophies are one type of symbiosis in which energetically unfavorable reactions are made more favorable by the close association of two or more microorganisms. In marine cold seeps and other subseafloor environments, for example, there are cell aggregations of archaea belonging to the *Methanosarcinales,* surrounded by, and often in direct physical contact with, sulfate-reducing bacteria related to the *Desulfosarcina* (Boetius et al., 2000). These organisms work together to achieve the thermodynamically unfavorable anaerobic oxidation of methane. In this association, the archaea are commonly located in the central core of each microbial aggregate and are surrounded by a shell of *Desulfosarcina* cells. The archaea are thought to mediate "reverse methanogenesis" while the bacterial partner is presumed to consume hydrogen, while reducing sulfate, thereby creating a favorable gradient for the conversion of methane to carbon dioxide.

An intriguing microbe-microbe symbiosis is between two thermophilic archaea *Ignicoccus hospitalis* and *Nanoarchaeum equitans* (Huber et al., 2002). *N. equitans* depends on *I. hospitalis* for growth, is only 400 nm in diameter, and has an extremely reduced genome lacking in many metabolic activities (Waters et al., 2003). However, the nature of the symbiotic relationship remains unclear since growth of *I. hospitalis* in pure culture is not affected (Huber et al., 2002).

Symbiont Synthesis of Secondary Metabolites

A variety of marine symbioses are involved in the synthesis of medically relevant secondary metabolites (review in Piel, 2006). One of the first identified was production of the antifungal compound isostatin by *Alteromonas* spp. isolated from shrimp eggs (Gil-Turnes et al., 1989). The antitumor metabolites bryostatins and pederins are synthesized by uncultured symbionts of bryozoans and sponges (and beetles), respectively (Piel et al., 2004; Sudek et al., 2007). Furthermore, a community of bacteria, many of which were subsequently cultivated, was found associated with the accessory nidamental gland, egg jelly, and sheaths of squid (including *E, scolopes*, see below) and is thought to aid in embryo development perhaps by providing a chemical defense (Kaufman et al., 1998).

Gutless Tubeworm *R. pachyptila*

R. pachyptila, the gutless giant hydrothermal vent tubeworm, was the first demonstration of an association between a chemoautotrophic bacterium and a marine invertebrate (Cavanaugh, 1981). Since its discovery, a number of other similar associations have been discovered in a variety of habitats including vents, cold seeps, sewage outfalls, eelgrass beds, and anoxic basins (Cavanaugh, 1994; Cavanaugh et al., 2006). *R. pachyptila* houses symbiotic γ-*Proteobacteria* in a unique internal organ known as the trophosome, which can account for up to 50% of the host volume and is highly vascularized (Cavanaugh et al., 1981; Cavanaugh, 1985). First observed via transmission electron microscopy (Cavanaugh et al., 1981), these intracellular bacteria were discovered to be sulfide-oxidizing and carbon dioxide-fixing symbionts solely responsible for the nutrition of the worm (Felbeck et al., 1981). Many important studies have been conducted on the metabolic needs of both partners as well as the nutritional dialogue between them. For example, it is now known that the host has evolved many biochemical adaptations for symbiont accommodation, including effective mechanisms to concentrate inorganic carbon internally, sulfide acquisition from the environment using specialized hemoglobin molecules, as well as unprecedented pH and ion regulation. The symbiont, in turn, provides the host with organic carbon, which enables the worm to grow rapidly, quickly dominating communities around newly established hydrothermal vents (Cavanaugh et al., 2006).

Wood-Boring Bivalve *Teredinibacter tumerae*

Wood-boring bivalves (family: *Teredinidae*), commonly known as shipworms, contain a cellulolytic nitrogen-fixing symbiotic bacterium, *T. tumerae,* allowing the bivalve to thrive on wood (Popham and Dickson, 1973; Waterbury et al., 1983; Lechene et al., 2007). Waterbury et al. cultivated the symbiont by dilution culture in mineral medium containing cellulose but lacking a nitrogen source (Waterbury et al., 1983). Like nitrogen fixing symbionts in termite guts (Lilburn et al., 2001), *T. tumerae* allows wood-boring bivalves to grow on nitrogen-limited wood. Recently, multi-isotope mass spectrometry was used to detect and quantify nitrogen fixation by symbiotic *T. tumerae* in situ (Lechene et al., 2007).

Epibiotic Associations

There are many examples of commensal or mutual associations between host organisms and symbionts on the external surface of the host ("epibiont"). For instance, a diverse number of marine hosts, including the *Stilbonematinae* nematodes, the *Alvinellidae* (polychaete worms), and the *Bresiliidae* (shrimp), have dense growth of epibionts often times consisting of highly ordered sulfide-oxidizing *ε-Proteobacteria* (Desbruyères et al., 1985; Van Dover et al., 1988; Ott, 1995). In all cases, the hosts physically bridge the gap between oxic and anoxic waters via movement, thereby exposing the symbionts to all necessary metabolites. The hosts are thought to graze on the symbionts as a source of nutrition.

E. scolopes–V. fischeri Light Organ

The *E. scolopes–V. fischeri* symbiosis is a well-developed model animal-bacterium symbiosis. The ability to isolate, cultivate, and genetically manipulate the symbiont, in addition to the ability to reestablish the association using aposymbitic squid, has contributed to the prominence of this symbiosis model. The Hawaiian bobtail squid, *E. scolopes,* inhabits shallow sand flats in the Hawaiian archipelago where it is uses a ventrally located light organ for counterillumination against the night sky during feeding (Jones and Nishiguchi, 2004). The bioluminescence of the light organ is due to the presence of *V. fischeri* at high cell densities inside the light organ. Hatchling squid have a light organ primordia and ciliated epithelia appendages, which are used to acquire symbiotic bacteria from the surrounding environment (i.e., horizontal transmission) (McFall-Ngai and Montgomery, 1990). As elegantly summarized by Nyholm and McFall-Ngai, a winnowing occurs to select for symbiotic *V. fischeri*, which is vastly outnumbered by other marine planktonic microorganisms (Nyholm and McFall-Ngai, 2004). After *V. fischeri* colonizes the light organ, more than 90% of *V. fischeri* are vented from the light organ before dawn. The remaining *V. fischeri* then repopulates the light organ during the day while the squid remains burrowed in sand on the seafloor. The diurnal venting of *V. fischeri* reinoculates the environment with symbiotic bacteria (Lee and Ruby, 1994).

Development of the mature light organ is coincident with colonization by symbiotic bacteria and requires reciprocal and sequential host-bacterial interactions (Nyholm and McFall-Ngai, 2004). Colonization involves mucus secretion, quorum sensing, biofilm formation, oxygen radical physiology, lipopolysaccharide and peptidoglycan signaling, apoptosis, and actin remodeling, many processes relevant to human health and disease. In fact, analysis of the completed genome sequences reveals striking similarities between *V. fischeri* and *Vibrio cholerae* (Ruby et al., 2005). These similarities include shared "virulence" factors and components of pathogenicity islands. One such island, TCP, is lacking certain phage sequences flanking it and some genes found in TCP in *V. cholerae* are distributed in the *V. fischeri* genome that may indicate a more recent ancient acquisition of TCP in *V. fischeri*. It will be of great interest to determine the role of these "virulence" factors in *V. fischeri*, e.g., homologs of the CTX proteins in *V. fischeri* might affect the actin cytoskeleton of host cells.

CHALLENGES AND BENEFITS OF CULTIVATION

With the application of molecular biological methods to microbial ecology it has been established that the majority of microorganisms have not been cultivated (Pace, 1997). Although some symbiotic marine microorganisms have been cultivated (Table 1), many have not. Metagenomic techniques have been used to discover the physiology of uncultured marine microbes (Beja et al., 2000) or to identify novel secondary metabolites from uncultured microbes (Gillespie et al., 2002). However, the ability to cultivate symbiotic microorganisms can lead to detailed genetic and physiologic analyses of the symbiont. Cultivation of the bioluminescent *V. fischeri* symbiont of squid has allowed this symbiosis to mature into a well-established model for animal-bacterial interactions, providing a deeper understanding of the processes and mechanisms employed to establish and maintain a mutualistic association (Nyholm and McFall-Ngai, 2004). Cultivation of symbiotic microbes that produce medically relevant secondary metabolites might also allow the production of these metabolites by sustainable and economical fermentative technologies as opposed to collection of marine organisms from the environment (Piel, 2006). Despite the advantages of cultivation, many challenges exist in the cultivation of marine symbionts.

Removal of Transient or Contaminating Microorganisms

Care should be made to eliminate transient or environmental microorganisms from symbiotic organisms or tissue, because they may lead to false conclusions regarding the identity and complexity of the symbiotic microbial community. Moreover, opportunistic microorganisms or microbial "weeds" can overgrow and outcompete the true symbionts during cultivation attempts.

Before analysis of and attempts to cultivate symbiotic marine microorganisms, transient and/or environmental microorganisms should be removed or killed. Symbiotic microorganisms are usually sequestered from the environment by being enclosed by host cells or tissue but detectable by microscopic techniques (Ciche and Goffredi, 2007). Transient or environmental microbes can be eliminated by washing whole organisms or symbiont tissue with sterile seawater or buffer and/or by treating with dilute bleach as described below. A similar approach was recently applied to remove contaminating microbes and DNA from Siberian mammoth hair, while preserving host, albeit thousands of years old, mitochondrial DNA (Gilbert et al., 2007). Alternatively, if symbiont tissue or cells are not enclosed by resistant surfaces or are surface exposed (i.e., epibionts), only extensive washes in sterile seawater or buffer can be used.

Surface Sterilization of Symbiotic Tissue with Dilute Bleach

1. Collect specimens thought to be symbiotic and dissect symbiont tissue of interest.

Note: Anesthesia of marine invertebrates prior to symbiont extraction is often desirable. A solution of 5% ethanol in filtered seawater works for many invertebrates, including cephalopods.

2. Wash specimens or tissues three times in (0.22 μm) filtered seawater (artificial or natural) or buffer.

3. Treat specimens with 1 to 2% commercial bleach (>0.1% sodium hypochlorite) for 5 min.

4. Wash specimens or tissues an additional three times.

5. Proceed to cultivation and molecular techniques for symbiont analysis.

After surface sterilization, whole specimens or tissue should be immediately processed or flash frozen in liquid nitrogen. Molecular methods should be used in conjunction with cultivation techniques (see chapter 8; Ciche and Goffredi, 2007).

Cultivation of Marine Symbionts

Since microorganisms symbiotic with marine organisms are diverse, there is no general solution for the development of a successful cultivation strategy. However, there are a few hints and techniques that can increase cultivation success (see chapter 10). One should use any knowledge of the habitat (e.g., temperature, pH, osmolarity, pressure) or function of the symbiotic microbe when attempting cultivation. For example, knowledge that shipworms live on cellulose lacking a fixed nitrogen source was used for the cultivation of the *T. tumerae* symbiont (Waterbury et al., 1983). Knowledge that soil contains large amounts of humic substrates led to increased cultivation of soil microorganisms (Stevenson et al., 2004). Lastly, addition of cAMP and acyl-homoserine lactone signaling molecules known to be present in some cells growing as colonies improved the cultivation of marine bacteria (Bruns et al., 2002). Sequencing of the symbiont 16S rDNA might also reveal useful information for which to develop a cultivation strategy if the physiology and cultivation methods are known for close relatives. However, one should always be cautious when hypothesizing symbiont physiology from phylogeny since these characteristics are not necessarily correlated.

A first approach for the cultivation of aerobic heterotrophic marine symbionts is to use one-half strength Marine Broth 2216 (Becton-Dickinson, Franklin Lakes, NJ) supplemented with one-half strength natural or artificial seawater, with 1.5% agar added when required. The next steps are to (i) homogenize symbiont-containing tissue aseptically, e.g., by using a pellet pestle or tissue homogenizer; (ii) dilute the homogenate by serial dilution technique to extinction using diluted Marine Broth or sterile seawater (note: visible counts of symbiotic bacteria can be performed using phase contrast or epifluorescence microscopy) (Patel et al., 2007); (iii) spread 0.1 ml of several dilutions onto 3 to 4 plates of dilute Marine Broth and incubate the plates at a temperature appropriate for the symbiotic tissue; (iv) incubate 2 to 3 weeks and observe colonies and microcolonies using a stereomicroscope; and (v) restreak isolated colonies onto diluted Marine Broth agar. If the isolated colonies grow upon restreak, freezer stock is made by scraping off the cells into Marine Broth and dropping dimethyl sulfoxide to 4.5%, flash freezing in a dry ice/ethanol bath or in liquid nitrogen, and storing at −80°C. Freezer stocks should be made promptly after adequate symbiont growth is achieved. Stock should be checked for viability and colony morphology after freezing.

The critical factors described in this approach are (i) surface sterilization and aseptic dissection and handling of symbiotic tissue to remove transient and/or opportunistic symbionts, which will complicate analyses of the symbiosis and might dominate cultivation attempts, (ii) dilution and plating to extinction to eliminate competition and or antagonism among strains, and (iii) patience and close examination for colonies and microcolonies that grow on the plates.

The diluted Marine Broth still is relatively rich, and increased diversity has been obtained from environmental samples simply using sterile natural or artificial seawater agar plates (Eilers et al., 2001). The high nutrient conditions present in rich microbial media may lead to the production of toxic metabolic products that inhibit the growth of some bacteria. Agar can also inhibit the growth of certain microorganisms and can be replaced with another gelling agent such as Gelrite. Additional supplements such as bacterial signal molecules can also result in the cultivation of a higher diversity of marine organisms. Recently, new approaches have been applied to cultivate a higher diversity of marine and nonmarine bacteria (Rappé et al., 2002; Connon and Giovannoni, 2002; Kaeberlein et al., 2002; Zengler et al., 2005). These approaches involve separation of the complex microbial community to single cells and sensitive methods to monitor microbial growth. One of these methods involves culturing marine bacteria in gel microdroplets initiated by a single cell but where metabolic exchange between the embedded microbes is possible (Zengler et al., 2002; Zengler et al., 2005). The gel microdroplets containing microcolonies were separated from gel microdroplets containing no or single cells by flow cytometry using differential light scattering. The gel microdroplets containing bacteria can be separated and grown in more nutrient-rich conditions where enough cells were grown to detect antifungal activity (Zengler et al., 2002).

Validation of Successful Symbiont Cultivation

The success of cultivation should be validated by molecular techniques (see chapter 8 and molecular criteria stated above) to determine whether the cultivated symbiont corresponds to a phylotype of the microbial symbiont present in host tissue. If so, one can attempt to fulfill Koch's postulates applied to symbiosis. Essential for this is the generation of aposymbiotic (i.e., without symbionts) organisms. For organisms that acquire the symbionts from the environment (i.e., horizontal transmission), eggs or freshly hatched larvae are often aposymbiotic and only need to be propagated in the absence of the symbiont(s). If host organisms are enclosed by resistant surfaces (e.g., egg case), these can be surface sterilized with dilute bleach as described above. If these techniques are not successful, one can attempt antibiotic treatment to eliminate symbiont bacteria, although symbiont cells present inside host cells and tissue might impermeable or resistant to antibiotics.

CONCLUSION

In summary, the ability to cultivate marine symbiotic microorganisms is a great advantage for the study of the symbiotic microorganism and the natural products it might produce. Cultivation attempts should always be performed in conjunction with molecular techniques to monitor whether the dominant member(s) of the symbiotic community are being isolated.

REFERENCES

Beja, O., L. Aravind, E. V. Koonin, M. T. Suzuki, A. Hadd, L. P. Nguyen, S. B. Jovanovich, C. M. Gates, R. A. Feldman, J. L. Spudich, E. N. Spudich, and E. F. DeLong. 2000. Bacterial rhodopsin: evidence for a new type of phototrophy in the sea. *Science* **289:**1902–1906.

Boetius, A., K. Ravenschlag, C. J. Schubert, D. Rickert, F. Widdel, A. Gieseke, R. Amann, B. B. Jørgensen, U. Witte, and O. Pfannkuche. 2000. A marine microbial consortium apparently mediating anaerobic oxidation of methane. *Nature* **407:**623–626.

Bruns, A., H. Cypionka, and J. Overmann. 2002. Cyclic AMP and acyl homoserine lactones increase the cultivation efficiency of heterotrophic bacteria from the central Baltic Sea. *Appl. Environ. Microbiol.* **68:**3978–3987.

Brussow, H., C. Canchaya, and W.-D. Hardt. 2004. Phages and the evolution of bacterial pathogens: from genomic rearrangements to lysogenic conversion. *Microbiol. Mol. Biol. Rev.* **68:**560–602.

Cary, S. C., W. Warren, E. Anderson, and S. J. Giovannoni. 1993. Identification and localization of bacterial endosymbionts in hydrothermal vent taxa with symbiont-specific polymerase chain reaction amplification and in situ hybridization techniques. *Mol. Mar. Biol. Biotechnol.* **2:**51–62.

Cavanaugh, C. M. 1985. Symbiosis of chemoautotrophic bacteria and marine invertebrates from hydrothermal vents and reducing sediments. *Bull. Biol. Soc. Wash.* **6:**373–388.

Cavanaugh, C. M. 1994. Microbial symbiosis: patterns of diversity in the marine environment. *Am. Zool.* **34:**79–89.

Cavanaugh, C. M., S. L. Gardiner, M. L. Jones, H. W. Jannasch, and J. B. Waterbury. 1981. Procaryotic cells in the hydrothermal vent tube worm *Riftia pachyptila* Jones: possible chemoautotrophic symbionts. *Science* **213:**340–342.

Cavanaugh, C. M., Z. P. McKiness, I. R. G. Newton, and F. J. Stewart. 2006. Marine chemosynthetic symbioses, p. 475–507. *In* M. Dworkin, S. Falkow, E. Rosenberg, and K.-H.

Schleifer (ed.), *The Prokaryotes: a Handbook on the Biology of Bacteria*, 3rd ed., vol. 1. Springer, New York, NY.

Ciche, T. A., and S. K. Goffredi. 2007. General methods to investigate microbial symbioses, p. 394–419. *In* C. A. Reddy, T. J. Beveridge, J. A. Breznak, G. A. Marzluf, T. M. Schmidt, and L. R. Snyder (ed.), *Methods for General and Molecular Microbiology*. ASM Press, Washington, DC.

Connon, S. A., and S. J. Giovannoni. 2002. High-throughput methods for culturing microorganisms in very-low-nutrient media yield diverse new marine isolates. *Appl. Environ. Microbiol.* **68:**3878–3885.

Currie, C. R., J. A. Scott, R. C. Summerbell, and D. Malloch. 1999. Fungus-growing ants use antibiotic-producing bacteria to control garden parasites. *Nature* **398:**701–704.

Davidson, S. K., S. W. Allen, G. E. Lim, C. M. Anderson, and M. G. Haygood. 2001. Evidence for the biosynthesis of bryostatins by the bacterial symbiont "*Candidatus* Endobugula sertula" of the bryozoan *Bugula neritina*. *Appl. Environ. Microbiol.* **67:**4531–4537.

de Bary, A. 1879. *Die Erscheinung der Symbiose.* Trubner, Strasburg, Germany.

Desbruyères, D., F. Gaill, L. Laubier, and Y. Fouquet. 1985. Polychaetous annelids from hydrothermal vent ecosystems: an ecological overview. *Bull. Biol. Soc. Wash.* **6:**103–116.

Distel, D. L., W. Morrill, N. MacLaren-Toussaint, D. Franks, and J. Waterbury. 2002. *Teredinibacter turnerae* gen. nov., sp. nov., a dinitrogen-fixing, cellulolytic, endosymbiotic gamma-proteobacterium isolated from the gills of wood-boring molluscs (Bivalvia: Teredinidae). *Int. J. Syst. Evol. Microbiol.* **52:**2261–2269.

Eilers, H., J. Pernthaler, J. Peplies, F. O. Glöckner, G. Gerdts, and R. Amann. 2001. Isolation of novel pelagic bacteria from the German bight and their seasonal contributions to surface picoplankton. *Appl. Environ. Microbiol.* **67:**5134–5142.

Enticknap, J. J., M. Kelly, O. Peraud, and R. T. Hill. 2006. Characterization of a culturable alphaproteobacterial symbiont common to many marine sponges and evidence for vertical transmission via sponge larvae. *Appl. Environ. Microbiol.* **72:**3724–3732.

Faruque, S. M., and J. J. Mekalanos. 2003. Pathogenicity islands and phages in *Vibrio cholerae* evolution. *Trends Microbiol.* **11:**505–510.

Felbeck, H., J. J. Childress, and G. N. Somero. 1981. Calvin-Benson cycle and sulfide oxidation enzymes in animals from sulfide-rich habitats. *Nature* **293:**291–293.

Fieseler, L., U. Hentschel, L. Grozdanov, A. Schirmer, G. Wen, M. Platzer, S. Hrvatin, D. Butzke, K. Zimmermann, and J. Piel. 2007. Widespread occurrence and genomic context of unusually small polyketide synthase genes in microbial consortia associated with marine sponges. *Appl. Environ. Microbiol.* **73:**2144–2155.

Fredericks, D., and D. Relman. 1996. Sequence-based identification of microbial pathogens: a reconsideration of Koch's postulates. *Clin. Microbiol. Rev.* **9:**18–33.

Gilbert, M. T. P., L. P. Tomsho, S. Rendulic, M. Packard, D. I. Drautz, A. Sher, A. Tikhonov, L. Dalen, T. Kuznetsova, P. Kosintsev, P. F. Campos, T. Higham, M. J. Collins, A. S. Wilson, F. Shidlovskiy, B. Buigues, P. G. P. Ericson, M. Germonpre, A. Gotherstrom, P. Iacumin, V. Nikolaev, M. Nowak-Kemp, E. Willerslev, J. R. Knight, G. P. Irzyk, C. S. Perbost, K. M. Fredrikson, T. T. Harkins, S. Sheridan, W. Miller, and S. C. Schuster. 2007. Whole-genome shotgun sequencing of mitochondria from ancient hair shafts. *Science* **317:**1927–1930.

Gillespie, D. E., S. F. Brady, A. D. Bettermann, N. P. Cianciotto, M. R. Liles, M. R. Rondon, J. Clardy, R. M. Goodman, and J. Handelsman. 2002. Isolation of antibiotics turbomycin a and B from a metagenomic library of soil microbial DNA. *Appl. Environ. Microbiol.* **68:**4301–4306.

Gil-Turnes, M. S., M. E. Hay, and W. Fenical. 1989. Symbiotic marine bacteria chemically defend crustacean embryos from a pathogenic fungus. *Science* **246:**116–118.

Haddad, A., F. Camacho, P. Durand, and S. C. Cary. 1995. Phylogenetic characterization of the epibiotic bacteria associated with the hydrothermal vent polychaete *Alvinella pompejana*. *Appl. Environ. Microbiol.* **61:**1679–1687.

Haygood, M. G., and D. H. Cohn. 1986. Luciferase genes cloned from the unculturable luminous bacteroid symbiont of the Caribbean flashlight fish, *Kryptophanaron alfredi*. *Gene* **45:**203–209.

Haygood, M. G., and S. K. Davidson. 1997. Small-subunit rRNA genes and in situ hybridization with oligonucleotides specific for the bacterial symbionts in the larvae of the bryozoan *Bugula neritina* and proposal of "*Candidatus* Endobugula sertula." *Appl. Environ. Microbiol.* **63:**4612–4616.

Haygood, M. G., and D. L. Distel. 1993. Bioluminescent symbionts of flashlight fishes and deep-sea anglerfishes form unique lineages related to the genus *Vibrio*. *Nature* **363:**154–156.

Hentschel, U., and M. Steinert. 2001. Symbiosis and pathogenesis: common themes, different outcomes. *Trends Microbiol.* **9:**585.

Huber, H., M. J. Hohn, R. Rachel, T. Fuchs, V. C. Wimmer, and K. O. Stetter. 2002. A new phylum of archaea represented by a nanosized hyperthermophilic symbiont. *Nature* **417:**63–67.

Jones, B. W., and M. K. Nishiguchi. 2004. Counterillumination in the Hawaiian bobtail squid, *Euprymna scolopes* Berry (Mollusca: Cephalopoda). *Mar. Biol.* **144:**1151–1155.

Kaeberlein, T., K. Lewis, and S. S. Epstein. 2002. Isolating "uncultivable" microorganisms in pure culture in a simulated natural environment. *Science* **296:**1127–1129.

Kaufman, M. R., Y. Ikeda, C. Patton, G. van Dykhuizen, and D. Epel. 1998. Bacterial symbionts colonize the accessory nidamental gland of the squid *Loligo opalescens* via horizontal transmission. *Biol. Bull.* **194:**36–43.

Lechene, C. P., Y. Luyten, G. McMahon, and D. L. Distel. 2007. Quantitative imaging of nitrogen fixation by individual bacteria within animal cells. *Science* **317:**1563–1566.

Lee, K.-H., and E. G. Ruby. 1994. Effects of the squid host on the abundance and distribution of symbiotic *Vibrio fischeri* in nature. *Appl. Environ. Microbiol.* **60:**1565–1571.

Lilburn, T. G., K. S. Kim, N. E. Ostrom, K. R. Byzek, J. R. Leadbetter, and J. A. Breznak. 2001. Nitrogen fixation by symbiotic and free-living spirochetes. *Science* **292:**2495–2498.

Lopanik, N., N. Lindquist, and N. Targett. 2004. Potent cytotoxins produced by a microbial symbiont protect host larvae from predation. *Oecologia* **139:**131–e139.

Margulis, L. 1970. *Origin of Eukaryotic Cells; Evidence and Research Implications for a Theory of the Origin and Evolution of Microbial, Plant, and Animal Cells on the Precambrian Earth.* Yale University Press, New Haven, CT.

McFall-Ngai, M. J., and M. K. Montgomery. 1990. The anatomy and morphology of the adult bacterial light organ of *Euprymna scolopes* Berry (Cephalopoda:Sepiolidae). *Biol. Bull.* **179:**33–339.

McFall-Ngai, M. J., and E. G. Ruby. 2000. Developmental biology in marine invertebrate symbioses. *Curr. Opin. Microbiol.* **3:**603–607.

Nyholm, S. V., and M. J. McFall-Ngai. 2004. The winnowing: establishing the squid *Vibrio*-symbiosis. *Nature Rev. Microbiol.* **2:**632–-642.

Orphan, V. J., C. H. House, K.-U. Hinrichs, K. D. McKeegan, and E. F. DeLong. 2001. Methane-consuming archaea revealed by directly coupled isotopic and phylogenetic analysis. *Science* **293:**484–487.

Ott, J. 1995. Sulfide symbioses in shallow sands, p. 143–147. *In* A. Eleftheriou, A. Ansell, and C. Smith (ed.), *Biology and Ecology of Shallow Coastal Waters.* Olsen & Olsen, Fredensborg, Denmark.

Pace, N. R. 1997. A molecular view of microbial diversity and the biosphere. *Science* **276:**734–740.

Paper, W., U. Jahn, M. J. Hohn, M. Kronner, D. J. Nather, T. Burghardt, R. Rachel, K. O. Stetter, and H. Huber. 2007. *Ignicoccus hospitalis* sp. nov., the host of *"Nanoarchaeum equitans." Int. J. Syst. Evol. Microbiol.* **57:**803–808.

Patel, A., R. T. Noble, J. A. Steele, M. S. Schwalbach, I. Hewson, and J. A. Fuhrman. 2007. Virus and prokaryote enumeration from planktonic aquatic environments by epifluorescence microscopy with SYBR Green I. *Nat. Protoc.* **2:**269–276.

Pichon, D., V. Gaia, M. Norman, and R. Boucher-Rodoni. 2005. Phylogenetic diversity of epibiotic bacteria in the accessory nidamental glands of squids (Cephalopoda: Loliginidae and Idiosepiidae). *Mar. Biol.* **147:**1323–1332.

Piel, J. 2006. Bacterial symbionts: prospects for the sustainable production of invertebrate-derived pharmaceuticals. *Curr. Med. Chem.* **13:**39–50.

Piel, J., D. Hui, G. Wen, D. Butzke, M. Platzer, N. Fusetani, and S. Matsunaga. 2004. Antitumor polyketide biosynthesis by an uncultivated bacterial symbiont of the marine sponge *Theonella swinhoei. Proc. Natl. Acad. Sci. USA* **101:**16222–16227.

Polz, M. F., and C. M. Cavanaugh. 1995. Dominance of one bacterial phylotype at a mid-Atlantic ridge hydrothermal vent site. *Proc. Natl. Acad. Sci. USA* **92:**7232–7236.

Polz, M. F., D. L. Distel, B. Zarda, R. Amann, H. Felbeck, J. A. Ott, and C. M. Cavanaugh. 1994. Phylogenetic analysis of a highly specific association between ectosymbiotic, sulfur-oxidizing bacteria and a marine nematode. *Appl. Environ. Microbiol.* **60:**4461–4467.

Popham, J. D., and M. R. Dickson. 1973. Bacterial associations in the teredo *Bankia australis* (Lamellibranchia: Mollusca). *Mar. Biol.* **19:**338–340.

Rappé, M. S., S. A. Connon, K. L. Vergin, and S. J. Giovannoni. 2002. Cultivation of the ubiquitous SAR11 marine bacterioplankton clade. *Nature* **418:**630–633.

Ruby, E. G. 1996. Lessons from a cooperative, bacterial-animal association: The *Vibrio fischeri-Euprymna scolopes* light organ symbiosis. *Ann. Rev. Microbiol.* **50:**591–624.

Ruby, E. G., M. Urbanowski, J. Campbell, A. Dunn, M. Faini, R. Gunsalus, P. Lostroh, C. Lupp, J. McCann, D. Millikan, A. Schaefer, E. Stabb, A. Stevens, K. Visick, C. Whistler, and E. P. Greenberg. 2005. Complete genome sequence of *Vibrio fischeri*: a symbiotic bacterium with pathogenic congeners. *Proc. Natl. Acad. Sci. USA* **102:**3004–-3009.

Sharp, K. H., B. Eam, D. J. Faulkner, and M. G. Haygood. 2007. Vertical transmission of diverse microbes in the tropical sponge *Corticium* sp. *Appl. Environ. Microbiol.* **73:**622–629.

Sipe, A. R., A. E. Wilbur, and S. C. Cary. 2000. Bacterial symbiont transmission in the wood-boring shipworm *Bankia setacea* (Bivalvia: Teredinidae). *Appl. Environ. Microbiol.* **66:**1685–1691.

Stevenson, B. S., S. A. Eichorst, J. T. Wertz, T. M. Schmidt, and J. A. Breznak. 2004. New strategies for cultivation and detection of previously uncultured microbes. *Appl. Environ. Microbiol.* **70:**4748–4755.

Sudek, S., N. B. Lopanik, L. E. Waggoner, M. Hildebrand, C. Anderson, H. Liu, A. Patel, D. H. Sherman, and M. G. Haygood. 2007. Identification of the putative bryostatin polyketide synthase gene cluster from "*Candidatus* Endobugula sertula," the uncultivated microbial symbiont of the marine bryozoan *Bugula neritina*. *J. Nat. Prod.* **70:**67–74.

Van Dover, C. L., B. Fry, J. F. Grassle, S. Humphris, and P. A. Rona. 1988. Feeding biology of the shrimp *Rimicaris exoculata* at hydrothermal vents on the mid-Atlantic ridge. *Mar. Biol.* **98:**209–216.

Walder, M. K., and J. J. Mekalanos. 1996. Lysogenic conversion by a filamentous bacteriophage encoding cholera toxin. *Science* **272:**1910–1914.

Waterbury, J. B., C. B. Calloway, and R. D. Turner. 1983. A cellulolytic-nitrogen fixing bacterium cultured from the gland of Deshayes in shipworms (Bivalvia: Teredinidae). *Science* **221:**1401–1403.

Waters, E., M. J. Hohn, I. Ahel, D. E. Graham, M. D. Adams, M. Barnstead, K. Y. Beeson, L. Bibbs, R. Bolanos, M. Keller, K. Kretz, X. Lin, E. Mathur, J. Ni, M. Podar, T. Richardson, G. G. Sutton, M. Simon, D. Soll, K. O. Stetter, J. M. Short, and M. Noordewier. 2003. The genome of *Nanoarchaeum equitans*: insights into early archaeal evolution and derived parasitism. *Proc. Natl. Acad. Sci. USA* **100:**12984–12988.

Zengler, K., G. Toledo, M. Rappé, J. Elkins, E. J. Mathur, J. M. Short, and M. Keller. 2002. Cultivating the uncultured. *Proc. Natl. Acad. Sci. USA* **99:**15681–15686.

Zengler, K., M. Walcher, G. Clark, I. Haller, G. Toledo, T. Holland, E. J. Mathur, G. Woodnutt, J. M. Short, and M. Keller. 2005. High-throughput cultivation of microorganisms using microcapsules. *Methods Enzymol.* **397:**124–430.

METHODS TO STUDY CONSORTIA AND MIXED CULTURES

Boran Kartal and Marc Strous

12

Koch's postulates are a central paradigm in microbiology. Originally formulated for the identification of pathogenic microbes, the postulates have since proved their value in environmental microbiology. Freely translated, they tell us that if we want to demonstrate that a certain microbe performs a certain task in nature, we have to isolate that microorganism in pure culture and show that the task can be reproduced. Now "isolation in pure culture" is not an experimentally neutral phrase that can be achieved by any means; it is an important aspect of the "culture" of our trade. It means that we must make a microbe proliferate on solid media and deposit it in culture collections such as the American Type Culture Collection or Deutsche Sammlung von Mikroorganismen und Zellkulturen.

It is not always possible, and perhaps it is even exceptional, for environmental microbes to withstand such domestication. Explanations for such "un- or-not-just-yet-cultivability" are not difficult to find: Microbes may depend on proximal contact with other microbes for essential but unknown growth factors (i.e., quorum-sensing agents). Microbes may form symbiotic consortia with other microbes or nonmicrobial hosts. Microbes may be sensitive to periods of environmental stress associated with the handling of cultures. Or microbes may simply grow too slowly. According to some studies, our view of the microbial world is severely biased toward exceptionally rapid growers (Teske, 2005). In the environment, most microbes may divide only once per month or even once per year. At this rate, growth of a colony on an agar plate would take several generations of Ph.D. students, and isolation via a dilution series would be scientific suicide when you are on a tenure track.

However, the culture collection is not the essence of Koch's postulates. The essence is that we demonstrate unambiguously that a certain microbe is responsible for a certain task. Nowadays many experimental approaches exist that one could not even imagine in Koch's days. Therefore we should carefully consider whether pure culture isolation is a fruitful exercise in a certain case—or not. An alternative approach that focuses on dealing with *undefined* populations of microbes in the laboratory is the scope of this chapter.

Boran Kartal and Marc Strous, Department of Microbiology, Faculty of Science, Radboud University Nijmegen, Toernooiveld 1, 6525 ED Nijmegen, The Netherlands.

Accessing Uncultivated Microorganisms: from the Environment to Organisms and Genomes and Back
Edited by Karsten Zengler

APPROACH

The approach aims to study a microbe *in molecular detail* without growing it in pure culture. "In molecular detail" means to identify and characterize the enzymes involved in central carbon and energy metabolism, to study its growth parameters quantitatively, for example, to measure growth rate and yield and affinity constants for the substrates. It means to know the range of substrates it can use and basic responses toward changing environmental conditions. In other words, in molecular detail means to define the niche of the organism to understand to what environmental conditions it is adapted and how those adaptations are realized in terms of the molecular machinery of the cell. Such detailed understanding is necessary when going back to the ecosystem to investigate the in situ importance. Only then it is possible to detect the abundance of the organism, for example, via fluorescence in situ hybridization, quantitative PCR, and specific biomarkers (i.e., membrane lipids specific for the organism). Then it is also possible to characterize and interpret its in situ activity (e.g., via quantitative PCR targeting the expression of its key genes). Subsequently, we can begin to understand why we find that particular organism under those particular circumstances and think about why the ecosystem behaves the way it does as a whole, for example, in response to climate change.

The approach consists of four steps. The first step is the definition of an ecological niche. In the second step the complement of the niche is engineered in a laboratory bioreactor, which is to become an enrichment culture. In the third step the culture is characterized in molecular detail. Finally, the detailed information is used to investigate the importance of the described processes and microbes in nature.

DEFINING THE ECOLOGICAL NICHE

Microbiology starts when a microbe makes itself known. Pathogens were among the first that attracted the interest of microbiologists, and in environmental microbiology nitrifiers and denitrifiers were also first noticed because of their specific (undesired) action, namely the removal of fertilizer from the fields. Microbes have been noticed because of their dominance in a given habitat (e.g., *Bifidobacterium* spp. among the gut bacteria) or because of their conspicuous morphology. Nowadays, microbes often spark the interest of microbiologists because of their dominance in clone libraries or an exotic position in a phylogenetic tree.

In the era of 16S rRNA gene sequencing we tend to forget that microbial species are really defined by their niche and not by their gene sequence. The 16S rDNA sequence is only an evolutionary expression of the niche. The niche definition includes the types of energy and carbon metabolism performed, the dependence on trace elements and cofactors, the temperature and pH range for growth, the resistance and sensitivity to toxic chemicals and free radicals, the dependencies on and interactions with other organisms, the responses to environmental change on different time scales (minutes, hours, days, seasons) and survival strategies, the defenses to viruses and predation by protists, the tendency to take up foreign DNA and exchange genes, the "plasticity" of the genome, the growth rate, the affinity for the substrates and much, much more.

Without a sense of at least some aspect of a niche, it is difficult to enrich or isolate a desired microorganism. Generally, we are clueless to what a position in a phylogenetic tree or a certain type of morphology means in terms of the niche. In the absence of meaningful information, a trial and error approach is possible by measuring the incorporation of a tracer in short incubations under specific conditions (Wuchter et al., 2006). Isotopic signatures can also be useful to tie a microbe to a specific carbon source (Hinrichs et al., 1999; Orphan et al., 2002). Alternatively, metagenomic approaches can link a 16S rRNA gene to a functional role. The discovery of proteorhodopsin and nitrification by crenarchaea are beautiful examples (Beja et al., 2000; Könneke et al., 2005).

It is also possible to simply define a theoretical niche without prior knowledge of the environmental significance of the associated microorganisms; they may not even exist in nature. For example, the existence of anaerobic ammonium oxidizers was "predicted" by Broda (Broda, 1977). Recently the idea was generalized in a "periodic system of one-carbon microorganisms" (Fig. 1). In this approach thermodynamic calculations showed which combinations of common electron donors and acceptors could in theory support growth on methane or carbon dioxide. That was only a small but sufficient aspect of a microbial niche (the energy and carbon metabolism). Ultimately microbes responsible for one of these processes, anaerobic methane oxidation coupled to denitrification, were enriched in the laboratory (Raghoebarsing et al., 2006). One might also focus on growth rate. Nobody knows which organisms would be specialized on aerobic respiration of glucose *at very low rates.* Or dynamic responses: Which organisms thrive when substrate supply is stopped every other day—and do those same organisms prevail when no substrate is supplied every other week? Does the option of growth in biofilms constitute part of a niche?

Unfortunately, isolation of an organism in pure culture does not only mean the separation of the organism from other organisms; it often also means the separation of the organism from its niche. In pure culture, the organism is forced to adapt to the rhythm of sequential transfer of cells from one colony to the next: alternating conditions of feast—when a cell first lands on a fresh plate—to famine—when the colony is fully grown. This way the organism experiences alternating periods of "transfer" stress, exponential growth, and stationary phase. There are no competitors to outcompete, no viruses and protists to outwit.

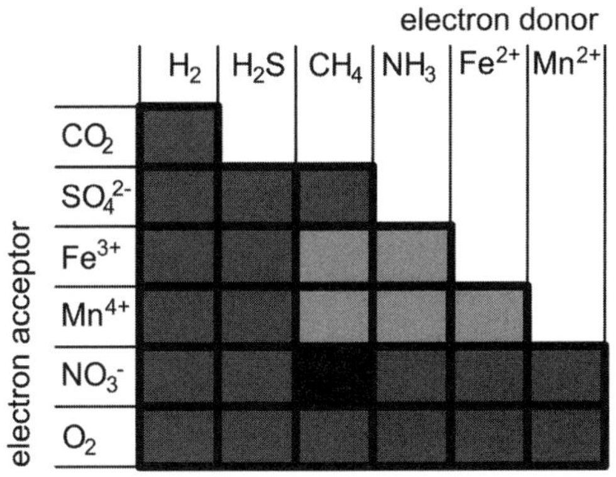

FIGURE 1 Periodic system of a number of one-carbon microorganisms.

Finally, if isolation proves difficult, the effort can also lead to the separation of the scientist and the problem; the problem that the organism refuses to grow on solid media is not necessarily the problem you would like to investigate. Let's say that with the intention to fix the car engine you notice the hood is stuck—and that becomes the focus for the remainder of a Saturday afternoon.

In the approach of this chapter, the target organism remains in continuous culture under chemo-spatio-temporal conditions that define its natural niche, in an open system that allows competition. The following section outlines how this can be achieved.

ENGINEERING THE COMPLEMENT OF THE NICHE IN THE LABORATORY

Continuous culture was "invented" by Leo Szilard, a physicist best known from his work on the nuclear chain reaction. In biology he contributed the concept of "repression" to bacterial gene regulation. In his paper (Novick and Szilard, 1950) on continuous culture he wrote: "We have developed a device for keeping a bacterial population growing at a reduced rate over an indefinite period of time. In this device, which we shall refer to as the chemostat, we have a vessel containing V ml of a suspension of bacteria..."

The chemostat has become a main concept in microbiological cultivation. It generally consists of a closed glass vessel with volume between 1 ml and a few liters. Fresh medium is supplied continuously with a peristaltic pump while a second pump removes liquid from the vessel at the same rate. This way a constant volume is maintained. When the vessel is inoculated with bacteria, they grow exponentially until all the nutrients in the vessel are depleted. From that point onward, they can only consume what is supplied by the influent pump and their growth rate (h^{-1}) is then defined by the ratio of the influent rate ($l\ h^{-1}$) and the vessel volume (l). As Szilard writes, this state can be maintained indefinitely. Among the substrates and growth factors in the medium one will limit growth; the others are present in surplus. Szilard referred to this compound as the "controlling substrate." The concentration of the controlling substrate determines the concentration of cells in the vessel and in the effluent. Apart from a small "maintenance energy" effect, the cell count is independent of the flow rate and this way the bacterial cell yield on the controlling substrate under the vessel conditions can be measured with precision. Generally, the controlling substrate is the energy or carbon source, but it can be chosen depending on the aim of the experiment.

The chemostat is an excellent way to cultivate microorganisms for experimental investigation because the conditions are defined and constant compared to batch cultures (i.e., shake flasks), enabling better reproducibility in experiments. Furthermore, the setup generates a continuous flow of microbial culture, much like your tap generates a continuous flow of clean water that can be conveniently harvested for further experiments.

The chemostat also has a few disadvantages: Because of the prolonged length of the experiment, it is more prone to contamination, and recently it has become clear that in long experiments (more than five volume changes) the strong selection pressure can generate a considerable degree of in vitro evolution (Jansen et al., 2005; Mashego et al., 2005). Second, chemostat equipment is available commercially but expensive compared to what is necessary for shaking flask experiments. Last, the experimental procedures and conceptual framework require some training and practice.

The chemostat has been used for pure cultures to study the kinetics of bacterial growth and also for more detailed -omics approaches (Elias et al., 2006). In microbial ecology, it has also been used for competition experiments. In such experiments two or three strains with comparable niches are released in the vessel under varying conditions, at high or low growth rates, at high or low oxygen concentrations, with or without growth factors (Kuenen and Veldkamp, 1973; Nogueira and Melo, 2006).

Application of the chemostat for the direct enrichment of organisms from the environment is not always possible. Many environmental organisms have a strong preference for attachment to surfaces, and this behavior makes great sense in the chemostat. Colonization of the vessel's surfaces means successful evasion of the effluent pump and so increases fitness compared to suspended competitors that are continuously removed from the vessel. Only when the substrate concentration becomes very low, growth in suspension starts to make evolutionary sense.

In our approach we make use of the preference of many microorganisms for biofilms by introducing a settler to the chemostat or by introducing time for settling (Fig. 2). The settler may be a separate vessel or engineered into the chemostat itself as a zone without mixing. In the settler aggregated organisms settle and return to the vessel, while suspended cells wash out. This way, aggregated biomass is retained in the chemostat, thus uncoupling the relationship between substrate concentration and cell counts. Without intervention, retained aggregates accumulate in the vessel indefinitely, leading to ever decreasing growth rates and high biomass concentration, for example, over 10 g of protein per liter at growth rates of less than one division per 10 weeks. Biomass concentration and growth rate can be maintained at a constant value by periodically removing a fixed amount of biomass aggregates from the system.

With such an experimental setup full control is achieved over the enrichment conditions and the ecological niche we would like to select for. For example, the medium may contain different electron acceptor/donor pairs, and gaseous substrates may be supplied by sparging the vessel with a gas of a desired composition; the enrichment may be performed in light or dark, with or without oxygen, or with and without oxygen by switching the gas supply on and off at given intervals. Like in the chemostat,

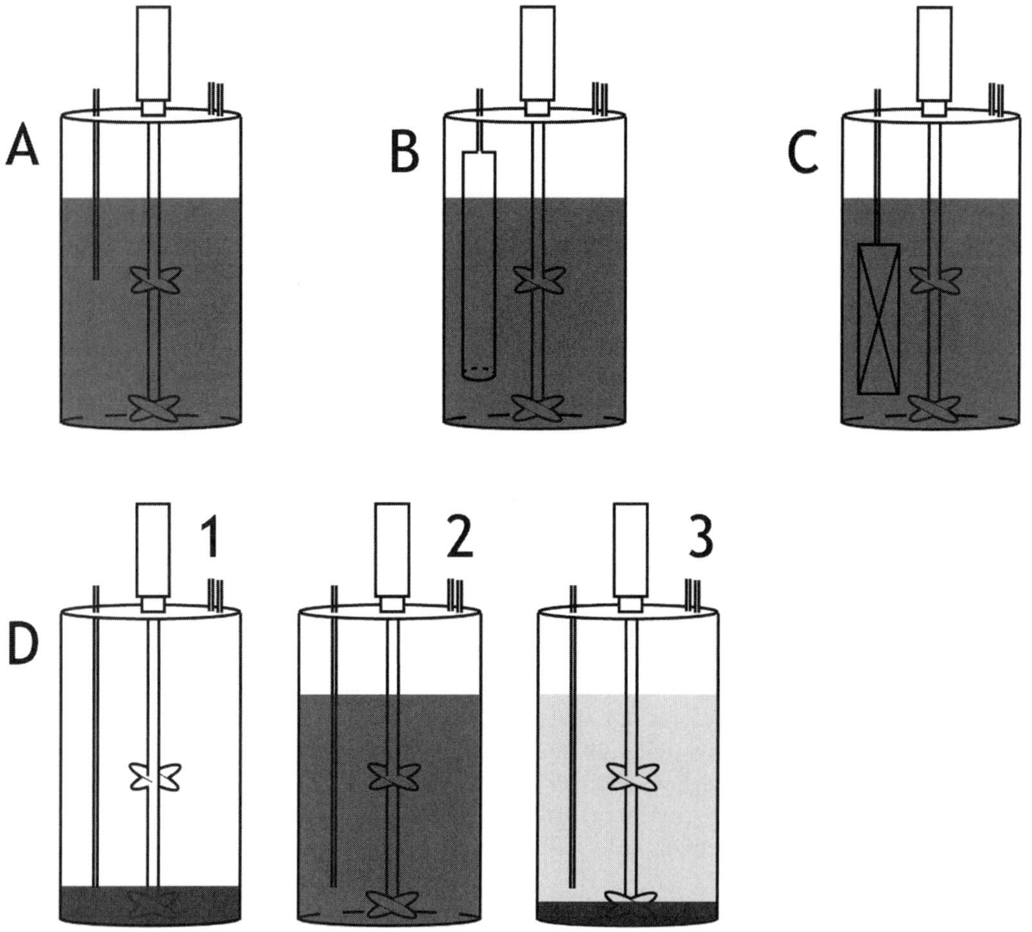

FIGURE 2 Experimental setups for enrichment in continuous culture. (A) Chemostat, (B) chemostat with an internal settler, (C) chemostat with a membrane unit, and (D) sequencing batch reactor.

the growth rate is an experimental parameter that can now be selected for.

When biomass retention is efficient, growth is almost exponential and because stable conditions can be maintained for a long time, very slowly growing microorganisms can be enriched and produced in large numbers for further study. For example, the anaerobic ammonium-oxidizing bacteria generally grow at doubling times over 2 weeks (Strous et al., 1999b).

The key to success with such slow-growing microorganisms is the prevention of experimental errors and equipment failures. For example, after several accidents, we completely banned the use of pH control for enrichment cultures. The pH probes tend to fail at least once per year and that is simply too frequent for organisms that take longer than a year to enrich for. Alternatively, pH can be maintained at a constant value by balancing the bicarbonate in the medium and the carbon dioxide concentrations in the gas supply.

Generally the stable end result of enrichment via this method is a culture dominated 70 to 80% by a target organism, the primary producer in the enrichment context, the "owner" of the niche defined by the enrichment conditions. The remaining 20 to 30% consists of a cryptic, heterogeneous by-population that may profit from leakage of metabolites, degradation of extracellular polymeric substances, or dead cells.

Compared to the original chemostat, the conditions experienced by the organisms are only known approximately. The chemical conditions in the bulk liquid (outside the microbial aggregates) can easily be measured, but one should be aware that gradients exist inside the biofilms. Therefore, microbiological parameters such as the affinity constant for the substrates cannot be measured as precisely as in a true chemostat.

In the future, membrane filters might be applied to overcome this problem and to apply the approach to suspended organisms. The filter could be mounted inside the reactor and would retain suspended cells. This way, the selection pressure for aggregated growth disappears, since suspended cells would also be retained and have higher substrate affinity than aggregated microorganisms. The fitting of chemostats with membranes has been described previously, and the system is known as the "retentostat" (Tappe et al., 1999). The challenge here is to engineer the filter in such a way that it is compatible with reliable performance. Filters are prone to frequent clogging (especially at higher solid concentrations, i.e., with muddy environmental inocula), and clogged filters rapidly lead to a flooded laboratory.

CHARACTERIZATION IN MOLECULAR DETAIL

It was stated before that the reason to "culture" a microorganism is to relate a biochemical reaction to a single microorganism. Once the microorganism is isolated, then the quest for revealing the secrets it is holding starts. In an enrichment culture, however, an organism is not isolated; it is still present in coculture with others. Then the question arises: How does one abide with Koch's postulates? The difficulty of fulfilling the central dogma of microbiology is the prime disadvantage of having an enrichment culture instead of a single isolate; actually all the other problems arise from it. The usual modus operandi when working with a single microorganism is to calculate the kinetic parameters, determine the intermediates, resolve the proteins responsible for the conversion of the intermediates, and connect the protein to the gene. Then the genes coding for the key enzymes may be knocked out, and the functionality may be ascertained. However, following these steps is not straightforward when working with an enrichment culture. For example, how does one work on kinetics in an enrichment culture? Or how does one figure out the metabolic intermediates or make a functional mutant?

Still, with many recently developed methods in environmental microbiology it is possible to characterize enrichment cultures in detail: PCR-based techniques (cloning, real-time PCR, denaturing gradient gel electrophoresis), different types of microscopy (e.g., epifluorescence and electron microscopy), physical isolation techniques (e.g, density gradient centrifugation), determination of microorganism-specific molecules (e.g., lipids), stable isotope tracing, the analysis of the metagenome from enrichment cultures, and proteomics approaches.

The first step in characterizing a mixed culture is to figure out which microorganisms are living in it. Constructing a general clone library with universal primers gives valuable preliminary information about the phylogenetic diversity in an enrichment culture. However, because of well-known biases of DNA extraction and PCR, a clone library might not always reflect the real diversity of an enrichment culture. Furthermore, with this technique it is not possible to quantify relative abundances of the different constituents of the culture.

A PCR-based, semiquantitative alternative to cloning is denaturing gradient gel electrophoresis (DGGE). In this technique the whole genomic DNA extract is amplified with a special forward primer with a GC-rich ending (GC clamp). When the complete amplificate is run through a denaturing acrylamide gel, the double-DNA strand melts until the GC clamp and anchors to the gel. Depending on their sequences, different DNA strands stop at different points in the gel, resulting in the separation of the DNA. Later the bands may be cut out of the gel and sequenced. The intensity of the bands directly relates to their quantity in the amplificates. In theory, the more dominant a band is, the more abundant is that species. Again, due to the accumulative bias arising from DNA extraction and PCR, quantification is at best semiquantitative. A disadvantage of this approach is that the amplified DNA fragments cannot be longer than 500 base pairs (Myers et al., 1985). Even though the sequencing of such a short fragment may indicate what kinds of microorganisms are present in the enrichment culture, it is far from conclusive. Moreover, the amount of amplificates that is separated by DGGE is limited by the size of the gel; at most 1% of the total community may be separated (Muyzer et al., 1993). Also there is the possibility of comigration in the polyacrylamide matrix, which then leads to multiple amplificates in one band. Still the method is very useful when monitoring an enrichment culture over a period of time or at different operation regimes. For example, Kowalchuk and coworkers used DGGE to study the ammonium-oxidizing bacteria in complex microbial systems in sand dunes (Kowalchuk et al., 1997). They found a correlation between pH and proximity to the sea with the occurrence of different types of ammonium-oxidizing bacteria. This study also shows the weaknesses of the DGGE approach such as comigration in the polyacrylamide matrix.

There are also PCR-based quantitative techniques such as real-time PCR. Real-time PCR is a minor modification to conventional PCR: A fluorescent DNA dye (e.g., SYBR green) is added to the reaction mix or a fluorescently labeled primer is used. The progress of the amplification is then monitored by a fluorescence reader. The intensity of the signal increases proportionally with the amount of the amplificate. With this approach and appropriate primers it is possible to specifically quantify a single organism or a clade of microorganisms. For example, the abundance of *Crenarchaeota* in soils was recently shown by real-time PCR of ammonium monooxygenase genes from archaea and bacteria (Leininger et al., 2006). However, the method does not provide any information on the activity of the detected constituents of the ecosystem. Moreover, it is not possible to have an overview of a natural system or an enrichment culture. Once a clone library is constructed from an enrichment culture, and certain microorganisms are expected to be major players, the real-time PCR method may be optimized toward those species. This strong point of the method seems to be one of its weaknesses because it brings in the bias of the researcher. Of course, one should not exclude the bias arising from DNA extraction and PCR.

Epifluorescence microscopy is of great value to visualize relative abundances determined with cloning or DGGE. The enrichment culture is a complex system; thus, markers that differentiate between different constituents of the culture are needed. Oligonucleotides labeled with fluorescent dyes (probes) are perfect as such markers. Fluorescence in situ hybridization (FISH) technique makes use of probes that target specific 16S rRNA of different microorganisms (Amann et al., 1990). Usually probes (e.g., kingdom, phylum, and genus specific) labeled with dyes with different fluorescence spectra are reacted with a sample. Then an epifluorescence microscope is used to count cells positive for the used probes to quantify the microorganisms. FISH does not provide information about the activity of the cells in the culture. Several approaches exist to overcome this problem.

It is possible to combine FISH with immunofluorescence: Antibodies targeting specific proteins are applied and made visible by a secondary antibody labeled with a fluorescent dye. This way, the expression of a certain protein may be visualized in the cell. For example, it was possible to localize the enzyme nitrite oxidoreductase of nitrite-reducing bacteria with monoclonal antibodies (Bartosch et al., 1999).

FISH-microautoradiography (FISH-MAR or STARFISH) can reveal the type of microorganism at the single-cell level (Lee et al., 1999; Ouverney and Fuhrman, 1999). In this method a radioactive substrate (e.g., $^{14}CO_2$ or ^{14}C-acetate) is fed to the culture. After labeling, the cells are transferred to a slide and a silver film is applied. After exposure, the cells that hybridize with a defined probe and cells that incorporate radiolabeled substrates are visualized with a confocal laser scanning microscope. If the fluorescence signal overlays with the radiation signal, it means that those cells were consuming the radioactive substrates. FISH-MAR is well suited to reveal complex interactions of the microbial community of an enrichment culture. For example, a recent application of this method on enrichment cultures revealed that *Chloroflexus*-like filamentous bacteria were dependent on nitrifying bacteria for organic matter supply in an "autotrophic" biofilm (Okabe et al., 2005).

Due to the limitations of light microscopy, it is not possible to determine cell structure with epifluorescence microscopy. Transmission electron microscopy (TEM) has a dramatically higher (5 to 10 nm) resolution; therefore, with this technique it is possible to determine the cell structure of the constituents of an enrichment culture. There is a technique that combines oligonucleotide probes with electron microscopy (Canto-Nogues et al., 2001). In this method instead of fluorescent dyes, gold particles are used to label the oligonucleotides. Then the gold particles are viewed via TEM. This way, it is possible to identify the type of the microorganism and visualize the structure of the cell. This electron microscopy in situ hybridization technique is not yet applied to environmental microbiology, but it looks promising for possible future applications because thin sections provide much more information about the cell and its envelope than fluorescent microscopy. Immunogold labeling may also be used conventionally to localize specific proteins in the cell. For example, the presence of hydroxylamine oxidoreductase of anammox bacteria in the anammoxosome and the DNA-rich region in the riboplasm of *Isosphaera pallida* were visualized by immunogold labeling and electron microscopy (Lindsay et al., 2001).

Results from clone libraries and microscopy reveal the types of microorganisms in the culture, and the conditions of the bioreactor reveal information about the niche of these organisms. From here on it may be possible to proceed with conventional isolation procedures. If this is not possible or difficult, it can be a very good option to physically separate target cells from the remainder of the community with cell density centrifugation. Density gradient centrifugation makes use of a dynamic gradient (such as Percoll) to separate particles according to their densities. It is possible to create gradients with very high resolutions to separate subcellular particles, cells, or viruses. This method has been used in microbiology to screen different types of microorganisms from a mixed cell culture. For example, use of density gradient centrifugation has become a cornerstone in the investigation of the anaerobic ammonium-oxidizing (anammox) bacteria (Strous et al., 1999a). The first anammox bacterium was identified after the cells were separated from other constituents of the enrichment culture by density gradient centrifugation (Fig. 3). With the help of physical separation it was possible to meet Koch's postulates; it was shown that one single organism was able to oxidize ammonium under anoxic conditions and incorporated radiolabeled carbon dioxide (Strous et al., 1999a). Now, through physical separation it is possible to gather unambiguous data related to anammox bacteria; it was of great value to study anammox physiology and essential for the assembly of the genome of the anammox bacterium *Kuenenia stuttgartiensis* from its enrichment culture (Kartal et al., 2007a; Strous et al., 1999a; Strous et al., 2006). Gradient centrifugation was also used to enrich for *Cenarchaeum symbiosum* cells before the genomic analysis of this species (Schleper et al., 1998).

It is also possible to identify microorganisms in an enrichment culture independent from molecular techniques, microscopy, or physical separation. Kingdom-, phylum-, or class-specific molecules, biomarkers, are very useful for determining the presence of certain microorganisms. Membrane lipids are especially well suited for this purpose. Crenarchaeol, a crenarchaeote-specific membrane lipid, has been used to detect these organisms in nature and in enrichment culture (Sinninghe

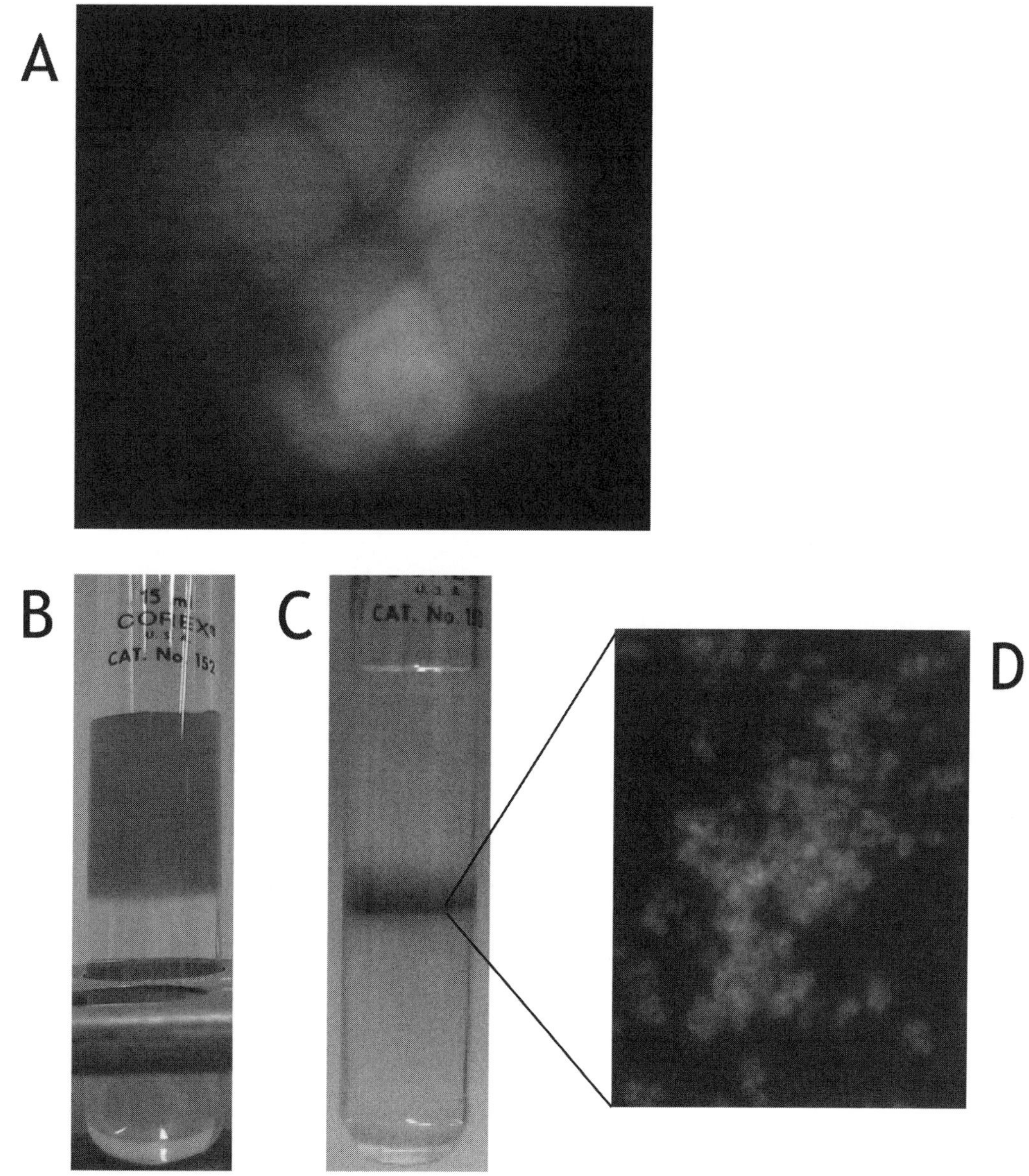

FIGURE 3 Percoll purification of anammox bacterium *K. stuttgartiensis* from an enrichment culture. (A) FISH micrograph of an anammox cluster (light gray) from an enrichment culture with other microorganisms. (B) Anammox cell suspension on Percoll gradient before centrifugation. (C) "Anammox band" that forms after centrifugation. (D) FISH micrograph of Percoll-separated *K. stuttgartiensis* cells.

Damsté et al., 2002a). Another example is the so-called ladderane lipids of the anammox bacteria (Sinninghe Damsté et al., 2002b). These lipids are made of concatenated cyclobutane rings and so far unique to anammox bacteria (Sinninghe Damsté et al., 2002b). The presence of these lipids in an enrichment culture or in nature shows the presence of anammox bacteria (Kartal et al., 2007b; Kuypers et al., 2003; Schmid et al., 2003). Moreover, different anammox species produce different types and amounts of these ladderane lipids; thus, it is possible to even determine what type of anammox bacteria is present in an enrichment culture solely based on their membrane lipid profiles. This technique can also be combined with natural fractionation studies, namely with the relative abundance of ^{12}C and ^{13}C carbon in membrane lipids. When ladderane lipids of anammox bacteria have been identified, it was discovered that these lipids are highly depleted in ^{13}C carbon (Schouten et al., 2004). When less-specific membrane lipids (such as hopanoids) were extracted from an anammox enrichment culture, they were also found to be highly depleted in ^{13}C carbon, indicating that they could also be produced by anammox bacteria (Schouten et al., 2004). This depletion in ^{13}C content of the lipids led to the hypothesis that anammox bacteria used the acetyl-CoA pathway for carbon fixation (Schouten et al., 2004). The hypothesis was later confirmed by the genome sequence of *Kuenenia stuttgartiensis*: the acetyl-CoA pathway was the only complete carbon fixation pathway present (Strous et al., 2006). Finally, one can use stable isotopes to trace specific biomarkers in a complex system. The enrichment culture may be fed a stable isotope (e.g., $^{13}CH_4$) and the incorporation of this compound to the cell material may be determined. One drawback of using biomarkers is that it is not always possible to find markers unique to species or even to phyla.

There are more clever ways to make use of isotopic labeling. In stable isotope probing the enrichment culture is incubated with a ^{13}C-labeled carbon source. Organisms that make use of the carbon source incorporate the ^{13}C into their DNA and RNA, and these nucleic acids can then be separated from their unlabeled counterparts by cesium chloride centrifugation (Radajewski et al., 2000). Then the heavy DNA may be further investigated by cloning or other genomic approaches. A recent exciting development is the implementation of secondary-ion mass spectrometry (SIMS) in biological sciences (Orphan et al., 2002); this method was well known for investigating isotopic composition in the chemical and materials sciences. The technique can be combined with FISH to determine the isotopic composition of the elements of a cell after identification by FISH. In other words, it is possible to determine the incorporation of a stable isotope into the cell in situ. The technique makes use of a laser beam that volatilizes the cell material where it hits the cell. The ionized particles are then detected by a mass spectrometer, and the incorporation of the stable isotope is determined. The resolution of the method is limited by the size of the laser beam. Currently high-resolution instruments are on the market and known as nano-SIMS; their resolution is currently 33 nm. It is highly likely that we will see this method used in many studies in near future (Lechene et al., 2006).

An overall problem of using stable isotopes in the study of complex cultures is the possibility of interconversion of these compounds. One may start an experiment with ^{13}C-acetate, but of course in a complex system the acetate may be oxidized to $^{13}CO_2$, and the $^{13}CO_2$ may subsequently be incorporated to cells, possibly contaminating the results. However, it might be possible to limit such problems with proper controls and shorter incubation times.

Nowadays the number of publicly available genome sequences increases by the day, and genome sequencing has become common practice. It is even possible to sequence genomes from natural habitats and enrichment cultures; bioinformatics makes possible the assembly of individual genomes from complex systems (Strous et al., 2006; Tyson et al., 2004). Genome sequences open new doors: When the genome information is available, Koch's postulates may sometimes even be satisfied in silico. In other words, it is possible to establish the existence of genes necessary to perform a biochemical reaction. Hypotheses that do not conform to the genes may be amended, new ideas may come up, and the research may take a wholly different path. For example, the metagenome of an acid-mine drainage community revealed that the community was dependent on a single organism able to fix dinitrogen (Tyson et al., 2004). That knowledge also directed the subsequent isolation of the nitrogen fixer (Tyson et al., 2005). The *K. stuttgartiensis* metagenome indicated the possibility that instead of hydroxylamine, nitric oxide was the most likely intermediate in central anammox catabolism (Strous et al., 2006). Without the genome sequence it would have taken several years to get to this point. Of course, inferences from genomes should always be validated experimentally.

Genomic data also make possible the identification of the major proteins of an organism by proteomics (two-dimensional gel electrophoresis combined with time of flight mass spectroscopy or digestion of the protein extract followed by liquid chromatography-mass spectroscopy). This analysis can be complemented by the analysis of the messenger RNA pool by RNA arrays that rapidly show which genes are expressed in a mixed culture. Once it is known that certain proteins are expressed, one can proceed with conventional biochemical purification to study their kinetics, reveal their sequences, construct crystal structures, and study the active sites. To attempt this usually a substantial amount of protein, i.e., a thick cell culture, is needed. Alternatively, target genes can sometimes be overexpressed in other hosts such as *Escherichia coli* and purified from there.

Naturally, the enrichment culture can be used to perform the conventional chemostat-type experiments to determine the maximum growth rate as a function of pH and temperature (Strous et al., 1999b). Competition experiments can also be performed between different enriched organisms with similar niches.

All these approaches are possible with enrichment cultures. Because the organisms grow under stable and more or less natural conditions, the results are likely to be better than for pure cultures. The only problem that arises is the ambiguity of the source of the activity, protein, or expressed gene. The interpretation depends on the correct assembly of the individual genomes from the metagenome. To overcome this point, immunogold and immunofluoresence approaches are useful. Antibodies can be raised with synthetic peptides designed for the target gene or by heterologous overexpression of proteins in other hosts (Huston et al., 2007; Lindsay et al., 2001).

The most important method that cannot be applied to enrichment cultures is genetic modification, for example, the construction of knockout mutants. If this is necessary, isolation is required. The knowledge gained so far may be used to direct such isolation (Tyson et al., 2005). However, in general, that should not be necessary, since with the methods described above it is possible not only to fulfill Koch's postulates but also to describe the niche of an enriched microorganism in molecular detail.

BACK TO NATURE

At this point we have a clear sense of many aspects of the niche of the enriched microorganism. Furthermore, we have set up a toolbox to investigate those aspects of the niche in more complex situations such as the natural habitat. Now comes the time to investigate whether our ideas also apply to the natural settings, to go back to nature and to find out whether the organism also fulfills its niche in nature.

The toolbox consists of a suite of FISH probes for detection of target cells in nature, target genes for real-time PCR quantification, labeling approaches to measure the growth and activity in short incubations of natural samples, biomarkers combined with knowledge of isotope fractionation effects, and the in silico probing of environmental metagenomes. For unambiguous results in the complex natural habitat it is absolutely essential to combine a number of these independent approaches in parallel, since all methods by themselves are prone to biases and systematic errors. In case of anammox bacteria biomarker analysis (detection of ladderanes), FISH and ^{15}N isotopic labeling have been combined successfully (Kuypers et al., 2003).

It is possible that the results are unexpected; the enriched organisms are not detected in habitats where the niche is really fulfilled. Apparently another organism exists that is a better fitter in the postulated niche. For a long time this was the case for betaproteobacterial aerobic ammonium oxidizers in the oceans. Ultimately, it appeared that *Crenarchaeota* were the major competitors for this niche in that environment (Könneke et al., 2005; Wuchter et al., 2006). With such an anomaly it is time to start again with a new definition of what the niche really means.

CONCLUSION

Enrichment of microorganisms in continuous culture is a mature technique that has been applied successfully to anaerobic ammonium oxidizers (Van de Graaf et al., 1995) and anaerobic methane-oxidizing denitrifiers (Raghoebarsing et al., 2006). For the anammox bacteria this approach was so successful that these bacteria are by far the best-understood organisms of their phylum (the *Planctomycetes*). Even though representatives of the phylum's other taxa are available in pure culture, we still have very limited understanding of their niche and metabolism. Every clade of anammox organisms detected in nature by molecular ecology approaches has at least one representative in enrichment culture (Fig. 4). We have a comparable understanding of their ecological

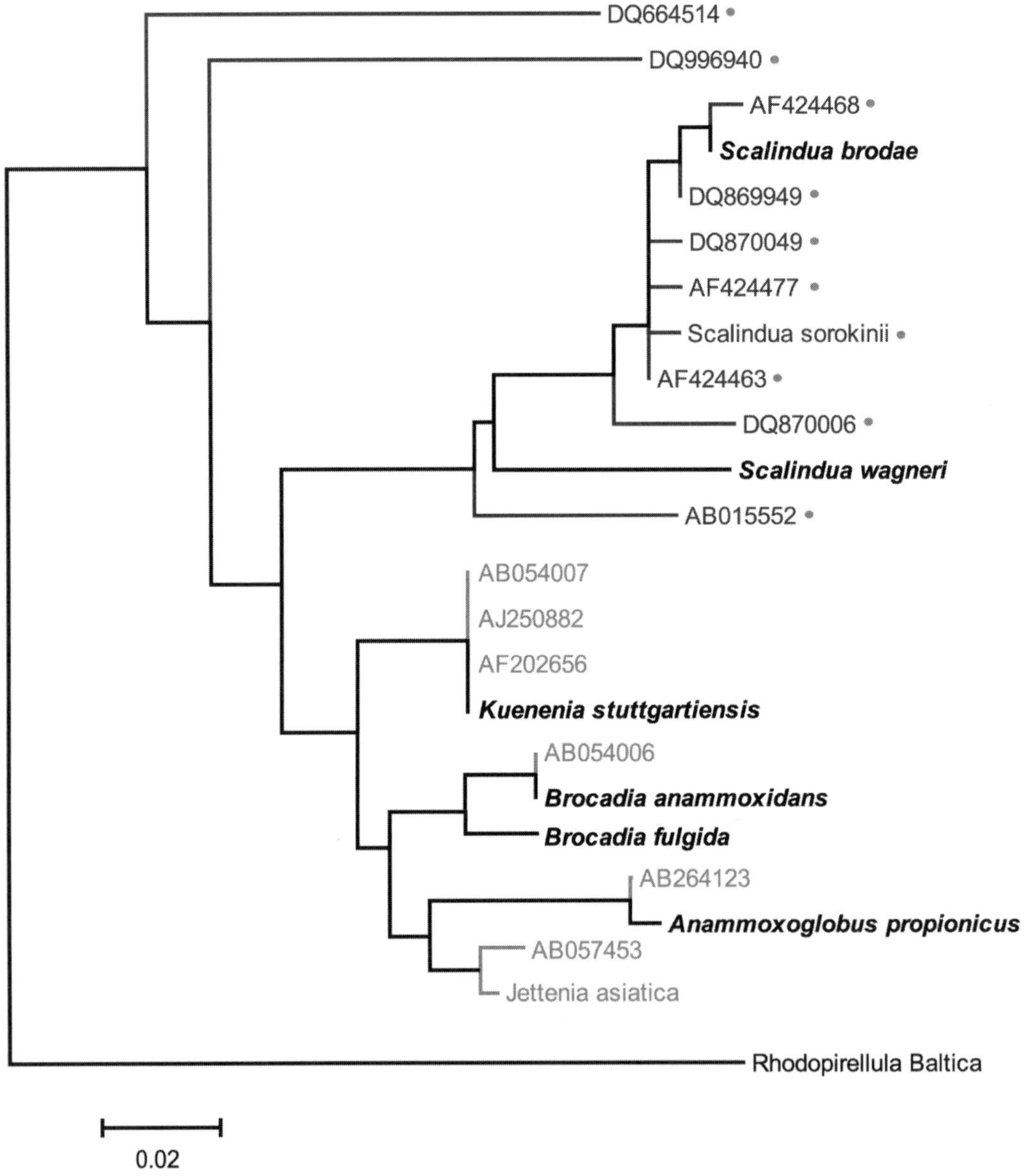

FIGURE 4 Phylogenetic tree showing cultivated and uncultivated anammox bacteria. Black, described species; dark gray with a dot, marine clone; light gray, reactor clone.

niche than we have for betaproteobacterial aerobic nitrifiers, which have been in pure culture for over a century. As indicated above, the study of microorganisms in undefined mixed cultures could be applied more generally and has great potential for environmental microbiology.

REFERENCES

Amann, R. I., B. J. Binder, R. J. Olson, S. W. Chisholm, R. Devereux, and D. A. Stahl. 1990. Combination of 16S ribosomal-RNA-targeted oligonucleotide probes with flow-cytometry for analyzing mixed microbial-populations. *Appl. Environ. Microbiol.* **56:**1919–1925.

Bartosch, S., I. Wolgast, E. Spieck, and E. Bock. 1999. Identification of nitrite oxidizing bacteria with monoclonal antibodies recognizing the nitrite oxidoreductase. *Appl. Environ. Microbiol.* **65:**4126–4133.

Beja, O., L. Aravind, E.V. Koonin, M.T. Suzuki, A. Hadd, L. P. Nguyen, S. Jovanovich, C. M. Gates, R. A. Feldman, J. L. Spudich, E. N. Spudich, and E. F. DeLong. 2000. Bacterial rhodopsin: evidence for a new type of phototrophy in the sea. *Science* **289:**1902–1906.

Broda, E. 1977. 2 Kinds of lithotrophs missing in nature. *Z. Allg. Mikrobiol.* **17:**491–493.

Canto-Nogues, C., D. Hockley, C. Grief, S. Ranjbar, J. Bootman, N. Almond, and I. Herrera. 2001. Ultrastructural localization of the RNA of immunodeficiency viruses using electron microscopy in situ hybridization and in vitro infected lymphocytes. *Micron* **32:**579–589.

Elias, D. A., M. E. Monroe, R. D. Smith, J. K. Fredrickson, and M. S. Lipton. 2006. Confirmation of the expression of a large set of conserved hypothetical proteins in *Shewanella oneidensis* MR-1. *J. Microbiol. Methods* **66:**223–233.

Hinrichs, K. U., J. M. Hayes, S. P. Sylva, P. G. Brewer, and E. F. DeLong. 1999. Methane consuming archaebacteria in marine sediments. *Nature* **398:**802–805.

Huston, W. M., H. R. Harhangi, A. P. Leech, C. S. Butler, M. S. M. Jetten, H. den Camp, and J. W. B. Moir. 2007. Expression and characterisation of a major c-type cytochrome encoded by gene kustc0563 from *Kuenenia stuttgartiensis* as a recombinant protein in *Escherichia coli*. *Protein Expr. Purif.* **51:**28–33.

Jansen, M. L. A., J. A. Diderich, M. Mashego, A. Hassane, J. H. de Winde, P. Daran-Lapujade, and J. T. Pronk. 2005. Prolonged selection in aerobic, glucose-limited chemostat cultures of *Saccharomyces cerevisiae* causes a partial loss of glycolytic capacity. *Microbiology-Sgm* **151:**1657–1669.

Kartal, B., M. M. M. Kuypers, G. Lavik, J. Schalk, H. J. M. Op den Camp, M. S. M. Jetten, and M. Strous. 2007a. Anammox bacteria disguised as denitrifiers: nitrate reduction to dinitrogen gas via nitrite and ammonium. *Environ. Microbiol.* **9:**635–642.

Kartal, B., J. Rattray, L. van Niftrik, J. van de Vossenberg, M. Schmid, R. I. Webb, S. Schouten, J. A. Fuerst, J. S. Sinninghe Damsté, M. S. M. Jetten, and M. Strous. 2007b. *Candidatus* "Anammoxoglobus propionicus" gen. nov., sp. nov., a new propionate oxidizing species of anaerobic ammonium oxidizing bacteria. *Syst. Appl. Microbiol.* **30:**39–49.

Könneke, M., A. E. Bernhard, J. R. de la Torre, C. B. Walker, J. B. Waterbury, and D. A. Stahl. 2005. Isolation of an autotrophic ammonia-oxidizing marine archaeon. *Nature* **437:**543–546.

Kowalchuk, G. A., J. R. Stephen, W. De Boer, J. I. Prosser, T. M. Embley, and J. W. Woldendorp. 1997. Analysis of ammonia-oxidizing bacteria of the beta subdivision of the class proteobacteria in coastal sand dunes by denaturing gradient gel electrophoresis and sequencing of PCR-amplified 16S ribosomal DNA fragments. *Appl. Environ. Microbiol.* **63:**1489–1497.

Kuenen, J. G., and H. Veldkamp. 1973. Effects of organic compounds on growth of chemostat cultures of *Thiomicrospira pelophila, Thiobacillus thioparus and Thiobacillus neapolitanus*. *Arch. Mikrobiol.* **94:**173–190.

Kuypers, M. M. M., A. O. Sliekers, G. Lavik, M. Schmid, B. B. Jørgensen, J. G. Kuenen, J. S. Sinninghe Damsté, M. Strous, and M. S. M. Jetten. 2003. Anaerobic ammonium oxidation by anammox bacteria in the Black Sea. *Nature* **422:**608–-611.

Lechene, C., F. Hillion, G. McMahon, D. Benson, A. M. Kleinfeld, J. P. Kampf, D. Distel, Y. Luyten, J. Bonventre, D. Hentschel, K. M. Park, S. Ito, M. Schwartz, G. Benichou, and G. Slodzian. 2006. High-resolution quantitative imaging of mammalian and bacterial cells using stable isotope mass spectrometry. *J. Biol.* **5:**30.

Lee, N., P. H. Nielsen, K. H. Andreasen, S. Juretschko, J. L. Nielsen, K. H. Schleifer, and M. Wagner. 1999. Combination of fluorescent in situ hybridization and microautoradiography—a new tool for structure-function analyses in microbial ecology. *Appl. Environ. Microbiol.* **65:**1289–1297.

Leininger, S., T. Urich, M. Schloter, L. Schwark, J. Qi, G. W. Nicol, J. I. Prosser, S. C. Schuster, and C. Schleper. 2006. Archaea predominate among ammonia-oxidizing prokaryotes in soils. *Nature* **442:**806–809.

Lindsay, M. R., R. I. Webb, M. Strous, M. S. Jetten, M. K. Butler, R. J. Forde, and J. A. Fuerst. 2001.

Cell compartmentalisation in planctomycetes: novel types of structural organisation for the bacterial cell. *Arch. Microbiol.* **175:**413–429.

Mashego, M. R., M. L. A. Jansen, J. L. Vinke, W. M. van Gulik, and J. J. Heijnen. 2005. Changes in the metabolome of *Saccharomyces cerevisiae* associated with evolution in aerobic glucose-limited chemostats. *FEMS Yeast Res.* **5:**419–430.

Muyzer, G., E. C. Dewaal, and A. G. Uitterlinden. 1993. Profiling of complex microbial-populations by denaturing gradient gel electrophoresis analysis of polymerase chain reaction amplified genes coding for 16s ribosomal-RNA. *Appl. Environ. Microbiol.* **59:**695–700.

Myers, R. M., S. G. Fischer, L. S. Lerman, and T. Maniatis. 1985. Nearly all single base substitutions in DNA fragments joined to a GC-clamp can be detected by denaturing gradient gel electrophoresis. *Nucleic Acids Res.* **13:**3131–3145.

Nogueira, R., and L. F. Melo. 2006. Competition between *Nitrospira* spp. and *Nitrobacter* spp. in nitrite oxidizing bioreactors. *Biotechnol. Bioeng.* **95:**169–175.

Novick, A., and L. Szilard. 1950. Description of the chemostat. *Science* **112:**715–716.

Okabe, S., T. Kindaichi, Y. Nakamura, and T. Ito. 2005. Ecophysiology of autotrophic nitrifying biofilms. *Water Sci. Technol.* **52:**225–232.

Orphan, V. J., C. H. House, K. U. Hinrichs, K. D. McKeegan, and E. F. DeLong. 2002. Multiple archaeal groups mediate methane oxidation in anoxic cold seep sediments. *Proc. Natl. Acad. Sci. USA* **99:**7663–7668.

Ouverney, C. C., and J. A. Fuhrman. 1999. Combined microautoradiography 16S rRNA probe technique for determination of radioisotope uptake by specific microbial cell types in situ. *Appl. Environ. Microbiol.* **65:**1746–1752.

Radajewski, S., P. Ineson, N. R. Parekh, and J. C. Murrell. 2000. Stable isotope probing as a tool in microbial ecology. *Nature* **403:**646–649.

Raghoebarsing, A. A., A. Pol, K. T. van de Pas-Schoonen, A. J. P. Smolders, K. F. Ettwig, W. I. C. Rijpstra, S. Schouten, J. S. Sinninghe Damsté, H. J. M. Op den Camp, M. S. M. Jetten, and M. Strous. 2006. A microbial consortium couples anaerobic methane oxidation to denitrification. *Nature* **440:**918–921.

Schleper, C., E. F. DeLong, C. M. Preston, R. A. Feldman, K. Y. Wu, and R. V. Swanson. 1998. Genomic analysis reveals chromosomal variation in natural populations of the uncultured psychrophilic archaeon *Cenarchaeum symbiosum*. *J. Bacteriol.* **180:**5003–5009.

Schmid, M., K. Walsh, R. Webb, W. I. C. Rijpstra, K. van de Pas-Schoonen, M. J. Verbruggen, T. Hill, B. Moffett, J. Fuerst, S. Schouten, J. S. Sinninghe Damsté, J. Harris, P. Shaw, M. Jetten, and M. Strous. 2003. *Candidatus* "Scalindua brodae", sp nov., *Candidatus* "Scalindua wagneri", sp nov., two new species of anaerobic ammonium oxidizing bacteria. *Syst. Appl. Microbiol.* **26:**529–538.

Schouten, S., M. Strous, M. M. M. Kuypers, W. I. C. Rijpstra, M. Baas, C. J. Schubert, M. S. M. Jetten, and J. S. Sinninghe Damsté. 2004. Stable carbon isotopic fractionations associated with inorganic carbon fixation by anaerobic ammonium-oxidizing bacteria. *Appl. Environ. Microbiol.* **70:**3785–3788.

Sinninghe Damsté, J. S., S. Schouten, E. C. Hopmans, A. C. T. van Duin, and J. A. J. Geenevasen. 2002a. Crenarchaeol: the characteristic core glycerol dibiphytanyl glycerol tetraether membrane lipid of cosmopolitan pelagic crenarchaeota. *J. Lipid Res.* **43:**1641–1651.

Sinninghe Damsté, J. S., M. Strous, W. I. C. Rijpstra, E. C. Hopmans, J. A. J. Geenevasen, A. C. T. van Duin, L. A. van Niftrik, and M. S. M. Jetten. 2002b. Linearly concatenated cyclobutane lipids form a dense bacterial membrane. *Nature* **419:**708–712.

Strous, M., J. A. Fuerst, E. H. M. Kramer, S. Logemann, G. Muyzer, K. T. van de Pas-Schoonen, R. Webb, J. G. Kuenen, and M. S. M. Jetten. 1999a. Missing lithotroph identified as new planctomycete. *Nature* **400:**446–449.

Strous, M., J. G. Kuenen, and M. S. M. Jetten. 1999b. Key physiology of anaerobic ammonium oxidation. *Appl. Environ. Microbiol.* **65:**3248–3250.

Strous, M., E. Pelletier, S. Mangenot, T. Rattei, A. Lehner, M. W. Taylor, M. Horn, H. Daims, D. Bartol-Mavel, P. Wincker, V. Barbe, N. Fonknechten, D. Vallenet, B. Segurens, C. Schenowitz-Truong, C. Medigue, A. Collingro, B. Snel, B. E. Dutilh, H. J. M. Op den Camp, C. van der Drift, I. Cirpus, K. T. van de Pas-Schoonen, H. R. Harhangi, L. van Niftrik, M. Schmid, J. Keltjens, J. van de Vossenberg, B. Kartal, H. Meier, D. Frishman, M. A. Huynen, H. W. Mewes, J. Weissenbach, M. S. M. Jetten, M. Wagner, and D. Le Paslier. 2006. Deciphering the evolution and metabolism of an anammox bacterium from a community genome. *Nature* **440:**790–794.

Tappe, W., A. Laverman, M. Bohland, M. Braster, S. Rittershaus, J. Groeneweg, and H. W. van Verseveld. 1999. Maintenance energy demand and starvation recovery dynamics of *Nitrosomonas europaea* and *Nitrobacter winogradskyi* cultivated in a retentostat with complete biomass retention. *Appl. Environ. Microbiol.* **65:**2471–2477.

Teske, A. P. 2005. The deep subsurface biosphere is alive and well. *Trends Microbiol.* **13:**402–404.

Tyson, G. W., J. Chapman, P. Hugenholtz, E. E. Allen, R. J. Ram, P. M. Richardson, V.

V. Solovyev, E. M. Rubin, D. S. Rokhsar, and J. F. Banfield. 2004. Community structure and metabolism through reconstruction of microbial genomes from the environment. *Nature* **428:**37–43.

Tyson, G. W., I. Lo, B. J. Baker, E. E. Allen, P. Hugenholtz, and J. F. Banfield. 2005. Genome-directed isolation of the key nitrogen fixer *Leptospirillum ferrodiazotrophum* sp nov from an acidophilic microbial community. *Appl. Environ. Microbiol.* **71:**6319–6324.

Van de Graaf, A. A., A. Mulder, P. Debruijn, M. S. M. Jetten, L. A. Robertson, and J. G. Kuenen. 1995. Anaerobic oxidation of ammonium is a biologically mediated process. *Appl. Environ. Microbiol.* **61:**1246–1251.

Wuchter, C., B. Abbas, M. J. L. Coolen, L. Herfort, J. van Bleijswijk, P. Timmers, M. Strous, E. Teira, G. J. Herndl, J. J. Middelburg, S. Schouten, and J. S. Sinninghe Damsté. 2006. Archaeal nitrification in the ocean. *Proc. Natl. Acad. Sci. USA* **103:**12317–12322.

DO WE HAVE TO CHANGE GEAR? NEW CULTIVATION APPROACHES AND NEW MOLECULAR APPROACHES COMBINED

IV

MICROBIAL CELL INDIVIDUALITY

Simon V. Avery

13

Individual cells within clonal microbial cultures exhibit marked phenotypic heterogeneity. This heterogeneity is manifest in a wide range of phenotypes, many of which are fundamental to organismal fitness and/or development. For example, individual cells of microbial pathogens display variable degrees of virulence, have varying degrees of dormancy and resistance to antimicrobial treatments and other stresses, and may exhibit differing propensities to differentiate and to express motility determinants. Heterogeneity also extends to variability between individual hyphae of filamentous microorganisms (Vinck et al., 2005), but that is beyond the scope of this chapter, which focuses on heterogeneity phenomena among single-celled organisms. In addition to, and underlying these phenomena, the principal control processes that regulate cell function (e.g., gene transcription, translation) can, at any moment in time, be differentially activated in different cells within genetically uniform populations. Heterogeneity at the single-cell level is typically masked in conventional studies of microbial populations, which rely on data averaged across thousands or millions of cells in a sample. However, recent interest in the processes governing cell-to-cell variability, fueled by methodological advances (Elowitz et al., 2002; Sumner et al., 2003; Brehm-Stecher et al., 2004; Colman-Lerner et al., 2005; Huang et al., 2007), has been changing such conventions and enhancing awareness of cell individuality.

It has generally been considered that phenotypic heterogeneity provides a dynamic source of diversity in addition to that derived from genotypic changes such as genome rearrangements and mutation (i.e., genotypic heterogeneity). Recent studies highlighted later in this chapter have substantiated this view. Microbial populations should benefit by the creation of variant subpopulations that have the potential to be better equipped to persist during perturbation and/or to exploit new niches (Blake et al., 2006; Bishop et al., 2007; Smith et al., 2007). A specific advantage of phenotypic over genotypic heterogeneity is that the former does not invoke an irreversible commitment to the new state (i.e., indefinite inheritability), so allowing rapid reversion to the original phenotype if appropriate. Also, heterogeneity may be evident as some cells in a population entering a dormant

Simon V. Avery, School of Biology, Institute of Genetics, University of Nottingham, University Park, Nottingham NG7 2RD, United Kingdom.

Accessing Uncultivated Microorganisms: from the Environment to Organisms and Genomes and Back
Edited by Karsten Zengler © 2008 ASM Press, Washington, DC

or uncultivatable state, which may enhance the mean stress resistance of a population. In each of the examples of cell-individualized phenotypes addressed in this chapter, the benefits of heterogeneity to population fitness can be readily envisaged. Until very recently, support for the hypothesis that heterogeneity confers benefits has largely been by inference, albeit quite convincing inference. Thus, modeling studies have predicted outcompetition of homogeneous cell populations by more heterogeneous populations (Thattai and van Oudenaarden, 2004; Kussell and Leibler, 2005), at least under certain conditions, and there is evidence for evolutionary selection of diversity-generating mechanisms in microorganisms (True and Lindquist, 2000; Fraser et al., 2004; Raser and O'Shea, 2004; Bar-Even et al., 2006). Moreover, the most recent studies have risen to the challenge of demonstrating fitness benefits of heterogeneity experimentally (Blake et al., 2006; Bishop et al., 2007; Smith et al., 2007), and these studies are discussed further below.

This chapter integrates recent studies to describe the molecular bases and consequences of heterogeneity for several of the key phenotypes characterized by variability in clonal microbial populations. Drivers of cell heterogeneity within microbial populations include progression through the cell cycle and biological rhythms, spontaneous epigenetic modifications that are metastably inherited, aging linked to cell division, mitochondrial activity, individual-cell growth rates, and stochastic gene expression (Davey and Kell, 1996; Sumner and Avery, 2002) (Table 1). As discussed later, these mechanisms are not necessarily independent of each other, although they operate over different timescales (Fig. 1).

STOCHASTIC GENE EXPRESSION AND ITS CONSEQUENCES

Much of the recent focus of research relevant to the field of cell individuality has been to uncover the drivers of diversity operating at the molecular level, in particular the contributions of stochasticity or noise to the processes of gene transcription and translation (other drivers are discussed later). Noise may distort the same information encoded in the same genome differently in different individual cells and, with amplification (e.g., via positive feedback), could give rise to cell subpopulations with differing phenotypes and developmental fates. Noise is particularly relevant to low-abundance molecules; a stochastic difference of one or two mRNAs or proteins becomes proportionally larger the smaller the pool of that molecular species. In addition to the variation that can occur in mRNA- or protein-turnover rates in individual cells, stochastic variation in the processes of both gene transcription and translation has been demonstrated (Ozbudak et al., 2002; Blake et al., 2003). It is predicted that a particular level of gene expression accomplished through high translation and low transcription yields higher levels of noise when compared to a similar level of expression attained through low translation and high transcription. Consequently, studies have found that key essential and regulatory proteins typically have high relative transcription and low relative translation rates in cells, consistent with the low noise observed with these primary functions (Ozbudak et al., 2002; Fraser et al., 2004). Conversely proteins with functions related to stress responses are characterized by high relative noise (Bar-Even et al., 2006; Newman et al., 2006). Genes other than those involved in stress responses or protein synthesis/degradation exhibit an expression variance that is approximately proportional to the mean, consistent with stochastic production/destruction of mRNA molecules (translational bursting) being a dominant source of noise (Bar-Even et al., 2006; Newman et al., 2006). Variation in bacterial transcription rates has less of an impact on gene expression noise than in the eukaryote *Saccharomyces cerevisiae* (Kaern et al., 2005). Although stochastic gene expression in bacteria was thought to be dominated by the influence of low-abundance mRNAs and translational bursting (Elowtiz et al., 2002; Ozbudak et al., 2002), a "transcriptional bursting" mechanism

TABLE 1 Major heterogeneously expressed microbial phenotypes and the relevant drivers of heterogeneity[a]

Phenotype	Comments	Underlying driver (and organisms investigated)
Heterogeneous antigenicity at the microbial cell surface	Different epigenetic mechanisms operate in different systems	Epigenetic (*P. falciparum, T. brucei, S. cerevisiae, E. coli*)[b]
Switching between morphologies in *Candida* spp.	White-opaque switching requires homo- or hemizygosity at the mating type locus. Histone deacetylases modulate switch frequency. Wor1 (Tos9) is a master regulator.	Probably stochastic
Resistance to metals and pro-oxidants	Several deterministic mechanisms of heterogeneity have been shown	Cell cycle, ultradian rhythm, aging (*S. cerevisiae*)
Resistance to heat stress	Variable expression of HSP-encoding genes	Cell cycle/stochastic? (*S. cerevisiae, S. enterica* serovar Typhimurium)
Resistance to DNA damage	Several mechanisms suggested	Aging/ultradian rhythm? (*S. cerevisiae*) Stochastic? (*E. coli*)
Antibiotic persistence	The proportion of persisters is influenced by the *hipBA* toxin-antitoxin module	Stochastic?[c] (*E. coli*)
Bacterial motility and chemotaxis	Variable activity in *E. coli* of the CheR methyltransferase, a key upstream component of the chemotaxis signaling network	Stochastic? (*E. coli, B. subtilis*)
Lysis/lysogeny in phage lambda	The decision hinges primarily on the activity of the cII activator protein	Probably stochastic
Competence development and sporulation	Heterogeneity requires bistable expression of the major transcriptional regulators, ComK and Spo0A	Probably stochastic (*B. subtilis*)

[a]Table adapted from Avery (2006) with permission.

[b]In many cases where spontaneous epigenetic modification is involved, there remains the underlying question as to what determines which cells will switch and when (see text).

[c]? indicates speculative, which is generally the case where stochasticity is indicated.

has also been reported (Golding et al., 2005) (as has also been described in *S. cerevisiae*) where slow transitions between promoter states make a significant contribution to noise (Blake et al., 2003; Raser and O'Shea, 2004).

Using a two-reporter assay system, two principal types of transcriptional variability have been identified: intrinsic and extrinsic (Elowitz et al., 2002). Intrinsic noise refers to the biochemical process of gene expression, and may be defined as the extent to which the activities of two identical copies of the same gene, in the same intracellular environment, fail to correlate. Extrinsic variability arises from factors that are global to a single cell, but that may vary from one cell to another, e.g., cell-cycle stage, cell age. The transcriptional and translational bursting phenomena described above refer to intrinsic noise, but other studies have indicated that extrinsic variability makes the larger contribution to total "noise" in gene expression (Elowitz et al., 2002; Raser and O'Shea, 2004; Colman-Lerner et al., 2005; Rosenfeld et al., 2005). Consistent with this proposal is the observation that intrinsic noise in protein production rates is calculated to occur over shorter timescales (≤10 min in *Escherichia coli*) than extrinsic variability (~40 min); in the latter case, accurate processing would require signal integration over inefficiently long timescales

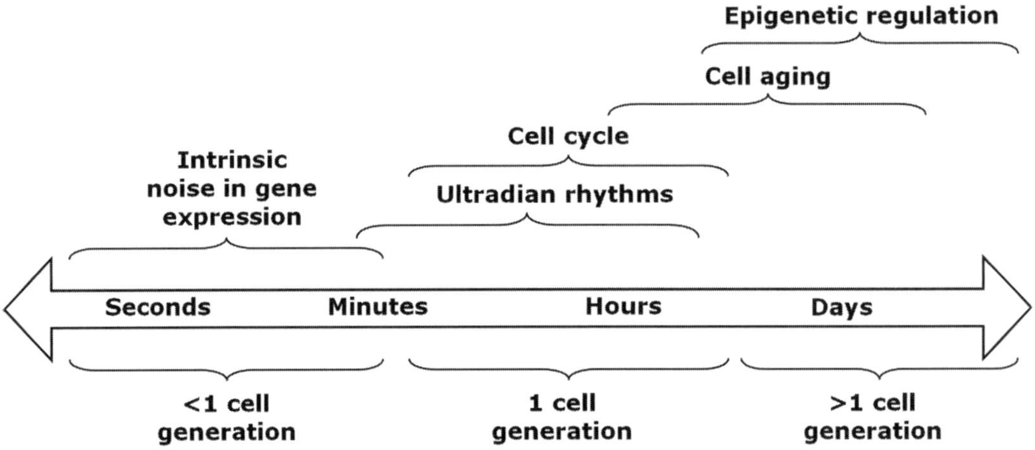

FIGURE 1 Hierarchy of timescales over which key drivers of cellular heterogeneity operate. Temporal information on the stability of heterogeneous phenotypes is generally lacking. However, the timescales of such stability (e.g., for how long is a particular phenotype maintained by individual cells?) can facilitate prediction of the source of the heterogeneity. The figure is intended to reflect the hierarchical nature of the timescales involved. The timescale values indicated are approximations and are generally organism and condition dependent. Note that while epigenetically regulated phenotypes may be sustained over many generations of cells, it could be stochastic events operating at the opposite end of the timescale that may with time trigger a switch spontaneously to an alternative epigenetic state. Adapted from Avery (2006) with permission.

(Rosenfeld et al., 2005). Extrinsic variability has been further subdivided into effects on yeast gene expression that are global, for example, linked to cell size, or that are pathway-specific. Both sources may contribute to variability, depending on the level of pathway induction (Colman-Lerner et al., 2005; Volfson et al., 2006).

Blake et al. (2006) have exploited such recent advances in our understanding of noisy gene expression to test the hypothesis that noise confers an adaptive advantage. They introduced mutations into the TATA box of the *GAL1* promoter, which allowed them to manipulate the kinetics and degree of noise in *GAL1* expression during induction. "Bursts" of gene expression enabled rapid responses in individual cells and led to increased cell-to-cell variability. Moreover, by manipulating the degree of noise while maintaining mean (culture-averaged) gene expression, Blake et al. (2006) were able to validate experimentally that rapid response and increased heterogeneity were beneficial traits in cell populations during acute stress.

HETEROGENEOUS ANTIGENICITY AT THE MICROBIAL CELL SURFACE

Bacterial Phase Variation

Many microorganisms (pathogens in particular) exhibit a considerable diversity of cell-surface properties and antigenicity due to spontaneous changes in the biosynthesis of surface-exposed structures such as flagella, fimbriae, pili, and outer lipopolysaccharides. In the classical example of bacterial "phase variation," such diversity typically arises from stochastic genetic alterations at specific loci in the genome, such as reversible changes in the lengths of short DNA sequence repeats, or microsatellites (Hallett, 2001; van der Woude and Baumler, 2004). These examples are outside the scope of this review as they require genotypic variation. However, certain specific types of microbial phase variation are primarily epigenetic in nature, i.e., they involve a heritable effect on gene expression that does not involve a change in DNA sequence. Switching of the pyelonephritis-associated pili (PAP) and switching of the outer membrane-located antigen-43 in *E. coli* are examples of the latter (Henderson et al., 1999;

Hernday et al., 2004a). The molecular regulation of PAP switching, together with the relevant environmental inputs, is becoming increasingly well understood (Hernday et al., 2004a, 2004b). The switch hinges on binding of a leucine-responsive regulatory protein (Lrp) to either proximal (transcription "off") or distal ("on") sites on the *pap* pilin promoter. An environmentally regulated PapI coregulator triggers the switch by altering the relative affinity of Lrp for the proximal or distal sites. Differential DNA methylation at these sites together with a positive feedback loop to *papI* transcription helps maintain the expression state epigenetically. The environmental nature of the signal in this case provides an example of a deterministic switch mechanism, although an element of stochasticity in initial Lrp binding (Hernday et al., 2003) could underlie some cell-to-cell heterogeneity.

Antigen Switching in Protozoal Parasites

The genome of *Trypanosoma brucei,* a member of the trypanosomes that cause African sleeping sickness, encodes hundreds of variant surface glycoprotein (VSG) genes organized in subtelomeric tandem arrays (Marcello and Barry, 2007). However, only one VSG is expressed dominantly at any time during the course of infection. The relevant control mechanism involves some genetic component in that DNA rearrangements move different VSGs into a single active expression site. However, it is transcriptional regulation that determines that only one expression site, out of the approximately 20 expression sites that occur in subtelomeric regions of the *T. brucei* genome, is active at any given time. The active expression site is known to be transcribed at a location outside the nucleolus, but the mechanisms dictating that all but one site are transcriptionally silent and that regulate switches between sites remain to be resolved fully. The involvement of DNA repair factors and/or chromatin structure has been proposed, while the possibility of a role specifically for telomere silencing is more controversial (Pays, 2005; Horn and Barry, 2005). Dreesen et al. (2007) have hypothesized that growth and breakage of telomeric repeats have an important role in regulating antigenic variation in trypanosomes.

Variable expression of the *var* antigen-encoding gene family in the malaria parasite *Plasmodium falciparum* is known to be under epigenetic control. The mechanism(s) generating such variability has attracted particular interest, not least because of the potential implications for antimalarial drug development. A recent model proposes that *var* genes are transcriptionally silenced by heterochromatin at the nuclear periphery. To establish transcription, the active *var* locus of a cell physically moves away from heterochromatic telomeric clusters that are maintained by Sir2-mediated histone deacetylation, and along the nuclear periphery toward euchromatin areas (Ralph and Scherf, 2005). It is thought that a localized structure dedicated to activation of a single *var* gene can be contained within a gene-dense chromosome end cluster that is otherwise transcriptionally silent (Marty et al., 2006). The active *var* gene adopts an open chromatin conformation that makes it more accessible to transcription complexes and that is thought to be maintained and stably inherited via active transcription (Ralph and Scherf, 2005). In a rare event, an alternative intronic promoter outcompetes the active *var* promoter for the transcription initiation complex. This produces a noncoding RNA species that tags the active locus for histone modification, heterochromatin formation, and subsequent silencing, with a switch in cell surface antigenicity. However, Voss et al. (2006) have recently discovered that activation of another *var* promoter by itself can be sufficient to cause such a switch. Either way, it is thought that the switch event from one *var* gene to another might be an entirely stochastic process (Ralph and Scherf, 2005).

Expression of Cell Wall Glycoproteins in Yeasts

Cell-to-cell variation in the expression of *FLO* genes in *S. cerevisiae* has also been found to have an epigenetic basis (Halme et al., 2004). The

FLO genes encode cell wall glycoproteins that are important for adherence and pseudohyphal formation, and have homology with the *ALS* and *EPA* gene families that encode adherence and virulence factors in pathogenic *Candida* spp. (Verstrepen and Klis, 2006). Under conditions of nitrogen starvation, pseudohypha formation by individual *S. cerevisiae* cells correlates closely with their *FLO11* expression states ("on" or "off"), which are inherited for ≥10 generations (Halme et al., 2004). The expression state of *FLO11* in individual cells is determined by epigenetic silencing dependent on the histone deacetylase Hda1p, which is also involved in *Candida* phenotypic switching (see below). Hda1p activity in turn is regulated both by chromosomal-position signals as well as promoter *(FLO11)*-specific signaling involving the Sfl1p transcription factor. Thus, metastable regulation of *FLO11* determines that some cells express *FLO11* and exhibit *FLO*-dependent phenotypes such as pseudohypha formation, whereas others do not. What is not fully clear is whether it is a deterministic or stochastic process that determines which individual cells will be the ones that express *FLO11*. In the same study, expression of the normally silent *FLO10* gene was altered by a genetic mechanism, involving high-frequency (10^{-3}) mutations in *IRA1* or *IRA2*, the yeast Ras GTPase-activating proteins. This resulted in heterogeneous *FLO10* expression and further cell surface variation. Observations of increased agar adhesion and flocculation in Ira^- cells were shown to depend on FLO10 (Halme et al., 2004).

The epigenetic transitions in cell surface properties of bacteria, protozoa, and yeasts described in this section provide a readily envisaged example of cell individuality serving to benefit the population, by allowing pathogenic microorganisms the advantage of variety in their strategies to overcome host defenses.

SWITCHING BETWEEN MORPHOLOGIES IN CANDIDA SPP.

The yeast pathogen *Candida albicans* is known to undergo high-frequency, spontaneous, and reversible phenotypic switching events that are manifest as alterations in colony morphology (Slutsky et al., 1985, 1987). The range of colony phenotypes exhibited is strain dependent and differs also in alternative *Candida* species such as *C. glabrata* (Brockert et al., 2003; Srikantha et al., 2005). One of the best-characterized switches observed in *C. albicans* is the so-called white-opaque transition (Slutsky et al., 1987). "White" and "opaque" describe the colony appearance, though the switch is also characterized by overt changes in cell morphology, patterns of gene expression, and significantly, virulence characteristics (Soll, 2004). Thus, switching in *Candida* spp. is considered to provide phenotypic variants that may enable rapid adaptation to different environments and host defenses (Odds, 1997; Brockert et al., 2003). The importance of this transition has been underscored by the discovery that only strains of diploid *C. albicans* that are homo- or hemizygous at the mating type locus, for either *MTLa* or *MTLα*, are able to switch to the opaque form, and only opaque phase cells are able to mate efficiently (Lockhart et al., 2002; Miller and Johnson, 2002). The Mtla1 and Mtlα2 proteins form a heterodimer that represses the switch from the white to the opaque form (Tsong et al., 2003). In strains of *C. albicans* that are heterozygous at the mating type locus, the principal mechanism by which repression of this white to opaque switch can be alleviated is thought to be mediated by chromosome loss followed by duplication (Wu et al., 2005).

Treatment with a histone deacetylase inhibitor or deletion of genes encoding histone deacetylases altered the frequency of white-opaque switching (Klar et al., 2001; Srikantha et al., 2001), indicating an involvement of metastable changes in chromatin structure. Furthermore, switching between different sets of colony types in *C. albicans* has been found to be modulated by *SIR2*-dependent silencing (Perez-Martin et al., 1999) or by inhibition of DNA methylation (G. Staniforth, E. Keeble, N. A. R. Gow, personal communication), consistent with epigenetic mechanisms of regulation. Recently, two studies have independently identified the transcriptional regulator Wor1 (Tos9) as a key regulator of the white-opaque switch (Srikantha et al., 2006; Zordan et al., 2006). Wor1 binds the a1-α2 corepressor complex and is expressed only in opaque cells where it localizes to the nucleus, and manipulations of Wor1 expression can be used to block cells in the white phase or induce conversion of white-phase cells to the opaque phase (Srikantha et al., 2006). Zordan et al. (2006) demonstrated that Wor1 forms a self-sustaining positive feedback loop (see "Bistable Phenotypes," below) to regulate its own expression. It was proposed that this feedback loop accounts, at least in part, for the heritability (epigenicity) of the opaque state. A model incorporating stochastic changes in Wor1 expression above and below a threshold level (presumably the threshold at which the feedback loop is activated or deactivated) was developed to explain spontaneous switching between the opaque state and the white state (Srikantha et al., 2006). This significant new understanding has helped to rationalize much of the previous experimental evidence in this area into a unifying mechanism of phenotypic switching. It will be interesting to see whether similar mechanisms underpin the other types of morphology switches seen in *Candida* spp. The contribution of bistable expression states (e.g., that described for Wor1) to heterogeneity in microbial populations is explored further under "Bistable Phenotypes," below.

HETEROGENEOUS RESISTANCE TO ENVIRONMENTAL AND CHEMICAL STRESSES

The observation that some cells of a microbial population are sensitive to an environmental stress, while others are resistant, is a readily observed manifestation of heterogeneity. Heterogeneous stress resistance is widely documented (though often not remarked on specifically) in reports dealing with virtually any stressor-microorganism combination. Such heterogeneous resistance can be confirmed to be noninheritable (i.e., not genotypic), as shown for acid- and osmotic-stress resistance in *E. coli* (Levina et al., 1999; Booth, 2002) and sorbic-acid and heat resistance in yeasts (Steels et al., 2000; Attfield et al., 2001).

In addition to mechanisms that tend to be stressor specific, it is appropriate to note two potential mechanisms of heterogeneous resistance that could exert influence more broadly. One example is epigenetic in nature and is prion based, resulting from aggregation of the Sup35 translation termination factor in *S. cerevisiae*. Spontaneous conversion from the normal form (psi^-) to the prion form (PSI^+) occurs with frequencies varying between 1 in 10^5 and 10^7 and can drive enhanced proteomic variation (Eaglestone et al., 1999; True and Lindquist, 2000; True et al., 2004). Linked to such changes at the proteome level, PSI^+ strains exhibit altered susceptibility to a wide array of stresses compared to coisogenic (psi^-) strains. However, the presence of prions in *S. cerevisiae* seems to be restricted to laboratory strains (Chernoff et al., 2000), and the frequencies of changes in phenotype of PSI^+ cells and of spontaneous conversion between normal and prion states (True and Lindquist, 2000) are lower than could account for much of the heterogeneous stress resistance evident within these cultures.

Heat shock proteins (HSPs) are involved in PSI^+ maintenance (Aertsen and Michiels, 2005), and one of these proteins, the chaperone Hsp90, is reported to function also as an "evolutionary capacitor," suppressing the expression of genotypic variation in a population (Rutherford and Lindquist, 1998). It is proposed that the susceptibility of this capacitor function to environmental perturbation serves to promote genotypic heterogeneity when it is most likely to be advantageous to the organism. This premise has been extended to nongenotypic heterogeneity; from modeling studies with yeast, it was suggested that a multitude of cellular gene products may act like capacitors by buffering variable expression of other genes in gene regulatory networks (Bergman and Siegal, 2003). However, this model did not explain at least some examples of enhanced nongenotypic heterogeneity (in stress resistance) in yeast deletion strains (Bishop et al., 2007).

Resistance to Metals and Pro-oxidants

Metals and pro-oxidants are major environmental stressors that are becoming increasingly well characterized in the context of heterogeneity. Similar mechanisms may underlie the toxicities of metals and pro-oxidants (Avery, 2001), so these agents are considered together here. All major cellular macromolecules (e.g., proteins, lipids, DNA) are susceptible to attack by reactive oxygen species. In *E. coli* cultures, variation in cell-cycle position has been suggested as a possible cause for differences between individual cells in their observed levels of accumulated oxidative protein damage (Desnues et al., 2003). Furthermore, cells with the higher levels of protein oxidation were largely nonculturable, indicating that variable oxidative deterioration could determine nonculturability versus culturability. Cell age in eukaryotes is a particularly well-characterized determinant of heterogeneous resistance to oxidative burden. In *S. cerevisiae*, oxidatively damaged proteins are preferentially retained in mother cells during cytokinesis in a manner apparently dependent on chromatin silencing and on the actin cytoskeleton (Aguilaniu et al., 2003). This mechanism presumably serves to exclude damaged components from daughter cells while contributing to the deterioration of older cells and to population heterogeneity. The degree of apparently age-related heterogeneity in the oxidative burden of yeast is more pronounced in mutants lacking antioxidant defenses, such as thioredoxin reductase (Drakulic et al., 2005).

The strategy of identifying gene products that influence phenotypic heterogeneity has enabled a detailed elucidation of the mechanistic basis for heterogeneous copper (Cu) resistance in yeast. It was initially shown that Cu resistance was cell cycle dependent and that the preexisting cellular oxidant status also was predictive of Cu sensitivity (Howlett and Avery, 1999). It was subsequently found that the cell cycle dependency was linked to the activity of the Cu,Zn-superoxide dismutase (Sod1p); Cu resistance was uncoupled from the cell cycle in *sod1Δ* deletion mutants (Sumner et al., 2003) (Fig. 2). The activity of Sod1p oscillated ~fivefold during the cell cycle, in approximate phase with peaks of Cu resistance. It was also established that Cu resistance diminished as cells progressed through successive rounds of cell division. This age-dependent resistance also required the activity of Sod1p, as well as the Cu metallothionein Cup1p (Sumner et al., 2003). The contribution of cell cycle-dependent SOD activity to heterogeneity was also suggested by Desnues et al. (2003) in their study on oxidative damage in *E. coli* (above). Thus, cell cycle regulation of SOD activity could have a widely conserved function in heterogeneity modulation. This proposal is consistent with the role of SOD enzymes in cellular protection against a wide range of common stresses (Avery et al., 2000; Touati, 2002; Wallace et al., 2005); however, in at least some of these cases, SOD does not influence heterogeneity in the same way as with Cu (Drakulic et al., 2005; Sumner et al., 2005). Associations between cell cycle position and pro-oxidant resistance have been demonstrated—the G_1 and G_2 stages of the *S. cerevisiae* cell cycle were identified as menadione- and H_2O_2-resistant phases, respectively (Flattery-O'Brien and Dawes, 1998). Leroy et al. (Leroy et al., 2001) also showed that H_2O_2 sensitivity of *S. cerevisiae* during the S-phase of the cell cycle could be a result of ineffective DNA base excision repair during this stage.

Another apparently deterministic fluctuation in gene expression that could drive variable protection of cells against metals and oxidative damage is the respiratory oscillation (often termed ultradian rhythm) of yeast, an oscillation that becomes synchronized during continuous culture. Linked to these oscillations, resistance to the pro-oxidants H_2O_2 and menadione as well as to heat cycled in phase with each other, i.e., resistances to these different stressors reached maxima or minima at the same times during the oscillation (Wang et al., 2000; Tonozuka et al., 2001). Resistance to an uncoupler of mitochondrial metabolism and to the toxic metal cadmium also oscillated, but slightly out of phase with the other stressors. Recent data have indicated that oscillation-driven fluctuations in the redox-state

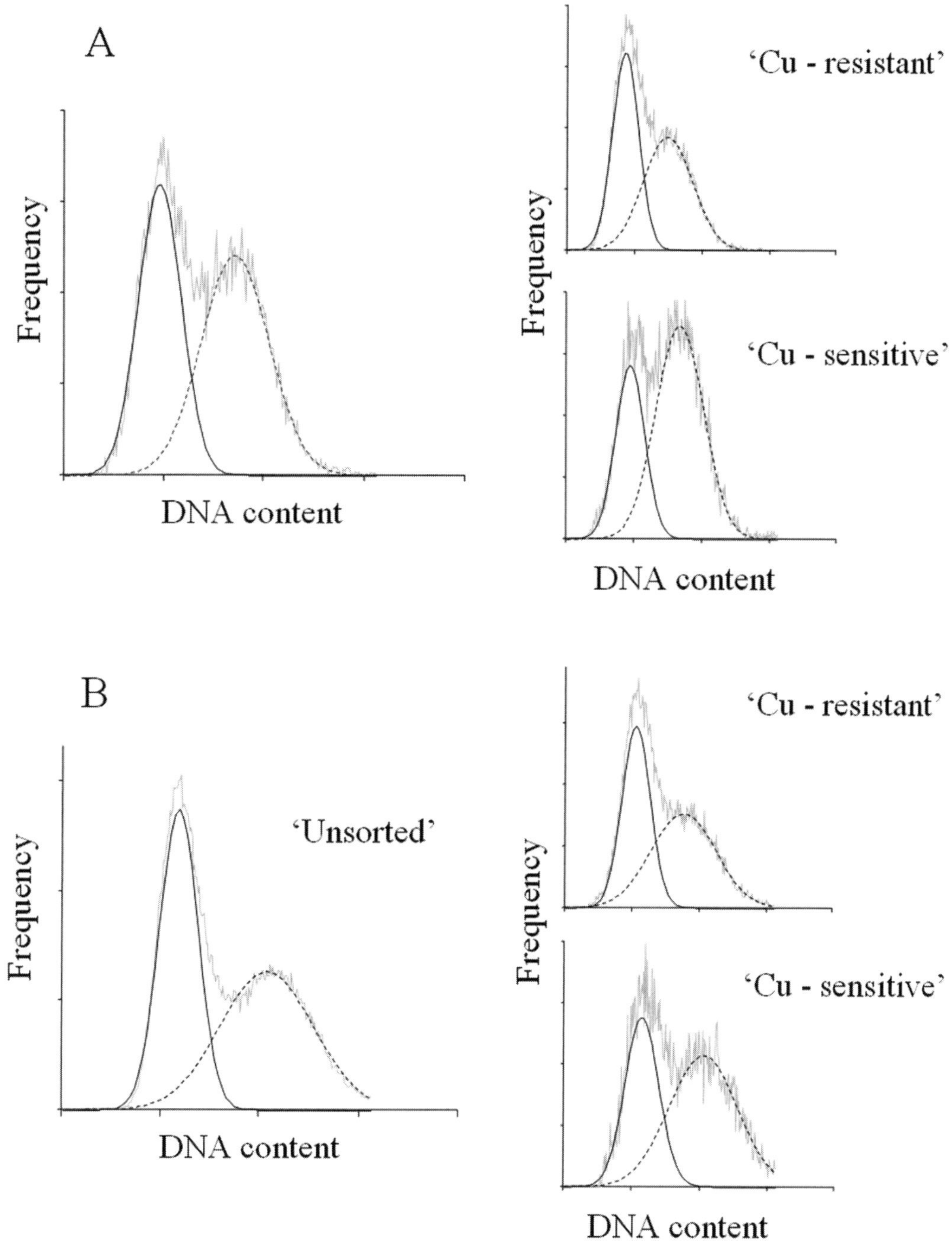

FIGURE 2 Cell cycle-dependent copper resistance in S. *cerevisiae* requires Sod1p. Wild-type (A) and *sod1Δ* (B) cells were exposed to $Cu(NO_3)_2$ for 10 min at concentrations that gave ~50% killing. With the aid of fluorescent viability dyes (Sumner et al., 2003), cells were sorted into Cu-resistant and -sensitive subpopulations. Unsorted cells (left panels) and the sorted subpopulations (right panels) were then stained for DNA content and reanalyzed with flow cytometry. Data were analyzed using SIMfit software to fit individual 1C (first peak) and 2C (second peak) DNA content plots. Cu-resistant and -sensitive wild-type subpopulations had markedly different cell cycle (1C and 2C DNA) distributions, whereas those of the *sod1Δ* subpopulations were similar. Adapted from Sumner et al. (2003) with permission.

of glutathione drive the variable cadmium resistances of oscillating yeast cells (Smith et al., 2007). Moreover, this insight was exploited to construct strains that exhibited dampened oscillations and so decreased heterogeneity in glutathione content and in cadmium resistance. Competition experiments between these strains have been used to demonstrate experimentally fitness advantages that depend on deterministic variation (i.e., ultradian rhythmicity) in gene expression within cell populations (Smith et al., 2007).

Another study examining metal and prooxidant resistance has recently provided independent confirmation of fitness benefits attributable to heterogeneity. Bishop et al. (2007) noted that each one of three different yeast mutants (defective for either *VMA3, CTR1,* or *SOD1*) were more heterogeneous than the wild type in their resistances to stressors such as nickel, copper, alkaline pH, menadione, or paraquat (N,N′-dimethyl-4,4′-bipyridinium dichloride). Thus, although the mutants were stress-sensitive at intermediate doses of stressor, enhanced heterogeneity meant that the resistances of individual cells spanned a broad range, and at high stress occasional cell survival in most of these populations overtook that of the wild type (Fig. 3). In the case of the Vma-dependent vacuolar H^+-ATPase, the Vma-dependent resistance mechanism was shown to suppress an influence of variable vacuolar pH on the metal resistances of individual wild-type cells (Bishop et al., 2007). The heterogeneity-specific advantage described in this study was particularly striking considering that it was prevalent in mutant populations that are conventionally considered to be disadvantaged.

Resistance to Heat Stress

HSPs are key determinants of cellular heat resistance. Evidence for single-cell variability in this regard has come from in situ reverse transcription-PCR studies with *Salmonella enterica* serovar Typhimurium, in which mRNA for the HSP GroEL displayed significant heterogeneity between individual cells (Holmstrom et al., 1999). It was suggested that this heterogeneity was related to the variability in the cell cycle position of individual cells in growing cultures. Cell cycle-dependent heat resistance has been reported in *S. cerevisiae* (Plesset et al., 1987). Further work revealed that individual cells within *S. cerevisiae* subpopulations exhibited a 1,500-fold variation in the induction of the *HSP104* promoter by mild heat shock (Attfield et al., 2001). The strength of *HSP104* promoter induction correlated with the cell-to-cell variation in resistance to a lethal heat stress, establishing a phenotypic consequence for the heterogeneous transcriptional response. The mechanistic basis for this heterogeneous response is not genotypic (the phenotypes were not inherited); however, the role of the cell cycle or other parameters that could function as the underlying driver of this heterogeneity was not investigated. One interesting observation that could be relevant is the short period (~3 to 6 min) shuttling to and from the nucleus exhibited by the transcriptional activators of the general stress response in *S. cerevisiae* (Msn2 and Msn4) during adaptation to mild stress (Jacquet et al., 2003). Superimposed on this cyclical behavior is the marked heterogeneity between individual cells in the patterns and timing of Msn2 and Msn4 shuttling. In continuous yeast cultures, nuclear localization of the redox-sensitive transcription factor Yap1p is also known to oscillate (Tu et al., 2005)

Resistance to DNA Damage

The respiratory oscillations in yeast described above are associated with a periodicity in expression of different classes of genes and metabolites, extending across most of the genome and metabolome (Murray et al., 2007). These have been resolved into clusters, largely linked to the reductive and respiratory phases of the oscillation (Klevecz et al., 2004; Tu et al., 2005). Synchronous bursts in DNA replication were restricted to the reductive phase, and this timing has been proposed to be an evolutionary adaptation for minimizing oxidative DNA damage during DNA replication. The replicative age of individual yeast cells (i.e., the number of times each has budded) has also been identified as an underlying cause of variable sensitivity

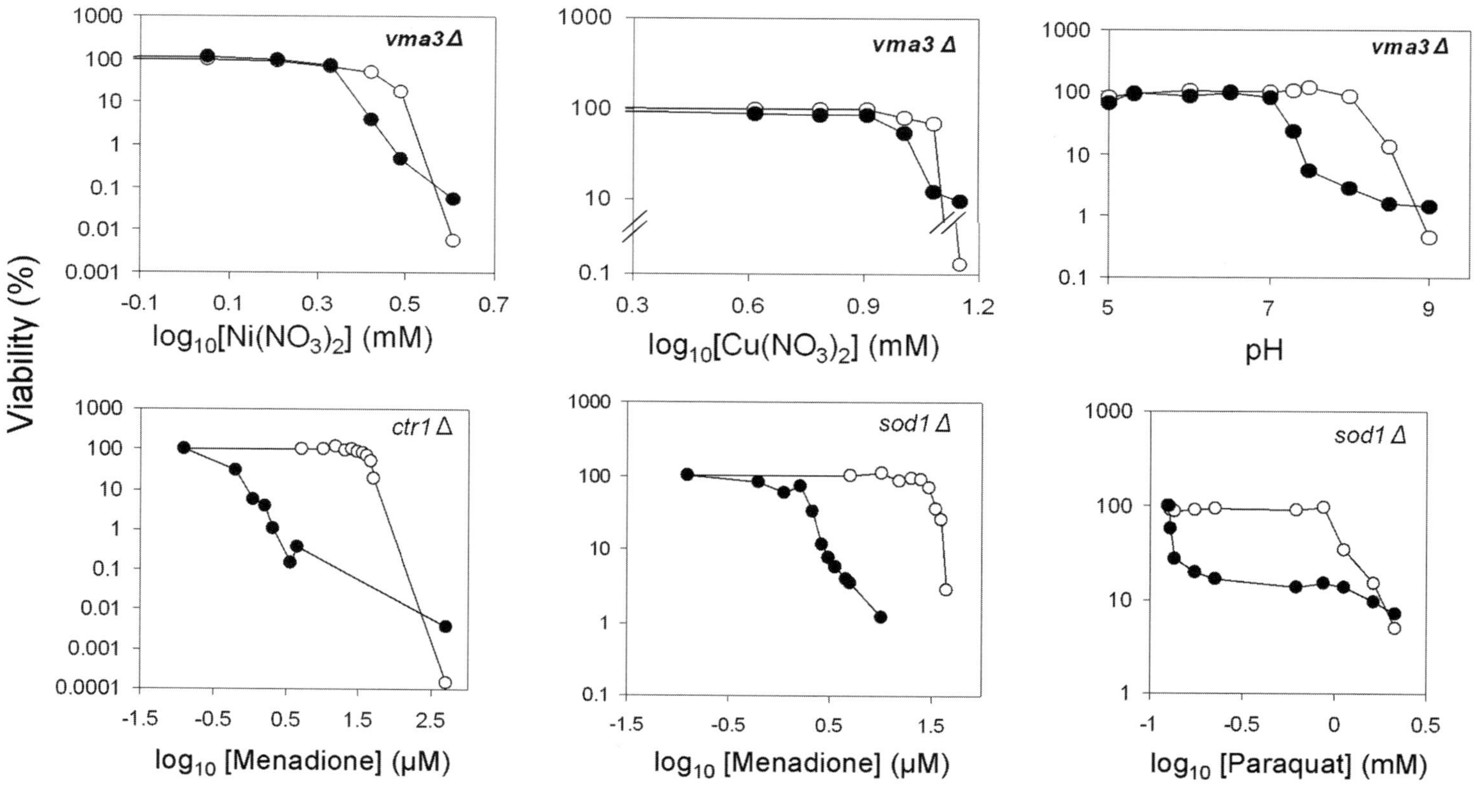

FIGURE 3 Release of phenotypic heterogeneity and enhanced survival under extreme stress in specific yeast mutants. Wild-type (○) or mutant (●) cultures of *S. cerevisiae* were plated onto YPD supplemented or not with either $Ni(NO_3)_2$, $Cu(NO_3)_2$, menadione, or paraquat, as indicated, or buffered to the specified pH values. Viability (colony formation) was determined after several days' incubation at 30°C, and converted to percentages by reference to growth in unsupplemented control medium. The gradients of the slopes reflect the degree of heterogeneity (Sumner et al., 2003). Data are shown for the deletion strains *vma3Δ*, *ctr1Δ*, and *sod1Δ*, as indicated. The points represent means from three replicate determinations. Typical results from one of at least two independent experiments are shown. Adapted from Bishop et al. (2007) with permission.

to DNA damage (Kale and Jazwinski, 1996). Susceptibility to an alkylating agent increased steadily with cell age, whereas resistance to UV irradiation peaked in "middle-aged" cells (~8 generations). The latter, biphasic, pattern of resistance was closely correlated with the level of *RAS2* mRNA; the Ras2p protein is involved in the repair of UV-induced DNA damage.

In *E. coli,* the SOS response to DNA damage is characterized by heterogeneity, which is described by a "two-population model" (McCool et al., 2004). Some variation around the mean SOS expression of subpopulations in that study was suggested to be due to stochasticity in gene transcription and translation. Moreover, while heterogeneity in SOS induction could give rise to considerable heterogeneity in cell phenotype (not tested), it should not be overlooked that heterogeneous SOS induction probably stems from preexisting heterogeneity in the population. Thus, McCool et al. (2004) hypothesized that variation in the DNA damage experienced by individual cells dictated whether and when those cells subsequently induced the SOS response.

Resistance to Antimicrobial Drugs

Antibiotic "persistence" is usually used to describe the small minority of cells within a genetically homogeneous microbial population that are able to survive a period of antibiotic treatment (Moyed and Bertrand, 1983). Phenotypic persisters were first reported over 60 years ago for *Staphylococcus* infections treated with penicillin, and similar phenomena have now been documented in microorganisms exposed to a wide range of antibiotics as well as other toxic agents (Wolfson et al., 1990; Harrison et al., 2005; Wiuff et al., 2005). Persistence can prevent clearance of infectious bacteria (particularly biofilm-forming cells) by administered antibiotics (Spoering et al., 2001; Lewis, 2005; Wiuff et al., 2005), so identification of the underlying mechanism is important to improve therapeutic efficacy and may suggest novel possible drug targets (Balaban et al., 2004).

Persistence-type effects have been reported in a variety of different microorganisms, including "heteroresistance" to antimicrobials in pathogens such as the yeasts *C. albicans* and *Cryptococcus neoformans* and the bacteria *Mycobacterium tuberculosis* and *Staphylococcus* spp. (Rinder, 2001). The degree of methicillin heteroresistance in *Staphylococcus epidermidis* has been linked to σ^B-dependent gene expression (Knobloch et al., 2005). In the yeasts, "resistant" cells isolated after drug treatment can take from one to many generations to regenerate heterogeneously resistant populations, and such heterogeneity is thought to stem from variability in the expression of drug-resistance determinants (Marr et al., 1998; Mondon et al., 1999). Recent work has revealed phenotypic variation in the drug resistance properties of *C. albicans* that is sensitive to the function of the molecular chaperone Hsp90; in this case, however, the variability is apparently of genotypic origin (Cowen et al., 2005).

Advances in our understanding of persistence have come largely from work on *E. coli.* This work is summarized only briefly here as it has been reviewed recently elsewhere (Dhar and McKinney, 2007; Lewis, 2007), articles to which the reader is referred for a detailed insight. It has been established that differences in the growth rates of individual cells of *E. coli* correlate with their subsequent abilities to persist during antibiotic (ampicillin) treatment: rare, nongrowing cells preexisting in the population are persisters, which, following antibiotic treatment, may then switch spontaneously and reversibly to a faster growth rate, reestablishing a majority-sensitive bacterial population (Sufya et al., 2003; Balaban et al., 2004). The nongrowing state of persister cells has been likened to that of unculturable bacterial cells, and so understanding of persistence may give insight to the problem of unculturability also (Lewis, 2007). Fitness loss of the population due to slow persister growth is outweighed by the benefits of this risk-reducing strategy (Kussell et al., 2005). In *E. coli*, the proportion of persisters in a stationary-phase population is influenced by the high persistence (*hipBA*) operon (Keren et al., 2004). In cultures carrying the *hipA7* allele, persisters appear in an arrested growth state at a rate orders of magnitude higher than in wild type (Moyed and Bertrand, 1983). Persisters also arise at a high rate in mutants at an alternative locus, *hipQ*. These bacterial cells do grow and divide, albeit at a rate much slower than nonpersistent cells, and they also produce semi-inheritable persistence (Balaban et al., 2004; Kussell et al., 2005). Evidence indicates that the *hipA7* persistence phenotype requires increased (p)ppGpp (3′,5′-bispyrophosphate) synthesis (Korch et al., 2003). Through analysis of the gene expression profiles of persister cells and other approaches, Keren et al. (2004) produced evidence consistent with the idea that the *hipBA* operon is a toxin-antitoxin module, with HipA serving as a "toxin" that inhibits the cellular function (e.g., translation) through which the drug acts, thus promoting persistence. Such reversible inhibition of an essential cellular function would also explain the correlation between single-cell growth rate and persister-cell formation.

Despite the above advances in our understanding of bacterial persistence, the underlying mechanism that variably drives the switch between persister and nonpersister states has not been elucidated. It has been proposed that stochastic fluctuations in gene expression (e.g., of *hipA*, to give occasional strong toxin-mediated inhibition of the drug-target function and hence persistence) are likely to be the cause (Keren et al., 2004; Wiuff et al., 2005; Lewis, 2007), but such possibilities have yet to be tested.

BACTERIAL MOTILITY AND CHEMOTAXIS

Flagella-driven motility in bacteria is characterized by intermittent reversals of the rotary direction of flagellar movement. In complex habitats with a sparse distribution of nutrients, variability among individual cells in the time periods between motility switches can be expected to provide an ideal search strategy (Viswanathan et al., 1999; Korobkova et al., 2004). When rotating anticlockwise the bacterium swims smoothly and with direction (a run), whereas clockwise flagellar rotation causes random reorientation ahead of the next run. Longer runs occur when cell movement is directed toward an increasing concentration gradient of attractant (Wadhams and Armitage, 2004). Variable responsiveness to chemoattractants could also help individual bacterial cells avoid predators, where attractant release is a mechanism for luring bacterial prey (Shi and Zusman, 1993).

Variability in the individual swimming behaviors of bacterial cells in a clonal population is not related to differences in nutritional state or cell cycle position (Spudich and Koshland, 1976). It was predicted that a standard deviation of 10% from the mean abundance of key proteins involved in the chemotaxis process would be sufficient to yield the observed variability in swimming behavior (Levin et al., 1998), with all proteins in the chemotaxis signal transduction pathway potentially contributing to phenotypic variability (Levin, 2003). Broad variability in the timescales of switching events of individual flagellar motors among *E. coli* cells has been reported (Korobkova et al., 2004). Cells manipulated to express stably the chemotaxis response regulator, CheY, exhibited less variability in flagellar switching than wild-type cells in which CheY was regulated by the chemotaxis network. Moreover, small increases in the concentration of one key upstream component of the chemotaxis signaling network, the methyltransferase CheR, were found to suppress variability. The slow methylation process catalyzed by CheR was suggested to help drive the variability and it was concluded that CheR-mediated variability in temporal chemotactic behavior is a selected trait. Due to signal amplification effects that occur downstream in the chemotaxis pathway (Cluzel et al., 2000; Sagi et al., 2003; Andrews et al., 2006), it could be inferred that any variability in CheR function is further accentuated at the phenotypic level. At the same time, other mechanisms have been proposed that could function to dampen any excessive cell-to-cell variability in motility or chemotactic behavior (Barak and Eisenbach, 2004; Kollmann et al., 2005). The underlying cause of variability in CheR function remains to be elucidated. Another recent study of heterogeneous bacterial motility was characterized by a binary phenotype (motility/sessility) (Kearns and Losick, 2005), and consequently is described in the next section.

BISTABLE PHENOTYPES

Various developmental transitions in microorganisms are characterized by a binary distribution of phenotypes among individual cells within populations, whereby each individual cell either does or does not follow a particular developmental pathway given a window of opportunity (Dubnau and Losick, 2006; Smits et al., 2006). Several examples have attracted attention as models of simple decision circuits. At the level of gene expression, binary (all or nothing) distributions are well documented (Siegele and Hu, 1997; Tolker-Nielsen et al., 1998). A graded gene expression pattern can be converted to a binary pattern by positive autoregulatory feedback and similar regulatory mechanisms (Becskei et al., 2001; Ferrell, 2002; Isaacs et al., 2003; Smits et al., 2006; Paliwal et al., 2007), increased input noise in regulatory cascades (Blake et al., 2003; Kaern et al., 2005), or changes in the rates of promoter state transitions (Kaern et al., 2005). In principle, such gene expression patterns could extrapolate downstream to whole-cell phenotype.

Motility/Sessility

Exponential phase *Bacillus subtilis* cultures are a mixture of either motile or sessile cell types, the type depending on whether the transcription factor for motility, σ^D, is in an "on" or "off" state (Kearns and Losick, 2005). A regulatory protein called SwrA was found to stimulate both transcription of the gene encoding σ^D and also σ^D-directed gene expression. Thus, SwrA activity biases the population toward motility. The phenotype was not cell cycle linked and, because the *swrA* gene is already known to be subject to a phase-variation switch, it was proposed that this bias component has a genetic basis. In contrast, it was also suggested that an epigenetic bistable switch ensures that the two cell types are formed. Generation of both types in this way is thought to allow the organism to exploit the current habitat while simultaneously exploring for new habitats (Kearns and Losick, 2005).

Lysis/Lysogeny

The "decision" by bacteriophage lambda whether to take the lytic (replication and release) or lysogenic (integration into host DNA) life cycle pathways remains one of the best-studied examples of a genetic switch mechanism. The option of lysis or lysogeny is thought to have evolved as an adaptation to fluctuations in bacterial-host availability. The principal molecular events that contribute to determine the outcome are well characterized, but there are continuing efforts to find the underlying regulator(s) of the switch. Lysogeny occurs when the lambda repressor protein, cI, is synthesized (Fig. 4). The unstable activator protein, cII, activates expression of the *cI* gene, and itself is stabilized by the cIII protein. However, cII and cIII synthesis is repressed by the Cro protein. Thus, if Cro activity is high, the bacteriophage becomes committed to the lytic pathway. Conversely, Cro synthesis is repressed by cI. As such, whether lysis or lysogeny is undertaken hinges on whether Cro or cI dominates the early regulatory events, with cII apparently a pivotal determinant of the decision (Folkmanis et al., 1977; Dodd et al., 2005). Repression by Cro and proteolysis determine cII activity, and it is incompletely defined effects of the host environment in addition to cIII activity that primarily dictate cII stability (Banuett et al., 1986). The issue of host environment adds further complexity to efforts to resolve further function(s) that may control variability in cII activity, and consequently in the decision circuit. Moreover, the fraction of infected cells found to proceed along one pathway versus the other fitted a stochastic kinetic model (Arkin et al., 1998), undermining the possibility of a deterministic master regulator of the decision. In lysogeny, the bacteriophage enters an epigenetic state that is very stable. This state is so stable, in fact, that transitions to lysis that are spontaneous

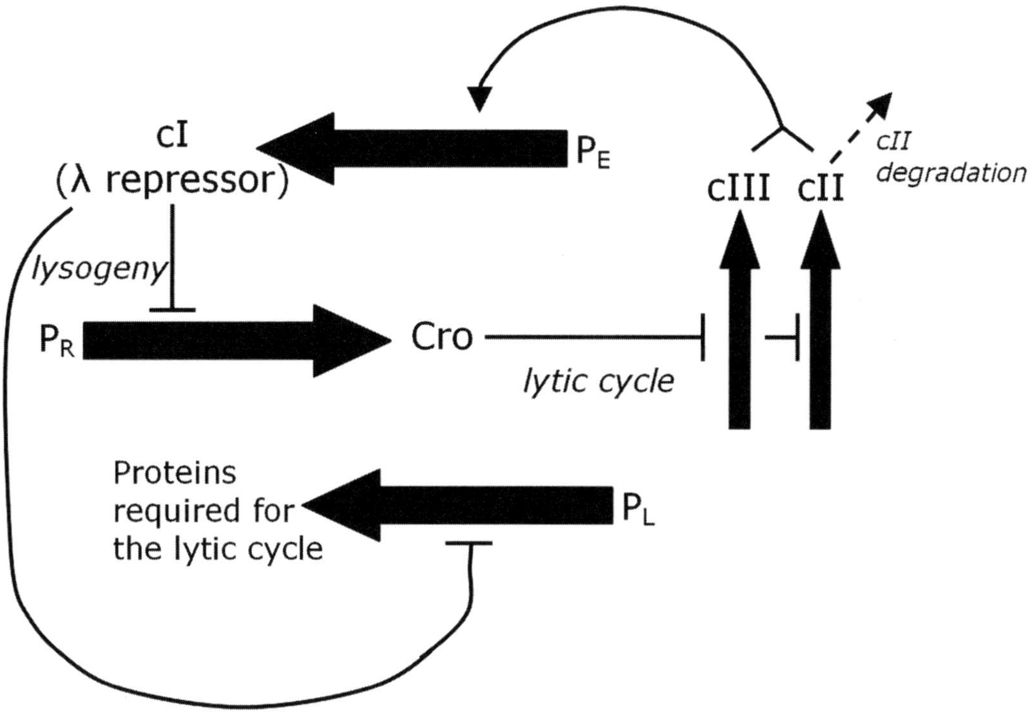

FIGURE 4 Lytic or lysogenic pathways in bacteriophage lambda. The figure summarizes some of the principal regulatory inputs that determine the "decision"'by bacteriophage lambda whether to take the lytic (replication and release) or lysogenic (integration into host DNA) life cycle pathways. Please see the main text for accompanying explanation. Adapted from Avery (2006) with permission.

(i.e., not a response to host DNA damage) arise more frequently due to mutation to a regulatory component of the decision circuit than to a change in epigenetic state (Baek et al., 2003).

Sporulation and Competence Development

The development of competence for transformation and decision to sporulate in *B. subtilis* are classic examples of binary phenotypes. Intrapopulation variability in competence permits individual cells to sample the potential benefits of DNA uptake, or continued growth and development. The *comK* gene encodes the master regulator of competence development, and an autostimulatory loop of *comK* expression has recently been found to be the only factor required for establishment of a bistable pattern of *comK* expression (Maamar and Dubnau, 2005; Smits et al., 2005). The trigger-point for *comK* self-activation is reached once the basal level of ComK (which is highly variable between cells) passes a critical threshold during the transition to stationary phase (Avery, 2005; Maamar and Dubnau, 2005; Smits et al., 2005). Leisner et al. (2007) showed that the mean basal ComK level increases during exponential phase to reach a peak before decreasing in the stationary phase. They proposed that due to noise in basal ComK expression, only those cells that are on the high end of ComK expression at peak basal expression trigger the autostimulatory loop of *comK* expression to become competent. The experimental manipulation of transcription factors that modulate ComK activity resulted in altered fractions of cells that developed competence, indicating that these factors can change the threshold ComK concentration required to initiate the autostimulatory loop (Smits et al., 2005).

Similar to the above, an autostimulatory loop of expression of the key transcriptional regulator of sporulation, Spo0A, is now known to be required to establish the bistability of sporulation gene expression (Veening et al., 2005). The simultaneous option of nonsporulation that this allows offers individual bacterial cells the opportunity of rapid outgrowth should nutrient availability suddenly increase again. Moreover, sporulating cells can use nonsporulating siblings that lyse as a nutrient source (Gonzalez-Pastor et al., 2003). The autostimulatory loop of Spo0A expression involves a multicomponent phosphorelay, which is also thought to regulate gradual increases in the levels of Spo0A protein and its activity as sporulation proceeds (Fujita and Losick, 2005). Bistability was abolished in mutants that lacked the external phosphatases that interface with this phosphorelay.

Regarding the fundamental question of what determines which cells reach the threshold for autostimulation of ComK or Spo0A expression, it has been speculated that there may be a stochastic basis to this phenomenon. In the case of *B. subtilis* sporulation, it has been shown that intrinsic transcriptional noise in *spoIIA* expression does not play a role (Veening et al., 2005). However, new studies have provided evidence consistent with stochastic bases for the initiation of competence and sporulation (Süel et al., 2006; Veening et al., 2006).

CONCLUDING REMARKS AND FUTURE DIRECTIONS

Recent research efforts in the area of phenotypic heterogeneity have served to advance significantly our understanding of its nature. The potential benefits of this form of heterogeneity to microbial cell populations, highlighted above, are now widely acknowledged and have recently been demonstrated experimentally. A challenge remains, in most cases, to delineate fully the pathway of heterogeneity from source to individual phenotype. As, in principle, it takes just one surviving variant cell to regenerate a population, a complete understanding of the molecular basis for individual cell-to-cell variation, and the resultant possibilities for its control, will have important ramifications for the food preservation industry, for understanding the impact of environmental perturbations on natural microbial populations, and for antimicrobial chemotherapy.

Recurrent themes apparent from the above cases are the confirmed roles of cell aging, the cell cycle, metabolic rhythms, and spontaneous epigenetic modifications in creating heterogeneity right up to the phenotypic level (Table 1). Stochasticity has been implicated as a cause of phenotypic heterogeneity, typically when heterogeneous phenotypes remain less well defined. Interestingly, it is the role of heterogeneity in stress resistance (rather than in normal developmental/physiological phenotypes) that primarily has been shown to be regulated by the more deterministic mechanisms. It will be interesting to see if such a trend persists as more mechanisms underlying heterogeneity become unravelled.

In several cases where epigenetic changes are involved in generating heterogeneity, the crucial question remains: what determines which individual cells will switch phenotype and when? It is tempting to speculate on a stochastic basis here. It is also plausible that stochastic events could impinge on otherwise deterministic parameters, such as rhythms or cell cycle progression to cause desynchronization, or such as slow growth rate in the case of antibiotic persistence, which may stem from stochastic dominance of a toxin over an antitoxin. The latter issue has been the focus of particular attention recently, and this angle could ultimately yield new insights also to the nature of unculturable microorganisms (Lewis, 2007).

Although some studies have shown that a significant contribution to gene expression noise comes from cell cycle heterogeneity (Raser and O'Shea, 2004; Colman-Lerner et al., 2005), it should not be assumed that the cell cycle and other commonly reported drivers underlie heterogeneity in most phenotypes. It may be no coincidence that these parameters are also among those that can be readily tested (Sumner et al., 2003). In contrast, whereas noise in gene expression is now well documented, the extrapolation and integration of these observations with whole-cell phenotype are less tractable. This may give a view of the biological relevance of stochastic gene expression that is currently distorted, and a major continuing onus should be to elucidate further the biological relevance of these observations.

ACKNOWLEDGMENTS

The author gratefully acknowledges the support of the NIH (R01 GM57945), the BBSRC (BB/C506656/1), and the NERC (NE/E005969/1) for heterogeneity-related work.

REFERENCES

Aertsen, A., and C.W. Michiels. 2005. Diversify or die: generation of diversity in response to stress. *Crit. Rev. Microbiol.* **31:**69–78.

Aguilaniu, H., L. Gustafsson, M. Rigoulet, and T. Nystrom. 2003. Asymmetric inheritance of oxidatively damaged proteins during cytokinesis. *Science* **299:**1751–1753.

Andrews, B. W., T. M. Yi, and P. A. Iglesias. 2006. Optimal noise filtering in the chemotactic response of *Escherichia coli*. *PLOS Comput. Biol.* **2:**1407–1418.

Arkin, A., J. Ross, and H. H. McAdams. 1998. Stochastic kinetic analysis of developmental pathway bifurcation in phage lambda-infected *Escherichia coli* cells. *Genetics* **149:**1633–1648.

Attfield, P. V., H. Y. Choi, D. A. Veal, and P. J. L. Bell. 2001. Heterogeneity of stress gene expression and stress resistance among individual cells of *Saccharomyces cerevisiae*. *Mol. Microbiol.* **40:**1000–1008.

Avery, S. V. 2001. Metal toxicity in yeasts and the role of oxidative stress. *Adv. Appl. Microbiol.* **49:**111–142.

Avery, S. V. 2005. Cell individuality: the bistability of competence development. *Trends Microbiol.* **13:**459–462.

Avery, S. V. 2006. Microbial cell individuality and the underlying sources of heterogeneity. *Nat. Rev. Microbiol.* **4:**577–587.

Avery, S. V., S. Malkapuram, C. Mateus, and K. S. Babb. 2000. Cu/Zn superoxide dismutase is required for oxytetracycline resistance of *Saccharomyces cerevisiae*. *J. Bacteriol.* **182:**76–80.

Baek, K., S. Svenningsen, H. Eisen, K. Sneppen, and S. Brown. 2003. Single-cell analysis of lambda immunity regulation. *J. Mol. Biol.* **334:**363–372.

Balaban, N. Q., J. Merrin, R. Chait, L. Kowalik, and S. Leibler. 2004. Bacterial persistence as a phenotypic switch. *Science* **305:**1622–1625.

Banuett, F., M. A. Hoyt, L. McFarlane, H. Echols, and I. Herskowitz. 1986. hflB, a new *Escherichia coli* locus regulating lysogeny and the level of bacteriophage lambda cII protein. *J. Mol. Biol.* **187:**213–224.

Barak, R., and M. Eisenbach. 2004. Co-regulation of acetylation and phosphorylation of CheY, a response regulator in chemotaxis of *Escherichia coli*. *J. Mol. Biol.* **342:**375–381.

Bar-Even, A., J. Paulsson, N. Maheshri, M. Carmi, E. O'Shea, Y. Pilpel, and N. Barkai. 2006. Noise in protein expression scales with natural protein abundance. *Nature Genet.* **38:**636–643.

Becskei, A., B. Seraphin, and L. Serrano. 2001. Positive feedback in eukaryotic gene networks: cell differentiation by graded to binary response conversion. *EMBO J.* **20:**2528–2535.

Bergman, A., and M. L. Siegal. 2003. Evolutionary capacitance as a general feature of complex gene networks. *Nature* **424:**549–552.

Bishop, A. L., F. A. Rab, E. R. Sumner, and S. V. Avery. 2007. Phenotypic heterogeneity can enhance rare-cell survival in "stress-sensitive" yeast populations. *Mol. Microbiol.* **63:**507–520.

Blake, W. J., G. Balazsi, M. A. Kohanski, F. J. Isaacs, K. F. Murphy, Y. Kuang, C. R. Cantor, D. R. Walt, and J. J. Collins. 2006. Phenotypic consequences of promoter-mediated transcriptional noise. *Mol. Cell* **24:**853–865.

Blake, W. J., M. Kaern, C. R. Cantor, and J. J. Collins. 2003. Noise in eukaryotic gene expression. *Nature* **422:**633–637.

Booth, I. R. 2002. Stress and the single cell: intrapopulation diversity is a mechanism to ensure survival upon exposure to stress. *Int. J. Food Microbiol.* **78:**19–30.

Brehm-Stecher, B. F., and E. A. Johnson. 2004. Single-cell microbiology: tools, technologies, and applications. *Microbiol. Mol. Biol. Rev.* **68:** 538–559.

Brockert, P. J., S. A. Lachke, T. Srikantha, C. Pujol, R. Galask, and D. R. Soll. Phenotypic switching and mating type switching of *Candida glabrata* at sites of colonization. *Infect. Immun.* **71:**7109–7118.

Chernoff, Y. O., A. P. Galkin, E. Lewitin, T. A. Chernova, G. P. Newnam, and S. M. Belenkiy. 2000. Evolutionary conservation of prion-forming abilities of the yeast Sup35 protein. *Mol. Microbiol.* **35:**865–876.

Cluzel, P., M. Surette, and S. Leibler. 2003. An ultrasensitive bacterial motor revealed by monitoring signaling proteins in single cells. *Science* **287:**1652–1655.

Colman-Lerner, A., A. Gordon, E. Serra, T. Chin, O. Resnekov, D. Endy, C. G. Pesce, and R. Brent. 2005. Regulated cell-to-cell variation in a cell-fate decision system. *Nature* **437:**699–706.

Cowen, L. E., and S. Lindquist. 2005. Hsp90 potentiates the rapid evolution of new traits: drug resistance in diverse fungi. *Science* **309:**2185–2189.

Davey, H. M., and D. B. Kell. 1996. Flow cytometry and cell sorting of heterogeneous microbial populations: the importance of single-cell analyses. *Microbiol. Rev.* **60:**641–696.

Desnues, B., C. Cuny, G. Gregori, S. Dukan, H. Aguilaniu, and T. Nystrom. 2003. Differential oxidative damage and expression of stress defence regulons in culturable and non-culturable *Escherichia coli* cells. *EMBO Rep.* **4:**400–404.

Dhar, N., and J. D. McKinney. 2007. Microbial phenotypic heterogeneity and antibiotic tolerance. *Curr. Opin. Microbiol.* **10:**30–38.

Dodd, I. B., K. E. Shearwin, and J. B. Egan. 2005. Revisited gene regulation in bacteriophage lambda. *Curr. Opin. Genet. Devel.* **15:**145–152.

Drakulic, T., M. D. Temple, R. Guido, S. Jarolim, M. Breitenbach, P. V. Attfield, and I. W. Dawes. 2005. Involvement of oxidative stress response genes in redox homeostasis, the level of reactive oxygen species, and ageing in *Saccharomyces cerevisiae*. *FEMS Yeast Res.* **5:**1215–1228.

Dreesen, O., B. B. Li, and G. A. M. Cross. 2007. Telomere structure and function in trypanosomes: a proposal. *Nat. Rev. Microbiol.* **5:**70–75.

Dubnau, D., and R. Losick. 2006. Bistability in bacteria. *Mol. Microbiol.* **61:**564–572.

Eaglestone, S. S., B. S. Cox, and M. F. Tuite. 1999. Translation termination efficiency can be regulated in *Saccharomyces cerevisiae* by environmental stress through a prion-mediated mechanism. *EMBO J.* **18:**1974–1981.

Elowitz, M. B., A. J. Levine, E. D. Siggia, and P. S. Swain. 2002. Stochastic gene expression in a single cell. *Science* **297:**1183–1186.

Ferrell, J. E., Jr. 2002. Self-perpetuating states in signal transduction: positive feedback, double-negative feedback and bistability. *Curr. Opin. Cell Biol.* **14:**140–148.

Flattery-O'Brien, J. A., and I. W. Dawes. 1998. Hydrogen peroxide causes *RAD9*-dependent cell cycle arrest in G_2 in *Saccharomyces cerevisiae* whereas menadione causes G_1 arrest independent of *RAD9* function. *J. Biol. Chem.* **273:**8561–8571.

Folkmanis, A., W. Maltzman, P. Mellon, A. Skalka, and H. Echols. 1977. The essential role of the *cro* gene in lytic development by bacteriophage lambda. *Virology* **81:**352–362.

Fraser, H. B., A. E. Hirsh, G. Giaever, J. Kumm, and M. B. Eisen. 2004. Noise minimization in eukaryotic gene expression. *PLoS Biol.* **2:**834–838.

Fujita, M., and R. Losick. 2005. Evidence that entry into sporulation in *Bacillus subtilis* is governed by a gradual increase in the level and activity of the master regulator Spo0A. *Genes Devel.* **19:**2236–2244.

Golding, I., J. Paulsson, S. M. Zawilski, and E. C. Cox. 2005. Real-time kinetics of gene activity in individual bacteria. *Cell* **123:**1025–1036.

Gonzalez-Pastor, J. E., E. C. Hobbs, and R. Losick. 2003. Cannibalism by sporulating bacteria. *Science* **301:**510–513.

Hallett, B. 2001. Playing Dr. Jekyll and Mr. Hyde: combined mechanisms of phase variation in bacteria. *Curr. Opin. Microbiol.* **4:**570–581.

Halme, A., S. Bumgarner, C. Styles, and G. R. Fink. 2004. Genetic and epigenetic regulation of the *FLO* gene family generates cell-surface variation in yeast. *Cell* **116:**405–415.

Harrison, J. J., R. J. Turner, and H. Ceri. 2005. Persister cells, the biofilm matrix and tolerance to metal cations in biofilm and planktonic *Pseudomonas aeruginosa*. *Environ. Microbiol.* **7:**981–994.

Henderson, I. R., P. Owen, and J. P. Nataro. 1999. Molecular switches—the ON and OFF of bacterial phase variation. *Mol. Microbiol.* **33:**919–932.

Hernday, A. D., B. A. Braaten, and D. A. Low. 2003. The mechanism by which DNA adenine methylase and papl activate the pap epigenetic switch. *Mol. Cell* **12:**947–957.

Hernday, A., B. Braaten, and D. Low. 2004a. The intricate workings of a bacterial epigenetic switch. *Adv. Exp. Med. Biol.* **547:**83–89.

Hernday, A. D., B. A. Braaten, G. Broitman-Maduro, P. Engelberts, and D. A. Low. 2004b. Regulation of the Pap epigenetic switch by CpxAR: phosphorylated CpxR inhibits transition to the phase ON state by competition with Lrp. *Mol. Cell* **16:**537–-547.

Holmstrom, K., T. Tolker-Nielsen, and S. Molin. 1999. Physiological states of individual *Salmonella typhimurium* cells monitored by in situ reverse transcription-PCR. *J. Bacteriol.* **181:**1733–1738.

Horn, D., and J. D. Barry. 2005. The central roles of telomeres and subtelomeres in antigenic variation in African trypanosomes. *Chrom. Res.* **13:**525–533.

Howlett, N. G., and S. V. Avery. 1999. Flow cytometric investigation of heterogeneous copper sensitivity in asynchronously-grown *Saccharomyces cerevisiae*. *FEMS Microbiol. Lett.* **176:**379–386.

Huang, B., H. K. Wu, D. Bhaya, A. Grossman, S. Granier, B. K. Kobilka, and R. N. Zare. 2007. Counting low-copy number proteins in a single cell. *Science* **315:**81–84.

Isaacs, F. J., J. Hasty, C. R. Cantor, and J. J. Collins. 2003. Prediction and measurement of an autoregulatory genetic module. *Proc. Natl. Acad. Sci. USA* **100:**7714–7719.

Jacquet, M., G. Renault, S. Lallet, J. De Mey, and A. Goldbeter. 2003. Oscillatory nucleocytoplasmic shuttling of the general stress response transcriptional activators Msn2 and Msn4 in *Saccharomyces cerevisiae*. *J. Cell Biol.* **161:**497–505.

Kaern, M., T. C. Elston, W. J. Blake, and J. J. Collins. 2005. Stochasticity in gene expression: from theories to phenotypes. *Nat. Rev. Genet.* **6:**451–464. (Comprehensive review on mechanisms of stochastic gene expression.)

Kale, S. P., and S. M. Jazwinski. 1996. Differential response to UV stress and DNA damage during the yeast replicative life span. *Dev. Genet.* **18:** 154–160.

Kearns, D. B., and R. Losick. 2005. Cell population heterogeneity during growth of *Bacillus subtilis. Genes Devel.* **19:**3083–3094.

Keren, I., D. Shah, A. Spoering, N. Kaldalu, and K. Lewis. 2004. Specialized persister cells and the mechanism of multidrug tolerance in *Escherichia coli. J. Bacteriol.* **186:**8172–8180.

Klar, A. J. S., T. Srikantha, and D. R. Soll. 2001. A histone deacetylation inhibitor and mutant promote colony-type switching of the human pathogen *Candida albicans. Genetics* **158:**919–924.

Klevecz, R. R., J. Bolen, G. Forrest, and D. B. Murray. 2004. A genomewide oscillation in transcription gates DNA replication and cell cycle. *Proc. Natl. Acad. Sci. USA* **101:**1200–1205.

Knobloch, J. K. M., S. Jager, J. Huck, M. A. Horstkotte, and D. Mack. 2005. *mecA* is not involved in the σ^B-dependent switch of the expression phenotype of methicillin resistance in *Staphylococcus epidermidis. Antimicrob. Agents Chemother.* **49:**1216–1219.

Kollmann, M., L. Lovdok, K. Bartholome, J. Timmer, and V. Sourjik. 2005. Design principles of a bacterial signalling network. *Nature* **438:**504–507.

Korch, S. B., T. A. Henderson, and T. M. Hill. 2003. Characterization of the *hipA7* allele of *Escherichia coli* and evidence that high persistence is governed by (p)ppGpp synthesis. *Mol. Microbiol.* **50:**1199–1213.

Korobkova, E., T. Emonet, J. M. G. Vilar, T. S. Shimizu, and P. Cluzel. 2004. From molecular noise to behavioural variability in a single bacterium. *Nature* **428:**574–578.

Kussell, E., R. Kishony, N. Q. Balaban, and S. Leibler. 2005. Bacterial persistence: a model of survival in changing environments. *Genetics* **169:**1807–1814.

Kussell, E., and S. Leibler. 2005. Phenotypic diversity, population growth, and information in fluctuating environments. *Science* **309:**2075–2078.

Leisner, M., K. Stingl, J. O. Radler, and B. Maier. 2007. Basal expression rate of *comK* sets a "switching-window" into the K-state of *Bacillus subtilis. Mol. Microbiol.* **63:**1806–1816.

Leroy, C., C. Mann, and M. C. Marsolier. 2001. Silent repair accounts for cell cycle specificity in the signaling of oxidative DNA lesions. *EMBO J.* **20:**2896–2906.

Levin, M. D. 2003. Noise in gene expression as the source of non-genetic individuality in the chemotactic response of *Escherichia coli. FEBS Lett.* **550:**135–138.

Levin, M. D., C. J. Morton-Firth, W. N. Abouhamad, R. B. Bourret, and D. Bray. 1998. Origins of individual swimming behavior in bacteria. *Biophys. J.* **74:**175–181.

Levina, N., S. Totemeyer, N. R. Stokes, P. Louis, M. A. Jones, and I. R. Booth. 1999. Protection of *Escherichia coli* cells against extreme turgor by activation of MscS and MscL mechanosensitive channels: identification of genes required for MscS activity. *EMBO J.* **18:**1730–1737.

Lewis, K. 2005. Persister cells and the riddle of biofilm survival. *Biochemistry (Mosc.)* **70:**267–274.

Lewis, K. 2007. Persister cells, dormancy and infectious disease. *Nat. Rev. Microbiol.* **5:**48–56.

Lockhart, S. R., C. Pujol, K. J. Daniels, M. G. Miller, A. D. Johnson, M. A. Pfaller, and D. R. Soll. 2002. In *Candida albicans*, white-opaque switchers are homozygous for mating type. *Genetics* **162:**737–745.

Maamar, H., and D. Dubnau. 2005. Bistability in the *Bacillus subtilis* K-state (competence) system requires a positive feedback loop. *Mol. Microbiol.* **56:**615–-624.

Marcello, L., and J. D. Barry. 2007. From silent genes to noisy populations—dialogue between the genotype and phenotypes of antigenic variation. *J. Eukaryot. Microbiol.* **54:**14–17.

Marr, K. A., C. N. Lyons, T. Rustad, R. A. Bowden, and T. C. White. 1998. Rapid, transient fluconazole resistance in *Candida albicans* is associated with increased mRNA levels of *CDR. Antimicrob. Agents Chemother.* **42:**2584–-2589.

Marty, A. J., J. K. Thompson, M. F. Duffy, T. S. Voss, A. F. Cowman, and B. S. Crabb. 2006. Evidence that *Plasmodium falciparum* chromosome end clusters are cross-linked by protein and are the sites of both virulence gene silencing and activation. *Mol. Microbiol.* **62:**72–83.

McCool, J. D., E. Long, J. F. Petrosino, H. A. Sandler, S. M. Rosenberg, and S. J. Sandler. 2004. Measurement of SOS expression in individual *Escherichia coli* K-12 cells using fluorescence microscopy. *Mol. Microbiol.* **53:**1343–1357.

Miller, M G., and A. D. Johnson. 2002. White-opaque switching in *Candida albicans* is controlled by mating-type locus homeodomain proteins and allows efficient mating. *Cell* **110:**293–302.

Mondon, P., R. Petter, G. Amalfitano, R. Luzzati, E. Concia, I. Polacheck, and K. J. Kwon-Chung. 1999. Heteroresistance to fluconazole and voriconazole in *Cryptococcus neoformans. Antimicrob. Agents Chemother.* **43:**1856–1861.

Moyed, H. S., and K. P. Bertrand. 1983. *HIPA*, a newly recognized gene of *Escherichia coli* K-12 that affects frequency of persistence after inhibition of murein synthesis. *J. Bacteriol.* **155:**768–775.

Murray, D. B., M. Beckmann, and H. Kitano. 2007. Regulation of yeast oscillatory dynamics. *Proc. Natl. Acad. Sci. USA* **104:**2241–2246.

Newman, J. R. S., S. Ghaemmaghami, J. Ihmels, D. K. Breslow, M. Noble, J. L. DeRisi, and J. S. Weissman. 2006. Single-cell proteomic analysis of *S.cerevisiae* reveals the architecture of biological noise. *Nature* **441:**840–846.

Odds, F. C. 1997. Switch of phenotype as an escape mechanism of the intruder. *Mycoses* **40:**9–12.

Ozbudak, E. M., M. Thattai, I. Kurtser, A. D. Grossman, and A. van Oudenaarden. 2002. Regulation of noise in the expression of a single gene. *Nat. Genet.* **31:**69–73.

Paliwal, S., P. A. Iglesias, K. Campbell, Z. Hilioti, A. Groisman, and A. Levchenko. 2007. MAPK-mediated bimodal gene expression and adaptive gradient sensing in yeast. *Nature* **446:**46–51.

Pays, E. 2005. Regulation of antigen gene expression in *Trypanosoma brucei*. *Trends Parasitol.* **21:**517–520.

Perez-Martin, J., J. A. Uria, and A. D. Johnson. 1999. Phenotypic switching in *Candida albicans* is controlled by a *SIR2* gene. *EMBO J.* **18:** 2580–2592.

Plesset, J., J. R. Ludwig, B. S. Cox, and C. S. McLaughlin. 1987. Effect of cell cycle position on thermotolerance in *Saccharomyces cerevisiae*. *J. Bacteriol.* **169:**779–784.

Ralph, S. A., and A. Scherf. 2005. The epigenetic control of antigenic variation in *Plasmodium falciparum*. *Curr. Opin. Microbiol.* **8:**434–440.

Raser, J. M., and E. K. O'Shea. 2004. Control of stochasticity in eukaryotic gene expression. *Science* **304:**1811–1814.

Rinder, H. 2001. Hetero-resistance: an under-recognised confounder in diagnosis and therapy? *J. Med. Microbiol.* **50:**1018–1020.

Rosenfeld, N., J. W. Young, U. Alon, P. S. Swain, and M. B. Elowitz. 2005. Gene regulation at the single-cell level. *Science* **307:**1962–1965.

Rutherford, S. L., and S. Lindquist. 1998. Hsp90 as a capacitor for morphological evolution. *Nature* **396:**336–342.

Sagi, Y., S. Khan, and M. Eisenbach. 2003. Binding of the chemotaxis response regulator CheY to the isolated, intact switch complex of the bacterial flagellar motor—lack of cooperativity. *J. Biol. Chem.* **278:**25867–25871.

Shi, W. Y., and D. R. Zusman. 1993. Fatal attraction. *Nature* **366:**414–415.

Siegele, D. A., and J. C. Hu. 1997. Gene expression from plasmids containing the *araBAD* promoter at subsaturating inducer concentrations represents mixed populations. *Proc. Natl. Acad. Sci. USA* **94:**8168–8172.

Slutsky, B., J. Buffo, and D. R. Soll. 1985. *High-frequency switching of colony morphology in Candida albicans. Science* **230:**666–669.

Slutsky, B., M. Staebell, J. Anderson, L. Risen, M. Pfaller, and D. R. Soll. 1987. White-opaque transition—a 2nd high-frequency switching-system in *Candida albicans*. *J. Bacteriol.* **169:** 189–197.

Smith, M. C., E. R. Sumner, and S.V. Avery. 2007. Glutathione and Gts 1p drive beneficial variability in the cadmium resistances of individual yeast cells. *Mol. Microbiol.* **66:**699–712.

Smits, W. K., C. C. Eschevins, K. A. Susanna, S. Bron, O. P. Kuipers, and L. W. Hamoen. 2005. Stripping *Bacillus*: ComK auto-stimulation is responsible for the bistable response in competence development. *Mol. Microbiol.* **56:**604–614.

Smits, W. K., O. P. Kuipers, and J.-W. Veening. 2006. Phenotypic variation in bacteria: the role of feedback regulation. *Nat. Rev. Microbiol.* **4:**259–271.

Soll, D. R. 2004. Mating-type locus homozygosis, phenotypic switching and mating: a unique sequence of dependencies in *Candida albicans*. *Bioessays* **26:**10–20.

Spoering, A. L., and K. Lewis. 2001. Biofilms and planktonic cells of *Pseudomonas aeruginosa* have similar resistance to killing by antimicrobials. *J. Bacteriol.* **183:**6746–6751.

Spudich, J. L., and D. E. Koshland. 1976. Non-genetic individuality–chance in single cell. *Nature* **262:**467–471.

Srikantha, T., A. R. Borneman, K. J. Daniels, C. Pujol, W. Wu, M. R. Seringhaus, M. Gerstein, S. Yi, M. Snyder, and D. R. Soll. 2006. *TOS9* regulates white-opaque switching in *Candida albicans*. *Eukaryot. Cell* **5:**1674–-1687.

Srikantha, T., L. Tsai, K. Daniels, A.J.S. Klar, and D.R. Soll. 2001. The histone deacetylase genes *HDA1* and *RPD3* play distinct roles in regulation of high-frequency phenotypic switching in *Candida albicans*. *J. Bacteriol.* **183:**4614–4625.

Srikantha, T., R. Zhao, K. Daniels, J. Radke, and D. R. Soll. 2005. Phenotypic switching in *Candida glabrata* accompanied by changes in expression of genes with deduced functions in copper detoxification and stress. *Euk. Cell* **4:**1434–1445.

Steels, H., S. A. James, I. N. Roberts, and M. Stratford. 2000. Sorbic acid resistance: the inoculum effect. *Yeast* **16:**1173–1183.

Süel, G. M., J. Garcia-Ojalvo, L. M. Liberman, and M. B. Elowitz. 2006. An excitable gene regulatory circuit induces transient cellular differentiation. *Nature* **440:**545–550.

Sufya, N., D. G. Allison, and P. Gilbert. 2003. Clonal variation in maximum specific growth rate and susceptibility towards antimicrobials. *J. Appl. Microbiol.* **95:**1261–1267.

Sumner, E. R., and S. V. Avery. 2002. Phenotypic heterogeneity: differential stress resistance among individual cells of the yeast *Saccharomyces cerevisiae. Microbiology* **148:**345–351.

Sumner, E. R., A. M. Avery, J. E. Houghton, R. A. Robins, and S. V. Avery. 2003. Cell cycle- and age-dependent activation of Sod1p drives the formation of stress-resistant cell subpopulations within clonal yeast cultures. *Mol. Microbiol.* **50:** 857–870.

Sumner, E. R., A. Shanmuganathan, T. C. Sideri, S. A. Willetts, J. E. Houghton, and S. V. Avery. 2005. Oxidative protein damage causes chromium toxicity in yeast. *Microbiology* **151:**1939–1948.

Thattai, M., and A. van Oudenaarden. 2004. Stochastic gene expression in fluctuating environments. *Genetics* **167:**523–530.

Tolker-Nielsen, T., K. Holmstrom, L. Boe, and S. Molin. 1998. Non-genetic population heterogeneity studied by in situ polymerase chain reaction. *Mol. Microbiol.* **27:**1099–1105.

Tonozuka, H., J. Q. Wang, K. Mitsui, T. Saito, Y. Hamada, and K. Tsurugi. 2001. Analysis of the upstream regulatory region of the *GTS1* gene required for its oscillatory expression. *J. Biochem.* **130:**589–595.

Touati, D. 2002. Investigating phenotypes resulting from a lack of superoxide dismutase in bacterial null mutants. *Methods Enzymol.* **349:**145–154.

True, H. L., I. Berlin, and S. L. Lindquist. 2004. Epigenetic regulation of translation reveals hidden genetic variation to produce complex traits. *Nature* **431:**184–187.

True, H. L., and S. L. Lindquist. 2000. A yeast prion provides a mechanism for genetic variation and phenotypic diversity. *Nature* **407:**477–483.

Tsong, A. E., M. G. Miller, R. M. Raisner, and A. D. Johnson. 2003. Evolution of a combinatorial transcriptional circuit: a case study in yeasts. *Cell* **115:**389–399.

Tu, B. P., A. Kudlicki, M. Rowicka, and S. L. McKnight. 2005. Logic of the yeast metabolic cycle: temporal compartmentalization of cellular processes. *Science* **310:**1152–1158.

van der Woude, M. W., and A. J. Baumler. 2004. Phase and antigenic variation in bacteria. *Clin. Microbiol. Rev.* **17:**581–611.

Veening, J. W., L. W. Hamoen, and O. P. Kuipers. 2005. Phosphatases modulate the bistable sporulation gene expression pattern in *Bacillus subtilis. Mol. Microbiol.* **56:**1481–1494.

Veening, J. W., W. K. Smits, L. W. Hamoen, and O. P. Kuipers. 2006. Single cell analysis of gene expression patterns of competence development and initiation of sporulation in *Bacillus subtilis* grown on chemically defined media. *J. Appl. Microbiol.* **101:**531–541.

Verstrepen, K. J., and F. M. Klis. 2006. Flocculation, adhesion and biofilm formation in yeasts. *Mol. Microbiol.* **60:**5–15.

Vinck, A., M. Terlou, W. R. Pestman, E. P. Martens, A. F. Ram, C. A. M. J. J. van den Hondel, and H. A. B. Wosten. 2005. Hyphal differentiation in the exploring mycelium of *Aspergillus niger. Mol. Microbiol.* **58:**693–699.

Viswanathan, G. M., S. V. Buldyrev, S. Havlin, M. G. E. da Luz, E. P. Raposo, and H. E. Stanley. 1999. Optimizing the success of random searches. *Nature* **401:**911–-914.

Volfson, D., J. Marciniak, W. J. Blake, N. Ostroff, L. S. Tsimring., and J. Hasty. 2006. Origins of extrinsic variability in eukaryotic gene expression. *Nature* **439:**861–864.

Voss, T. S., J. Healer, A. J. Marty, M. F. Duffy, J. K. Thompson, J. G. Beeson, J. C. Reeder, B. S. Crabb, and A. F. Cowman. 2006. A *var* gene promoter controls allelic exclusion of virulence genes in *Plasmodium falciparum* malaria. *Nature* **439:**1004–1008.

Wadhams, G. H., and J. P. Armitage. 2004. Making sense of it all: bacterial chemotaxis. *Nat. Rev. Mol. Cell Biol.* **5:**1024–1037.

Wallace, M. A., S. Bailey, J M. Fukuto, J. S. Valentine, and E. B. Gralla. 2005. Induction of phenotypes resembling CuZn-superoxide dismutase deletion in wild-type yeast cells: an in vivo assay for the role of superoxide in the toxicity of redox-cycling compounds. *Chem. Res. Toxicol.* **18:**1279–1286.

Wang, J. Q., W. D. Liu, T. Uno, H. Tonozuka, K. Mitsui, and K. Tsurugi. 2000. Cellular stress responses oscillate in synchronization with the ultradian oscillation of energy metabolism in the yeast *Saccharomyces cerevisiae. FEMS Microbiol. Lett.* **189:**9–13.

Wiuff, C., R. M. Zappala, R. R. Regoes, K. N. Garner, F. Baquero, and B. R. Levin. 2005. Phenotypic tolerance: antibiotic enrichment of noninherited resistance in bacterial populations. *Antimicrob. Agents Chemother.* **49:**1483–1494.

Wolfson, J. S., D. C. Hooper, G. L. McHugh, M. A. Bozza, and M. N. Swartz. 1990. Mutants of *Escherichia-coli* K-12 exhibiting reduced killing by both quinolone and beta-lactam antimicrobial agents. *Antimicrob. Agents Chemother.* **34:**1938–1943.

Wu, W., C. Pujol, S. R. Lockhart, and D. R. Soll. 2005. Chromosome loss followed by duplication is the major mechanism of spontaneous mating-type locus homozygosis in *Candida albicans. Genetics* **169:**1311–-1327.

Zordan, R. E., D. J. Galgoczy, and A. D. Johnson. 2006. Epigenetic properties of white-opaque switching in *Candida albicans* are based on a self-sustaining transcriptional feedback loop. *Proc. Natl. Acad. Sci. USA* **103:**12807–12812.

NANOMECHANICAL METHODS TO STUDY SINGLE CELLS

Ramya Desikan, Laurene Tetard, Ali Passian, Ram Datar, and Thomas Thundat

14

Fundamental understanding of the structure and functions of a cell is central to biology. The understanding of many complex biological processes of living organisms would be greatly advanced by analyzing the content of their constituent single cells. The cell, consisting of an enclosing membrane, the cytoplasm, and various organelles (and a nucleus in the case of eukaryotes), has been acknowledged as one of the fundamental building blocks of life. Furthermore, individual cells also have discrete molecular, metabolic, and proteomic identities. Microbial cells such as bacteria, fungal spores, and yeasts, as well as single-cell eukaryotes like protists, have rigid cell walls that play a vital role in protecting the cytoplasm from the outer environment, providing the cell with a robust structure, determining its shape, and controlling its adhesion phenomena. Microbial adhesion processes have major consequences in natural environments (symbiotic interactions, biofouling), medicine (infections), and biotechnology (bioremediation, immobilized cells in bioreactors). Understanding the structure, properties, and functions of cell surfaces is, therefore, of great significance for both fundamental and applied research. Investigating the functions and properties of a single cell requires developing novel techniques with higher resolution and sensitivity. Efforts to understand the role of a given cell in an organism often entails defining the boundaries of the cell in order to investigate its properties within those boundaries, or isolating a cell from its environment in preparation for further study.

Since the characteristic size of a single cell is typically in the micrometer range, optical imaging is routinely used for low-resolution studies. Optical microscopy has long been used for visualizing microbial cells in their native state. Light microscopy is useful for counting the cells, identifying them, and describing their general morphological details. Cells can be distinguished from each other if the cells display unique physical characteristics. Optical techniques also allow identification and localization of specific cellular components using fluorescent tags where the sample is illuminated with UV light (Rizzuto et al., 1995).

Isolation of single cells from the tissues of an organism and their microscopic analysis using molecular probes such as antibodies can provide details about the structure as well as

Ramya Desikan, Laurene Tetard, Ali Passian, Ram Datar, and Thomas Thundat, Biosciences Division, Oak Ridge National Laboratory, 1 Bethel Valley Road, Oak Ridge, TN 37831-6123.

Accessing Uncultivated Microorganisms: from the Environment to Organisms and Genomes and Back
Edited by Karsten Zengler

possible functional role of the cells. However, since the resolution of light microscopes is limited by diffraction, it is difficult to obtain structural information at the nanometer level with far-field microscopes.

High-resolution imaging is routinely carried out using electron microscopy techniques that use energetic electron beams for image formation. The use of freeze-fracture and surface replica techniques in transmission electron microscopy makes it possible to visualize cell surface structures at high resolution. However, these approaches are limited by the requirement of vacuum during the analysis, i.e., living cells cannot be directly investigated in their native environments. In addition, electron microscopy cannot provide information on the biophysical properties of the cell surface.

Despite the enormous advances in imaging cells and biomolecules using conventional electron microscopy techniques, high-resolution imaging techniques and high-sensitivity characterization techniques that are compatible with the aqueous environment are highly desired. In recent years, many novel techniques based on scanning probe microscopies have been developed that have myriad applications in studying cells. Here, we introduce scanning probe microscopy and related techniques that can have application in high-resolution imaging of cells and real-time monitoring of multiple cellular components in a multimodal fashion.

SCANNING PROBE MICROSCOPY

Scanning probe microscopes trace their origin back to scanning tunneling microscope (STM) invented by Binnig and Roher in the early eighties (Binnig et al., 1982 1986). The working principle of an STM is that the electron tunneling occurs between a conducting sample and a sharp tip positioned a subnanometer distance away from the sample, and the magnitude of the tunneling current varies exponentially with tip-sample distance, such that scanning the tip over an area produces a current image that is directly related to surface morphology. Due to the exponential dependence of the tunneling current with tip-sample separation distance, surface features of electrically conducting surfaces are greatly amplified. STM is capable of achieving even atomic-scale resolution. However, STM requires electrically conducting samples, and is therefore not well suited for imaging biological samples.

Initially, STM was confined to applications under ultrahigh vacuum conditions. Hansma et al. soon introduced a simpler STM that can operate in air and under solution (Giambattista et al., 1987; Hansma et al., 1988). The ability of STM to operate under these conditions for real-time imaging of surface features with nanometer resolution spurred a sudden and rapid growth in the development of many novel scanning probe microscopes and related techniques. Today, there exist a variety of scanning probe techniques that include STM, atomic force microscopy (AFM), near-field scanning optical microscopy (NSOM), photon scanning tunneling microscopy, scanning electrochemical microscopy, among others. All these techniques are based on scanning a probe over a sample surface at nanometer or subnanometer separation distance, and all offer ways of imaging different physical properties of the sample.

Recent advances in the scanning probe microscopy techniques offer unprecedented opportunities for visualizing cells at nanometer resolution. Since STM requires conducting samples, its application for cellular imaging is very limited. The NSOM and photon scanning tunneling microscope that use near-field optics can be used to obtain images with a resolution of a few nanometers to tenths of nanometers, depending on the size and shape of the probe and the wavelength of the illumination. The AFM, on the other hand, offers subnanometer resolution for topographical imaging. The AFM is more versatile and widely used for imaging biological samples, including surfaces of cells.

ATOMIC FORCE MICROSCOPE

As explained earlier, AFM was a natural outgrowth of ideas catalyzed by the invention of STM. The AFM utilizes the force between a probe and the surface of a sample to map the topographical details of the surface (Fig. 1). Unlike the STM, the AFM can be used for imaging both conductors and insulators since the AFM operating principle is based on the interaction forces between an object and the AFM probing tip. The probing tip, which is located at the free end of a soft cantilever, is brought in close proximity to the surface by piezoelectric means. The cantilevers with a probe tip are microfabricated from silicon through photolithographic techniques. The force constant of a cantilever used in an AFM is very small, usually around 0.1 N/m, which is 2 orders of magnitude smaller than the force constants between atoms in a solid. These cantilevers have typical dimensions of 100 to 400 μm length, 20 to 40 μm width, and 1 to 2 μm thickness. The deflection of the cantilever is monitored by focusing the beam of a laser diode onto the free end of the cantilever, while monitoring the reflected beam with a position sensitive detector (PSD). A change in the PSD signal denotes bending of the cantilever. Here, it is assumed that the cantilever deflection is proportional to the normal force acting between the tip and the sample. In fact, the signal from PSD really measures the curvature of the cantilever, which is assumed to be proportional to the cantilever deflection. As the cantilever is brought close to the sample, usually by using a

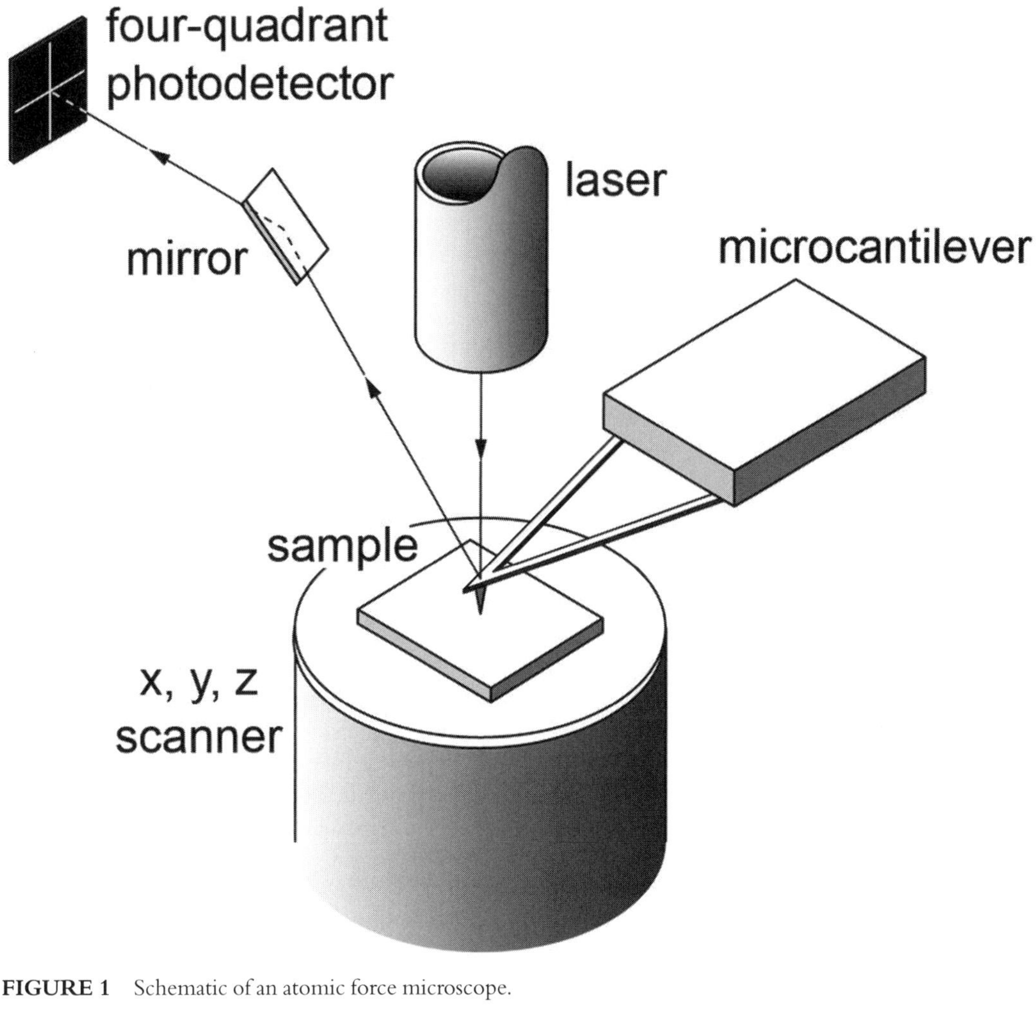

FIGURE 1 Schematic of an atomic force microscope.

piezoelectric element, at a separation distance of a few nanometers, the tip jumps into contact with the surface due to van der Waals force, which is operative at these distances. The interaction force, which is acting between the tip of the cantilever and the sample surface, deflects the cantilever. As the separation distances decrease, the attractive force is replaced with short-range repulsive forces because of the Pauli Exclusion Principle. The strong repulsive force increases rapidly with decreasing separation distance between the tip and the surface. This nonlinearity of the repulsion force is responsible for the high resolution observed in AFM images.

Cantilever displacement as small as a fraction of a nanometer can be measured with this method. The force is calculated from knowing the cantilever displacement and the spring constant of the cantilever. The spring constant of the cantilever can be determined by different methods, for example, from the knowledge of its resonance frequency and mass. Typical forces are in the nano-Newton range.

By scanning the probe location in the x-y direction in a systematic fashion, and noting the cantilever deflection as z-coordinate, one acquires high-resolution images of the surface. Since the van der Waals forces in z-direction fall off rapidly with separation distance, the resolution in z-direction is high enough to see atomic steps on a surface. AFM imaging can therefore reveal atomic arrangements on clean surfaces. The x-y imaging size depends on the piezoelectric scanners used for scanning probe (or sample). In general, a single piezoelectric tube is used, whose outer electrode is divided into four quadrants. Electrodes in one set of opposing quadrants are used in push-pull geometry to create deflections for scanning in x-axis (fast scan axis). At the end of the scan, the other two electrodes are used to deflect and hold the tube in y-axis (slow scan axis) for a small distance. This process is repeated until the probe images the entire surface. Piezoelectric scanners with scan ranges varying from a few nanometers to 100 microns are commercially available. The operation of the AFM, therefore, is analogous to the operation of a phonograph, where a sharp needle is scanned over a grooved surface, although with a different scanning pattern.

Modes of Operation

The two main modes of operation of an AFM are the static (contact) mode and the dynamic mode. In the static mode of operation, the static tip deflection is used as a feedback signal. In the dymanic mode, the change in frequency and amplitude of a resonating cantilever with the tip-sample distance is registered. The often-used dynamic mode, also known as the tapping mode, is based on a decrease in the amplitude of the resonating cantilever. The contact mode uses soft, low-frequency cantilevers, while tapping mode relies on rigid, high-frequency cantilevers. The lateral (x-y) resolution is higher for contact mode operation than for tapping mode. Tapping mode, however, overcomes issues associated with friction, electrostatic forces, adhesion, and other difficulties that often plague contact mode AFM. Since the contact mode is based on the measurement of the contact force between the tip and the sample, low-stiffness (low-spring constant) cantilevers are used for detecting displacements. The measurement of a static signal is often prone to noise and drift. Also, large capillary forces are present due to condensation of humidity around the tip-sample contact area. Since the contact forces are controlled only in the z-direction, lateral forces can cause problems during scanning. However, the cantilevers are geometrically designed in such a way that lateral forces are minimized. In contact mode, the force between the tip and the surface is maintained constant during scanning by keeping a constant deflection.

In the dynamic mode of operation, the cantilever is externally oscillated at a frequency equal to or close to its resonance frequency. The phase, amplitude, and resonance frequency of oscillation are modified by tip-sample interaction forces. The changes in resonance response due to tip-sample interaction with respect to the external reference oscillation produce information about the topography of the sample. Usually, changes in the amplitude of the cantilever oscillation are used for obtaining topography. The phase variation contains information on the viscoelastic properties of the sample and is commonly used for obtaining high contrast when imaging soft materials. The ability of AFM to produce images under ambient conditions or in solution makes it an ideal tool for physical studies of biological specimens under physiological conditions. In contact-mode, constant-force AFM imaging, one could show processes induced by viral infection on live cells and cellulose microfibrils that could be imaged with atomic-scale resolution. In recent years, high-resolution AFM images of bacterial membrane proteins have also been published (Scheuring et al., 2002; Horber et al., 2003). AFM imaging of cells by using contact mode of operation often results in rupturing the cell membrane. Tapping mode or intermittent-contact mode is therefore usually preferred in studies involving high-resolution imaging of subcellular structures. The main complication that arises in the tapping mode of operation is the damping caused by the surrounding liquid environment, making it essential to develop an appropriate method to enhance contrast and improve the quality of images.

Another major application of the AFM is cell studies involving real-time monitoring of live cells' dynamics such as exocytosis of viral particles from an infected cell in real-time, intercellular interactions and functions, and the cellular response to internal and external perturbations (Butt et al., 1990; Häberle et al., 1991; Kasas et al., 1995; Dufrêne et al., 1999, 2001; Van der Aa et al., 2001). The main issue in monitoring the dynamic behavior of the cell is to reduce the cantilever perturbation while scanning and maintaining optimal environmental conditions in terms of temperature and pH changes. Another technical difficulty that needs to be overcome relates to the requirement to achieve a higher temporal resolution; the acquisition time for a full scan of a living cell often exceeds the timescale of the dynamic processes being investigated. It is possible to reduce undesirable cellular stimulation by slightly modifying the tapping mode of operation in liquid. This could also be achieved by developing a new technique in which much lower cantilever-loading forces are needed or by designing novel cantilever probes that are biochemically and mechanically compatible with biological samples. One possible solution is to make the temporal resolution higher to speed up the scan rate, though often at the expense of spatial resolution. The current AFM apparatus and techniques not only allow us to monitor certain dynamic cellular processes, such as cell growth, exocytotic and endocytotic events that are fairly slow and do not require high spatial resolution, but also provide the ability to study the cell morphology in real time in the presence of growth factors, hormones, and other biological reagents. With the development of higher-scan-rate AFMs, it could be possible to monitor the processes that occur at the cell membrane during receptor-ligand binding, vesicle transfer, channel blocking or gating, etc., and to obtain information on the delivery of a specific drug with molecular resolution. Information about micromechanical properties is important for cellular systems as it helps to understand the cell architecture and its functions. Local elastic properties of a cell can be quantitatively derived from the force versus distance (F-S) curves obtained at fixed surface points by using AFM.

Figure 2 shows atomic-scale images of a freshly cleaved mica surface acquired with an AFM. The image consists of 512 lines, and each line has 512 pixels. The image shows raw data without any signal processing or pixel averaging. The white areas show a higher force between the atoms at the end of the cantilever tip and the molecular groups on the surface of mica. Hexagonal close packing of molecular arrangements of the mica surface is clearly visible in the image. The image is collected using a tip-surface interaction force of 1 nN.

In addition to atomic and molecular arrangements, nanometer-size objects can be easily visualized with an AFM. Figure 3 shows an image of double-stranded plasmid DNA adsorbed on a mica surface acquired in air (Thundat et al., 1993). Imaging of nanometer-

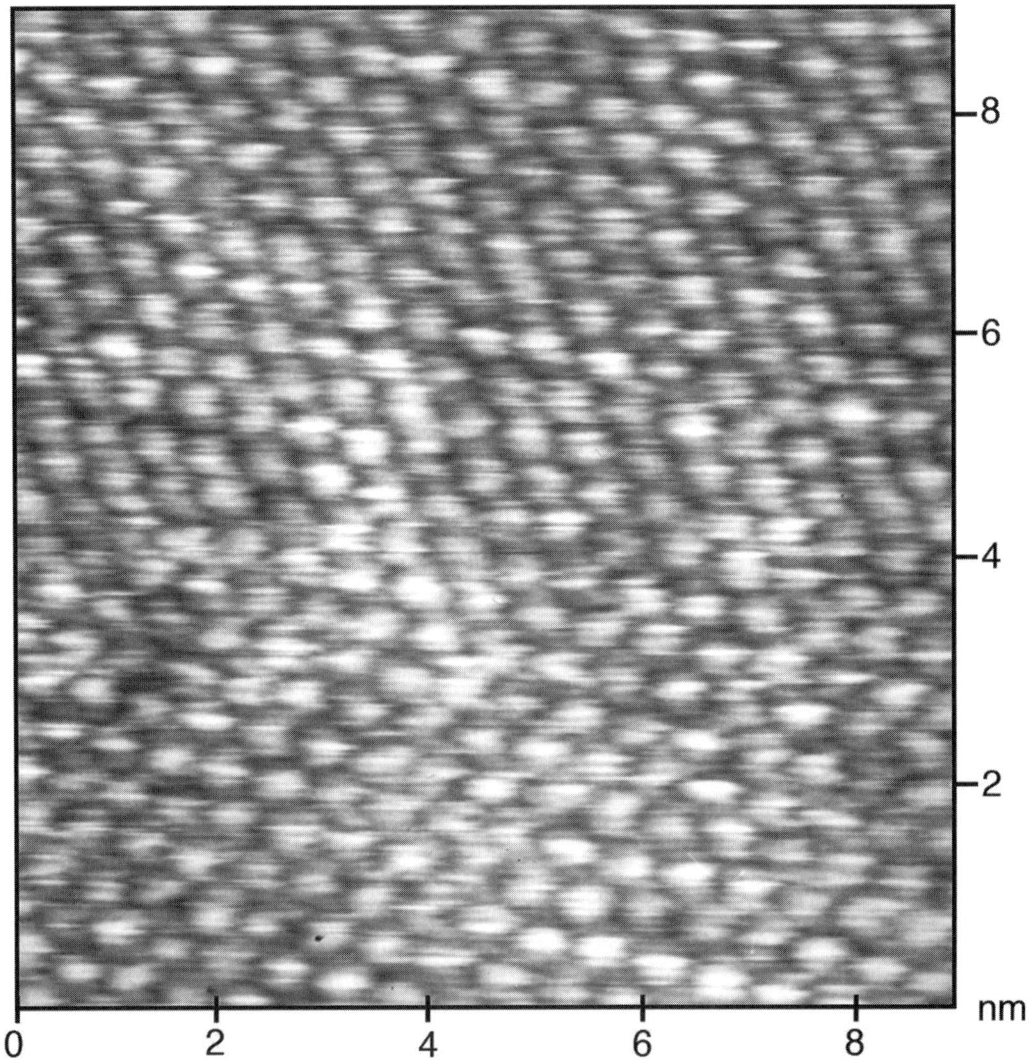

FIGURE 2 Atomic resolution images on a mica surface.

sized objects, such as nanoparticle or a molecule of DNA adsorbed on flat surface, results in geometrical broadening of the image due to finite size of the object (Fig. 3). The apparent width, w, observed in an AFM is approximately, $w = 4\sqrt{Rr}$, where R is the radius of the probe and r is the radius of the object. Typical radius of a micromachined probe tip is around 10 to 15 nm. The z-height, which is related to the van der Waals force between the tip and the surface, does not show any artificial enlargement. In some samples the z-height, however, will be influenced by the presence of other forces such as electrostatic or magnetic forces.

In ambient conditions when the tip comes in contact with the surface, capillary condensation creates a meniscus force, which has an approximate value of $F_c = 4\pi R\gamma\cos\theta$, where γ is the surface tension of water (72 mN/m) and θ is the contact angle. This produces a force of 9 nN for a tip radius of 10 nm and a contact angle of 0. Therefore, operation of an AFM in high humidity conditions results in loss of resolution due to uncontrolled forces of capillary action.

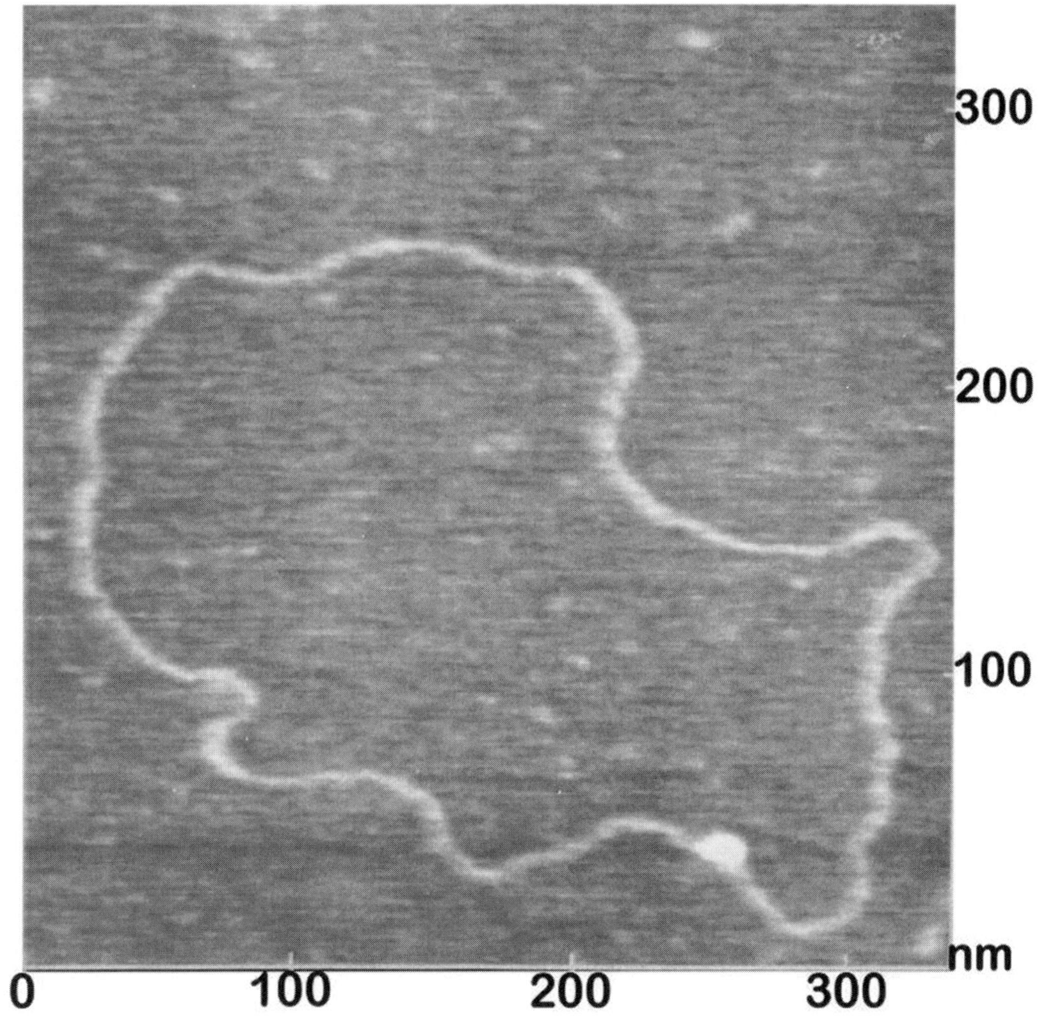

FIGURE 3 Image of double-stranded DNA adsorbed on a mica surface (Thundat et al., 1993).

Imaging surfaces under dry nitrogen or hydrogen improves the image quality due to much reduced capillary forces.

The introduction of the intermittent-contact mode or the tapping mode was an important development because it greatly reduced the shear forces on the specimen and, as a dynamic technique, also offered the possibility of phase imaging. At each point in an image, the interaction between the tip of the probe and the specimen is related to the phase difference between the driving force supplied to the cantilever and the response of the cantilever. The relationship between the nature of the tip-specimen interaction and phase shift is fairly complicated; however, images constructed from the phase information are not useful in identifying regions of a similar chemical nature. The tapping mode was initially developed for operation in air, and its eventual application in liquid brought the technique in the realm of biological environments. However, oscillation of the cantilever in liquid environments results in damping of the cantilever motion by the liquid. Higher forces are needed to drive the cantilever in liquid, and the damping of the motion results in decreasing the quality factor (Q). To overcome this, the resonant amplitude peak of the cantilever is broadened so that the value of Q ranges from a few hundred in air to around 1 in liquid. This results in greatly reducing the sensitivity to the nature of the tip-specimen interaction, resulting in a phase contrast, which is much weaker. The recent development involving active resonance control when applied to the oscillating cantilever in tapping mode in liquid enables an increase in the effective value of the Q so that tip-specimen forces are drastically reduced. The use of this active-resonance technique significantly reduces the deformation of soft biological and organic specimens, resulting in images that have greater resolution or would otherwise be unattainable; the phase contrast in such images is also tremendously improved. Advantages of AFM include its ability to image insulators, soft samples such as biological materials, and its ability to operate in air, in vacuum, and under solution. Since an AFM image depends on the interaction force between the tip and the sample, specially designed tips can be used for magnetic and electrostatic imaging of certain samples. It is also possible to make the AFM tip similar to a thermocouple for imaging the thermal properties of samples (Wang, 2004; Haeberle et al., 2006).

In summary, the AFM uses a microfabricated cantilever beam that is an extremely sensitive force sensor and detects the force between the atoms at the end of the cantilever tip and the cell surface. The ability of a cantilever to measure forces as small as subpiconewtons makes it an ideal tool for measuring interaction forces between molecules. The cantilever in an AFM allows application of precise quantifiable forces to single cells in a site-specific manner. Cells also respond to mechanical stress and strain. Understanding the single-cell response to these forces will open a new and exciting area for cell biology. AFM when combined with fluorescence microscopy allows biochemical events in the cell to be correlated with mechanical changes in the cell, making the AFM/fluorescence microscopy a novel tool in the study of mechanotransduction in single cells (Kassies et al., 2005).

The AFM has opened exciting new avenues in microbiology and biophysics for probing microbial cells (Horber et al., 2003). The unprecedented capabilities of AFM include the potential to measure local physical properties such as elasticity and adhesion forces and imaging the surface topography to the nanometer scale under different physiological conditions. With the scope for topographic imaging, one can directly visualize cell surface nanostructures and the changes of cell surface morphology occurring during various physiological processes (such as germination, division) (Hoh et al., 1992; Nagao and Dvorak, 1998). More improvements in the current sample preparation techniques, instrumentation, and experimental conditions could bring subnanometer-level resolution to these living cells for monitoring molecular conformational changes, as is already the case with reconstituted microbial surface layers (Fotiadis et al., 2002; Bahatyrova et al., 2004).

AFM produces not only high-resolution imaging of cellular structures below the optical limit, which is quite "natural" for this method, but also has the potential to study the micromechanical properties of the cell and the ability to monitor cell dynamics and intracytoplasmic processes in real time (Radmacher et al., 1996; Ricci et al., 1997). With AFM, cells can be imaged with practically little or no sample pretreatment, which is noteworthy in most native physiological media such as aqueous solutions. In addition to several advantages over conventional microscopic techniques, AFM can also be combined with other methods such as electron microscopy, scanning near-field optical microscopy, and others for further improvement (Vesenka et al., 1995; Haydon et al., 1996; Proksch et al., 1996; Langer et al., 1997). Direct imaging of fixed or living cells and subcellular structures gives us important information on the structure and features of the membrane, organelles, and cytoskeleton of cells. The AFM also has the potential to image, localize, and identify integral membrane proteins at the surface of living cells (Lal and John, 1994; Bao and Suresh, 2003).

Photonic Force Microscopy

AFM and related techniques are based on sensing normal forces by using a cantilever beam that is not capable of obtaining information from sidewalls. Since living cells exist in three dimensions, cantilever-based AFM cannot be used as a tool for obtaining three-dimensional images. The AFM is a surface tool with performance that is directly coupled to the flatness of the underlying surface examined. Imaging sidewalls or inside cells is impossible because of the instrument's mechanical connection to the imaging tip. Hence, a scanning probe microscope without a mechanical connection to the tip, working with extremely minute loading forces, would be an ideal complementary technique for the study of live cells with the AFM. Photonic force microscope (PFM) is such an imaging tool recently developed at the European Molecular Biology Laboratory in Heidelberg that can be used for three-dimensional imaging (Pralle et al., 2000).

The PFM utilizes a micron- or submicron-sized bead in an optical trap as the imaging probe. Unlike in the AFM, where the tip is attached to a cantilever anchored on a rigid mass, the bead is essentially free floating and can probe sidewalls. The bead is trapped in the three-dimensional trapping potential of a focused laser beam. Trapping and manipulating micrometer-sized beads with a laser beam focused in a fluid were originally described by Ashkin et al. (1986). The depth and shape of the trapping potential can be determined from the difference in the refractive index between the bead and the fluid medium, the bead diameter, and the laser intensity and the beam profile. The trapped bead executes Brownian motion inside the trap as if the bead is attached to an invisible spring with a spring constant 3to 4 orders of magnitude smaller than the cantilever used in an AFM. The effective spring constant of the trapped bead can be tuned by varying the intensity of the laser beam. The beads commonly used have a radius of 50 nm. However, it is also possible to use even smaller beads (~10 nm) provided the refractive index of the bead material is sufficiently high, for example, metals.

PFM uses a three-dimensional detection system for determining bead position with respect to trapping potential. This allows measurement of the force (magnitude and direction) acting on the bead with subpiconewton precision on a timescale of microseconds. The Brownian envelope of the bead motion, which is much larger than the bead diameter, is used for probing the three-dimensional shape of the object. Beads used as scanning probe tips can be moved along a surface to make interaction force measurements between the tip and the surface in the pico- and subpiconewton force range. If latex or glass beads are used as tips, many standard chemical surface modifications are commercially available with a bead size of 200 to 400 nm, for which the objective used to focus the laser provides a good optical control for the readings.

The image resolution obtained with these beads by scanning them across a surface while measuring the interaction forces is limited by the interaction region between bead and sample and is also limited by thermal fluctuations, which can be up to 100 nm, depending on the trapping potential. With a detection system giving a spatial resolution of better than 1 nm and a time resolution of 1 μs, this limitation can be easily overcome by using the thermal fluctuations as a random scan generator for the exploration (within milliseconds) of a small three-dimensional volume of several tenths of nanometers. Such a method opens up many new applications because the position probability measured for a certain volume reflects the presence of other objects in this volume, the interaction potential with these objects, and the interaction with the surrounding medium. By applying such a technique, three-dimensional polymer networks can be imaged, the mechanical properties of single molecules binding the bead to a surface can be measured, and the viscosity of the membrane of living cells in areas smaller than 100 nm in diameter can be determined if beads are linked to single-membrane components.

Scanning Acoustic Holography

AFM can image surface features with subnanometer resolution in the *z*-direction since the force used as a signal is nonlinear. The nonlinear relation results in amplification of signal in *z*-direction. The signals in the *x*-y directions have no amplification. One of the disadvantages of AFM is that it cannot obtain subsurface information. All images obtained with an AFM are surface features. In 2005 Shekhawat and Dravid developed a scanning probe technique called scanning near-field ultrasonic holography (SNFUH) (Skekhawat and Dravid, 2005). SNFUH is a modification of contact-mode AFM. In this modification the sample is vibrated at megahertz frequencies by attaching a piezoelectric crystal to the sample holder. The acoustic waves traveling through the sample vibrate the cantilever at the same frequency. Since this high-frequency vibration of the cantilever cannot be detected by optical beam deflection, a beat frequency technique is utilized to detect cantilever motion. The contact-mode cantilever of the AFM is excited at a frequency close to the frequency of sample vibration. This superposition of two frequencies makes the cantilever vibrate at a beat frequency, which is the difference between the two excitation frequencies. The excitation frequencies can be adjusted in such a way that the beat frequency lies below 800 kHz, which can be detected with electronics used for optical beam deflection. The DC deflection of the cantilever can be used as the topographic image of the sample. The phase of the vibration at the beat frequency as a function of *x*-*y* scan results in phase image of the sample. For properly tuned frequencies the phase image shows subsurface features that cannot be seen in topographs.

SNFUH could be an ideal technique for imaging cells where subsurface information is highly desired, for example, investigating the nucleus of a cell or imaging microtubules. SNFUH, however, is so new that it has not been applied to many exciting problems. Shekhawat and Dravid have employed SNFUH to image malarial parasites within red blood cells (Skekhawat and Dravid, 2005). Figure 4 shows AFM topography of red blood cells on a mica surface. Figure 5 shows an image of a macrophage from a mouse lung.

Near-Field Scanning Optical Microscopy

For a quantum particle, simultaneous measurement or knowledge of its position and momentum is impossible (Heisenberg, 1927). When trying to resolve small details of a sample by light microscopy, one experiences a similar measurement limitation. Abbe observed in 1873 that the smallest distance that can be resolved between two lines (spatial resolution) by optical microscopy is limited (Abbe, 1873). Therefore, objects that are closer than about one-third of the wavelength of the illuminating light cannot be distinguished. Breaking this diffraction limit has been at the center of many efforts (Stelzer, 2002). Based on the original idea of Synge in 1928, subwavelength resolution optical microscopy was demonstrated in 1986 by Pohl and Betzig, who used an optical fiber in the near-field (Pohl et al., 1984; Betzig et al., 1986). Since then, many efforts have been made to use NSOM for the study of biological samples and investigation of single molecules (Edidin, 2001; Kulzer and Orrit, 2004). Fluorescence microscopy provides a noninvasive method for cell biology (Stephens and Allan, 2003), and NSOM beats the diffraction limit; thus the combination of the two provides a powerful platform to obtain structural information. Room temperature detection of single-molecule fluorescence by NSOM was reported in 1993 (Heinz and Hoh, 1999). In recent years, another fusion of features from two powerful microscopies has emerged. Replacing its probe with a microcantilever that has an aperture through which light would be conducted may augment an NSOM. This complex probe allows force measurements on the same point on the sample from where an optical signal is also collected (Wissenschaftliche Instrumente und Technologie GmbH, Ulm, Germany).

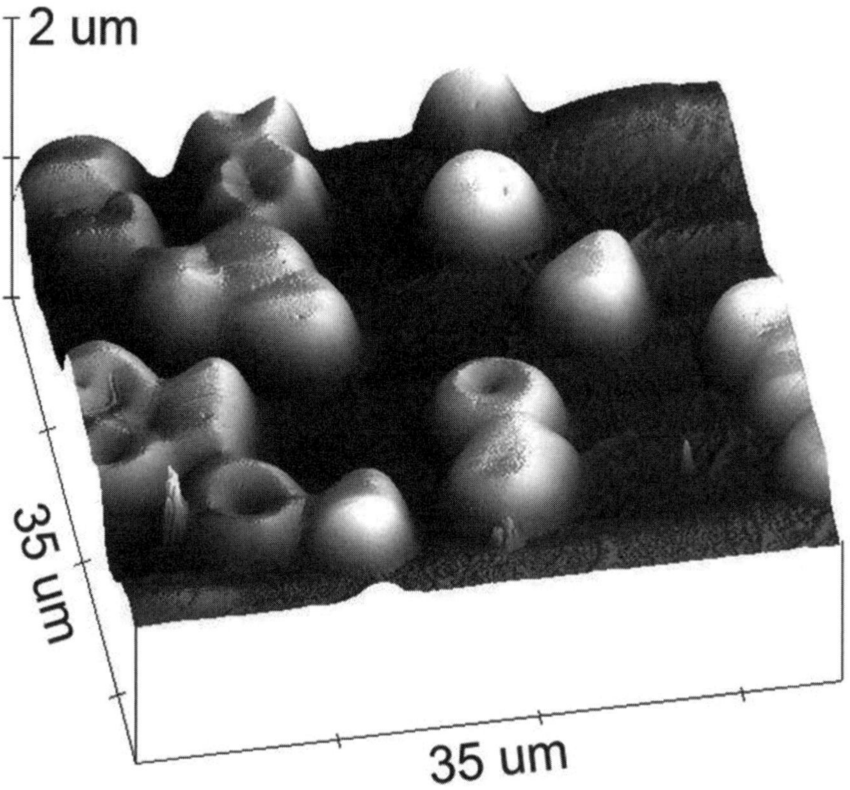

FIGURE 4 AFM (contact-mode) images of erythrocytes from mice. Blood samples were diluted in phosphate-buffered saline, centrifuged onto freshly cleaved mica by using a cytospin, and fixed with methanol.

CANTILEVER-BASED SPECTROSCOPY

Force Spectroscopy

Mechanical forces and molecular conformations play vital roles in the function of biological organisms. Force spectroscopy is a powerful dynamic analytical technique that allows the study of the mechanical properties of large molecules and the properties of chemical bonds. In general, AFMs are used for imaging at the nanoscale, but they also have found potential applications in nanomanipulation and as force sensors for detecting forces between DNA strands and interacting forces between antibodies and antigens; other examples are cited in the literature (Lee et al., 1994; Clausen-Schaumann et al., 2000). This single-molecule analytic technique allows much finer control of the molecule under study. Single molecular force spectroscopy offers a new way to measure directly the strength of single covalent bonds. Force-distance curves (Fig. 6) provide complementary information on surface forces, interatomic forces, adhesion, and nanomechanics, yielding new insight into the mechanisms of biological events such as cell adhesion and aggregation.

There are several ways to manipulate single molecules accurately. The two most common methods are the optical or magnetic tweezers and the AFM cantilevers. The force sensor is usually a micrometer-sized bead or a cantilever with displacements that can be measured to determine the force. In all of these techniques, a

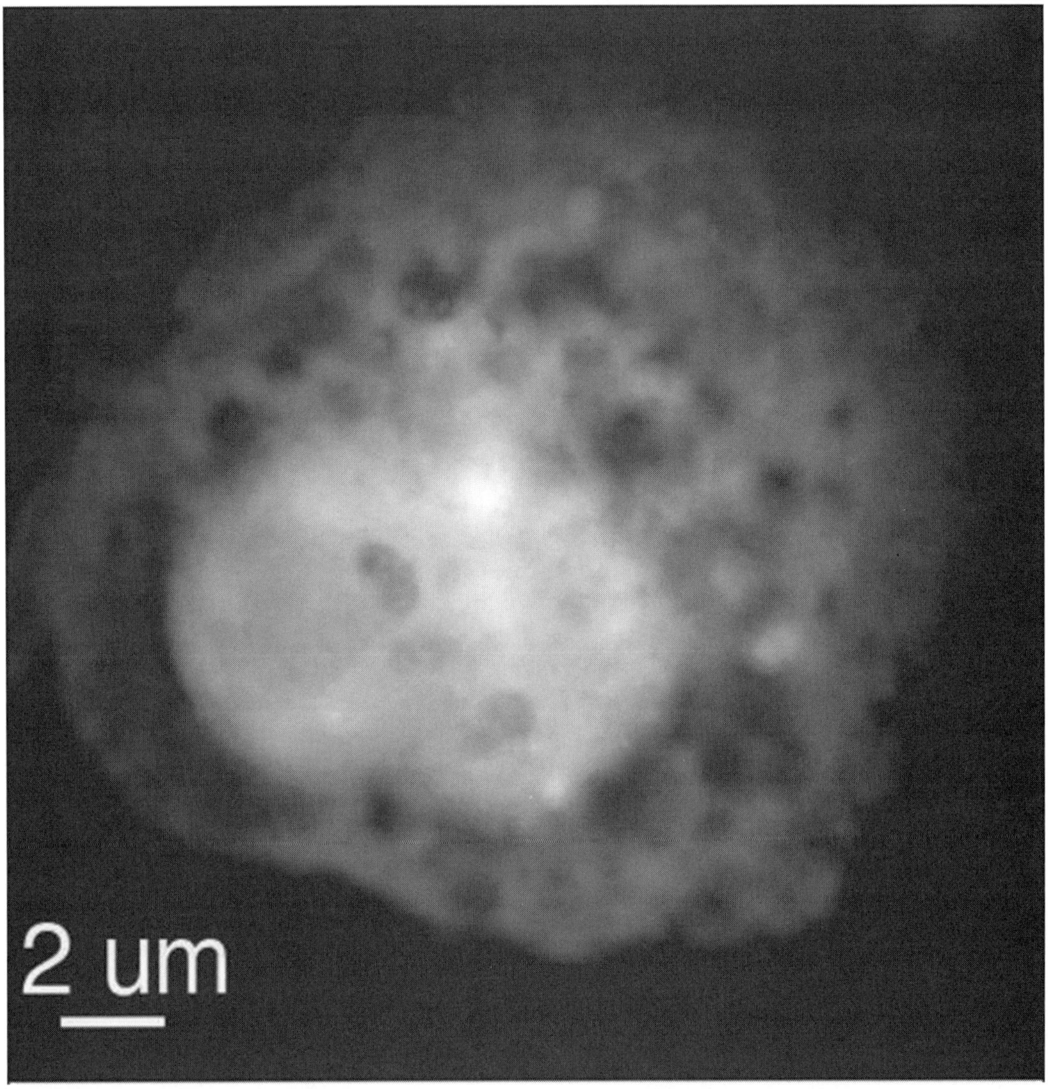

FIGURE 5 AFM image of a macrophage from mouse lungs. Cells were centrifuged onto freshly cleaved mica by using a cytospin and fixed with methanol.

biomolecule, such as protein or DNA, has one end adhered to a surface and the other to a force sensor. Force spectroscopy is mainly used to make measurements of elasticity, especially biopolymers such as RNA and DNA. Force spectroscopy is also used to unravel details on protein unfolding by making the proteins adsorb onto a gold surface and then by stretching it. The unfolding is carefully observed, and the characteristic pattern is plotted in a force versus elongation graph. Valuable information about protein elasticity and the unfolding pattern can be obtained with this technique. Force spectroscopy is also applied in the study of mechanical resistance of chemical bonds.

Photothermal Spectroscopy

Microcantilevers may be used to obtain unique spectroscopic information. A silicon cantilever interacts with an incident beam of photons through a number of channels. The net effect is a mechanical response that can be measured, for

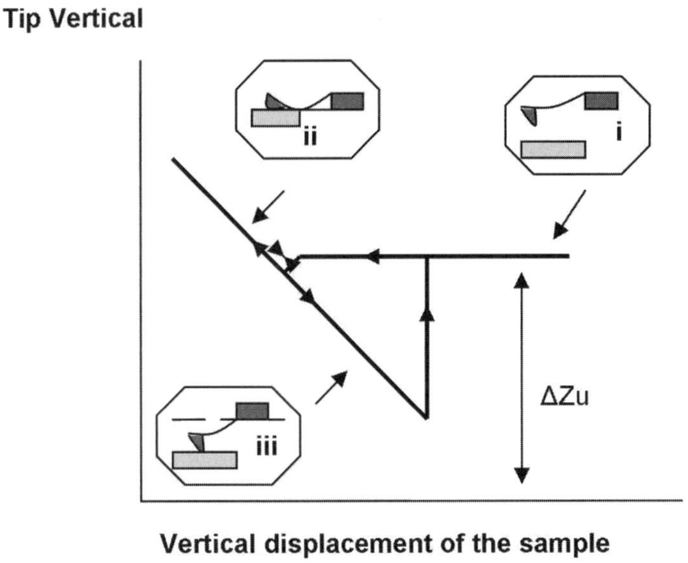

FIGURE 6 Cantilever deflection in an AFM due to contact with a surface. (i) The system is in equilibrium; (ii) the tip is in contact with the sample and the cantilever is compressed; and (iii) the force due to the extended cantilever equals the adhesive force, and it snaps back into the equilibrium position. ΔZu is the vertical displacement.

example, by using lock-in detection. Understanding the involved mechanisms for the observed response is ongoing research and is well beyond the scope of this chapter. Bypassing the fundamental physics associated with the scattering of photons in sensing applications, one may coat a cantilever with a solution that contains an analyte of interest whereby a spectrum may be acquired by illuminating the coated cantilever with a broadband light source that can be scanned with a monochromator. The analyte may also be adsorbed on the cantilever without a need for coating. Through such explicit sensing methodology to implement photothermal spectroscopy, first infrared spectra of nanogram quantities of *Bacillus anthracis* and *Bacillus cereus* were reported in 2006 (Wig et al.). In comparing the *B. anthracis* results with those of *B. cereus* by monitoring the relative differences observed in the photothermal deflections spectra (Fig. 7), it was concluded that the system could be used for chemical identification. Furthermore, when the data obtained with the cantilever-based sensor were compared with traditional spectroscopic techniques, similar results were obtained, suggesting possible uses of the microcantilevers as sensitive detectors of minute quantities of bioagents. Clear benefits of a faster, simpler sample preparation that requires a much lower concentration of spores and the low-cost platform highlight the advantage of nanomechanical sensing, where molecules undergoing transitions modify the dynamic or static state of the cantilever (Wig et al., 2006). We note that a spectrum from the mechanical response of the cantilever is also observed in the absence of any molecules, and thus any analyte-based spectral data acquired by such a sensing platform have to account for the "natural" spectrum of the cantilever (Wig et al., 2004).

CANTILEVER-BASED SENSING

Adsorption-Induced Response

The AFM microcantilever can be used as a physical, chemical, or biological sensor. The cantilevers are extremely sensitive force sensors. They are usually microfabricated from silicon by using conventional photolithographic masking and etching techniques. Typical dimensions of a cantilever are 100 μm in length, 40 μm in width, and 1 μm in thickness. Silicon and silicon nitride cantilevers and cantilever arrays that utilize optical beam deflection for signal transduction are commercially available. Piezoresistive cantilever arrays are also commercially available. The deflection of a piezoresistive can-

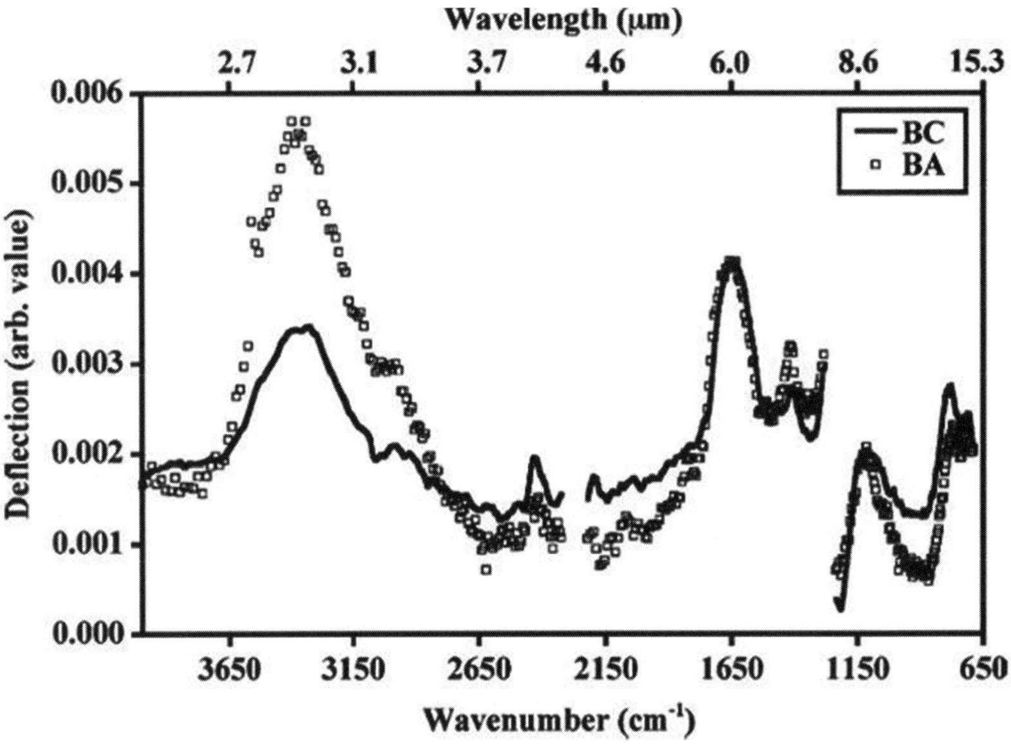

FIGURE 7 Deflections of two cantilevers coated with *B. anthracis* (BA) and *B. cereus* (BC) when exposed to infrared light. The deflection response as a function of infrared wavelength resembles infrared absorption spectra of the adsorbed material (Wig et al., 2006).

tilever is detected as a change in its resistivity without need for optical components.

Microcantilever sensors can be operated in dynamic or static modes. In the dynamic mode, mass loading due to molecular adsorption is detected by monitoring the variation in resonance frequency of the cantilever. In the static mode of operation, molecular adsorption on one side of the cantilever results in a differential surface stress that can be detected by monitoring the bending of the cantilever. The dynamic mode, where resonance frequency is monitored, is very similar to the operation of other gravimetric sensors, such as quartz crystal microbalance and surface acoustic wave transducers (Ballantine et al., 1996; Liu et al., 2003). The sensitivity of the dynamic mode of operation is directly related to the frequency of the cantilever; the higher the frequency, the higher the sensitivity. In the static mode with cantilevers with low frequency and spring constant, large deflections are observed due to adsorption-induced forces. Figure 6 shows a schematic diagram of cantilever bending due to differential molecular adsorption. Microcantilever-based sensing satisfies many requirements of an ideal biosensor in the sense that the microcantilevers can be operated under solution, are capable of simultaneously detecting many analytes, and the overall sensor can have a small footprint, hence potentially portable. Microcantilever-based sensors often utilize receptors that are immobilized on the cantilever surface for molecular recognition.

The microcantilever is an ideal displacement sensor. The ability to detect motion of a cantilever beam with nanometer precision makes the cantilever ideal for measuring bending. Cantilever bending can be related to adsorption/desorption of molecules through adsorption forces. As molecular reactions on a surface are ultimately driven by free energy reduction of the surface, the free energy reduction leads to a change in surface stress. Although they would produce no observable macroscopic change on the surface of a bulk solid, the adsorption-induced surface stresses are sufficient to bend a cantilever if the adsorption is confined to one surface. Adsorption-induced forces, however, should not be confused with bending due to dimensional changes such as swelling of thicker polymer films on cantilevers. The sensitivity of adsorption-induced stress sensors can be orders of magnitude higher than those of frequency-variation mass sensors (for resonance frequencies in the range of tenths of kilohertz).

Microcantilever deflection changes as a function of adsorbate coverage when adsorption is confined to a single side of a cantilever (or when there is differential adsorption on opposite sides of the cantilever). Since we do not know the absolute value of the initial surface stress, we can only measure its variation. A relation can be derived between cantilever bending and changes in surface stress from Stoney's formula and equations that describe cantilever bending (Stoney, 1909). Specifically, a relation can be derived between the radius of curvature of the cantilever beam and the differential surface stress:

$$\frac{1}{R} = \frac{6(1-\nu)}{Et^2}\,\delta s$$

where R is the cantilever's radius of curvature, ν and E are Poisson's ratio and E is Young's modulus for the substrate, respectively, t is the thickness of the cantilever, and $\delta s = \Delta\sigma_1 - \Delta\sigma_2$ is the differential surface stress. Surface stress, σ, and surface free energy, γ, can be related using the Shuttleworth equation (Shuttleworth, 1950):

$$\sigma = \gamma + \left(\frac{\partial\gamma}{\partial\epsilon}\right)$$

where σ is the surface stress. The surface strain, $\partial\epsilon$, is defined as the ratio of change in surface area, $\partial\epsilon = \frac{dA}{A}$. Since the bending of the cantilever is very small compared to the length of the cantilever, the strain contribution is only in the part-per-million (10^{-6}) range. Therefore, one can possibly neglect the contribution from surface strain effects and equate the free energy change to surface stress variation (Butt, 1996). By using equation (2), a relationship between the cantilever deflection, h, and the differential surface stress, δs, is obtained as:

$$h = \frac{3L^2(1-\nu)}{Et^2}\,\delta s$$

where L is the cantilever length. Therefore, the deflection of the cantilever (Fig. 8) is directly proportional to the adsorption-induced differential surface stress. Surface stress has units of N/m or J/m^2. Equation (3) shows a linear relation between cantilever bending and differential surface stress. Adsorption-induced forces are applicable only for monolayer films and, as mentioned above, should not be confused with bending due to dimensional changes such as swelling of thicker polymer films. It should also not be confused with deflection due to weight of the adsorbed molecules. The deflection due to weight is extremely small, for example, for a cantilever with a spring constant of 0.1 N/m; the bending due to weight of 1 ng of adsorbed material will be 0.1 nm.

The minimum detectable signal for cantilever bending depends on the geometry and the material properties of the cantilever. For a silicon nitride cantilever that is 200 microns long and 0.5 micron thick, with the Young's modulus $E = 8.5 \times 10^{10}$ N/m^2 and the Poisson's ratio, $\nu = 0.27$, a surface stress of 0.2 mJ/m^2 will result in a deflection of 1 nm at the end. Because a cantilever's deflection strongly depends on geometry, the surface stress change, which is directly related to molecular adsorption on the cantilever surface, is a more convenient quantity of the reactions for comparison of various measurements. Changes in free energy density in biomolecular reactions are usually in the range of 1 to 50 mJ/m^2 but can be as high as 900 mJ/m^2.

Microcantilever-Based Physical Sensors

Microcantilever can be used as a physical sensor for measuring changes in temperature, flow

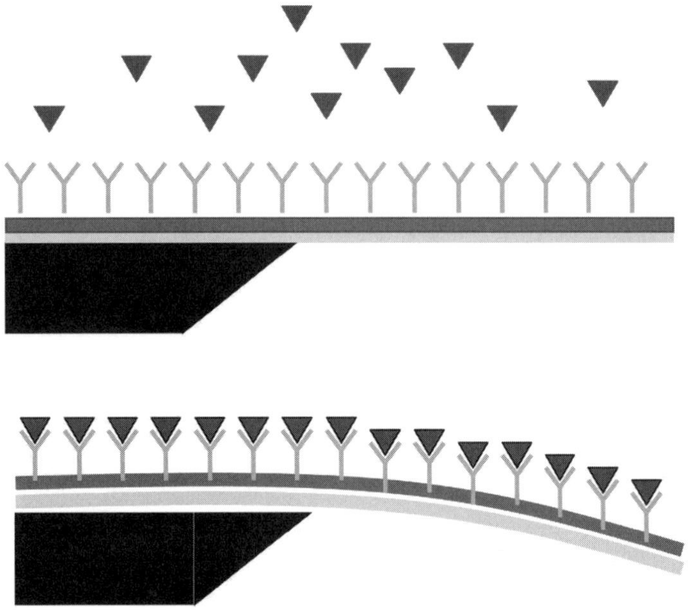

FIGURE 8 A cartoon of cantilever bending caused by binding of target molecules to immobilized probe molecules present on the surface of the cantilever. The probes are immobilized only on one side, and target binding induces a differential stress on the cantilever.

rate, pH, etc. (Datskos et al., 1998; Ji et al., 2001; Muralidharan et al., 2001; Mehta et al., 2001; Cherian and Thundat, 2002). The cantilever needs to be tuned for particular applications. For example, using cantilever as a temperature sensor requires making the cantilever bimetallic by depositing a thin layer of metal on the silicon cantilever. Change in temperature makes the cantilever bend due to a differential thermal expansion of metal and silicon. Developing a microcantilever pH sensor requires coating one side of the cantilever with a material that accumulates surface charge proportional to the pH of the surrounding solution. For example, silicon nitride has many groups that ionize, depending on the pH of the solution. It is also possible to immobilize self-assembled monolayers on one side of the cantilever, which can respond to variation in pH. A properly fashioned cantilever can be an ideal flow rate sensor where cantilever bending varies as a function of flow rate.

Microcantilever-Based Biosensors

Being a very simple structure, the microcantilever beam has the potential of being a highly effective sensing element offering numerous transduction applications. Micromachined cantilevers, in addition to their ability in characterizing surface features in AFM, also lend themselves well to numerous chemical and biological sensing applications wherein the presence of an analyte is manifested mechanically in cantilever deflection and/or a resonance frequency change. The selectivity, sensitivity, compactness, cost, low power consumption, and versatility of microcantilever sensors make them highly suitable for biological sensing. Biosensors are sensors in which biomolecular interactions are used as sensing reactions. Biomolecular interactions, when combined with a microcantilever platform, can produce an extremely powerful biosensing design (Thundat et al., 1997; Raiteri et al., 2001; Wu et al., 2001a, 2001b; Hansen et al., 2001; Ming et al., 2003). The resonance frequency of a microcantilever shifts sensitively due to mass loading. Since the resonance response is very broad under solution, mass detection sensitivity is lower for operation under solution. However, adsorption-induced bending is not affected by presence of solution. Therefore, cantilever bending is usually used as a sensor signal for sensing under solution. The specificity of the cantilever sensor originates from the specificity of probe-target interactions. Nonspecific adsorption of target molecules does not cause the cantilever to bend. Figure 9 shows bending of a cantilever with immobilized antibodies when exposed to a solution of *Francisella tularensis* (1 × 10^6 cells/ml in 0.1M phosphate-buffered saline solution) (Ji et al., 2004). The cantilever without any immobilized antibodies does not show any appreciable bending from injection of sample solution. The relatively small bending observed with the reference cantilever is most probably due to change in ionic concentration of the solution. This result was observed with a setup where cantilever bending was monitored with an optical beam. The optical beam deflection technique is influenced by concentration of ionic species in the solution. The adsorption of organisms on the antibody-covered cantilevers causes a large deflection compared to that of a bare cantilever. The continued bending of the cantilever observed in Fig. 9 is probably due to movement of the organisms on the cantilever surface.

Figure 10 shows cantilever bending due to adsorption of thiolated single-stranded DNA (ssDNA) on a cantilever with a thin layer of gold coating on one side. In the case of ssDNA

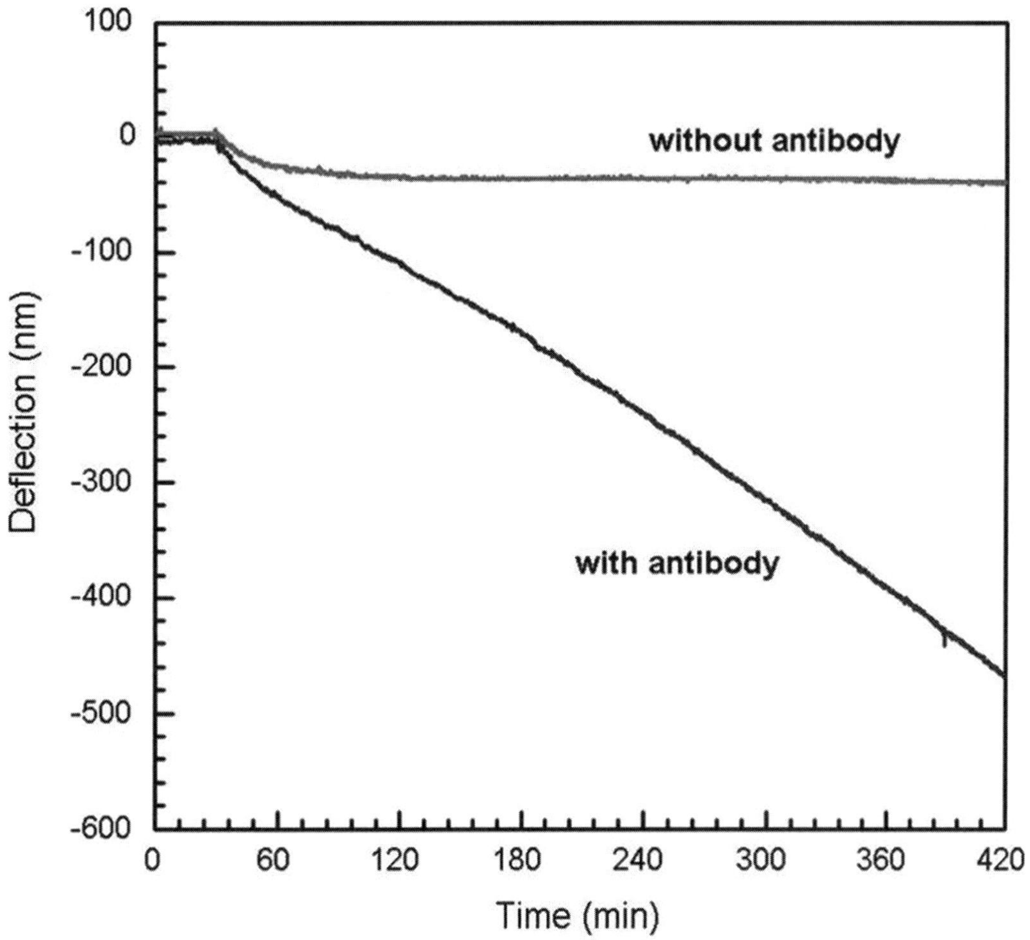

FIGURE 9 Cantilever deflection as a function of exposure to tularemia in solution. Cantilever with immobilized antibodies shows bending due to antibody-antigen interaction. The small deflection observed with uncoated cantilever is due to changes in ionic concentration of the solution.

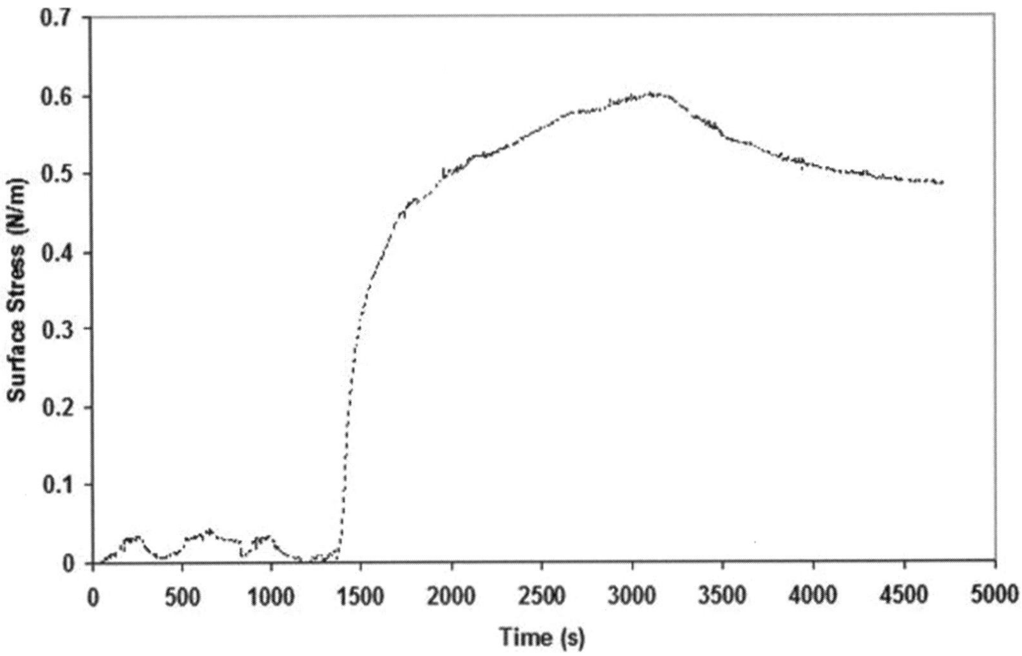

FIGURE 10 Surface stress (proportional to cantilever bending) observed on a cantilever as a function of time during thiolated ssDNA immobilization on the cantilever surface.

adsorption, the cantilever reaches a steady state due to saturation of available adsorption sites on the cantilever surface. The cantilevers used in the ssDNA experiments were piezoresistive cantilevers where the cantilever resistance changes as a function of cantilever bending.

The effect of interferents such as changes in ionic concentrations, pH, temperature, and flow rate can be overcome by using reference cantilevers (Boisen et al., 2000; Thaysen et al., 2001; Shekhawat et al., 2006). Also, it is possible to fabricate multiple cantilevers in an array for detection of multiple targets. Recently there have been many advances in fabrication of cantilevers with electronic signal transduction. Cantilevers with electronic readout have the inherent advantages over the optical beam deflection approach. In addition, electronic readout is compatible with array arrangement and packaging into a small integrated system.

CONCLUSIONS

We have described a variety of techniques based on scanning probe microscopy that have become available in recent years and can be effectively extended to sensing, physiological studies, and diagnostics in biological and microbiological analyses. These truly interdisciplinary developments have immense potential to transcend academic and industrial barriers and are expected to allow significant advancements in nanoscale studies in biological systems. Certainly, as the probes of the discussed microscopes, that is, the microcantilever, the bead, the optical fibers, etc., improve, thanks to recent developments in material research and processing, manufacturing, and the necessary understanding of the involved physics, important knowledge is expected to transpire with regard to biological functions. In particular, reduction in size of the probes and alteration of the mechanical, electronic, and optical properties have been witnessed to provide higher resolution, operation on a wider range of sample material in more realistic environments, and richer interpretation of the collected data. Although the presented material provides the basic framework of the current cellular and biomolecular imaging techniques, the future probes may be envisioned to be hybrid and multifunctional such that reproducible and robust data may be collected on a large number of biological samples.

REFERENCES

Abbe, E. 1873. Beitrage zur theorie des mikroskops und der mikroskopischen wahrnehmung. *Arch. Mikrosk. Anat. Entwicklung Mech.* **9:**413–468.

Ashkin, A., J. M. Dziedzic, J. E. Bjorkholm, and S. Chu. 1986. Observation of a single-beam gradient force optical trap for dielectric particles. *Opt. Lett.* **11:**288–290.

Bahatyrova, S., R. N. Frese, C. A. Siebert, J. D. Olsen, K. O. Van der Werf, R. Van Grondelle, R. A. Niederman, P. A. Bullough, C. Otto, and C. N. Hunter. 2004. The native architecture of a photosynthetic membrane. *Nature* **430:**1058–1061.

Ballantine, D. S., R. M. White, S. J. Martin, A. J. Ricco, G. C. Frye, E. T. Zellers, and Y. H. Wohltijen. 1996. *Acoustic Wave Sensors.* Academic Press, London, United Kingdom.

Bao, G., and S. Suresh. 2003. Cell and molecular mechanics of biological materials. *Nat. Mater.* **2:**715–725.

Betzig, E., A. Lewis, A. Harootunian, M. Isaacson, and E. Kratschmer. 1986. Near field scanning optical microscopy (NSOM): development and biophysical applications. *Biophys. J.* **49:**269–279.

Binnig, G., C. F. Quate, and C. H. Gerber. 1986. Atomic force microscope. *Phys. Rev. Lett.* **56:**930–933.

Binnig, G., H. Roher, C. H. Gerber, and E. Weibel. 1982. Surface studies by scanning tunneling microscopy. *Phys. Rev. Lett.* **49:**57–61.

Boisen, A., J. Thaysen, H. Jensenius, and O. Hansen. 2000. Environmental sensors based on micromachined cantilevers with integrated read-out. *Ultramicroscopy* **82:**11-16.

Butt, H. J. 1996. A sensitive method to measure changes in the surface stress of solids. *J. Colloid Interface Sci.* **179:**251–260.

Butt, H. J., E. K. Wolff, S. A. C. Gould, B. D. Northern, C. M. Peterson, and P. K. Hansma. 1990. Imaging cells with the atomic force microscope. *J. Struct. Biol.* **105:**54–61.

Cherian, S., and T. Thundat. 2002. Determination of adsorption-induced variation in the spring constant of a microcantilever. *Appl. Phys. Lett.* **80:**2219–2221.

Clausen-Schaumann, H., M. Seitz, R. Krautbauer, and H. E. Gaub. 2000. Force spectroscopy with single biomolecules. *Curr. Opin. Chem. Biol.* **4:**524–530.

Datskos, P. G., S. Rajic, and I. Datskou. 1998. Photoinduced and thermal stress in silicon microcantilevers. *Appl. Phys. Lett.* **73:**2319–2321.

Dufrêne, Y. F. 2001. Application of atomic force microscopy to microbial surfaces: from reconstituted cell surface layers to living. *Micron* **32:**153–165.

Dufrêne, Y. F., C. J. P. Boonaert, P. A. Gerin, M. Asther, and P. G. Rouxhet. 1999. Direct probing of the surface ultrastructure and molecular interactions of dormant and germinating spores of *Phanerochaete chrysosporium. J. Bacteriol.* **181:**5350–5354.

Edidin, M. 2001. Near-field scanning optical microscopy, a siren call to biology. *Traffic* **2:**797–803.

Fotiadis, D., S. Scheuring, S. A. Muller, A. Engel, and D. J. Muller. 2002. Imaging and manipulation of biological structures with the AFM. *Micron* **33:**385–397.

Giambattista, B., W. W. McNairy, C. G. Slough, A. Johnson, L. D. Bell, R. V. Coleman, J. Schneir, R. Sonnenfeld, B. Drake, and P. Hansma. 1987. Atomic resolution images of solid-liquid interfaces. *Proc. Natl. Acad. Sci. USA* **84:**4671–4674.

Häberle, W., J. K. H. Hörber, and G. Binnig. 1991. Force microscopy on living cells. *J. Vac. Sci. Technol. B.* **9:**1210–1213.

Haeberle, W., M. Pantea, and J. K. H. Hoerber. 2006. Nanometer-scale heat-conductivity measurements on biological samples. *Ultramicroscopy* **106:**678–686.

Hansen, K. M., H. F. Ji, G. Wu, R. Datar, R. Cote, A. Majumdar, and T. Thundat. 2001. Cantilever-based optical deflection assay for discrimination of DNA single-nucleotide mismatches. *Anal. Chem.* **73:**1567–1571.

Hansma, P. K., V. B. Elings, O. Marti, and C. E. Bracker. 1988. Scanning tunneling microscopy and atomic force microscopy: application to biology and technology. *Science* **242:**209–216.

Haydon, P. G., R. Lartius, V. Parpura, and S. P. Marchese-Ragona. 1996. Membrane deformation of living glial cells using atomic force microscopy. *J. Microsc.* **182:**114–120.

Heinz, W. F., and J. H. Hoh. 1999. Spatially resolved force spectroscopy of biological surfaces using the atomic force microscope. *Tibtech* **17:**143–150.

Heisenberg, W. 1927. Ueber den anschaulichen inhalt der quantentheoretischen kinematik and mechanic. *Z. Phys.* **43:**172–198.

Hoh, J. H., and P. K. Hansma. 1992. Atomic force microscopy for high-resolution imaging in cell biology. *Trends Cell. Biol.* **2:**208–213.

Horber, J. K. H., and M. J. Miles. 2003. Scanning probe evolution in biology. *Science* **302:**1002–1005.

Ji, H. F., K. M. Hansen, Z. Hu, and T. Thundat. 2001. Detection of pH variation using modified microcantilever sensors. *Sens. Actuators B.* **72:**233–238.

Ji, H. F., X. Yan, J. Zhang, and T. Thundat. 2004. Molecular recognition of biowarfare agents using micromechanical sensors. *Expert Rev. Mol. Diagn.* **4:**859–866.

Kasas, S., and A. Ikai. 1995. A method for anchoring round shaped cells for atomic force microscope imaging. *Biophys. J.* **68:**1678–1680.

Kassies, R., K. O. Van Der Werf, A. Lenferink, C. N. Hunter, J. D. Olsen, V. Subramaniam, and C. Otto. 2005. Combined AFM and confocal fluorescence microscope for applications in bio-nanotechnology. *J. Microsc.* **217:**109–116.

Kulzer, F., and M. Orrit. 2004. Single molecule optics. *Ann. Rev. Phys. Chem.* **55:**585–611.

Lal, R., and S. A. John. 1994. Biological applications of atomic force microscopy. *Am. J. Physiol.* **266:**C1–C21.

Langer, M. G., W. Offner, H. Wittmann, H. Flosser, H. Schaar, W. Haberle, A. Pralle, J. P. Ruppersberg, and J. K. H. Horber. 1997. A scanning force microscopy for simultaneous force and patch-clamp measurements on living cell tissues. *Rev. Sci. Instrum.* **68:**2583–2590.

Lee, G. U., D. A. Kidwell, and R. J. Colton. 1994. Sensing discrete streptavidin-biotin interactions with atomic force microscopy. *Langmuir* **10:**354–361.

Liu, T., J. Tang, M. Han, and L. Jiang. 2003. A novel microgravimetric DNA sensor with high sensitivity. *Biochem. Biophys. Res. Commun.* **304:**98–100.

Mehta, A., S. Cherian, D. Hedden, and T. Thundat. 2001. Manipulation and controlled amplification of brownian motion of microcantilever sensors. *Appl. Phys. Lett.* **78:**1637–1639.

Ming, S., S. Li, and V. P. Dravid. 2003. Microcantilever resonance-based DNA detection with nanoparticle probes. *Appl. Phys. Lett.* **82:**3562–3564.

Muralidharan, G., A. Mehta, S. Cherian, and T. Thundat. 2001. Analysis of amplification of thermal vibrations of a microcantilever. *J. Appl. Phys.* **89:**4587–4591.

Nagao, E., and J. A. Dvorak. 1998. An integrated approach to the study of living cells by atomic force microscopy. *J. Microsc.* **191:**8–19.

Pohl, D. W., W. Denk, and M. Lanz. 1984. Optical stethoscopy: image recording with resolution 20. *Appl. Phys. Lett.* **44:**651–653.

Pralle, A., E. L. Florin, E. H. K. Stelzer, and J. K. H. Hörber. 2000. Photonic force microscopy: a new tool providing a new method to study membranes at the molecular level. *Single Molec.* **1:**129–133.

Proksch, R., R. Lal, P. K. Hansma, D. Morse, and G. Stucky. 1996. Imaging the internal and external pore structure of membranes in fluid: tapping mode scanning ion conductance microscopy. *Biophys. J.* **71:**2155–2157.

Radmacher, M., M. Fritz, C. M. Kacher, J. P. Cleveland, and P. K. Hansma. 1996. Measuring the viscoelastic properties of human platelets with the atomic force microscope. *Biophys. J.* **7:**556–567.

Raiteri, R., M. Grattarola, H. J. Butt, and P. Skladal. 2001. Micromechanical cantilever-based biosensors. *Sens. Actuators B.* **79:**115–126.

Ricci, D., M. Tedesco, and M. Grattarola. 1997. Mechanical and morphological properties of living 3T6 cells probed via scanning force microscopy. *Microsc. Res. Tech.* **36:**165–171.

Rizzuto, R., M. Brini, P. Pizzo, M. Murgia, and T. Pozzan. 1995. Chimeric green fluorescent protein as a tool for visualizing subcellular organelles in living cells. *Curr. Biol.* **5:**635–642.

Scheuring, S., D. J. Müller, H. Stahlberg, A. Engel, and A. Engel. 2002. Sampling the conformational space of membrane protein surfaces with AFM. *Eur. Biophys. J.* **31:**172–178.

Shekhawat, G., S. H. Tark, and V. P. Dravid. 2006. MOSFET-embedded microcantilevers for measuring deflection in biomolecular sensors. *Science* **311:**1592–1595.

Shekhawat, G. S., and V. P. Dravid. 2005. Nanoscale imaging of buried structures via scanning near-field ultrasound holography. *Science* **310:**89–92.

Shuttleworth, R. 1950. The surface tension of solids. *Proc. Phys. Soc. (London), Ser. A* **63:**444–457.

Stelzer, E. H. K. 2002. Light microscopy: beyond the diffraction limit? *Nature* **417:**806–807.

Stephens, D. J., and V. J. Allan. 2003. Light microscopy techniques for live cell imaging. *Science* **300:**82–86.

Stoney, G. G. 1909. The tension of metallic films deposited by electrolysis. *Proc. R. Soc. London Ser. A* **82:**172–175.

Synge, E. H. 1928. A suggested method for extending the microscopic resolution into the ultramicroscopic region. *Phil. Mag.* **6:**356–362.

Thaysen, J., R. Marie, and A. Boisen. 2001. Cantilever-based bio-chemical sensor integrated in a microliquid handling system, p. 401–404. *Technical Digest. MEMS: Proc. 14th IEEE International Conference on Micro Electro Mechanical Systems.*

Thundat, T., D. P. Allison, R. J. Warmack, M. J. Doktycz, K. B. Jacobson, and G. M. Brown. 1993. Atomic force microscopy of single-and double-stranded deoxyribonucleic acid. *J. Vac. Sci. Technol. A.* **11:**824–828.

Thundat, T., P. I. Oden, and R. J. Warmack. 1997. Chemical, physical, and biological detection using microcantilevers, p. 179–187. *Proceedings of the Third International Symposium on Microstructures and Microfabricated Systems.*

Van der Aa, B. C., R. M. Michel, M. Asther, M. T. Zamora, P. G. Rouxhet, and F. Dufrêne. 2001.

Stretching cell surface macromolecules by atomic force microscopy. *Langmuir* **17:**3116–3119.

Vesenka, J., C. Mosher, S. Schaus, L. Ambrosio, and E. Henderson. 1995. Combining optical and atomic force microscopy for life sciences research. *Biotechniques* **19:**240–253.

Wang, C. 2004. The principle of micro thermal analysis using atomic force microscope. *Thermochimica Acta* **423:**89–97.

Wig, A., E. T. Arakawa, A. Passian, T. L. Ferrell, and T. Thundat. 2006. Photothermal spectroscopy of *Bacillus anthracis* and *Bacillus cereus* with microcantilevers. *Sens. Actuators B.* **114:**206–211.

Wig, A., A. Passian, E. Arakawa, T. L. Ferrell, and T. Thundat. 2004. Optical thin-film interference effects in microcantilevers. *J. Appl. Phys.* **95:**1162–1165.

Wu, G., R. H. Datar, K. M. Hansen, T. Thundat, R. J. Cote, and A. Majumdar. 2001a. Bioassay of prostate-specific antigen (PSA) using microcantilevers. *Nat. Biotechnol.* **19:**856–860.

Wu, G., H. Ji, K. M. Hansen, T. Thundat, R. Datar, R. Cote, M. F. Hagan, A. K. Chakraborty, and A. Majumdar. 2001b. Origin of nanomechanical cantilever motion generated from biomolecular interactions. *Proc. Natl. Acad. Sci. USA* **98:**1560–1564.

SINGLE-CELL GENOMICS

Martin Keller, Christopher W. Schadt, and Anthony V. Palumbo

15

Current environmental genomic studies largely originated from Pace's cultivation-independent survey approach to studying natural microbial populations (Pace and Marsh, 1985; Olsen et al., 1986) and have developed into single-cell genomics. Pace's approach has evolved and expanded to include large genome fragment analyses for the characterization of uncultivated, indigenous microbes (Stein et al., 1996). We have seen an explosion in total genome and metagenomic projects based on a shotgun sequencing approach (Tringe et al., 2005; Tyson et al., 2004). In addition, the repertoire of tools for "postenvironmental genomics" is also expanding, with microarray, proteomic, and metabolomic experiments following the environmental genomic discoveries (DeLong, 2004). At the same time, the use of these molecular techniques has led to the realization that the microbial diversity found in almost all investigated environments is much larger than ever anticipated.

The presence of this tremendous diversity in combination with the finding of significant lateral gene transfer within these environments challenges the conventional understanding and definition of a microbial species and of evolution of microbes. As discussed by Goldenfeld and Woese (2007), "The emerging picture of microbes as gene-swapping collectives demands a revision of such concepts as organisms, species and evolution itself. Similarly, the convergence of fresh theoretical ideas in evolution and the coming avalanche of genomic data will profoundly alter our understanding of the biosphere—and is likely to lead the revision of concepts such as species, organisms and evolution." It could be argued that the molecular reductionisms that dominated the last century will be replaced by an interdisciplinary approach that will take all collective phenomena into account. Questions arise: are genomes really discrete or do genomes change and adapt to the need and pressure induced through the specific environment?

These findings and new concepts raise the question of how significant it is to study single microbial strains in isolation and outside the content of the specific microbial environment. Goldenfeld and Woese (2007) suggest that "there is a continuity of energy flux and informational transfer from the genome up through cells, community, virosphere and envi-

Martin Keller, Christopher Schadt, and Anthony Palumbo, Biosciences Division, Oak Ridge National Laboratory, 1 Bethel Valley Road, Oak Ridge, TN 37831-6035.

Accessing Uncultivated Microorganisms: from the Environment to Organisms and Genomes and Back
Edited by Karsten Zengler © 2008 ASM Press, Washington, DC

ronment." Nesbo et al. (2006) characterized regions of genomes of eight members of the hyperthermophilic genus *Thermotoga* that differ from each other physiologically and occupy physically distinct environments. The data they presented suggest a compelling reason to believe that there is recombination between different "ecotypes" or "species" across vast distances. On the basis of these findings, it could be argued that no single bacterial and/or archaeal species concept might be able to capture the uncoupling of ecotypic and genetic aspects of cohesion and diversity. However, it cannot be ruled out that the differences found among the investigated strains are due to a biased and limited isolation of these specific strains with all the other strains present at the same time in the same environment (Nesbo et al., 2006). Therefore, it is uncertain whether the differences came from organisms at global scales or rather from inadequate sampling of the local environments. All these compared strains are cultivated representatives and therefore it cannot be ruled out that the different "species" were not present within the same environment.

More extensive work with increased sampling will be needed to understand the genetic diversity of species within and between locations (Staley, 2006). Almost all microbial diversity surveys are based on 16S rRNA gene sequencing (Brennan et al., 2004). However, recent studies have shown that at different geographical locations, vast numbers of diverse genotypes can exist within a single species (Whitaker et al., 2003). Current shotgun sequencing and in silico assembly algorithms are challenged by this inherent intraspecies genetic complexity (Green and Keller, 2006), assuming that the metagenomic data will be assembled and assigned to specific species or ecotypes.

The fact that we are trying to understand complex microbial environments and the function and contribution of certain microbial ecotypes makes the binning of sequences an important and challenging task. The microbial assembly and binning of sequences of any group of microbial microorganisms are hard to predict based on the variation of the needed genome coverage, especially if underrepresented species are of interest. The current metagenomic technologies used to cover environmental assemblages reveal enormous biodiversity and demonstrate tremendous amounts of novel genes. However, these approaches suffer from two major related drawbacks: (i) the difficulty of assembling contigs into discrete genomes and (ii) biased sampling toward abundant species (Zhang et al., 2006).

As demonstrated by Zhang et al. (2006) and Podar et al. (2007), the ability to sequence an entire genome from a single bacterial cell might overcome some limitations of current metagenomic approaches, enable genomic analyses such as the characterization of genetic heterogeneity in a population of cells, and reveal the *cis*-relationships between sequences that are more than 200 kb apart. Individual microorganisms, even those in a clonal population, may differ widely from each other in terms of their genetic composition, physiology, biochemistry, or behavior (Brehm-Stecher and Johnson, 2004). Microbial genomes are capable of substantial changes within very short periods of time and genetic heterogeneity in individual microorganisms can arise from a number of random, semirandom, or programmed events and can be subject to selection on relatively short timescales. Modes and mechanisms of genetic variability include spontaneous point mutations, random transcription events, phage-related phenomenon, chromosomal duplications, gene amplification, and the presence, absence, and copy number of mobile genetic elements such as plasmids and transposons (Bridson and Gould, 2000; Davey and Kell, 1996; Elowitz et al., 2002; Papadopoulos et al., 1999). Large metagenomic studies demonstrate the central theme, the microbial heterogeneity of individual microbial cells, which will drive single-cell genomics and microbiology. Bulk-scale measurements made on heterogeneous populations, or even worse, complete microbial environments, show only averaged values for the population or the environment. The contribution of single microbes is normalized, and only the average population trait is visible.

Single-cell genomics, the total genome sequence from a single or a very few cells, is becoming reality and will allow insight into many of these phenomena. This chapter gives an overview of the current state of the art and highlights the different technological steps necessary to achieve the total genome sequence from a single microbial cell.

This chapter addresses (i) methods to isolate single bacterial cells, (ii) DNA isolation and amplification from a single microorganisms, (iii) DNA sequencing, and (iv) applications and future outlook.

METHODS TO ISOLATE SINGLE BACTERIAL CELLS

The method to ensure cultivation of a pure strain is to start cultivation from a single cell resulting in an actively growing colony created by clonal growth. This goal has been the driving force behind the development of techniques based on micromanipulation for isolation of single bacterial and eukaryotic cells throughout the last century.

Many methods can be used to create single cells as starting material for DNA isolation and amplification, and the easiest approach to achieve single cells is through serial dilution. For example, serial dilution was used to isolate *Escherichia coli* and *Prochlorococcus* cells following DNA amplification (Zhang et al., 2006). Although this technology is very easy to apply, it has significant limitations. The selection of single cells is random and thus, when applied to the environment, serial dilution will give access to only the most abundant cells. In addition, dilution factors can be off, and therefore more than one cell or a mixture of two or more different cell types or species can still be present in the final dilution.

Other cell separation technologies are based on the use of optical instrumentation to visualize the specific microorganisms coupled with a method to manipulate the cells. Through the development of micromanipulation, two principal methods, laser manipulation and mechanical manipulation, have dominated. For example, Huber et al. (1995) used a focused laser beam to capture and move archaeal cells within a capillary filled with fresh medium from one side of the capillary to the other side. Because cells are visualized during the process, cells can be chosen on morphological criteria and even phylogenetic criteria if combined with fluorescent in situ hybridization (FISH). The capillary, containing the separated cell at the end, is cut, and the individual cell can be used for further cultivation or theoretically for DNA isolation. The second approach is based on microinjectors in combination with the precision of a servo-powered micromanipulator, enabling the easy handling of a single microbial cell. Fröhlich and König (1999) applied such a system using prefabricated Bacto-tips to isolate and cultivate single cells from the natural environment. Ishey et al. (2006) further improved this technology and developed a system that is based on an inverted microscope, a microinjector, and a micromanipulator. The isolated cell is captured in a microcapillary pipet and transferred into an appropriate growth medium. They demonstrated that the micromanipulator system was able to isolate a single cell from a mixed culture without contamination by any of the nontarget cells from the surrounding media. However, micromanipulation of individual microorganisms can be challenging. Thomsen et al. (2004) used the combination of micromanipulation of FISH-labeled filaments or microcolonies in activated sludge samples with reverse transcription-PCR for further identification. The applied method was more successful on filamentous bacteria than on microcolonies. The filamentous bacteria were easier to separate from the floc material in comparison to the microcolonies. However, it was not possible to isolate microcolonies consisting of only one single bacterial species due to the presence of several different species within a microcolony or the contamination of the selected microcolony by surrounding free bacteria (Thomsen et al., 2004). Kvist et al. (2007) combined the micromanipulator system with FISH followed by DNA amplification. *Methanothermobacter thermautotrophicus* was used as a test organism, followed by the isolation of an archaeal strain closely related to the crenarchaeotal BAC clone SCA1170 obtained from a soil sample.

Another technology, widely used to separate microorganisms, is flow cytometry and this approach offers many advantages. This technology is commonly used in many medical applications to analyze and separate eukaryotic cells but is gaining more popularity for bacterial applications. Flow cytometry coupled with cell sorting is a powerful fluorescence-based cell diagnostic and separation tool that enables the rapid analysis of entire cell populations on the basis of single-cell characteristics. Thus characteristics can be collected simultaneously and include cell count, cell size or content, and responses to fluorescent probes as diagnostics of cell function. Cells in a liquid sample are passed individually in front of an intense light source (laser or laser diode), and data on light scattering and fluorescence are collected and saved as a data file (Davey and Kell, 1996; Vives-Rego et al., 2000). Bacteria can be separated and sorted based on the forward and side scatter effect alone even without the use of any fluorescence dyes. Flow cytometry can be combined with FISH, which will allow the combination of identification and cell separation (Sekar et al., 2004; Kalyuzhnaya et al., 2006). Raghunathan et al. (2005) used flow cytometry to separate single, unstained *E. coli* cells and amplified isolated DNA 5-billion-fold. Podar et al. (2007) combined flow cytometry with FISH and multiple displacement amplification (MDA) to amplify DNA from five sorted bacterial cells belonging to the candidate division TM7, an uncultured division of microorganisms commonly found in different soils. The rapid development in laser technology and nanotechnology has recently led to the miniaturization of flow cytometry and the development of an integrated microfabricated cell sorter using multilayer soft lithography (Fu et al., 2002, 1999). This integrated cell sorter can be incorporated with various microfluidic functionalities, including peristaltic pumps, dampers, switch-valves, and input and output wells, to perform cell sorting in a coordinated and automated fashion with extremely low fluidic volumes. Due to the simple fabrication process and inexpensive materials, these devices can be made disposable so that the user can eliminate any cross-contamination from previous runs. Thus, they would be ideal for downstream molecular methods that require extreme sterility and to avoid any contaminating DNA in single cell culturing or genomics applications (Fu et al., 1999). This benefit might compensate for the much lower sorting speed compared to conventional fluorescent-activated cell sorting (FACS) machines (Fu et al., 2002). In addition, the cell sorting device can be directly combined with chemical or enzymatic reactions, such as cell lysis and DNA amplification methods (MDA, PCR).

DNA ISOLATION AND AMPLIFICATION FROM SINGLE MICROBIAL CELLS

Many applications of single-cell genomics require careful DNA amplification to overcome several problems and potential limitations. A typical bacterial chromosome contains a few femtograms (10^{-15} g) of DNA. Consequently, single-cell genomics using current sequencing methods require amplification of the DNA of a single cell by factors of about 10^9 to obtain the microgram amounts that are required for these technologies (Hutchison and Venter, 2006). The lyses of bacterial cells have to be extremely gentle to avoid breakage of large chromosomal DNA. Amplification of DNA cannot occur across breaks and therefore linkage information can be lost with DNA breakage. Also, contamination can be a significant issue and steps must be taken to minimize chances for contamination (Hutchison and Venter, 2006).

A variety of techniques have been used to avoid DNA breaks and contamination. The most applied genomic DNA isolation method from single bacterial cells includes a cell treatment with lysozyme followed by a denaturation step through an alkaline solution (Dean et al., 2001, 2002; Zhang et al., 2006). Podar et al. (2007) used a similar alkaline lysis buffer with the modification to perform cell lyses in a volume of 20 μl to keep DNA as concentrated as possible. Small reaction volumes have a significant positive effect in reducing the possibility of contamination. Additionally, the reduced reaction volume increases the effective concentration ratio of template to enzymatic reaction components and may reduce the conditions that are responsible for spurious background amplification (Hutchison et al., 2005). Using submicroliter reaction volumes, Hutchinson et al. amplified single circular DNA molecules to give amounts of DNA that can easily be visualized by gel electrophoresis and can be sequenced. The 600-nl reactions used by Hutchison et al. (2005) can be prepared by using conventional manual pipetting devices. However, preliminary results indicate that it may be advantageous to reduce volumes even further. Microfluidic chips present the opportunity to decrease reaction volumes and combine DNA isolation and amplification. For example, Hong et al. (2004) fabricated a microfluidic chip that can sequentially process nanoliter volumes to isolate variable numbers of microbial cells, lyse them, and purify their DNA or mRNA. All of the steps in the process were carried out on a single microfluidic chip without any pre- or postsample treatment (Hong et al., 2004).

Because a key issue for single-cell genomics is the ability to perform high-quality DNA amplifications to obtain enough DNA molecules for DNA sequencing (whole-genome amplification [WGA]), novel DNA amplification methods are being developed and applied. Although, new DNA sequencing methods that offer higher sequencing rates and lower initial DNA concentrations are developing very rapidly (Ronaghi et al., 1998), it is not yet feasible to sequence single DNA molecules. Therefore, WGA is required to obtain enough DNA for sequencing.

At present three primary forms of WGA have been developed: primer extension preamplification (PEP) (Zhang et al., 1992), degenerate oligonucleotide primed (DOP) PCR (Telenius et al., 1992), and MDA (Dean et al., 2002). PEP creates multiple copies of the DNA sequences present in a single cell by multiple rounds of extension with the *Taq* DNA polymerase and a random mixture of 15-base oligonucleotides as primers. DOP PCR allows the statistical and representative amplification of an uncharacterized or unknown DNA template. The reaction utilizes a universal primer possessing a 6-nucleotide-long degenerated region, which statistically represents all possible 6 nucleotide combinations. MDA may currently represent the most promising reaction for DNA amplification. MDA uses the DNA polymerase encoded by phage Phi 29. Application of this technique for WGA was first reported in 2002 by Lasken and colleagues, who used it to amplify human genomic DNA (Dean et al., 2002). Phi 29 polymerase was used by Lasken in combination with random primers to amplify circular DNA directly from cells or plaques, called multiply primed rolling-circle amplification (RCA), generating high-quality template for use in DNA sequencing, probe generation, or cloning. This amplification method is simple, optimally performed at 30°C, and is able to achieve a 10,000-fold amplification. On average each enzyme copy has the possibility to perform strand displacement DNA synthesis for more than 70,000 nucleotides without dissociation from the template. In multiply primed RCA, the use of multiple primers annealed to a circular template DNA generates multiple replication forks. RCA proceeds by displacing the non-template strand. In this way, product strands are "rolled off" the template as tandem copies of the circle. Random priming allows synthesis of both strands, resulting in double-stranded product. A cascade of priming events results in exponential amplification (Dean et al., 2001).

Pinard et al. (2006) compared all three methods (PEP, DOP, and MDA) to assess bias introduced through WGA. Each reaction started with 25 ng of input DNA. Within the MDA, two different methods—REPLI-G (DNA amplification kit, Qiagen, Germantown, MD) and GenomiPhi Whole Genome Amplification kit (GE Healthcare Life Sciences, Uppsala,

Sweden)—were tested. The two methods, respectively, generated a 2,100-fold and 640-fold amplification of the input DNA. In this test, PEP amplified the starting material 120-fold and DOP 92-fold. A sharp contrast was evident between the yields derived from the MDA-based (REPLI-G and GenomiPhi) amplifications versus the PCR-based (PEP and DOP) approaches. The higher yield with the MDA-based approaches can be attributed to the differences between the highly processive strand displacement activities of Phi 29, the polymerase used in both MDA reactions, and the *Taq*-like enzymes used in the PCR-based reactions. Due to the strand-displacing capabilities of Phi 29, MDA reactions do not require repetitive cycles of denaturation and annealing temperatures. By utilizing isothermal reactions the MDA methods are able to preserve enzyme functionality for a full 16-h reaction, and generate substantially more DNA in the process by its hyperbranched mechanism (Pinard et al., 2006) of DNA amplification (Fig. 1). Also, the two MDA methods, GenomiPhi and REPLI-G, introduced the lowest amplification bias. Thus, the MDA method potentially has many applications in microbial ecology. Microbial assessment of natural biodiversity is usually achieved through PCR amplification. DNA sequences from natural samples are often difficult to amplify because of the presence of PCR inhibitors or the low number of copies of specific sequences.

These DNA amplification methods have also been used for sequencing genomes from single cells of cultured bacteria to near completion and for preliminary characterization of relatives of cultured species. Podar et al. (2007) combined the use of taxon-specific separation of microbial cells by flow cytometry with WGA to gain access to a low-abundance soil bacterium from the candidate TM7 division. This was the first targeted isolation and partial genomic sequencing of cells representing a completely uncultured group of organisms. A significant fraction of the genome of an uncultured bacterium was obtained, starting with only a few (~5) cells selectively isolated from a complex environmental sample. Because environmental bacterial populations representing a "species" are not clonal, and their genomes may contain sequence and genetic map polymorphisms not reflected at the level of the rRNA genes (Whitaker and Banfield, 2006), the computational burden in resolving the polymorphisms and assembling a "pan-genome" increases with the number of pooled cells. Therefore, it is desired to keep the number of cells at a minimum while providing a sufficient input template for WGA. Thus these two powerful methods used in microbial ecology (FISH and cell separation by flow cytometry) were linked to WGA and sequencing. This approach combines high specificity derived from the stringent hybridization of oligonucleotide probes to target rRNA in a taxonomically

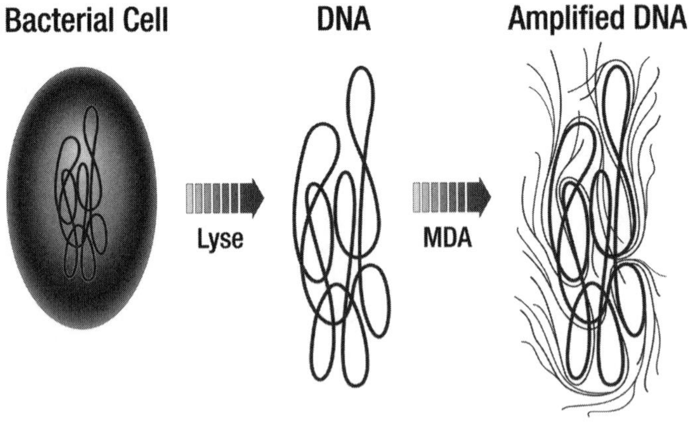

FIGURE 1 Schematic of DNA amplification through MDA. Bacterial cells will be lysed followed by DNA amplification through MDA using the strand displacement enzyme Phi 29.

predefined cellular population, with high sensitivity and throughput from the detection and separation of labeled cells from complex mixtures of organisms generated by flow cytometry. The isolated cells are then subject to MDA and further analysis (Fig. 2).

The current MDA technique has limitations when applied to a single or few cells for application to uncultured bacterial cells for genome recovery and characterization. There is a significant sequence bias during amplification, causing variations in genome coverage. As observed by Zhang et al. (2006), the distribution of bias appeared to be random and is probably due to the limited number of initial replication initiation events and relatively short amplified DNA fragments. When the starting amount of DNA was relatively high (nanogram range), the observed bias was not as severe (Pinard et al., 2006; Stein et al., 1996).

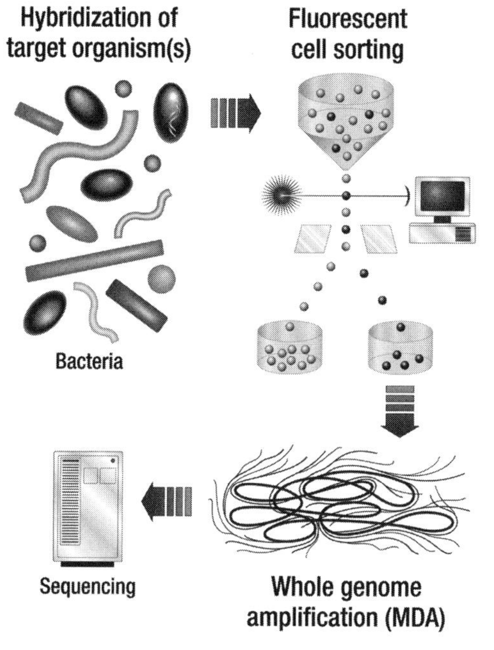

FIGURE 2 Targeted access to microbial cells directly from the environment. This approach combines high specificity derived from the stringent hybridization of oligonucleotide probes to target rRNA in a taxonomically predefined cellular population, with high sensitivity and throughput from the detection and separation of labeled cells from complex mixtures of organisms generated by flow cytometry. The isolated cells are then subject to WGA MDA, followed by DNA sequencing.

Amplification bias could also partially be reduced by pooling multiple separate reactions. However, in the case of environmental bacteria, this will result in an increased chance of heterogeneity due to the nonclonal nature of the population. Another current limitation of MDA is the formation of chimeric structures, which result in fragmented genes and difficulties in assembling large genomic contigs. Podar et al. (2007) determined empirically that the MDA of five cells achieved the lowest level of contamination with nontarget cells while providing sufficient template to result in a lower fraction of chimeric fragments, as judged by assembly artifacts and fragmented genes. The low level of sequences from contamination, a common problem in cell separation using flow cytometry, was mitigated by G+C content and taxonomic sequence binning (Podar et al., 2007).

Although there are valuable uses for MDA in single-cell analysis and other applications, the amplification bias and other problems must be considered. The amplification bias for single cells will result in the loss of some sequences and will also require more effort in preparation and interpretation of DNA libraries constructed from the amplified DNA (Raghunatghan et al., 2005). Detter et al. (2002) constructed a DNA library of MDA-amplified DNA, isolated from 1,000 cells of the bacterium *Xylella fastidiosa* and sequenced to a depth of approximately sevenfold. These data demonstrated that it may be necessary to sequence to an even greater depth for single cells due to the amplification bias introduced through MDA. Also, some microbes are expected to be more difficult to lyse and will create challenges to isolate DNA without introducing DNA strand breakages. In addition, the degree to which MDA may result in sequence rearrangements and chimeric sequences is not fully determined. The hyperbranched nature of the Phi 29 product may inhibit efficient transformation into *E. coli* sequencing vectors or result in chimeric clone products. Zhang et al. (2006) were able to solve these problems by digesting the amplified products with S1 nuclease enzyme that cleave the DNA at the branch junctions. This library showed a significant improvement in comparison to the library constructed without S1 nuclease treatment (Zhang et al., 2006). However, the remaining chimera still limited the quality of genome assemblies. An improved assembly with longer contigs was obtained by computationally splitting these chimeric sequences at their junction points based on a reference genome (Zhang et al., 2006) available in the case of *Prochlorococcus*. In addition, if the chimeric structures are produced during library construction, it is possible that novel sequencing approaches such as the 454 sequencing technology (454 Life Science, Branford, CT), which requires no DNA cloning, will help to overcome these problems. Finally, MDA will amplify all DNA present within the sample, including contaminating DNAs, multicopy sequences, and plasmids, which will thus make sequence annotation more challenging.

MDA techniques are also being used in microbial ecology for strategies beyond single-cell genomics applications, which should accelerate their refinement and adoption. For example, Gonzalez et al. (2005) proposed a nonspecific preamplification procedure to overcome the presence of inhibitors and to increase the number of copies before carrying out standard amplification by PCR. The pre-PCR step is carried out through an MDA technique using random hexamers as priming oligonucleotides and Phi 29 DNA polymerase in an isothermal, WGA reaction. Results from the natural community showed successful amplifications using a two-step protocol proposed in the described study, whereas standard, direct PCR amplification most often resulted in no amplification product. In addition, MDA techniques have recently been applied to characterize the structure of communities from highly contaminated sites, where the amount of biomass was below standard detection levels. These studies have applied both shotgun sequencing (Abulencia et al., 2006) and microarray technologies (Wu et al., 2006; He et al., 2007) for downstream characterizations of these communities after MDA. Other studies have also recently used MDA for characterization of mixed populations of methanotrophs enriched by FISH followed by FACS (Kvist et al., 2007). The current and potential uses of MDA warrant further work on addressing the potential biases and problems with its application.

SEQUENCING OF DNA AMPLIFIED FROM SINGLE BACTERIAL CELLS

Sanger sequencing has served as the cornerstone for genome sequencing, including microbial sequencing, for over a decade (Goldberg et al., 2006), but there are new developments that may displace or supplement Sanger sequencing. Continued improvements in DNA sequencing techniques, bioinformatics, and data analysis over the past few years have helped reduce the cost and time associated with sequencing a genome. However, it still costs an estimated \$8,000 to \$10,000 per megabase pair to produce a high-quality microbial genome draft sequence (Goldberg et al., 2006). There is a need for a more efficient and cost-effective approach for genome sequencing that can maintain the high quality of data produced by conventional Sanger sequencing. One of the most promising new sequencing technologies is the 454 sequencing platforms (454 Life Science, Branford, CT). For example, the GS20 system is a high parallel noncloning pyrosequencing-based (Ronaghi et al., 1996, 1998) system capable of sequencing 100 times faster than current state-of-the-art Sanger sequencing and capillary electrophoresis platforms. However, this technology still has issues that hinder the full replacement of Sanger sequencing at this stage. Short read length, the lack of pair end reads, and the accuracy of individual reads limit the assembly of individual sequencing reads into larger scaffolds required for high-quality draft sequences and gap closure. Therefore, currently, pyrosequencing is a complement to Sanger sequencing rather than a replacement (Goldberg et al., 2006). However, this can rapidly change with the further development of this technology, especially if longer read length will be available as promised in the 454 GSFX instruments. Other high-throughput systems also are being developed. Margulies et al. (2005) describe an integrated system whose throughput routinely enables applications requiring millions of bases of sequencing information, including whole-genome sequencing. This was achieved through the development of an emulsion-based method to isolate and amplify DNA fragments in vitro, and development of a fabricated substrate and instrument that performs pyrophosphate-based sequencing in picoliter-sized wells. This technology generates random libraries of DNA fragments by shearing an entire genome and isolating single DNA molecules by limiting dilution. Specialized common adaptors are then added to the pieces of the randomly fragmented genome followed by the capture of these fragments into an individual bead and the clonal amplification of the individual fragment within an emulsion droplet.

OUTLOOK

These new types of sequencing technologies will likely further revolutionize microbial ecology. Sequencing will become very inexpensive, and it will be routine to sequence whole genomes and diverse environmental samples. In addition, these new methods will allow much greater sequencing coverage and will be essential to single microbial cell genomics. However, because of the inherent complexities and dramatic diversity of many microbial communities, cheaper and faster sequencing alone is not likely to enable the assembly of all the genomes within such samples. Targeted approaches, such as single-cell genomics, could, however, allow access to minority genomes in such environments using intelligent selection of "one at a time" targets (as opposed to random, shotgun approaches). New technologies such as microfluidics, single-cell separation, targeted access through FISH staining in combination with WGA from single cells will enable access to the total genome sequence from organisms separated directly from the environment without cultivation (Fig. 2). These approaches will not replace new cultivation efforts as described in Keller and Zengler (2004). However, the total genomic information of these so far uncultured microorganisms will potentially enable insights into the metabolisms and might give hints about the growth conditions. Targeted metagenomics of individual microorganisms will complement environmental metagenomic data and will allow assigning individual genes to the corresponding microorganisms.

Another target metagenomics approach was used by Ottesen et al. (2006) with microfluidic digital PCR to perform a multigene analysis of individual environmental bacteria. Microfluidic devices will allow control and manipulation of small volumes of liquid, allowing rapid separation and partitioning of single cells from a complex parent sample. Single, partitioned cells can serve as templates for individual genome sequencing or multiplex PCR reactions using primers and probes for simultaneous amplification of both 16S rRNA and metabolic genes of interest. In this first experiment Ottesen et al. (2006) described rapid colonization of two genes (encoding 16S rRNA and key metabolic enzymes) to single-genome templates, along with determination of the fraction of cells within the community that encoded them.

The next logical step following metagenomic DNA sequencing initiated through lower sequencing costs and increased sequencing capacity will be the analysis of global RNA expression directly from the environment and eventually from a single bacterial cell. Microarray-based analysis of microbial community RNAs is a powerful tool for viewing the expression of thousands of genes simultaneously in a single experiment. While this technology was initially designed for transcriptional profiling of a single species, its applications have been dramatically extended to environmental applications in recent years (Gao et al., 2007). One of the greatest challenges in using microarrays for analyzing environmental samples is the low detection sensitivity of microarray-based hybridization in combination with the low biomass often present in samples from environmental settings. Gao et al. (2007) developed a novel method for randomly amplifying whole-community RNAs. In this approach a T7 RNA promoter sequence is attached to a random hexamer, which is then used for reverse transcription of RNAs. This method produced sufficient mRNA from environmental samples for microarray analysis. The current bottleneck in single microbial cell RNA profiling lies within a sufficient method to amplify mRNA. Marcus et al. (2006) demonstrated the first quantitative calibration for microfluidic mRNA isolation and cDNA synthesis, sensitive enough to detect medium- and low-copy number transcripts in single eukaryotic cells. All five steps (cell capture, cell lysis, mRNA purification, cDNA synthesis, and cDNA purification) were implemented in a microfluidic assay on one integrated device. Novel RNA amplification methods in combination with the next-generation sequencing technology will open the field of single microbial RNA profiling. This will generate important data to further foster our understanding of how single bacterial species or ecotypes contribute within intricate microbial communities. Single-cell microbiology will allow us to take complex microbial communities apart, to analyze the contributions of individual ecotypes, and then to model and integrate this information back to further our understanding of the function of the specific ecotype within the community. In these ways single-cell microbiology will reveal many answers toward our ultimate goal of understanding complex environments and the roles that individual ecotypes play within them.

ACKNOWLEDGMENTS

We thank Meghan S McNeilly for critical reading and editing and Brett Hopwood for the graphical design of the figures.

REFERENCES

Abulencia, C. B., D. L. Wyborski, J. A. Garcia, M. Podar, W. Q. Chen, S. H. Chang, H. W. Chang, D. Watson, E. L. Brodie, T. C. Hazen, and M. Keller. 2006. Environmental whole-genome amplification to access microbial populations in contaminated sediments. *Appl. Environ. Microbiol.* **72:**3291–3301.

Brehm-Stecher, B. F., and E. A. Johnson. 2004. Single-cell microbiology: tools, technologies, and applications. *Microbiol. Mol. Biol. Rev.* **68:**538–559.

Brennan, Y., W. N. Callen, L. Christoffersen, P. Dupree, F. Goubet, S. Healey, M. Hernandez, M. Keller, K. Li, N. Palackal, A. Sittenfeld, G. Tamayo, S. Wells, G. P. Hazlewood, E. J. Mathur, J. M. Short, D. E. Robertson, and B. A. Steer. 2004. Unusual microbial xylanases from insect guts. *Appl. Environ. Microbiol.* **70:**3609–3617.

Bridson, E. Y., and G. W. Gould. 2000. Quantal microbiology. *Lett. Appl. Microbiol.* **30:**95–98.

Davey, H. M., and D. B. Kell. 1996. Flow cytometry and cell sorting of heterogeneous microbial populations: the importance of single-cell analyses. *Microbiol. Rev.* **60:**641–696.

Dean, F. B., S. Hosono, L. H. Fang, X. H. Wu, A. F. Faruqi, P. Bray-Ward, Z. Y. Sun, Q. L. Zong, Y. F. Du, J. Du, M. Driscoll, W. M. Song, S. F. Kingsmore, M. Egholm, and R. S. Lasken. 2002. Comprehensive human genome amplification using multiple displacement amplification. *Proc. Natl. Acad. Sci. USA* **99:**5261–5266.

Dean, F. B., J. R. Nelson, T. L. Giesler, and R. S. Lasken. 2001. Rapid amplification of plasmid and phage DNA using phi29 DNA polymerase and multiply-primed rolling circle amplification. *Genome Res.* **11:**1095–1099.

DeLong, E. F. 2004. Microbial population genomics and ecology: the road ahead. *Environ. Microbiol.* **6:**875 878.

Detter, J. C., J. M. Jett, S. M. Lucas, E. Dalin, A. R. Arellano, M. Wang, J. R. Nelson, J. Chapman, Y. Lou, D. Rokhsar, T. L. Hawkins, and P. M. Richardson. 2002. Isothermal strand-displacement amplification applications for high-throughput genomics. *Genomics* **80:**691–698.

Elowitz, M. B., A. J. Levine, E. D. Siggia, and P. S. Swain. 2002. Stochastic gene expression in a single cell. *Science* **297:**1183–1186.

Fröhlich, J., and H. König. 1999. Rapid isolation of single microbial cells from mixed natural and laboratory populations with the aid of a micromanipulator. *Syst. Appl. Microbiol.* **22:**249–257.

Fu, A. Y., H. P. Chou, C. Spence, F. H. Arnold, and S. R. Quake. 2002. An integrated microfabricated cell sorter. *Anal. Chem.* **74:**2451–2457.

Fu, A. Y., C. Spence, A. Scherer, F. H. Arnold, and S. R. Quake. 1999. A microfabricated fluorescence-activated cell sorter. *Nat. Biotechnol.* **17:**1109–1111.

Gao, H. C., Z. M. K. Yang, T. J. Gentry, L. Y. Wu, C. W. Schadt, and J. Z. Zhou. 2007. Microarray-based analysis of microbial community RNAs by whole-community RNA amplification. *Appl. Environ. Microbiol.* **73:**563–571.

Goldberg, S. M. D., J. Johnson, D. Busam, T. Feldblyum, S. Ferriera, R. Friedman, A. Halpern, H. Khouri, S. A. Kravitz, F. M. Lauro, K. Li, Y. H. Rogers, R. Strausberg, G. Sutton, L. Tallon, T. Thomas, E. Venter, M. Frazier, and J. C. Venter. 2006. A Sanger/pyrosequencing hybrid approach for the generation of high-quality draft assemblies of marine microbial genomes. *Proc. Natl. Acad. Sci. USA* **103:**11240–11245.

Goldenfeld, N., and C. Woese. 2007. Biology's next revolution. *Nature* **445:**369–369.

Gonzalez, J. M., M. C. Portillo, and C. Saiz-Jimenez. 2005. Multiple displacement amplification as a pre-polymerase chain reaction (pre-PCR) to process difficult to amplify samples and low copy number sequences from natural environments. *Environ. Microbiol.* **7:**1024–1028.

Green, B. D., and M. Keller. 2006. Capturing the uncultivated majority. *Curr. Opin. Biotech.* **17:**236–240.

He, Z., T. J. Gentry, C. W. Schadt, L. Wu, J. Liebich, S. C. Chong, W. M. Wu, B. Gu, P. Jardine, C. S. Criddle, and J.-Z. Zhou. 2007. GeoChip: a novel comprehensive microarray for investigating biogeochemical and environmental processes. *Int. Soc. Microb. Ecol. J.* **1:**67–77.

Hong, J. W., V. Studer, G. Hang, W. F. Anderson, and S. R. Quake. 2004. A nanoliter-scale nucleic acid processor with parallel architecture. *Nat. Biotechnol.* **22:**435–439.

Huber, R., S. Burggraf, T. Mayer, S. M. Barns, P. Rossnagel, and K. O. Stetter. 1995. Isolation of a hyperthermophilic archaeum predicted by in-situ RNA analysis. *Nature* **376:**57–58.

Hutchison, C. A., H. O. Smith, C. Pfannkoch, and J. C. Venter. 2005. Cell-free cloning using phi 29 DNA polymerase. *Proc. Natl. Acad. Sci. USA* **102:**17332–17336.

Hutchison, C. A., and J. C. Venter. 2006. Single-cell genomics. *Nat. Biotechnol.* **24:**657–658.

Ishey, T., T. Kvist, P. Westermann, and B. K. Ahring. 2006. An improved method for single cell isolation of prokaryotes from meso-, thermo- and hyperthermophilic environments using micromanipulation. *Appl. Microbiol. Biotechnol.* **69:**510–514.

Kalyuzhnaya, M. G., R. Zabinsky, S. Bowerman, D. R. Baker, M. E. Lidstrom, and L. Chistoserdova. 2006. Fluorescence in situ hybridization-flow cytometry-cell sorting-based method for separation and enrichment of type I and type II methanotroph populations. *Appl. Environ. Microbiol.* **72:**4293–4301.

Keller, M., and K. Zengler. 2004. Tapping into microbial diversity. *Nat. Rev. Microbiol.* **2:**141–150

Kvist, T., B. K. Ahring, R. S. Lasken, and P. Westermann. 2007. Specific single-cell isolation and genomic amplification of uncultured microorganisms. *Appl. Microbiol. Biotechnol.* **74:**926–935.

Marcus, J. S., W. F. Anderson, and S. R. Quake. 2006. Microfluidic single-cell mRNA isolation and analysis. *Anal. Chem.* **78:**3084–3089.

Margulies, M., M. Egholm, W. E. Altman, S. Attiya, J. S. Bader, L. A. Bemben, J. Berka, M. S. Braverman, Y. J. Chen, Z. T. Chen, S. B. Dewell, L. Du, J. M. Fierro, X. V. Gomes, B. C. Godwin, W. He, S. Helgesen, C. H. Ho, G. P. Irzyk, S. C. Jando, M. L. I. Alenquer, T. P. Jarvie, K. B. Jirage, J. B. Kim, J. R. Knight, J. R. Lanza, J. H. Leamon, S. M. Lefkowitz, M. Lei, J. Li, K. L. Lohman, H. Lu, V. B. Makhijani, K. E. McDade, M. P. McKenna, E. W. Myers, E. Nickerson, J. R. Nobile, R. Plant, B. P. Puc, M. T. Ronan, G. T. Roth, G. J. Sarkis, J. F. Simons, J. W. Simpson, M. Srinivasan, K. R. Tartaro, A. Tomasz, K. A. Vogt, G. A. Volkmer, S. H. Wang, Y. Wang, M. P. Weiner, P. G. Yu, R. F. Begley, and J. M. Rothberg. 2005. Genome sequencing in microfabricated high-density picolitre reactors. *Nature* **437:**37–380.

Nesbo, C. L., M. Dlutek, and W. F. Doolittle. 2006. Recombination in thermotoga: implications for species concepts and biogeography. *Genetics* **172:**759–769.

Olsen, G. J., D. J. Lane, S. J. Giovannoni, N. R. Pace, and D. A. Stahl. 1986. Microbial ecology and evolution—a ribosomal-RNA approach. *Annu. Rev. Microbiol.* **40:**337–365.

Ottesen, E. A., J. W. Hong, S. R. Quake, and J. R. Leadbetter. 2006. Microfluidic digital PCR enables multigene analysis of individual environmental bacteria. *Science* **314:**1464–1467.

Pace, N. R., and T. L. Marsh. 1985. RNA catalysis and the origin of life. *Origins Life Evol. B.* **16:**97–116.

Papadopoulos, D., D. Schneider, J. Meier-Eiss, W. Arber, R. E. Lenski, and M. Blot. 1999. Genomic evolution during a 10,000-generation experiment with bacteria. *Proc. Natl. Acad. Sci. USA* **96:**3807–3812.

Pinard, R., A. de Winter, G. J. Sarkis, M. B. Gerstein, K. R. Tartaro, R. N. Plant, M. Egholm, J. M. Rothberg, and J. H. Leamon. 2006. Assessment of whole genome amplification-induced bias through high-throughput, massively parallel whole genome sequencing. *BMC Genomics* **7:**21.

Podar, M., C. B. Abulencia, M. Walcher, D. Hutchison, K. Zengler, J. A. Garcia, T. Holland, D. Cotton, L. Hauser, and M. Keller. 2007. Targeted access to the genomes of low-abundance organisms in complex microbial communities. *Appl. Environ. Microbiol.* **73:**3205–3214.

Raghunathan, A., H. R. Ferguson, C. J. Bornarth, W. M. Song, M. Driscoll, and R. S. Lasken.

2005. Genomic DNA amplification from a single bacterium. *Appl. Environ. Microbiol.* **71:**3342–3347.

Ronaghi, M., S. Karamohamed, B. Pettersson, M. Uhlen, and P. Nyren. 1996. Real-time DNA sequencing using detection of pyrophosphate release. *Anal. Biochem.* **242:**84–89.

Ronaghi, M., M. Uhlen, and P. Nyren. 1998. A sequencing method based on real-time pyrophosphate. *Science* **281:**363–365.

Sekar, R., B. M. Fuchs, R. Amann, and J. Pernthaler. 2004. Flow sorting of marine bacterioplankton after fluorescence in situ hybridization. *Appl. Environ. Microbiol.* **70:**6210–6219.

Staley, J. T. 2006. The bacterial species dilemma and the genomic-phylogenetic species concept. *Philos. T. Roy. Soc. B.* **361:**1899–1909.

Stein, J. L., T. L. Marsh, K. Y. Wu, H. Shizuya, and E. F. DeLong. 1996. Characterization of uncultivated prokaryotes: isolation and analysis of a 40-kilobase-pair genome fragment front a planktonic marine archaeon. *J. Bacteriol.* **178:**591–599.

Telenius, H., N. P. Carter, C. E. Bebb, M. Nordenskjöld, B. A. Ponder, and A. Tunnacliffe. 1992. Degenerate oligonucleotide-primed PCR: general amplification of target DNA by a single degenerate primer. *Genomics* **13:**718–725.

Thomsen, T. R., J. L. Nielsen, N. B. Ramsing, and P. H. Nielsen. 2004. Micromanipulation and further identification of FISH-labelled microcolonies of a dominant denitrifying bacterium in activated sludge. *Environ. Microbiol.* **6:**470–479.

Tringe, S. G., C. von Mering, A. Kobayashi, A. A. Salamov, K. Chen, H. W. Chang, M. Podar, J. M. Short, E. J. Mathur, J. C. Detter, P. Bork, P. Hugenholtz, and E. M. Rubin. 2005. Comparative metagenomics of microbial communities. *Science* **308:**554–557.

Tyson, G. W., J. Chapman, P. Hugenholtz, E. E. Allen, R. J. Ram, P. M. Richardson, V. V. Solovyev, E. M. Rubin, D. S. Rokhsar, and J. F. Banfield. 2004. Community structure and metabolism through reconstruction of microbial genomes from the environment. *Nature* **428:**37–43.

Vives-Rego, J., P. Lebaron, and G. Nebe-von Caron. 2000. Current and future applications of flow cytometry in aquatic microbiology. *FEMS Microbiol. Rev.* **24:**429–448.

Whitaker, R. J., and J. F. Banfield. 2006. Population genomics in natural microbial communities. *Trends Ecol. Evol.* **21:**508–516

Whitaker, R. J., D. W. Grogan, and J. W. Taylor. 2003. Geographic barriers isolate endemic populations of hyperthermophilic archaea. *Science* **301:**976–978.

Wu, L., X. Liu, C. W. Schadt, and J. Zhou. 2006. Microarray-based analysis of subpicogram quantities of microbial community DNA. *Appl. Environ. Microbiol.* **72:**4931–4941.

Zhang, K., A. C. Martiny, N. B. Reppas, K. W. Barry, J. Malek, S. W. Chisholm, and G. M. Church. 2006. Sequencing genomes from single cells by polymerase cloning. *Nat. Biotechnol.* **24:**680–686.

Zhang, L., X. F. Cui, K. Schmitt, R. Hubert, W. Navidi, and N. Arnheim. 1992. Whole genome amplification from a single cell - implications for genetic-analysis. *Proc. Natl. Acad. Sci. USA* **89:**5847–5851.

HOW MANY GENES DOES A CELL NEED?

Hamilton O. Smith, John I. Glass, Clyde A. Hutchison III, and J. Craig Venter

16

Ever since the 1930s when a handful of physicists, chemists, and biologists banded together into the phage school led by Max Delbrück, there have been efforts to understand life at its simplest and most fundamental level. In the ensuing decades up to the present, we have achieved a complete understanding of the genetic and chemical structure of a number of viruses and know in many cases the role of all of their genes. The same cannot be said for cells. The simplest cells are bacteria, but they generally contain thousands of genes. In 1995 the first complete sequence of a bacterial cell was determined. Since then hundreds of bacteria, archaea, and eukaryotes have been sequenced. It is now possible to ask the fundamental questions: How many genes does a cell need? To what extent can one reduce the size of a cell's genome by consecutive deletions and still have a viable cell? What is the minimal set of genes or functions necessary to sustain the cell under ideal laboratory conditions? These are the fundamental questions we address in this chapter.

Hamilton O. Smith, John I. Glass, Clyde A. Hutchison III, and J. Craig Venter, The J. Craig Venter Institute, 9704 Medical Center Drive, Rockville, MD 20850.

DEFINING A MINIMAL CELL

A cell can be defined as minimal under ideal laboratory conditions if (i) it can be grown in pure (axenic) culture and (ii) each and every one of its genes is essential. This is a straightforward and apparently rigorous definition. However, consider the example of a cell that has 300 essential genes plus gene X and gene Y. Genes X and Y may be redundant for the same essential function, or they could have different functions. In the first case, if we knock out X, the cell is still viable. The same is true if we knock out Y. However, if we knock out both X and Y, the cell no longer grows because the essential function is missing. This leaves us with two solutions to the minimal cell: 300 + gene X or 300 + gene Y.

In the second case, genes X and Y have different functions. Let us suppose that each is required for optimal growth. If we knock out gene X, the cell grows slowly but survives, and the gene is said to be dispensable. The same is true for gene Y. However, if we knock out both X and Y simultaneously, the cell no longer grows within a reasonable period of observation, and again we are left with two solutions to the minimal cell: 300 + gene X or 300 + gene Y. From an operational standpoint, if a cell with a disrupted gene grows too slowly, it

Accessing Uncultivated Microorganisms: from the Environment to Organisms and Genomes and Back
Edited by Karsten Zengler © 2008 ASM Press, Washington, DC

appears to be nonviable during the period of observation, and the gene is classed as essential. However, if one is patient and extends the period of observation, growth could be observed, and the gene is then called nonessential. If one requires reasonable growth rates, our minimal cell is likely to need more genes than for a very feeble cell. In fact, if we insist on a growth rate similar to that of the original cell from which the minimal cell is derived by genome reduction, then any genes that decrease the overall fitness or growth rate would be listed as essential. From the foregoing, it appears that the exact meanings of "essential" and "minimal cell" must be considered somewhat unclear.

Minimal cells probably do not exist in nature because natural environments tend to be adverse and variable. Organisms in their natural habitats must adapt to changing environments or nutrient sources in order to survive. This requires that they carry genes for necessary adaptive regulatory shifts and stress responses, for DNA repair, for membrane transport, for metabolic pathways to allow use of a variety of compounds and energy sources, and for biosynthetic capacity to make essential compounds missing in the environment. The hardiest organisms, for example, *Escherichia coli* (Kolisnychenko et al., 2002), tend to have large genomes with many redundant functions and a variety of biosynthetic pathways. In the laboratory under favorable and constant conditions, these organisms could presumably dispense with a large fraction of their genes. In the most extreme case, it might be possible to reduce their genomes to just the core functions for running the cell.

The minimal set of genes clearly depends on the environment. For example, a minimal gene complement for a poor medium lacking in amino acids will be larger than the minimal gene complement for a rich medium that supplies these metabolites. We are primarily interested in determining a minimal set of genes that can sustain a cell under ideal conditions in which the medium contains all of the necessary factors and small molecules and in which adverse conditions such as excessive heat, cold, ultraviolet radiation, etc., are avoided.

There is probably no unique set of minimal genes. For example, there are several choices for carbon source, energy metabolism, or envelope structure and cell architecture. It is sufficient, however, to determine one minimal set. We are particularly interested in determining the core set of cellular functions that constitute the basic machinery of the cell. These functions include such basic processes as DNA replication, RNA transcription, protein translation and folding, nucleotide synthesis, membrane transport, energy metabolism, cell division, and cell envelope synthesis. Presumably a minimal cell contains the core functions and little else.

When we talk about a minimal set of genes, it should be kept in mind that we implicitly mean a minimal set of functions. Ideally one gene equals one function. However, gene fusions occur during evolution, resulting in multimodular proteins in which two or more independent proteins become connected, so it is quite common for a gene to produce a protein with more than one function. For example, the *E. coli* protein (gi1787250) has histidinol-phosphatase activity in its N-terminal domain and imidazoleglycerol-phosphate dehydralase activity in its C-terminal domain. In *Clostridium acetobutylicum* each of these activities is on separate proteins (Serres and Riley, 2005). Enzymes may also acquire relaxed specificity for their substrates so that a single enzyme can exhibit two or more related activities that are usually carried out by two separate enzymes. A case in point is the malate/lactate dehydrogenase in *Mycoplasma genitalium* (Cordwell et al., 1997). An arginine residue was found in the active site of lactate dehydrogenase, which is different from the hydrophobic residue usually found. This is thought to allow the enzyme to act on both lactate and malate. Because related functions often tend to be joined together into a single protein during evolution, and because some proteins can acquire relaxed specificity, the minimal set of genes will generally be smaller than the minimal set of functions.

WHERE DOES ONE START?

To determine a minimal set of genes, we need a starting point. We are not currently knowledgeable enough to sit with a table of genes and design a minimal cell from scratch (Gil et al., 2004). We need to find a natural bacterium with a very small set of genes. This naturally existing approximation to a "minimal cell" is likely to be found in an environment that is nutrient rich and relatively invariant. Thus one might expect to find good candidates for minimal cells among symbiotic or parasitic animal pathogens. In 1980, a mycoplasma species was discovered in the urethra of a man diagnosed with nongonococcal urethritis (Tully et al., 1981) and was successfully cultured in highly nutritious SP4 medium (Tully et al., 1979). It was immunologically distinct from other known mycoplasmas and was named *M. genitalium* strain G-37 (Tully et al., 1983).

It is a tiny flask-shaped cell with a tiplike attachment organelle. It has a cell diameter of around 300 nm and is surrounded by a cytoplasmic membrane. There is not a cell wall. It contains a single circular chromosome, 580 kb in size. It was the second bacterial genome ever to be sequenced (Fraser et al., 1995). Sequencing has become more accurate in the past 10 years since the appearance of automated capillary sequencers so the genome was resequenced in 2005. Errors were found at 34 sites. Careful annotation of the resequenced genome showed 482 protein-coding genes and 43 RNA genes. Recently 3 additional small protein-coding genes have been found, giving a revised total of 485 protein-coding genes. This highlights the difficulty of accurately determining all of the genes, particularly the small ones. *M. genitalium* has no insertion sequences and only a few repeated DNA sequences (Peterson et al., 1995). Unlike many bacteria, only 6% of its genes are in paralogous families (paralogous genes are homologous genes arising by gene duplication within an organism) (Glass et al., 2006). Overall, it has a very compact, gene-rich genome. Its complement of genes is the smallest known for any cellular organism capable of independent growth, and this establishes an upper bound for the minimal set.

M. genitalium belongs to the class *Mollicutes* (more generally called the mycoplasmas). These wall-less bacteria are thought to have evolved from more typical gram-positive bacteria in the *Firmicutes* taxon by undergoing massive genome reduction (Maniloff, 2002). *M. genitalium* lives symbiotically in its natural environment attached closely to epithelial cells of the urethral tract and probably requires most of its genes for survival.

The primary aim of this chapter is to review, discuss, and propose studies aimed at determining how many *M. genitalium* genes are really necessary when cells are grown under ideal laboratory conditions. This question has been approached in two ways: (i) by comparative genomics and ii) by exhaustive transposon mutagenesis. As we shall see in the following sections, not all of its protein-coding genes are necessary—perhaps 100 or more can be dispensed with—but all of its RNA genes appear to be essential. It is likely that we will not know a true minimal gene set until a minimal cell can be synthesized and tested in the laboratory.

COMPARATIVE GENOMICS APPROACH TO DETERMINING A MINIMAL SET OF GENES FOR *M. GENITALIUM*

The publication of the *Haemophilus influenzae* Rd (Fleischman et al., 1995) and *M. genitalium* G-37 (Fraser et al., 1995) genome sequences in 1995 led to the first attempt to determine a minimal set of genes. Mushegian and Koonin at the National Center for Biotechnology Information reasoned that since the two bacteria must both contain the minimal set required for a living cell, a comparison of the genes of the two bacteria would produce that set (Mushegian and Koonin, 1996). They did indeed find 240 genes in *M. genitalium* with orthologous counterparts in *H. influenzae* Rd. Orthologous genes are by definition related by vertical descent from a common ancestor, have the same functions, and have protein products that align through most of their length with higher similarity than with any other proteins in the respective organisms. They hypothesized that these were essential genes since they were conserved in these two bacteria for an estimated 1.5 billion years of evolution since their last common ancestor. So a minimal set of genes could number as few as 240. However, on inspection, it became apparent that enzymes for several intermediate steps in essential pathways were missing.

To account for these missing essential functions, they invoked the phenomenon of nonorthologous gene displacement (NOD) (Koonin et al., 1996). NODs are paralogous or unrelated genes that supply the missing functions. It appears that over millions to billions of years of evolution, a gene of an unrelated sequence or a member of a paralogous gene family can evolve to supply the function, and the original gene is eventually lost. Nonorthologous displacements have turned out to be a common feature in comparing two or more organisms and add a layer of complexity to comparative analyses (Galperin and Koonin, 1998; Pollack et al., 2002). In fact, the larger the set of bacteria that one compares, the smaller the number of orthologs becomes and the greater the number of NODs becomes (Koonin, 2000).

Returning to *M. genitalium,* one example of a nonorthologous gene displacement is the enzyme nucleoside diphosphate kinase. The gene for nucleoside diphosphate kinase is present in *H. influenzae*, but not in *M. genitalium*. However, there are two genes, MG264 and MG268, in *M. genitalium* that contain sequence motifs for nucleoside and nucleotide kinases, and these are candidates for nonorthologous displacement of the missing ortholog (Mushegian and Koonin, 1996). Several other genes have been found in *M. genitalium* that are possible NODs for other missing orthologs. Mushegian and Koonin postulated that 22 NODs were required to provide missing essential enzymes. Thus their minimal set was 240 + 22 = 262 (Fig. 1). They further refined this set by removing 6 genes from the original 240 orthologs that appeared to be "functionally redundant or parasite-specific" (Mushegian and Koonin, 1996), leaving a total of 256 genes in their minimal set (Table 1). Inspection of the functions of the 256 genes reveals that most of them are ones that would be anticipated to be essential. The question is whether an actual cell with just these genes plus essential RNA genes would be viable. One would have to build such a cell to test the hypothesis.

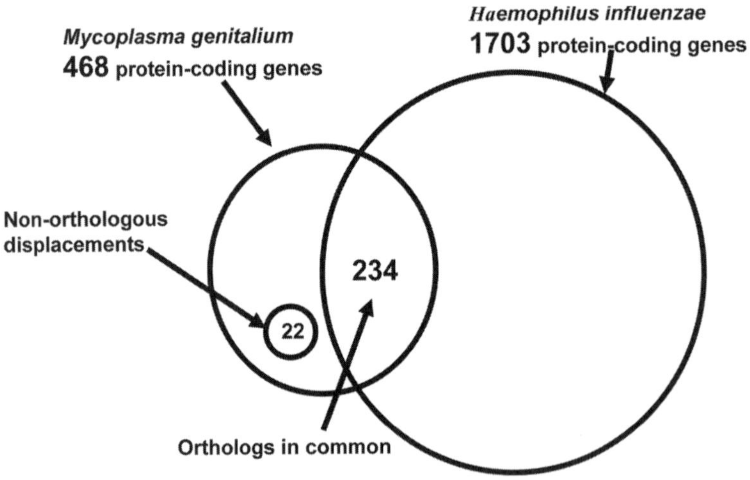

FIGURE 1 Comparison of orthologous genes in *M. genitalium* and *H. influenzae* Rd (Mushigian and Koonin, 1996). The total number of protein-coding genes in each bacterium is based on the original 1995 annotations, which have been modified several times in the intervening years.

TABLE 1 Minimal protein-coding gene set defined by comparing *M. genitalium* and *H. influenzae* Rd (Mushegian and Koonin, 1996)

Function	No. of genes
Translation	95
Replication	18
Transcription	9
Recombination and repair	8
Chaperone functions	13
Nucleotide metabolism	23
Amino acid metabolism	7
Lipid metabolism	6
Energy	34
Coenzyme metabolism and utilization	8
Exopolysaccharides	8
Uptake of inorganic ions	5
Secretion, receptors	5
Other conserved proteins	18
Total protein coding genes	256

One criticism of the above analysis is that *H. influenzae* and *M. genitalium* are structurally very different, the former being gram-negative with a cell envelope consisting of inner and outer membranes and a peptidoglycan cell wall, and the latter possessing only a cholesterol-containing cytoplasmic membrane. *M. genitalium* may require certain genes with no counterparts in *H. influenzae* and vice versa. The probable existence of mutually exclusive sets of genes coupled with the long period of divergence from the last common ancestor, which increases the likelihood of extensive nonorthologous gene displacements, weakens the comparative approach. Furthermore, many NODs are hard to detect because they are not involved in well-characterized pathways.

Perhaps comparing 13 sequenced mycoplasmas (Fig. 2) could be more informative. These are similar organisms that diverged from their last common ancestor only a few hundred million years ago. There are 173 orthologous genes common to all 13. However, if one leaves out *Phytoplasma asteris,* which is an obligate

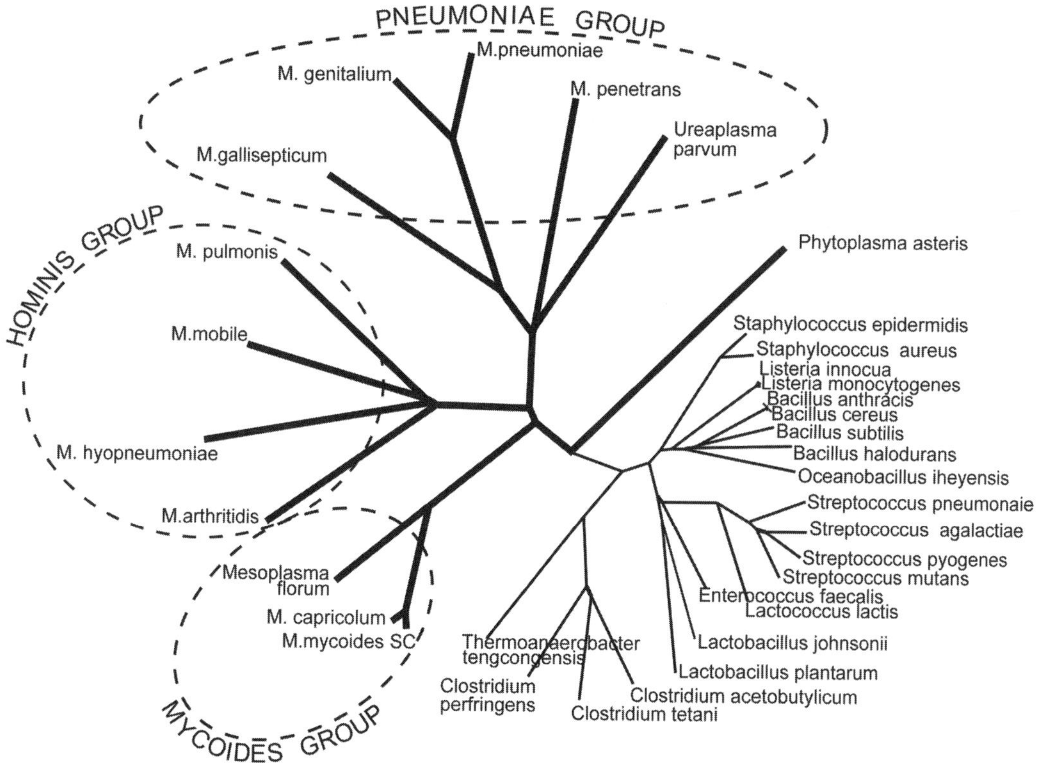

FIGURE 2 Evolutionary tree showing the relatedness of 13 mycoplasmas. *P. asteris* is an obligate intracellular organism and is thus distinct from others.

intracellular organism, the number of orthologs jumps to 220. To this, one must add an estimated 90 NODs. Thus the minimal set derived by comparing the 12 mycoplasmas is estimated to be at least 310 genes (J. Glass, unpublished data).

Generally, these genomic comparisons are not satisfying and do not give one much confidence in the validity of the results. There are simply too many gaps in our knowledge of the genetic functions in bacterial cells. Nearly a quarter of the genes in *M. genitalium* have unknown functions. And in comparing a large number of bacterial genomes, the number of orthologs common to all drops below 80 genes because of the NOD problem (Koonin, 2000). Experiments seem to offer the only route to an accurate answer.

EXPERIMENTAL DETERMINATION OF A MINIMAL SET OF GENES

The above comparative studies suggest that only a subset of the genes in *M. genitalium* may be required for growth in the laboratory. To test for dispensable genes, one needs to be able to knock out or remove genes in some systematic fashion. Individual genes or in some cases multiple genes can be targeted for disruption or deletion using in vivo recombination methods (Dhandayuthapani et al., 1999; Burgos et al., 2006), although these methods are probably of limited applicability in *M. genitalium*. Another approach is random mutagenesis using frameshift mutagens, or random transposon insertions. The latter is the most useful because one can determine the precise insertion site of the transposon by sequencing across the transposon/chromosome junction using primers originating near the transposon termini.

Transposons are widely used for global genome mutagenesis. A transposon, bearing a selectable antibiotic resistance marker, is introduced into a cell on a nonreplicating plasmid by transformation or electroporation. It then mobilizes and inserts into more or less random locations depending on the sequence specificity requirements of the particular element (Fig. 3). If it inserts into and disrupts a gene, and the cell survives, then the gene is probably nonessential.

The 1999 Global Transposon Mutagenesis Study

In 1999, Hutchison and colleagues carried out global transposon mutagenesis of both *M. genitalium* (580 kb) and *Mycoplasma pneumoniae* (816 kb) to determine which genes were nonessential (Hutchison et al., 1999). *M. pneumoniae* is the closest known relative of *M. genitalium* and contains orthologs to nearly every one of the 485 genes in *M. genitalium* plus an additional 197 genes. They argued that the 485 genes in common would contain all of the essential genes and the 197 extra genes would be largely dispensable in *M. pneumoniae* when grown in SP4 medium in the laboratory. Transposon Tn*4001*, carrying gentamycin (Gm) resistance, was introduced into *M. genitalium* by electroporation and the cells were grown in SP4 medium plus Gm for 2 to 4 weeks. During this time, cells with transposon insertions increased in number by nearly a billion-fold. Genomic DNA, which was now predominantly from cells containing transposons, was isolated, digested with DraI restriction enzyme, which cuts the mycoplasma DNA frequently but does not cut Tn*4001*, and the fragments were circularized by

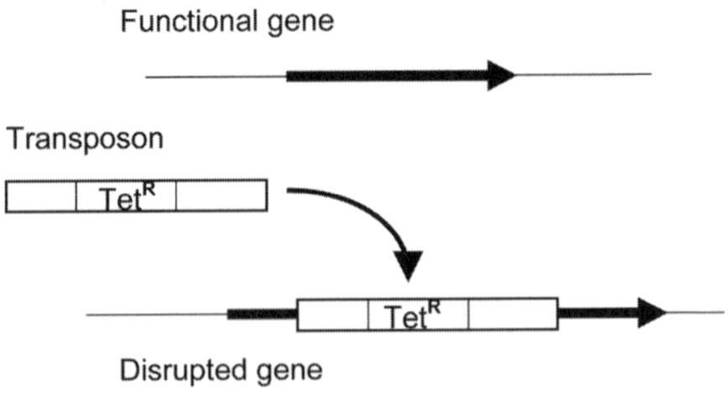

FIGURE 3 Diagram illustrating transposon mutagenesis. The transposon carries a transposase enzyme, an antibiotic resistance gene (Tet^R), and terminal sequences recognized by the transposase during transposition. The transposon is generally carried on a plasmid that does not replicate in the cell to be mutagenized. Antibiotic resistance is acquired by the cells only when the transposon jumps into the cell chromosome. A gene is disrupted when the transposon lands inside a gene.

ligation. Transposon junctions were amplified by inverse PCR using primers specific for the Tn*4001* ends. After cloning of the amplified DNA using EcoRI and HindIII sites located in the primers, the junctions were sequenced.

Transposon insertions were found in 140 genes in *M. genitalium* and 179 genes in *M. pneumoniae* (Table 2). However, not every insertion necessarily inactivates a gene. Tn*4001* has an outwardly directed promoter that could transcribe DNA downstream of the transposon-chromosome junction, resulting in a gene product if a translational start site is near the junction. Insertions within the 9 bp duplicated by Tn*4001* when it inserts were considered nondisruptive. In addition, insertions within the last 20% of the 3′ end of the gene were considered nondisruptive because the COOH-terminal region of a protein is often not necessary for function. Genes deemed to be disrupted were thus reduced to 93 in *M. genitalium* and 150 genes in *M. pneumoniae*. One of the more interesting results was the extensive disruption of genes in the species-specific portion of *M. pneumoniae*. Of the 150 genes disrupted, 93 were among the 197 species-specific genes and only 57 were in the remaining 485 *M. genitalium* orthologs. This strongly supports the notion that the 485 orthologs comprise the majority of the essential genes.

The study had several shortcomings. Mutants were selected and processed in batches. There was no attempt to isolate individual mutants to confirm their validity. Furthermore, the study was not exhaustive. Thus one could only estimate the total number of dispensable genes by extrapolation. Orthologs that were disrupted in *M. pneumoniae* but not in *M. genitalium* were included among the final count of dispensable genes. This makes sense, but one could be misled if the extra genes in *M. pneumoniae* result in substantial functional redundancy. If two genes have the same essential function, each can be individually disrupted. Only if both are simultaneously knocked out does one discover that the function they encode is essential. It is possible that this could explain some of the unexpected results. The authors concluded, "proof of dispensability of any specific gene requires cloning and detailed characterization of a pure population carrying the disrupted gene."

TABLE 2 Summary of global transposon mutagenesis insertion site data

Parameter	*M. genitalium* 1999	*M. pneumoniae* 1999	*M. genitalium* 2006
Junctions sequenced	1,291	918	2,462
Distinct sites	685	669	600
Intergenic sites	199	261	96
Sites in genes	484	408	504
Different genes	140	179	107
"Disrupted" genes	93	150	100

The 2006 Global Transposon Mutagenesis Study

In follow-up work to address some of the problems in the 1999 study, saturation transposon mutagensis of *M. genitalium* was carried out and individual clones were isolated and examined (Glass et al., 2006). Tn*4001*tetM was introduced into *M. genitalium* cells by electroporation and the cells were plated. Individual colonies were picked from tetracycline plates and the transposon junctions were sequenced. Only 62% of these colonies produced useful sequence. This was because of the tendency of mycoplasma cells to clump into aggregates, leading to colonies with mixtures of mutants. Subcolonies, obtained by replating dispersed colonies, yielded useful sequence at a rate of 82%. Fifty-nine percent of the subcolonies yielded insertions at positions different from those of the primary colony. In total, 3,321 colonies and subcolonies were sequenced. A total of 2,462 transposon insertion sites were identified and mapped onto the genome. Of these, 600 were at distinctly different sites (Table 2). Multiple insertions in "hot spots" accounted for the remainder. A total of 504 (84%) of the insertions were in protein-coding genes and 96 were intergenic (Fig. 4). None

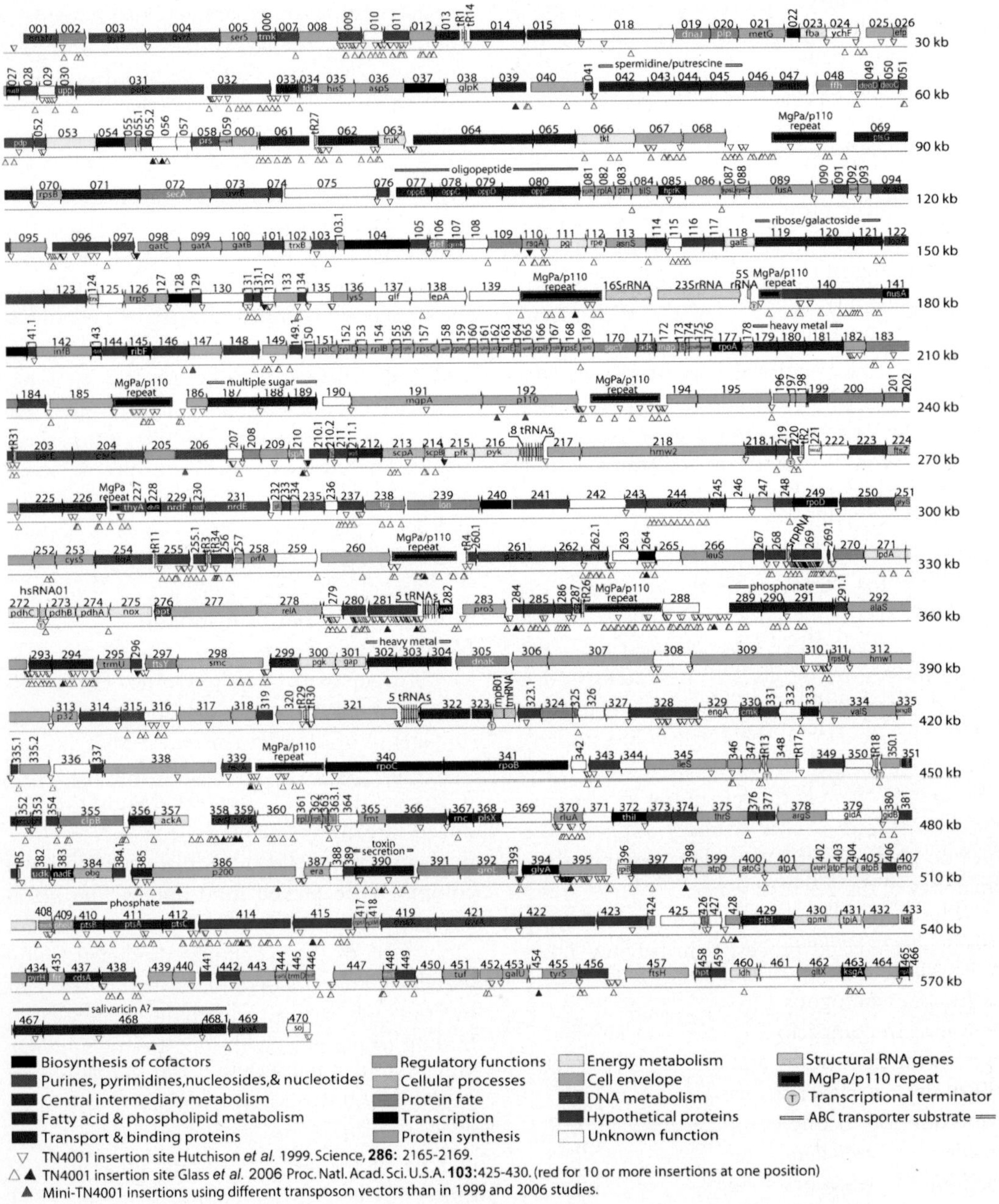

FIGURE 4 *M. genitalium* genome map showing the location of genes and transposon insertion sites.

were in known RNA genes. Using the same rules as in the 1999 study for categorizing insertions as disruptive, 100 genes were classed as disrupted, nonessential genes. The number of new genes hit changed only slightly as the number of distinct sites approached 600, suggesting that most of the dispensable genes had been hit.

The experimentally derived minimal gene set for *M. genitalium* (385) is substantially larger than that arrived at in the two comparative genomics studies (256 and 310). However, uncertainty remains as to the validity of the 385-gene set. The set was determined by single, one-at-a-time, gene knockouts. It seems unlikely that one could simultaneously remove the entire set of 100 dispensable genes. A number of the single-knockout clones showed phenotypic changes such as loss of adherence (MG185), slower growth rates, slightly faster growth rates (MG414, MG415, MG460), growth in chains of clumped cells (MG066), or membrane fragility (MG066), suggesting that many of the knockouts have slightly lower fitness. If one were to attempt to make a strain simultaneously lacking several of these genes, it might not be viable. Do we define a minimal cell as one that is robust and with near normal overall phenotype, or can it be feeble and slow growing? If the latter, then we can say that the essential core machinery to maintain the cell is there, but other genes that contribute to overall fitness under ideal laboratory conditions are not there. Furthermore, there may be certain essential sequences in the genome, which have not been annotated simply because annotation is still not an exact science. Some essential small genes, nested genes, small RNA genes, or regulatory signals may be present that are difficult to discover. We only recover transposon knockouts when they are in nonessential sequences. Thus, we would not detect such sequences.

In the 1999 study 130 genes were counted as disrupted in *M. genitalium* if one includes both the 93 actual disruptions in *M. genitalium* and the 37 orthologous genes disrupted in *M. pneumoniae.* In comparing the two sets, i.e., the 100 from the 2006 study and the 130 from the 1999 study, it is rather surprising that only 67 are in common. If one compares just the disruptions in *M. genitalium,* there are only 57; the remaining 10 were orthologous gene hits in *M. pneumoniae.* These results are perhaps not too surprising because transposon mutagenesis was not done exhaustively in the 1999 study. What is puzzling and hard to explain is that, in the 1999 study, 63 additional genes were disrupted. Of these, 37 were disrupted in *M. genitalium* and 26 were orthologs disrupted only in *M. pneumoniae.*

The Nonessential Genes of *M. genitalium*

The 100 *M. genitalium* protein-coding genes disrupted in the 2006 study are listed in Table 3 (this table also includes the putative essential genes that remain after removing the nonessential genes). Using the The Institute for Genomic Research (TIGR) functional classification scheme, the 100 genes can be broken into 15 groups based on their main roles. The two largest groups are hypothetical proteins with 30 members and proteins of unknown function with 13 members. The hypothetical proteins are homologous to proteins found in other species, i.e., they have been conserved in evolution but have no identified function. The proteins with unknown functions have known motifs and domains or can be assigned to protein families, but their specific functions are unknown. These two groups constitute 43% of the disrupted genes but only 27% of all genes in *M. genitalium.* This enrichment in the disrupted group is perhaps not surprising, but it is unexpected that there are 86 genes without functional assignments remaining in the essential gene list.

Among other functional groups, there are seven cell envelope proteins that are classed as putative lipoproteins or membrane proteins. The 35 cell envelope proteins remaining in the list of essential genes are presumably redundant for the functions of these 7 or are all that are needed. Most noteworthy is the small number of disrupted genes among those involved in protein synthesis and transcription. This is not unexpected since transcription and translation are critical functions for a cell and are among the most conserved processes in nature (Forster and Church, 2006).

Noteworthy is the finding that the phosphonate ABC transporter gene cluster (MG289–91) and the phosphate ABC transporter gene cluster (MG410–12) were both disrupted. The phoU (MG409) regulatory protein gene, however, was not disrupted and thus may be essential for both systems. Since phosphate transport across the membrane is an essential function, one must assume that either of these

TABLE 3 The essential and nonessential genes of *M. genitalium* are listed by main role and subroles using the TIGR functional classification system[a]

Main role	Subrole	Essential genes	Nonessential genes
DNA metabolism	DNA replication, recombination, repair	**001, 003, 004,** 007, 031, **073, 091, 094, 097, 122,** 199, **203, 204, 206,** 235, **250, 254, 261,** 419, **421, 469**	**244, 262.1,** 315, **339,** 352, **358, 359**
	Restriction/modification	184	438
	Chromosome-associated protein	**353**	213, 214, 298
	Degradation of DNA	186, **262**	**009**
Protein synthesis	Ribosomal proteins	055.1, **070, 081, 082, 087, 088, 090, 092, 093, 150–169, 174–176, 178, 197, 198,** 210.2, **232, 234, 257,** 311, **325, 361–363, 363.1, 417, 418, 424, 426, 444, 446, 466**	
	Translation factors	**026, 089, 142, 173, 196, 258, 433, 435, 451**	
	tRNA synthases	**005, 021, 035, 036,** 098–100, **113, 126, 136, 194, 195, 251, 253,266, 283, 292, 334, 345, 365, 375, 378, 455, 462**	
	tRNA/rRNA modification	**008,** 084, **182, 209, 295, 347,** 372, **445**	**012, 252, 346,** 370, **463**
	Other	**059, 083**	110
Transcription	RNA processing	**143, 465**	**367**
	RNA polymerase	022, **177, 340, 341**	
	Transcription factors	027, **054, 141, 249, 282**	
	RNA degradation	**104**	
Regulatory functions	DNA or protein interactions	**127,** 205	428
Transport and binding proteins	AAs, peptides, amines	**042–045, 077–080,** 225	226
	Anions	409	289–291, **410, 411,** 412
	Carbs, alcohols, acids	041, 069, 085, **119, 120,** 429	033, 062, 121
	Cations, iron	**071,** 179, **180,** 181, 302–304, **322,** 323	
	Unknown substrates	014, **015,** 064, **065, 187,** 188, 189, 467, 468, 468.1	061, 294, 390
Cofactors, prosthetic groups	Folic acid	**228, 394**	
	Pyridine nucleotides	037, 128, 240, **383**	
	Riboflavin, FMN, FAD	**145**	
	Pantothenate, coA		264
Cell envelope	Synthesis and degradradation of surface polysaccharides	025, **060,** 335.2, **453**	
	Other	068, 095, 133,135, 217, **247,** 277, 306, 307, 309, 313, 320, 321, 338, 348, 350.1, 395, 432, 439, 440, 443, 447, 464	040, 067, 147, 149, 185, 260, 452
	Surface structures	075, 191, 192, 218, 312, 317, 318, 386	
Cellular processes	Adaptations	**278**	
	Cell division	**224, 335, 384, 387, 457**	
Protein fate	Degradation	020, 046, 208, **239,** 324, **391**	183, **355**

TABLE 3 *Continued*

Main role	Subrole	Essential genes	Nonessential genes
	Secretion, trafficking	103.1, **048,** 055, **072, 170, 297**	**210**
	Folding, stabilization	**019,** 200, **201, 305, 392, 393**	002, **238**
	Modification and repair	**086, 106,** 109, **172, 270, 448**	**408**
Purines, pyrimidines	DNA nucleotide metabolism	**229–231**	**227**
	Interconversions	**006,** 034, **107, 171, 330, 434**	
	Salvage	**030, 049, 052, 058, 276, 382, 458**	051
Energy metabolism	Aerobic (NADH oxidase)	**275**	
	Anaerobic		**460**
	ATP synthase	**399–405**	**398**
	Electron transport	**102, 124**	
	Fermentation	**299, 357**	
	Glycolysis/gluconeogen	**023, 111, 215, 216, 300, 301, 407, 430, 431**	**063**
	Other	**038, 050**	
	Pentose phosphate path	396	**066, 112**
	Pyruvate dehydrogenase	**272–274**	**271**
	Sugars	**053, 118**	
Lipid metabolism	Biosynthesis	**114,** 211.1, **212, 287, 333,** 356, 368	039, **437**
	Degradation		**293,** 385
Central metabolism	1-carbon metabolism	**245**	
	Other	013, **047**	
	Phosphorus compounds	**351**	
Hypothetical proteins	Conserved	028, 055.2, 074, 076, 101, 105, 117, 123, 129, 141.1, 144, 146, 148, 202, 210.1, 211, 218.1, 219, 233, 241, 243, 260.1, 267, 269.1, 291.1, 296, 314, 319, 323.1, 331, 335.1, 337, 349, 354, 366, 373, 374, 376, 377, 381, 384.1, 389, **406,** 422, 423, 441, 442, 459	011, 032, 096, 103, 116, 131, 131.1, 134, 140, 149.1, 220, 237, 248, 255, 255.1, 256, **268,** 269, 280, 281, 284–286, 328, 343, 397, 414, 415, 449, 456
Unknown functions	Enzymes of unknown specificity	108, **125,** 137, 139, 190, **222,** 246, 263, **265,** 308, 310, 327, **336,** 344, 369, **425**	018, 207, **380**
	General	029, 057, 130, 132, **221,** 236, 242, 259, 326, 329, **332,** 342, 350, 364, 371, **379,** 388, 427, 450, 461, 470	010, **024, 056,** 115, **138,** 279, 288, 316, 360, 454

[a]The essential genes predicted by comparison of *M. genitalium* and *H. influenzae* Rd (Mushegian and Koonin, 1996) are indicated in bold type. (Note: MG420 is no longer in the gene list.)

transporters can import phosphate, perhaps because of relaxed specificity. This may be an example of duplicated functions such that either can be disrupted, but not both at the same time.

Seven DNA repair and recombination genes were disrupted, including *recA*, which is one of the most ubiquitous proteins found in nature. In addition, *recU* and two Holliday junction DNA helicases, *ruvA* and *ruvB*, were disrupted. Apparently, in the laboratory, recombination is not a critical function. But disruption of these genes decreases cell fitness and slows cell growth. Growth in the laboratory was only observed over short time spans. Recombination might be important for long-term survival. On the other hand, several repair genes are essential. These include the *uvrA*, *uvrB*, and *uvrC* genes for repair of ultraviolet radiation damage and the genes for apurinic endonuclease and uracil-DNA glycosylase.

The Essential Genes of *M. genitalium*

At least 385 genes are essential (Table 3), and the number may be substantially greater since no one has tested multiple simultaneous gene disruptions. Currently 86 genes classed as essential have no assigned functions. Determining the functions of these genes will be one of the most critical areas for further investigation if scientists are to fully understand the requirements for a minimal cell. The remaining essential genes have well or relatively well-documented functions, and most of these are involved in obviously vital cellular processes or structures.

The biggest single category is the 95 genes involved in protein synthesis. This includes 53 protein components of the small and large ribosomal subunits, 23 tRNA aminoacylation proteins, 8 tRNA or rRNA modification enzymes, and 9 translation factors involved in initiation, elongation, or release of the polypeptide chains. Although not protein-coding genes, one must also include in the protein synthesis group the 36 genes for the set of tRNAs; the 3 genes for the 5S, 16S, and 23S ribosomal RNAs; and the 4 genes for the small RNAs (srp01, hsRNA01, rnpB01, and tmRNA1). There are also 24 genes that play roles in protein fate. These include the chaperonins, kinases, signal recognition proteins, translocases, and peptidases.

Another large category is genes involved in cell envelope synthesis and structure. There are 35 genes in this group. Another large category is the 35 genes that specify components of ABC transporters and membrane-binding proteins. Some 25 genes are involved in DNA metabolism. This group includes genes for DNA replication, repair, topoisomerases, gyrases, and DNA modification. A total of 29 genes are involved in energy production, which includes a complete glycolytic pathway and all the subunits of the standard ATP synthase. There are seven genes for cofactor and prosthetic group synthesis, six for cellular processes, and four for central intermediary metabolism. Seven genes are involved in lipid metabolism. Sixteen genes are designated for purine and pyrimidine metabolism. Transcription includes five genes for subunits of the DNA-directed RNA polymerase, four for elongation and termination, two ribonucleases, and ribosome-binding factor A. Only two genes are listed for a main role in regulation.

It is interesting that of the 256 genes predicted by Mushegian and Koonin (1996) to be essential based on comparative analysis (listed in bold type in Table 3), 31 were disrupted by transposon mutagenesis and appear to be dispensable. The remaining 225 genes are in agreement with the assignments obtained by transposon mutagenesis.

Making a Minimal Cell

A minimal set of genes cannot be convincingly determined by either comparative genomics or global transposon mutagenesis. However, these studies do identify more than 200 genes for core functions such as DNA replication, transcription, protein translation, energy metabolism, transport, lipid metabolism, nucleotide metabolism, and protein fate, which one has confidence will be represented in the minimal cell. However, a number of apparently essential genes have been identified in *M. genitalium* that have not been assigned to these basic processes. For example, there is uncertainty regarding the many hypothetical genes and genes of unknown function, which appear to be essential. And the question remains as to how many putative nonessential genes can be eliminated simultaneously without decreasing cell fitness to the point of nonviability. These questions highlight the gaps in our knowledge.

An inescapable conclusion is that one must actually construct a minimal cell and demonstrate that every one of its genes is essential. There are three possible approaches to making a minimal cell: (i) cumulative inactivation of genes using mutagens that produce frameshifts, (ii) sequential genome reduction using recombineering (*recombi*nation-mediated genetic engi*neering*) methods, and (iii) chemical synthesis of a minimal genome and installation into a receptive cell cytoplasm.

INACTIVATION OF MULTIPLE GENES BY ACCUMULATION OF FRAMESHIFT MUTATIONS

Introducing frameshifts into multiple genes is not a tidy way to arrive at a minimal cell since the genes are inactivated but not removed. The genome is not changed in size, and one is left with a substantial amount of superfluous, inactive DNA. However, this approach could yield much useful information as to what combinations of genes are simultaneously dispensable (Peterson and Fraser, 2001).

Mutagens such as acridine orange, ethidium bromide, and proflavine that intercalate between adjacent base pairs in DNA can produce single-base additions and deletions. If a frameshift occurs within an open reading frame, the codons are shifted out of register and the downstream amino acids of the expressed protein will be changed. Not infrequently, one of the out-of-frame codons will become a nonsense triplet, thus producing premature chain termination. Frameshifts generally totally inactivate genes, and such mutations are not readily revertible.

If a bacterial culture is serially propagated for many generations in the presence of a frameshift mutagen, then multiple gene inactivations can occur, and after a prolonged period of growth with the mutagen, including several serial transfers, and further growth with the mutagen, surviving cells will accumulate mutations in nonessential genes, but not in essential genes. Survivors of a long course of mutagenesis can be sequenced to determine which combinations of gene inactivations can be tolerated. It would be a rare event to produce a cell in which the full set of nonessential genes is inactivated. A more likely outcome would be a number of different cells, each with some of its genes inactivated. The primary value of this approach would be to identify groups of genes that are simultaneously dispensable.

GENOME REDUCTION BY RECOMBINEERING

Some bacteria have active recombination systems that permit sequential deletion of genes. Several methods have been developed for use in *E. coli*. At least three of these methods use the phage lambda red system to accomplish the initial deletion. The inducible red *exo, beta,* and *gamma* genes that comprise the red system are introduced into the cell on a plasmid. The exo protein is a 5′-exonuclease that generates 3′ overhangs on linear double-stranded DNA that has been electroporated into *E. coli* cells. The beta protein binds to the 3′-single-stranded tails and mediates invasion and recombination with homologous chromosomal DNA. The gamma protein inhibits recBCD protein that would normally digest the incoming linear DNA.

Recombineering generally works in the following way (see, for example, Warming et al., 2005). A target region on the chromosome is selected for deletion. Flanking 50-bp sequences to either side of the region are the sites for the recombination process that will delete the region. A linear fragment carrying an antibiotic resistance gene is subjected to PCR with primers containing the 50-bp flanking sequences. This linear fragment is then transformed into the cell. The lambda red system carries out recombination in the 50-bp flanking homologies, thus deleting the chromosomal region and substituting the antibiotic marker. The second step accomplishes the removal of the marker gene, leaving a clean deletion (Fig. 5). The various methods differ primarily in how the marker is removed.

Blattner and colleagues carried out extensive reduction of *E. coli* by their method of "scarless" deletions (Kolisnychenko et al., 2002). They first replaced the region to be deleted by an antibiotic resistance marker, which is flanked by SceI meganuclease sites. The resistance marker is then excised by introducing a plasmid carrying the inducible *SceI* gene. RecA-mediated double-strand break recombination within flanking homologies

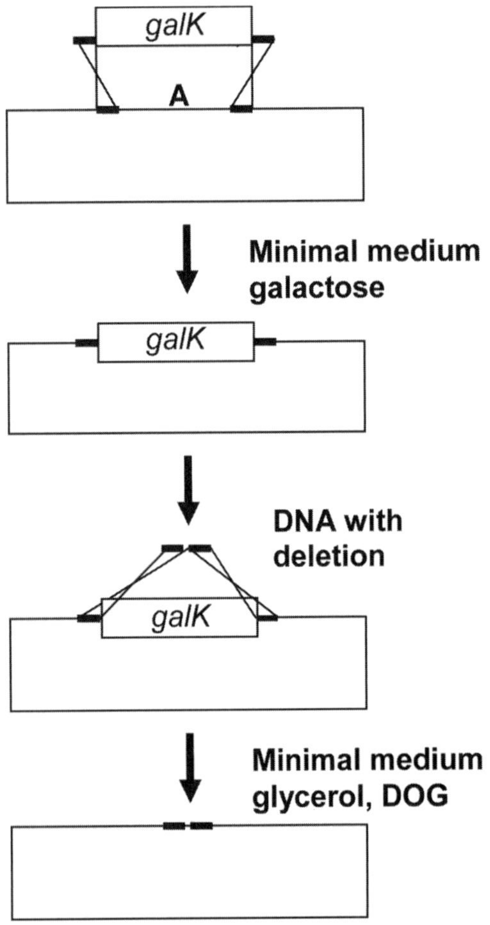

FIGURE 5 Deletion of a gene by recombineering using the lambda red system. Gene A is flanked by 50 bp of sequence on each side that are the targets for a linear piece of DNA that consists of an expressible *galK* gene flanked by the two 50-bp sequences. After introduction of the *galK* DNA fragment by electroporation, recombination occurs within the two 50-bp homologous segments, thus deleting gene A and replacing it with *galK*. Subsequent steps, as described in the text, counterselect for removal of the *galK* gene, leaving a clean ("scarless") deletion. The process can be repeated many times to remove other genes.

then restores the continuity of the genome. Details are given in Kolisnychenko et al. (2002). With this method, 11 regions were excised, resulting in 9.3% reduction in the gene count, removal of 24 of 44 transposable elements, and a removal of 8.7% of the genome. The resulting *E. coli* strain grows normally. This group recently achieved a 15% genome reduction of their *E. coli* strain while preserving good growth characteristics. In addition, they found unanticipated improvements in electroporation ability and stability of recombinant genes and plasmids when propagated in the reduced strain (Pósfai et al., 2006).

A Japanese group (Hashimoto et al., 2005) has accomplished an even more extensive reduction of *E. coli*. They used three markers, one to positively select for the initial deletion, and the other two to negatively select against the markers. They cleanly deleted a total of 16 segments of the genome, amounting to 1,377,172 bp, or 29.7% of the chromosome. The resulting cells, however, grew more slowly, exhibited changes in cell shape, and were altered in cell nucleoid organization.

An efficient bacterial artificial chromosome (BAC) recombineering system using *galK* positive/negative selection was recently described (Warming et al., 2005). The *galK* gene can replace any segment of a BAC insert, from a single base to longer regions (Fig. 5). This is accomplished using short flanking homologies, the lambda red system, and positive selection on galactose plates. The *galK* gene is then removed by negative selection on 2-deoxy-galactose (DOG) plates. The 2-deoxy-galactose is metabolized to 2-deoxy-galactose-1-phosphate, which accumulates and is toxic to the cells, so only cells that have eliminated the *galK* gene survive, resulting in the desired "clean" DNA deletion (or substitution). Reduction of the *E. coli* chromosome or reduction of any chromosome or chromosomal fragment that can be propagated in *E. coli* could be done using this method.

Single genes have been replaced with antibiotic markers using the recombination system of *M. genitalium* (J. Glass, personal communication; Dhandayuthapani et al., 1999), but this is very inefficient and no method has been developed for subsequent removal of the antibiotic marker. Thus, recombineering is of limited utility in *M. genitalium*. However, recombineering can be efficiently done in yeast. Any DNA that can be cloned in yeast can be genetically engineered using yeast recombination mechanisms

(Larionov et al., 1996; Raymond et al., 2002; Leem et al., 2003).

SYNTHESIS OF A MINIMAL GENOME

The *M. genitalium* genome is small enough to be chemically synthesized in its entirety although until recently this would be too technically challenging and expensive to be seriously contemplated. In the past several years, chemically synthesized oligonucleotides have become readily available at only a few cents per nucleotide. A number of companies now synthesize hundreds of genes per month for customers on delivery schedules of no more than a few days or weeks. The chemical synthesis of the 580-kb *M. genitalium* genome is thus feasible and relatively affordable in a research laboratory.

One strategy is to divide the chromosome into a hundred or so sections of around 6 kb, a size that is readily clonable in *E. coli*. *M. genitalium* DNA should be well tolerated in *E. coli* because in mycoplasmas UGA is used to code for tryptophan; thus most proteins are truncated when expressed in *E. coli*, and toxic gene products are less likely. The 6-kb sections are not made from oligonucleotides in one reaction but are instead assembled from several smaller pieces of 1,000 bp or less. The smaller pieces are synthesized from overlapping oligonucleotides, 40 to 60 bases in length by ligation (Smith et al., 2003) or by PCR (Stemmer et al., 1995). The error rate in commercial oligonucleotides is of the order of 1 per 200 bp of assembled product. Thus, most of the initially synthesized pieces will contain several errors. To lower the rate of errors, the double-stranded synthetic DNA products can be melted and reannealed such that an error in one strand will usually be paired to the correct sequence in the other strand, creating a mismatch. The mismatch products can then be eliminated by binding to mismatch recognizing proteins such as the mutS protein (Carr et al., 2004), or the mismatches can be cleaved by mismatch-recognizing endonucleases such as T7 endonuclease I (Young and Dong, 2004), and the error removed by excising a few bases adjacent to the break point. This lowers the error rate so that when cloned in a suitable *E. coli* vector such as pUC19 or pBR322, the majority will be correct.

Assembly of 6-kb sections from the smaller pieces typically uses TypeIIS restriction enzymes that produce degenerate 4-base overhangs. Since these enzymes cleave several bases to one side of their 5- to 7-bp recognition sites, the sites can be placed outside the cloned synthetic pieces with orientations such that cleavage occurs just within the inserts at both ends. Adjacent smaller pieces that make up the larger assembly are cleaved such that complementary overlaps are produced. In this way, adjacent pieces with complementary 4-base overhangs can be joined and ligated in correct order to complete the 6-kb segments (Mandecki and Bolling, 1988).

Larger (>10 kb) assemblies are made in stages. Small assemblies are joined to make moderate-sized assemblies, and these assemblies are joined to make even larger assemblies, and so forth. This method works very well for assemblies of a few kilobases but becomes limiting as the assembled units increase in size. This is because of the increased difficulty in finding TypeIIS enzymes that do not cleave within the pieces that are being assembled into larger structures. To address this problem, one would need to discover new enzymes with larger recognition sites or find ways to increase the size of the recognition sites.

One other problem that arises is the frequent difficulty in cloning large pieces in *E. coli*. Yeast (Burke et al., 1987; Kouprina et al., 1998) can support the cloning of megabase-sized DNAs. *Bacillus subtilis*, which is more closely related to the mycoplasmas than is *E. coli*, might also make a suitable cloning host (Keggins et al., 1978), although it has not been widely used and BAC-type vectors are not available. However, the *B. subtilis* genome can be used as a vector for cloning very large DNAs, and these DNAs could be recovered as linear molecules if suitable flanking restriction sites are available (Itaya et al., 2005).

In Vitro Assembly of the *M. genitalium* Chromosome from 6-kb Sections

To construct the entire circular 580-kb *M. genitalium* chromosome, the assembly must be carried out in stages. Assembly of the 100 6-kb sections is difficult using the TypeIIS method. However, if the 6-kb sections are designed so as to overlap each other by 60 bp or so, then it should be possible to join them at the homologous overlaps by an in vitro recombination method.

One such method has been described for use in ligase-independent cloning (Kuijper et al., 1992). The vector is amplified by PCR using primers that contain 5′-sequences that overlap the DNA to be cloned. The double-stranded insert DNA and the vector with the overlapping ends are then subjected to a 3′ chew-back reaction using the 3′ exonuclease proofreading activity of T4 polymerase. The exposed complementary ends are then annealed, and the DNA is used for cloning in *E. coli* without repair. This method joins two DNA molecules, a vector, and an insert together to form a circle. Very recently, a similar in vitro method has been used to join a vector and several tandem inserts (Li and Elledge, 2007). The essential features of these in vitro methods are illustrated in Fig. 6.

The overlap method could be used to assemble the entire *M. genitalium* chromosome, but it would need to be done in stages. For example, four of the 5- to 6-kb pieces could be joined together at one time and cloned in a BAC to make approximately 25 assemblies, each around 20 to 25 kb in length. The 25 assemblies could then be joined, three at a time, via overlaps into approximately 70- to 75-kb sections, each representing about one-eighth of the *M. genitalium* chromosome. The one-eighth molecules could be joined to make quarter molecules, then half molecules, and finally, the intact chromosome. Cloning of assemblies at each stage would be necessary to obtain DNA for the next stage and for sequence verification.

There are several difficulties with this scheme. Ability to clone bigger and bigger pieces may become limiting. No one has reported the cloning of BAC inserts over around 300 kb. The electroporation efficiency of large BACs is poor. Large BACs may not replicate well, and they have an increased likelihood of producing toxic gene products. Large pieces of DNA are more sensitive to shear so that handling becomes more difficult. And circularization of long pieces of DNA becomes less efficient. The final full-sized chromosome

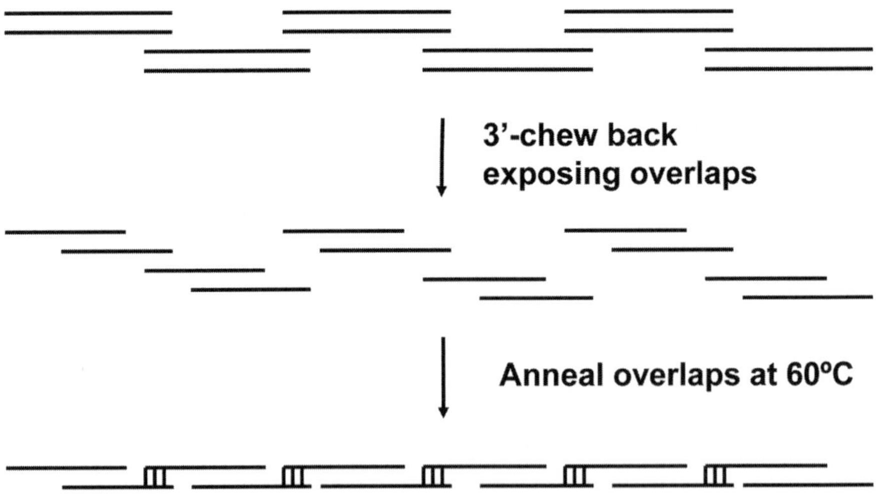

FIGURE 6 Assembly of multiple overlapping pieces of DNA by in vitro recombination. If vector DNA is included in the reaction and it overlaps sequences at each end of the assembly, then a circular DNA is formed, which can be cloned in a suitable host cell.

must be cloned in order to confirm the sequence, and most importantly, to allow multimicrogram preparations suitable for use in transplantation into a cell cytoplasm where the synthetic genome can "boot-up" and create a new living cell. To accomplish late cloning steps, other cloning systems such as yeast might need to be employed.

In Vivo Assembly

It is possible that the whole *M. genitalium* genome might not be successfully assembled in vitro as described in the previous section because of either the limitations of assembling large DNA molecules in vitro or because of the limitations of cloning in *E. coli*. It is possible that the cloning limitations could be overcome by using yeast. Yeast artificial chromosomes have been used to clone very large, megabase-sized pieces of DNA (Burke et al., 1987). Like yeast chromosomes, yeast artificial chromosomes are linear so that the final assembled chromosome would need to be circularized. However, small circular yeast vectors are also available for cloning and have been used to clone relatively large pieces of DNA (Kouprina et al., 1998). It may be possible to clone assembled *M. genitalium* quarters, halves, or the entire chromosome in one of these small vectors.

Linear DNA molecules transformed into yeast become highly recombinogenic. This has led to the development of an efficient transformation-associated recombination cloning system that operates on homologous regions as small as approximately 60 bp in length. When yeast is cotransformed with linear yeast vector and insert DNA that share overlapping homologous sequences of 60 or so base pairs, recombinants are formed and propagated under appropriate selection. This lays open the possibility to carry out some of the late steps in the assembly of the *M. genitalium* genome by transformation-associated recombination cloning (Larionov et al., 1996; Kouprina et al., 1998; Leem et al., 2003). Based on the high efficiency of transformation-associated recombination cloning, there is a reasonable probability that the two in vitro synthesized, overlapping halves of the *M. genitalium* chromosome could be joined and cloned in combination with yeast vector DNA. If so, then an extract of DNA from the yeast would be a source of the completely assembled chemically synthesized *M. genitalium* chromosome, perhaps in sufficient quantity for transplantation (see next section).

GENOME TRANSPLANTATION: INSTALLING A SYNTHETIC GENOME

Chemically synthesizing an entire chromosome would be a substantial achievement, but this alone is not sufficient. One must introduce the chromosome into a recipient cell cytoplasm and establish a viable cell containing the synthetic chromosome. Donor DNA can be transformed into some competent recipient bacterial cells by the natural mechanism of transformation. For example, using competent *B. subtilis* cells, cotransformation of genetic markers spread over more than 30% of the 4.2 megabase pairs genome was demonstrated using nucleoid DNA isolated from gently lysed *B. subtilis* protoplasts (Akamatsu and Taguchi, 2001). *E. coli* is usually transformed with DNA using artificial laboratory methods such as electroporation or the use of competent calcium chloride-treated cells. BACs as large as 300 kb in size have been transformed into *E. coli* (Shizuya et al, 1992). However, transformation of intact bacterial chromosomes has not been reported.

For the synthetic genomics field to progress, scientists must be able to transplant chemically synthesized chromosomes into supporting cytoplasms such that the new chromosomes can become established. By analogy to computers, the genome can be considered as the software that runs a cell, and the cytoplasm is the hardware (Fig. 7). The cytoplasm must support the initial boot-up of the instructions in the synthetic chromosome. Since *M. genitalium* and other mycoplasmas use UGA as a tryptophan codon, the supporting cytoplasm should be that of a mycoplasma species. It is also important that restriction enzyme barriers be removed or inactivated. To lower the chance of recombination with the recipient genome on entrance of

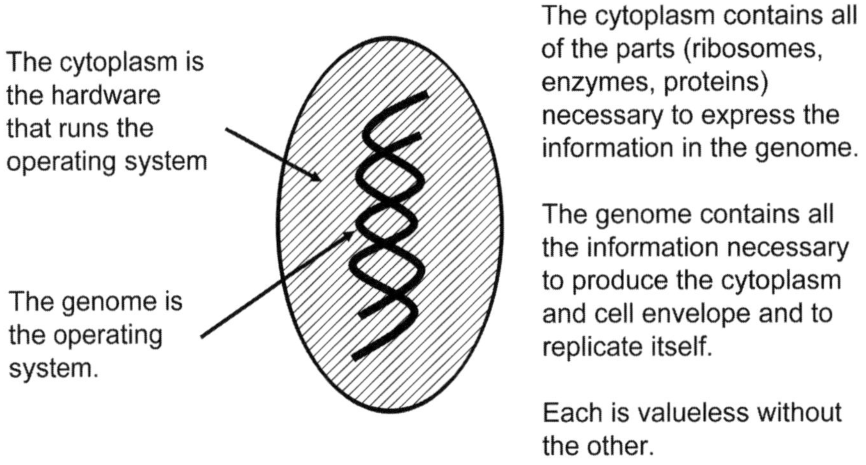

FIGURE 7 A diagram illustrating the basic tenet of synthetic biology. In analogy to computers, the genome of a cell is the operating system and the cytoplasm of the cell is the hardware that runs the operating system.

the synthetic genome, the recipient should preferably be *recA*$^-$ or its genome should be nonhomologous to the donor DNA. The donor needs to have at least one antibiotic resistance marker for selection and several genetic differences from the resident chromosome in order to convincingly demonstrate that the new genome has been installed.

Since *M. genitalium* lacks a cell wall and has a cholesterol-containing cytoplasmic membrane, it is likely that DNA can be introduced by methods that are commonly employed for mammalian cells. These include treatment with $CaCl_2$, polyethylene glycol, and electroporation. Recently, Glass and colleagues reported polyethylene glycol-mediated transplantation of isolated whole naked DNA genomes of *Mycoplasma mycoides* large colony into *Mycoplasma capricolum* recipient cells (Lartigue et al., 2007). The authors speculated that after the donor genome enters a recipient cell, a transient diploid state probably results. As the transformed cell begins to divide on an agar plate under antibiotic selection for the donor genome, daughter cells initially contain a recipient genome, a donor chromosome, or both. Those with only a recipient genome would cease to grow whereas those carrying the donor genome would survive. Eventually cells carrying only the donor genome would overgrow and produce a colony. The exception would be if the diploid state were stable for a number of generations. In that case, substantial recombination might occur between the two genomes. However, they found no evidence of stable diploids or recombination between the donor and recipient genomes. The transplanted cells contained only the donor genome and lacked any recipient DNA sequences.

MINIMIZING THE SYNTHETIC *M. GENITALIUM* GENOME

A recombineering approach (Fig. 5) can be used, either in *E. coli* or yeast, to delete genes or sets of genes from the synthetic chromosome. Alternatively, relevant sections of the synthetic chromosome can be resynthesized with some of the genes left out, followed by reassembly. Removal of putatively dispensable genes should be done in steps, testing the viability of the deleted, reduced genome each time by transplantation.

Possibly as many as 100 genes could be removed. However, this seems unlikely because several of the genes disrupted by transposon insertions are phenotypically altered and/or have reduced fitness as demonstrated by slower growth. A reasonable strategy would be to initially select, as candidates for removal, some of the clusters of genes that were disrupted multiple times in both the 1999 and 2005 studies. The best candidates are: genes MG009–011 (and possibly MG012), MG255–255.1, MG279–281, MG284–286, MG293–294, and MG410–415. In addition, several single genes are good candidates for deletion, for example, MG032, MG067, MG110, MG140, MG237, MG268, MG269, MG288, MG316, MG370, and MG390. Care should be taken that no RNA genes are included in a cluster since all of the RNA genes appear to be essential. Other vital *cis* information such as the origin and termination of replication and transcriptional and translational signals must also be preserved. Our knowledge of this information is still imperfect so every deletion must be tested. A minimal cell will be achieved when transposon mutagenesis no longer yields gene disruptions.

THE FUTURE OF SYNTHETIC GENOMICS

Synthetic genomics is still in its infancy, but methods to synthesize DNA are rapidly improving. It has become routine to synthesize individual genes on demand. Larger synthetic DNAs are now possible. Several complete viral genomes have been synthesized, for example, polio virus (Cello et al., 2002), the 1918 strain of influenza virus (Tumpey et al., 2005), and bacteriophage phiX174 (Smith et al., 2003). However, to date, no one has reported the synthesis of a bacterial chromosome.

One of the major goals of the new field is to design a minimal cell. This will give scientists fundamental knowledge of the essential genes, parts, organization, and machinery of a living cell. In nature, cells have many more genes than are minimally required to run the cell. It is the core machinery, the bare bones machinery, that synthetic biologists seek to define. We will then know in simplest terms the essence of a living system.

The minimal set of genes (functions) will serve as the basis for computational design and logical organization of cells. It may be possible to design cell genomes more logically than is done by nature. Evolution follows a crooked and convoluted path that often leaves genetic remnants and appears to defy intelligent design. Synthetic genomics has the power to design genomes logically. Related gene pathways and structures can be arranged in clusters. Various cellular processes can be supplied as modules in much the same way as electronic equipment is assembled from integrated circuits. Cells can in the future be assembled from a collection of genetic modules (Chan et al., 2005). However, modular design can only be correctly done when we understand not just the essential genes, but also *cis* information encoded in intergenic regions. The genetic parts list for a minimal cell must specify the *cis* sequence elements such as the replication origin and termination sites and the location of appropriate transcription signals and termination signals. In addition, gene design requires that codons be selected in such a way that the gene sequence contains information to regulate translation rates and includes appropriate translational pauses to facilitate efficient folding of proteins.

Currently, the set of genes for a minimal *M. genitalium* cell includes more than 80 genes without functional assignments. Scientists must decipher the functions of these genes in order to fully understand the requirements of the minimal cell.

Once a minimal cell has been achieved, layers of complexity can be added to allow the cell to adapt to various environments and do useful things. Biochemical pathways can be added and hardier cells can be built from the ground up. An industry of cell fabrication will be possible in the same way that integrated circuits are designed in today's world. Synthetically, one can design and make genomes that would be virtually impossible to make from existing organisms by genetic or recombinant DNA techniques. We expect that in a few years the cost of such constructions will greatly decrease and the methodologies will improve dramatically.

A minimal genome may not by itself be useful in a practical sense, but one can envision in the not too distance future the ability to design and synthesize microbes for useful purposes such as production of pharmaceutical products, industrial compounds, and fuels. Stripped down organisms outfitted with useful biosynthetic pathways could provide the basis for new industrial processes.

REFERENCES

Akamatsu, T., and H. Taguchi. 2001. Incorporation of the whole chromosomal DNA in protoplast lysates into competent cells of *Bacillus subtilis*. *Biosci. Biotechnol. Biochem.* **65:**823–829.

Burgos, R., O. Q. Pich, M. Ferrer-Navarro, J. B. Baseman, E. Querol, and J. Pinol. 2006. *Mycoplasma genitalium* P140 and P110 cytadhesins are reciprocally stabilized and required for cell adhesion and terminal-organelle development. *J. Bacteriol.* **188:**8627–8637.

Burke, D. T., G. F. Carle, and M. V. Olson. 1987. Cloning of large segments of exogenous DNA into yeast by means of artificial chromosome vectors. *Science* **236:**806–812.

Carr, P. A., J. S. Park, Y.-J. Lee, T. Yu, S. Zhang, and J. M. Jacobson. 2004. Protein-mediated error correction for de novo DNA synthesis. *Nucleic Acids Res.* **32:**e162.

Cello, J., A. V. Paul, and E. Wimmer. 2002. Chemical synthesis of poliovirus cDNA: generation of infectious virus in the absence of natural template. *Science* **297:**1016–1018.

Chan, L. Y., S. Kosuri, and D. Endy. 2005. Refactoring bacteriophage T7. *Mol. Sys. Biol.* **1:**1–10.

Cordwell, S. J., D. J. Basseal, J. D. Pollack, and I. Humphery-Smith. 1997. Malate/lactate dehydrogenase in mollicutes: evidence for a multienzyme protein. *Gene* **195:**113–120.

Dhandayuthapani, S., W. G. Rasmussen, and J. B. Baseman. 1999. Disruption of gene *mg218* of *Mycoplasma genitalium* through homologous recombination leads to an adherence-deficient phenotype. *Proc. Natl. Acad. Sci. USA* **96:**5227–5232.

Fleischmann, R. D., M. D. Adams, O. White, R. A. Clayton, E. F. Kirkness, A. R. Kerlavage, C. J. Bult, J. F. Tomb, B. A. Dougherty, J. M. Merrick, et al. 1995 Whole-genome random sequencing and assembly of *Haemophilus influenzae* Rd. *Science.* **269:**496–512.

Forster, A. C., and G. M. Church. 2006. Towards synthesis of a minimal cell. *Mol. Syst. Biol.* **2:**45.

Fraser, C. M., J. D. Gocayne, O. White, M. D. Adams, R. A. Clayton, R. D. Fleischmann, C. J. Bult, A. R. Kerlavage, G. Sutton, J. M. Kelley, R. D. Fritchman, J. F. Weidman, K. V. Small, M. Sandusky, J. Fuhrmann, D. Nguyen, T. R. Utterback, D. M. Saudek, C. A. Phillips, J. M. Merrick, J. F. Tomb, B. A. Dougherty, K. F. Bott, P. C. Hu, T. S. Lucier, S. N. Peterson, H. O. Smith, C. A. Hutchison, III, and J. C. Venter. 1995. The minimal gene complement of *Mycoplasma genitalium*. *Science* **270:**397–403.

Galperin, M. Y., and E. V. Koonin. 1998. Sources of systematic error in functional annotation of genomes: domain rearrangement, non-orthologous gene displacement, and operon disruption. *In silico Biology* **1:**55–67.

Gil, R., F. J. Silva, J. Pereto, and A. Moya. 2004. Determination of the core of a minimal bacterial gene set. *Microbiol. Mol. Biol. Rev.* **68:**518–537.

Glass, J. I., N. Assad-Garcia, N. Alperovich, S. Yooseph, M. R. Lewis, M. Maruf, C. A. Hutchison III, H. O. Smith, and J. C. Venter. 2006. Essential genes of a minimal bacterium. *Proc. Natl. Acad. Sci. USA* **103:**425–430.

Hashimoto, M., T. Ichimura, H. Mizoguchi, K. Tanaka, K. Fujimitsu, K. Keyamura, T. Ote, T. Yamakawa, Y. Yamazaki, H. Mori, T. Katayama, and J. Kato. 2005. Cell size and nucleoid organization of engineered *Escherichia coli* cells with a reduced genome. *Mol. Microbiol.* **55:**137–149.

Hutchison, C. A., S. N. Peterson, S. R. Gill, R. T. Cline, O. White, C. M. Fraser, H. O. Smith, and J. C. Venter. 1999. Global transposon mutagenesis and a minimal *Mycoplasma* genome. *Science* **286:**2165–2169.

Itaya, M., K. Tsuge, M. Koizumi, and K. Fujita. 2005. Combining two genomes in one cell: stable cloning of the *Synechocystis* PCC6803 genome in the *Bacillus subtilis* 168 genome. *Proc. Natl. Acad. Sci. USA* **102:**15971–15976.

Keggins, K. M., P. S. Lovett, and E. J. Duvall. 1978. Molecular cloning of genetically active fragments of *Bacillus* DNA in *Bacillus subtilis* and properties of the vector plasmid pUB110. *Proc. Natl. Acad. Sci. USA* **75:**1423–1427.

Kolisnychenko, V., G. Plunkett III, C. D. Herring, T. Fehér, J. Pósfai, F. R. Blattner, and G. Pósfai. 2002. Engineering a reduced *Escherichia coli* genome. *Genome Res.* **12:**640–647.

Koonin, E. V. 2000. How many genes can make a cell: the minimal-gene-set concept. *Annu. Rev. Genomics Hum. Genet.* **1:**99–116.

Koonin, E. V., A. R. Mushegian, and P. Bork. 1996. Non-orthologous gene displacement. *Trends Genet.* **12:**334–336.

Kouprina, N., L. Annab, J. Graves, C. Afshari, J. C. Barrett, M. A. Resnick, and V. Larionov. 1998. Functional copies of a human gene can be directly isolated by transformation-associated recombination cloning with a small 3′ end target sequence. *Proc. Natl Acad. Sci. USA* **95:**4469–4474.

Kuijper, J. L., K. M. Wiren, L. D. Mathies, C. L. Gray, and F. S. Hagen. 1992. Functional cloning vectors for use in directional cDNA cloning using cohesive ends produced with T4 DNA polymerase. *Gene* **112:**147–155.

Larionov, V., N. Kouprina, J. Graves, X. N. Chen, J. R. Korenberg, and M. A. Resnick. 1996. Specific cloning of human DNA as yeast artificial chromosomes by transformation-associated recombination. *Proc. Natl. Acad. Sci USA* **93:**491–496.

Lartigue, C., J. I. Glass, N. Alperovich, R. Pieper, P. P. Parmar, C. A. Hutchison III, H. O. Smith, and J. C. Venter. 2007. Genome transplantation in bacteria: changing one species to another. *Science* **317:**632–638.

Leem, S.-H., V. N. Noskov, J.-E. Park, S. I. Kim, V. Larionov, and N. Kouprina. 2003. Optimum conditions for selective isolation of genes from complex genomes by transformation-associated recombination cloning. *Nucleic Acids Res.* **31:**e29.

Li, M. Z., and S. J. Elledge. 2007. Harnessing homologous recombination in vitro to generate recombinant DNA via SLIC. *Nat. Methods* **4:**251–256.

Mandecki, W., and T. J. Bolling. 1988. FokI method of gene synthesis. *Gene* **15:**101–107.

Maniloff, J. 2002. Phylogeny and evolution, p. 31–43. *In* S. Razin and R. Hermann (ed.), *Molecular Biology and Pathogenicity of Mycoplasmas*. Kluwer Academic/Plenum Publishers, New York, NY.

Mushegian, A. R., and E. V. Koonin. 1996. A minimal gene set for cellular life derived by comparison of complete bacterial genomes. *Proc. Natl. Acad. Sci. USA* **93:**10268–10273.

Peterson, S. N., C. C. Bailey, J. S. Jensen, M. B. Borre, E. S. King, K. F. Bott, and C. A. Hutchison III. 1995. Characterization of repetitive DNA in the *Mycoplasma genitalium* genome: possible role in the generation of antigenic variation. *Proc. Natl. Acad. Sci. USA* **92:**11829–11833.

Peterson, S. N., and C. M. Fraser. 2001. The complexity of simplicity. *Genome Biol.* **2:**COMMENT2002. [Epub 2001 Feb 8.]

Pollack, J. D., M. A. Myers, T. Dandekar, and R. Herrmann. 2002. Suspected utility of enzymes with multiple activities in the small genome *Mycoplasma* species: the replacement of the missing "household" nucleoside diphosphate kinase gene and activity by glycolytic kinases. *Omics* **6:**247–258.

Pósfai, G., G. Plunkett, T. Fehér, D. Frisch, G. M. Keil, K. Umenhoffer, V. Kolisnychenko, B. Stahl, S. S. Sharma, M. de Arruda, V. Burland, S. W. Harcum, and F. R. Blattner. 2006. Emergent properties of reduced-genome *Escherichia coli*. *Science* **312:**1044–1046.

Raymond, C. K., E. H. Sims, and M. V. Olson. 2002. Linker-mediated recombinational subcloning of large DNA fragments using yeast. *Genome Res.* **12:**190–197.

Serres, M. H., and M. Riley. 2005. Gene fusions and gene duplications: relevance to genomic annotation and functional analysis. *BMC Genomics* **6:**33–47.

Shizuya H, B. Birren, U.-J. Kim, V. Mancino, T. Slepak, Y. Tachiri, and M. Simon. 1992. Cloning and stable maintenance of 300-kilobase-pair fragments of human DNA in *Escherichia coli* using F-factor-based vector. *Proc. Natl. Acad. Sci. USA* **89:**8794–8797.

Smith, H. O., C. A. Hutchison III, C. Pfannkoch, and J. C. Venter. 2003. Generating a synthetic genome by whole genome assembly: phiX174 bacteriophage from synthetic oligonucleotides. *Proc. Natl. Acad. Sci. USA* **100:**15440–15445.

Stemmer, W. P., A. Crameri, K. D. Ha, T. M. Brennan, and H. L. Heyneker. 1995. Single-step assembly of a gene and entire plasmid from large numbers of oligonucleotides. *Gene* **164:**49–53.

Tully, J. G., D. L. Rose, R. F. Whitcomb, and R. P. Wenzel. 1979. Enhanced isolation of *Mycoplasma pneumoniae* from throat washings with a newly modified culture medium. *J. Infect. Dis.* **139:**478–482.

Tully, J. G., D. Taylor-Robinson, R. M. Cole, and D. L. Rose. 1981. A newly discovered mycoplasma in the human urogenital tract. *Lancet* **1:**1288–1291.

Tully, J. G., D. Taylor-Robinson, D. L. Rose, R. M. Cole, and J. M. Bove. 1983. *Mycoplasma genitalium*, a new species from the human urogenital tract. *Int. J. Syst. Bacteriol.* **33:**387–396.

Tumpey, T. M., C. F. Basler, P. V. Aguilar, H. Zeng, A. Solórzano, D. E. Swayne, N. J. Cox, J. M. Katz, J. K. Taubenberger, P. Palese, and A. García-Sastre. 2005. Characterization of the reconstructed 1918 Spanish influenza pandemic virus. *Science* **310:**77–80.

Warming, S., N. Costantino, D. L. Court, N. A. Jenkins, and N. G. Copeland. 2005. Simple and highly efficient BAC recombineering using *galK* selection. *Nucleic Acids Res.* **33:**e36.

Young, L., and Q. Dong. 2004. Two-step total gene synthesis method. *Nucleic Acids Res.* **32:**e59.

INDEX

Acanthamoeba, 70
Acantharia, 75
Acidimicrobidae, 177
Acidobacteria, 174–178, 182, 183, 184
Actinobacteria, 14, 15, 18, 20, 174, 177, 178, 183, 184
Actinobacteridae, 175
Acylhomoserine lactones, 183
Adsorption-induced response, 257–259
Agar-based media, nonselective, 176–177
Akkermansia muciniphilia, 16
Alteromonas species, 197
Alveolates, 80
  protistan taxa within, major clades of, 81
Alvinellidae, 198
Amoebophrya, 76, 80, 81
Amoebozoa, 73
Amplified ribosomal intergenic spacer analysis, 139, 140
Anammox bacteria, 214–215, 216
Anapore system, 17
Antigen switching, in protozoal parasites, 227
Antimicrobial drugs, resistance to, 234–235
Antonie van Leeuwenhoek, 11, 67
Archaeoglobus fulgidus, 154
Aristotle, classification system of, 118
Ascomycota, 56, 57, 58–59, 62–63
Assemblage fingerprinting, 139–140
Assembling the Fungal Tree of Life initiative, 56
Atomic force microscope, 246–252, 254, 255, 256
Aureococcus anophagefferens, 72, 73

Bacillus anthracis, 257, 258
Bacillus cereus, 257, 258
Bacillus subtilis, 154, 236, 237, 293, 295
Bacteria, flagellum-driven motility of, 235–236
  immunological approaches to characterize and count, 134
  soil. *See* Soil bacteria
Bacterial cells, single, methods to isolate, 269–270
  sequencing of DNA amplified from, 274–275
Bacterial colony formation on plates, carbon, nitrogen, phosphorus, and sulfur for, 177
Bacterial cultures, Robert Koch and, 3
Bacterial motility, and chemotaxis, 235–236
Bacterial phase variation, 226–227
Bacteroides, 17, 18, 20
Bacteroides dorei, 15
Bacteroides thetaiotaomicron, 15–16
Bacteroidetes, 14, 15, 18, 20, 23, 174
Basidiomycota, 56, 63
Bifidobacterium species, 16, 18, 20, 23, 206
Biodiversity, metagenomics to study, 153–169
  total, of ecosystem, 131
Biodiversity studies, terms used in, 7
Biogeography, island, equilibrium theory of, 104–105, 106
  microbial, 95–115
    history of, 96
    incorporation of phylogenetics into, 110
    shaping of, environmental or historical, 105–108
    study of, future directions in, 108–111
Bolidophyceae, 82–84
Bresiliidae, 198
Butyrate producers, cultivation targeting, 16

Campylobacter species, 20
Candida, 58
Candida albicans, 228, 234
Candida glabrata, 228

Candida species, switching between morphologies in, 228–229
Cantilever-based sensing, 257–262
  adsorption-induced response to, 257–259
Cantilever-based spectroscopy, 255–257
Carbon and energy source, to culture soil bacteria, 178
CARD FISH, 143
Cell wall glycoproteins, expression in yeasts, 227–228
Cells, genes necessary for, 279–299
  minimal, definition of, 279–280
    environment and, 280
  single, DNA isolation and amplification, 270–274
    isolation using antibodies, 245–246
    methods to isolate, 269–270
    nanomechanical methods to study, 245–265
Cenarchaeum symbiosum, 212
Cercozoa, 74
Chao's estimator, 38–40, 41, 42, 47, 146
Chemostat, disadvantages of, 208
  for pure cultures, 208
  in microbiological cultivation, 207–208
  membrane filters and, 210
  with settler, 208, 209
Chemotaxis, bacterial motility and, 235–236
Chloroflexi, 174, 175, 177, 184, 212
Chroococcales, 14
Chthoniobacter flavus, 175
Chytridiomycota (chytrids), 56, 57, 60–62
Chytriomyces angularis, 60–62
Ciliates, 73
Clade, 126
Clone libraries, pitfalls of, 57
Cloning, versus fingerprints, 140
Clostridium, 24
Clostridium acetobutylicum, 280
Clostridium difficile, 20
Clostridium glycyrrhizinilyticum, 15
Clostridium histolyticum, 23
Cluster, 126
Clustered regularly interspaced short palindromic repeats, 154
Clusters of Orthologous Genes database, 162
Collinsella spp., 18
Collozoum longiforme, 67
Colonies, detection of, 185–186
Communities of individuals, distribution of taxon abundances in, 37, 38
Community metagenomics, 143–145
Community proteomics (metaproteomics), 7
Community sequencing (metagenomics), 7
Community structure, of natural microbial communities, methods to assess, 133
"Compound Poisson" distribution, 46
Comprehensive Microbial Resource, 161–162
Consortia and mixed cultures, methods to study, 205–219
Continuous culture, 207
Copper resistance, in yeast, 230
Crenarchaea, nitrification by, 206
Crenarchaeol, 212–213
Crenarchaeota, 215
Crohn's disease, 17
  microbiota in, 23
Cryptococcus neoformans, 234
Cryptosporidium, 70
Cultivation, challenges and benefits of, 199–201
Culture-independent analysis, 17, 25–26
Culturing, growth conditions for, 182–183
  to obtain microbial diversity information, 132–133
Cyanobacteria, immunological approaches, 134
Cylindrotheca species, 72, 73

Delbrück, Max, 279
Delft school, in isolation of microorganisms, 3
Denaturing gradient gel electrophoresis, 121–123, 140, 210–211
Desulfosarcina, 197
Diarrhea, microbes causing, 20
Dinobryon, 71
Dinoflagellates, parasitic, 76
  photosynthetic, 71–73
Dinophysis, 72, 73, 75, 76
Dinophysis acuminata, 75
Distribution. *See also* Taxon-abundance distribution
  posterior, 45
  shape of, 41–42
    fitting using larger samples, 44–47
    inferring by DNA reassociation, 43–44
Diversity, components of, 99, 131
  in microbial communities, 98–103
    locally and globally, 103–104
  microbial. *See* Microbial diversity
  nonparametric estimation of, 37–51
  observed, extrapolations from, 145–146
  of free-living protists seen and unseen, cultured and uncultured, 67–93
  of natural microbial communities, methods to assess, 133
  total genetic, sampling to capture, 50–51
Diversity analysis, 131–132
Diversity-disturbance relationships, 111
Diversity estimates, studies of, 51–52
Diversity estimators, nonparametric, 38–41
    use by simulation, 40–41
Diversity indices, 37–39
Diversity-productivity relationships, 111
DNA, community, "fingerprinting" of, 135
  fungal, sequencing from soil, 56
  isolation and amplification from single microbial cells, 270–274
  overlapping pieces of, assembly by in vitro recombination, 294–295
    in vivo assembly of, 294–295
  reassociation of, inferring shape of distribution by, 43–44

ribosomal, serial analysis of, 155
single-sample, analysis of reassociation rates of, 134–135
DNA amplification, through multiple displacement amplification, 272
DNA and RNA, fecal, analysis of, 18, 19
DNA approaches, non-sequence-based, 134–135
DNA damage, resistance to, 232–234
DNA-DNA hybridization, 5, 134–135
DNA-DNA reassociation experiments, 119–120
DNA sequencing, new methods of, 271

Ebria, 76
Ecosystem, biodiversity of, 131
Ecotype, and geotype, 125
Electron transport processes, in oxygen, 184
Electrophoresis, denaturing gradient gel, 210–211
Entamoeba, 70
Enterobacteriaceae, 20, 23
Enterococcus caccae, 14
Enterococcus species, 23
Environmental and chemical stresses, heterogeneous resistance of, 229–235
Environmental complexity hypothesis, 101
Environmental heterogeneity, influence on microbial communities, 106–107
Environmental microbiology, "black box" approach to, 4
"bottom up" approach to, 4–5
metagenomic approach to, 4
methods in, 4
Epifluorescence microscopy, 211, 212
with fluorescent oligonucleotide probes, 142
Equilibrium theory of island biogeography, 104–106
Eremothecium, 58
Escherichia coli, 154, 159, 215, 225, 226, 230, 234, 235, 269, 270, 273–274, 280, 291–292, 293, 295, 296
Escherichia coli position 1492, 137–138
Escovopsis, 196
Eubacterium, 17
Euglena gracilis, 72, 73
Eukarya, 55
Eumycota, 55
Euprymna scolopes-Vibrio fischeri symbiosis, 198–199
Euprymna scolopes, 195

Faecalibacterium, 16
Ferroplasma species, 159
Fingerprints, cloning versus, 140
Firmicutes, 14, 15, 20, 21, 23, 24, 174, 183, 281
FISH. *See* Fluorescent in situ hybridization
Fish, quantitative probes of, 142–143
Flagellum-driven motility of bacteria, 235–236
Flagellate, 74
Flow cytometric sorting, 85
Flow cytometry, 142
Fluorescent in situ hybridization, 17, 25–26, 85–86
and microautoradiography, 211–212
techniques of, 134
with immunofluorescence, 211
with whole-genome amplification, 275
Force spectroscopy, 255–256, 257
454 Life Sciences technology, 157
Fragment-based analyses, 139
Frameshift mutations, inactivation of multiple genes by accumulation of, 291
Francisella tularensis solution, 260–262
Function-driven analyses, 24
Fungal sequences, gene library of, 56
Fungus(i), 6
diversity in soil, 55–56
missing, culture-independent molecular studies of soil for, 55–65
novel, in soil, 57–62
Fusobacterium prausnitzii, 16

Gastrointestinal tract, disorders of, microbiotas in, 17–18
human, approaches for isolation of microbes from, 16
dimensions of, and microbiota density, 12
disorders of, factors complicating studies of, 21
inaccessibility of, 12–13
microbiota of, 12–13
and disease, 20–23
diversity of, 13–15
functions of, 12
in healthy individuals, 17–20, 22
studies of, 14
novel cultivation strategies in research on, 15–17
microbes of, meta-analysis of, 23–25
Gaussian distribution, inverse, 46–47, 48
Gelling agents, for culturing of soil bacteria, 180
Geminigera cryophila, 75
Gemmatimonadetes, 174, 175, 177, 184
Gene expression, stochastic, 224–226
Genes, diversity of, sampling to capture, 48–50
minimal set of, 280, 290–291
comparative, 281–284
determination of, 281
experimental determination of, 284–291
multiple, inactivation of, by accumulation of frameshift mutations, 291
necessary for cell, 279–299
nonessential, of *Mycoplasma genitalium*, 287–290
Genome(s), changes in short time, 268
complete, sequencing of, 8–9
estimates of number of, 51
first, insight provided by, 154
minimal, synthesis of, 293–295
reduction by recombineering, 291–293
sequencing from single bacterial cell, 268
synthetic, installation of, for transplantation, 295–296
transplantation of, 295–296

Genomic and metagenomic comparisons, tools and databases for, 161–163
Genomic coherence, 121–122
Genomics, 86, 87
comparative, approach to determining minimal set of genes, 281–284
history of, 153–157
"postenvironmental," 267
single-cell, 8, 267–278
definition of, 269
synthetic, future of, 297–298
Genotype, 126
Geotype, and ecotype, 125
Giardia, 70
Giardia lamblia, 11
Global Ocean Survey, 44, 47, 51
Global Transposon Mutagenesis Study—1999, 284–285, 287
Global Transposon Mutagenesis Study—2006, 285–287
Globigerinoides sacculifer, 72, 73
Glomeromycota, 56
Glycoproteins, cell wall, expression in yeasts, 227–228
Growth conditions, 182–183
Gutless giant hydrothermal vent tubeworm, 197–198
Gymnodinium beii, 73

Haemophilus influenzae, 153, 281, 282, 283
Heat shock proteins, 229
Heat stress, resistance to, 232
Helicobacter pylori, 20, 163
Hematodinium, 80, 81
Hermesinum, 76
Heterogeneous antigenicity, at microbial cell surface, 226–228
High-throughput phenotyping screening, 24, 25
High-throughput sequence analysis, 138–139
Histone deacetylase inhibitor, 228
Holography, scanning acoustic, 254, 255, 256
Hooke, Robert, 67
Human Gut Microbiome Initiative, 25
Human intestinal microbiota, and impact on health, 11–32
Human Intestinal Tract Chip, 25
Hyaloraphidium curvatum, 60

Ignicoccus hospitalis, 197
Immunofluorescence, FISH with, 211
Immunological approaches, to characterize and count bacteria and cyanobacteria, 134
Inoculum, and interactions, 180–182
Integrated Microbial Genomes data management system, 162
Intestinal bowel syndrome, 17, 21
Island biogeography, equilibrium theory of, 104–105, 106
Isosphaera pallida, 212

K-strategists, 174, 176, 179
Koch, Robert, 3, 194–195
Koch's postulates, 195, 205, 214
adapted for symbiosis, 195
applied to symbiosis, 194–196
Kuenenia stuttgartiensis, 212, 214–215
Kyoto Encyclopedia of Genes and Genomes database, 162

Laboratory results, and environment, 5
Lactobacillus, 23
Ladderane lipids, 213–214
Lambda repressor protein, 236, 237
Latitudinal diversity gradient, 110–111
Length heterogeneity PCR, 139, 140
Lentisphaera, 15
Lentisphaerae, 16
Leptospirillum ferrodiazotrophum, 9
Leptospirillum species, 159
Lineages, 126
Lingulodinium polyedrum, 72, 73
Linnaeus, taxonomic schema of, 118
Lipid anlysis, 134
Listeria monocytogenes, 20
Lobulomycetaceae, 62
Lobulomycetales, 62
"Long-tail" distribution, 99, 100
Low-molecular-weight RNA profiles, 134
Lunn's estimator, 41
Lysozyme, cell treatment with, DNA breaks and, 270–271

Maillard reaction, 177
Marine broth, diluted, 200
Marine microbial plankton data set, 47–48
Marine microbial symbioses, 194, 238
Marine symbionts, cultivation of, 200–201
Marine symbioses, bacteria and phage, 197
diversity and significance of, 196–199
epibiotic associations, 198
gutless tubeworm, 197–198
microbe-microbe, 197
of secondary metabolites, 197
wood-boring bivalve, 198
Marine symbiotic microorganisms, cultivation of, 193–204
MetaGene, open reading frames of, 163
Metagenome, to evaluate transcriptome, 164
Metagenomic and genomic comparisons, tools and databases for, 161–163
Metagenomic clone library achievement, 121–122
Metagenomic data, analysis of, wet lab tools for, 163
assembly and processing of, 161
Metagenomic sequencing, communities subject to, 158–160
Metagenomic studies, methods for performing, 155
Metagenomics, 23–25, 86

community, 143–145
definition of, 153
from gastrointestinal tract samples of humans, 24
limitations of, 24
potential of, 164–165
to study biodiversity, 153–169
viral, 160–161
Metals and pro-oxidants, resistance to, 230–232
Metaproteomics, 163–164
Methanococcus jannaschii, 154
Methanosarcinales, 197
Methanothermobacter thermautotrophicus, 269–270
Michaelis-Menten equation, 48–49
Microautoradiography, combined with fluorescence in situ hybridization, 122
Microbes, as gene-swapping collectives, 267
body size and spatial scaling of, 101–102
defining ecological niche of, 206–207
dispersal and colonization of, 96–98
diversity of, locally and globally, 103–104
engineering complement of niche of, 207–210
environmental complexity of, 101
speciation and extinction rates of, 102
"storage effect" and, 102–103
study in molecular detail, 206
Microbial biogeography, 95–115
history of, 96
incorporation of phylogenetics into, 110
shaping of, environmental or historical, 105–108
study of, future directions in, 108–111
Microbial cell individuality, 8, 223–243
Microbial cell surface, heterogeneous antigenicity at, 226–228
Microbial cells, single, DNA isolation and amplification from, 270–274
targeted access to, from environment, 273, 275
Microbial communities, diversity in, 7, 98–103
history of studying, 154–156
Microbial diversity, 4. *See also* Diversity
sequence-based surveys of, 110
studying of, 51–52
Microbial diversity studies, based on 16S rRNA gene sequencing, 268
Microbial diversity surveys, scope of, 6
Microbial ecology, community diversity and, 36
Microbial eukaryotes, community structure of, 77, 78
Microbial habitat, characterization at microbe scale, 111
Microbial "species," definition of, 105
Microbiota, human intestinal, and disease, 20–23
and impact on health, 11–32
in healthy individuals, 17–20, 22
Microcantilever, 257–259
Microcantilever-based biosensors, 260–262
Microcantilever-based physical sensors, 259–260
Microcantilever sensors, 258
Micrococcus luteus, 183
Microdiversity, 146
Micromonas, 82
Microorganisms, characterization of, in molecular detail, 210–215
cultivation of, 3
interaction with human body, 5
isolation of, 4
marine symbiotic, cultivation of, 193–204
one-carbon, periodic system of, 207
role in environment, 4
terrestrial, 173–192
transient contaminating, removal of, 199
visible colonies on plates and, 3
Microscope, atomic force, 246–252, 254, 255, 256
modes of operation of, 248–252
Microscopy, epifluorescence, 211, 212
near-field scanning optical, 254
optical, 245
photonic force, 253
scanning probe, 246
to obtain microbial diversity information, 132
Mixed culture, characterization of, 210
Mixotrophic algae, 76
Molecular techniques, microbial diversity and, 3
patterns of microbial diversity and, 6–7
Mollicutes, 281
Monoblepharidales, 60
mRNA, for microarray analysis, 276
Multilocus sequence typing, 5
Multiple displacement amplification, 269–270
amplification bias and, 273
DNA amplification through, 272
following targeted access to cells, 273
pre-PCR step and, 274
Multiply primed rolling-circle amplification, 271
Mycobacterium tuberculosis, 234
Mycoplasma capricolum, 9, 296
Mycoplasma genitalium, 280, 281, 282, 283, 284–285, 286–287, 292–293, 295
essential and nonessential genes of, 288–289, 290
minimal cell, set of genes for, 297–298
nonessential genes of, 287–290
Mycoplasma genitalium chromosome, from 6-kb sections, in vitro assembly of, 294–295
Mycoplasma genitalium genome, synthetic, minimizing of, 292, 296–297
Mycoplasma genitalium genome map, 285–286
Mycoplasma mycoides, 9, 296
Mycoplasma pneumoniae, 284–285
Myrionecta rubra, 75

Nanoarchaeum equitans, 197
National Research Council for a Global Metagenomics Initiative, 165
Near-field scanning optical microscopy, 254
Neolecta, 58

Nephroselmis, 77
Non-sequence-based DNA approaches, 134–135
Nucleotide identity, average, 5

Ochromonas, 71
Oligonucleotide probes, 141–142
One-carbon microorganisms, periodic system of, 207
Operational taxonomic units, 99, 100
Optical microscopy, 245
Ornithocercus, 75
Ornithocercus species, 72, 73
Ostreococcus species, 67

Pace's cultivation-independent survey, 267
Paleogenomics, 158
"Paradox of scale," 106–107
Paraphysomonas, 71, 103–104
Paraphysomonas imperforata, 77–79
PCR, biases introduced by, 140–141
  chimeras or heteroduplexes formed during, 141
  length heterogeneity, 139, 140
  quantitative probes and, 141–143
  real-time quantitative, 142, 211
PCR-based cloning, 137–138
PCR-based sequencing, 138
Pelagophyceae, 82–84
Pezizomycotina, 58–59, 63
Phaeocystis antarctica, 72, 73, 77
Phenotypes, bistable, 236–238
    lysis/lysogeny of, 236–237
    motility-sessility of, 236
    sporulation amd competence development of, 237–238
Phenotypic heterogeneity, 223–224
Photonic force microscopy, 253
Photosynthetic flagellate taxa, 76
Photothermal spectroscopy, 256–257, 258
Phylogenetic microassays, 18, 25
Phylogenetics, incorporation into microbial biogeography, 110
Phylotypes, and ribotypes, 123–125
Phytoplankton, 77
Phytoplasma asteris, 283
Picobiliphytes, 82–84
Piezoresistive cantilever, 257
Pinguiophyceae, 82–84
Planctomycetes, 7, 174, 175, 176, 177, 184
Planktonic foraminifera, 75
Plasmodium, 70
Plasmodium falciparum, 227
"Polony" sequencing, 164
"Postenvironmental" genomics, 267
Posterior means, 45
Probe/microautoradiography combination, 143
Prochlorococcus, 141, 269, 274
Prokaryote taxonomy, 118
Pro-oxidants and metals, resistance to, 230–232
Protein-coding gene set, minimal, 282, 283
Protein coding genes, community analysis from, 145
Proteins, heat shock, 229
Proteobacteria, 15–16, 23, 159, 174, 183, 197, 198
Proteorhodopsin, 206
Protista, 67–93
Protistan phylogeny, reorganization of, 68–69
Protistan taxa, within alveolates, major clades of, 81
  within Tree of Life, phylogenetic schemes for, 69
Protistan taxonomy, 86
Protists, 6
  flagellated, 74
  free-living, seen and unseen, cultured and uncultured, diversity of, 67–93
    present knowledge on, 69–74
  rhizarian, 74
  sarcodine, 74
Protozoal parasites, antigen switching in, 227
Pseudolysogeny, 182
Pseudomonas, 101
Pyramimonas species, 72, 73
Pyrosequencing, 138, 157–158
  for metagenomic analysis, 145

r-strategists, 174, 180
Radiolaria, 75
rDNA sequence analysis, 5
Real-time quantitative PCR, 142, 211
Recombineering, genome reduction by, 291–293
Respiratory oscillation, of yeast, 230–232
Restriction fragment length polymorphism analysis, 139
Ribosomal Database Project, 155
Ribosomal RNA (rRNA), 11
  small-subunit, 11, 23–24, 25, 119, 154–155
Ribotypes, and phylotypes, 123–125
Riftia pachyptila, 195–196, 197–198
RNA profiles, low-molecular-weight, 134
Roseburia spp., 16, 21
rRNA, 11
  small-subunit, 11, 23–24, 25, 119, 154–155
    in DNA databases, 11
Rubrobacteridae, 177, 178, 184
Ruminococcus, 17

16S rDNA genes, 35–36
16S rRNA gene sequencing, 206
  microbial diversity studies based on, 268
16S rRNA genes, 155
  PCR cloning and sequencing of, 137–138
16S rRNA sequences from biomass, 136–137
Saccharomyces cerevisiae, 224, 225, 227, 229–232
Saccharomycotina, 58, 59
Sargasso Sea dataset, 162

Salinibacter ruber, 5
Salmonella enterica, 232
Sanger dideoxynucleotide chain termination method, 157
Sanger sequencing method, 153
Scanning acoustic holography, 254, 255, 256
Scanning near-field ultrasonic holography, 254
Scanning probe microscopy, 8, 246
  techniques in microbiological analyses, 262
Scrippsiella nutricula, 72, 73
Secondary-ion mass spectrometry, 214
SEED database, 162–163
Sequencing, future directions in, 164
Sequencing communities, technologies for, 157–158
Serial analysis of ribosomal DNA, 155
Shannon index, 37–39
Shotgun sequencing approach, 153
Simpson's index, 37–39
Single-cell genomics, 8, 267–278
  definition of, 269
  outlook for, 275–276
Single-genome sequencing projects, 162
Slow-growing bacteria, detection of, 185–186
Soil, fungal diversity in, 55–56
  microbial diversity in, 35, 36
  novel fungi in, 57–62
Soil bacteria, 173
  cultivation efficiency of, signal compounds for, 183
  culturing of, 173–175
    choice of carbon and energy source for, 178
    gelling agents in, 180
    inoculum and interactions in, 180–182
    medium choice for, 176–178
    medium components for, 179–180
  isolation of, intention of, 175
  mean growth rate of, 184
  transition to growth, 182–183
  "unculturable," characteristics of, 175–176
Soil Clone Group I, 58
  in rDNA clone libraries in soils, 58, 60
Soil extract agar, cold and hot extracted, 177
Spartobacteria, 175, 181
Species, and taxa, 123
  as term, 117–118, 119
  classification of new, 120
  genomic coherence of, 121–122
Spectrometry, secondary-ion mass, 214
Spectroscopy, cantilever-based, 255–257
  force, 255–256, 257
  photothermal, 256–257, 258
Sphingopyxis alaskensis, 175
Spumella, 71
Staphylococcus epidermidis, 234
Staphylococcus infections, antibiotic resistance of, 234
Stilbonematinae, 198
"Storage effect," 102–103
Stramenopiles, 71, 80, 82, 84
Streptococcus species, 18
Sulfolobus, 107
Suppression subtractive hybridization, 163
Symbiont cultivation, successful, validation of, 201
Symbiont synthesis, of secondary metabolites, 197
Symbiosis, 8
  biochemical and molecular criteria for, 195–196
  definition of, 193–194
  Koch's postulates adapted for, 195
  Koch's postulates applied to, 194–196
  prevalence of, 196
Symbiotic microorganisms, marine, cultivation of, 193–204
Symbiotic tissue, surface sterilization of, using dilute bleach, 199–200
Syndiniales, 80, 82, 83
Synechocystis PCC6803, 164
Synthetic biology, basic tenet of, 296
Synthetic genomics, future of, 297–298
Szilard, Leo, 207

T-RFLP, 139, 140
Taphrinomycotina, 58, 59
Taxa, and species, 123
Taxon-abundance distribution, 38, 41–42, 46, 50–51
  and parameter set, likelihood function for, 45
Taxon-area relationships, 104–105
Taxon-time relationships, 109
Taxonomic unit(s), observed in environment, 117
  operational, 125–126
  species, etymology of, historical perspective on, 117–120
    new, pure cultures in, 120–121
    recognition among noncultured organisms, 121–123
  species or operational, 117–130
Taxonomy, definition of, 118
Telonemia, 82
Teredinibacter tumerae, 198, 200
Terrestrial microorganisms, new cultivation strategies for, 173–192
Tetrahymena, 70–71
Thalassicola nucleata, 82
Thermal gradient gel electrophoresis, 140
Thermotoga, 267
Thiolated single-stranded DNA, 261–262
Toxoplasma, 70
Transcriptional regulator Wor1, 229
Transcriptome, metagenome to evaluate, 164
Transmission electron microscopy, 211–212
Transposon mutagenesis, 284
Tree of Life, phylogenetic schemes for protistan taxa within, 69
Trophosome, 197

Trypanosoma, 70
Trypanosoma brucei, 227
Trypticase soy agar, 176
Two-reporter assay system, 225

Ulcerative colitis, 17, 21
  microbiota in, 22–23
Uronema marina, 72, 73
U.S. Department of Energy Joint Genome Institute, 162

van Leeuwenhoek, Antonie, 11, 67
Verrucomicrobia, 15, 16, 174, 175, 176, 177, 178, 179–180, 181, 184, 186
Vibrio fischeri, 195, 198–199
Victivallis vadensis, 16
Viral metagenomics, 160–161
Viruses, as abundant biological entity, 6
  taxonomic diversity of, 109–110

Whole-genome amplification, 271
  assessment of bias introduced through, 271–272
  FISH staining with, 275
  three primary forms of, 271
Whole-genome shotgun sequencing, 160
Winogradsky's salt solution agar, 177
Wood-boring bivalve, 197

Xylella fastidiosa, 273

Yeast mutants, phenotypic heterogeneity in, 232, 233
Yeasts, expression of cell wall glycoproteins in, 227–228

Zygomycota, 56, 60–62